Lecture Notes in Computer Science   9996

Commenced Publication in 1973
Founding and Former Series Editors:
Gerhard Goos, Juris Hartmanis, and Jan van Leeuwen

Quanyan Zhu · Tansu Alpcan
Emmanouil Panaousis · Milind Tambe
William Casey (Eds.)

Decision and
Game Theory for Security

7th International Conference, GameSec 2016
New York, NY, USA, November 2–4, 2016
Proceedings

 Springer

Editors
Quanyan Zhu
New York University
New York, NY
USA

Milind Tambe
University of Southern California
Los Angeles, CA
USA

Tansu Alpcan
The University of Melbourne
Melbourne, VIC
Australia

William Casey
Carnegie Mellon University
Pittsburgh, NY
USA

Emmanouil Panaousis
University of Brighton
Brighton
UK

ISSN 0302-9743   ISSN 1611-3349   (electronic)
Lecture Notes in Computer Science
ISBN 978-3-319-47412-0   ISBN 978-3-319-47413-7   (eBook)
DOI 10.1007/978-3-319-47413-7

Library of Congress Control Number: 2016953220

LNCS Sublibrary: SL4 – Security and Cryptology

Printed on acid-free paper

This Springer imprint is published by Springer Nature
The registered company is Springer International Publishing AG
The registered company address is: Gewerbestrasse 11, 6330 Cham, Switzerland

Preface

Communication and information technologies have evolved apace. Recent advances feature greater ubiquity and tighter connectivity for systems exchanging increasingly larger amounts of social, personal, and private information. Indeed, cyberspace, constructed on top of these technologies, has become integral to the lives of people, communities, enterprises, and nation states.

Yet protecting the various assets therein to ensure cybersecurity is a difficult challenge. First, and no differently from physical security, a wide variety of agent utilities abound, including adversarial and antithetical types. Second, being constructed upon heterogeneous, large-scale, and dynamic networks, cyberspace is fairly complex, offering adversaries a large attack surface and ample room for evasive maneuvers, even within carefully designed network and software infrastructure. Nonetheless, security is critical and warrants novel analytic, computational, and practical approaches to thought, planning, policy, and strategic action so we can protect systems and the critical assets they contain, minimize risks and maximize investments, and ultimately provide practical and salable security mechanisms. Collectively our aim is to enhance the trustworthiness of cyber-physical systems.

Recently the analytic and modeling framework of modern game theory has yielded powerful and elegant tools for considering security and the effects of non-cooperative and adversarial types. The problems of security and cybersecurity by necessity must confront the challenging adversarial and worst-case outcomes. To address these, researchers have brought to bear diverse methodologies from control, mechanism design, incentive analysis, economics, and data science to co-evolve advances in game theory, and to develop solid underpinnings of a science of security and cybersecurity.

The GameSec conference brings together academic, industry, and government researchers to identify and discuss the major technical challenges and present recent research results that highlight the connections between and among game theory, control, distributed optimization, and economic incentives within the context of real-world security, trust, and privacy problems. The past meetings of the GameSec conference took place in Berlin, Germany (2010), College Park Maryland, USA (2011), Budapest, Hungary (2012), Fort Worth Texas, USA (2013), Los Angeles, USA (2014), and London, UK (2015). GameSec 2016, the 7th Conference on Decision and Game Theory for Security took place in New York, USA, during November 2–4, 2016. This year we extended the two-day format to a three-day program, allowing GameSec to expand topic areas, include a special track and a poster session.

Since its first edition in 2010, GameSec has attracted novel, high-quality theoretical and practical contributions. This year was no exception. The conference program included 18 full and eight short papers as well as multiple posters that highlighted the research results presented. Reviews were conducted on 40 submitted papers. The selected papers and posters were geographically diverse with many international and transcontinental authorship teams. Whith the geographical diversity underscoring the global concern for and significance of security problems, the papers this year demonstrated several international efforts formed to address them.

The themes of the conference this year were broad and encompassed work in the areas of network security, security risks and investments, decision-making for privacy, security games, incentives in security, cybersecurity mechanisms, intrusion detection, and information limitations in security. The program also included a special track on "validating models," which aims to close the gap between theory and practice in the domain, chaired by Prof. Milind Tambe. Each area took on critical challenges including the detection/mitigation problems associated with several specific attacks to network systems, optimal and risk-averse management of systems, the increased concern of data integrity, leakage, and privacy, strategic thinking for/against adversarial types, adversarial incentives and robust and novel designs to counter them, and acting/decision making in partially informed adversarial settings.

Collectively the conference presents many novel theoretical frameworks and impacts directly the consideration of security in a wide range of settings including: advanced persistent threat (APT), auditing elections, cloud-enabled internet of controlled things, compliance, crime and cyber-criminal incentives, cyber-physical systems, data exfiltration detection, data leakage, denial of service attacks (DOS), domain name service (DNS), electric infrastructures, green security, Internet of Things (IoT), intrusion detection systems (IDS), patrolling (police and pipeline), privacy technology, routing in parallel link networks, secure passive RFID networks, social networking and deception, strategic security investments, voting systems, and watermarking.

We would like to thank NSF for its continued support for student travel, which made it possible for many domestic and international undergraduate and graduate students to attend the conference. We would also like to thank Springer for its continued support of the GameSec conference and for publishing the proceedings as part of their *Lecture Notes in Computer Science* (LNCS) series. We hope that not only security researchers but also practitioners and policy makers will benefit from this edition.

November 2016

Quanyan Zhu
Tansu Alpcan
Emmanouil Panaousis
Milind Tambe
William Casey

Organization

Steering Board

Tansu Alpcan   The University of Melbourne, Australia
Nick Bambos   Stanford University, USA
John S. Baras   University of Maryland, USA
Tamer Başar   University of Illinois at Urbana-Champaign, USA
Anthony Ephremides   University of Maryland, USA
Jean-Pierre Hubaux   EPFL, Switzerland
Milind Tambe   University of Southern California, USA

Organizers

General Chair

Quanyan Zhu   New York University, USA

TPC Chairs

Tansu Alpcan   University of Melbourne, Australia
Emmanouil Panaousis   University of Brighton, UK

Publication Chair

William Casey   Carnegie Mellon University, USA

Special Track Chair

Milind Tambe   University of Southern California, USA

Local Arrangements and Registration Chair

Raquel Thompson   New York University, USA

Publicity Chairs

Mohammad Hossein   Isfahan University of Technology, Iran
  Manshaei
Stefan Rass   Universität Klagenfurt, Austria
Charles Kamhoua   US Air Force Research Laboratory, USA

Web Chair

Jeffrey Pawlick   New York University, USA

Technical Program Committee

TPC Chairs

Tansu Alpcan   University of Melbourne, Australia
Emmanouil Panaousis   University of Brighton, UK

TPC Members

Habtamu Abie   Norsk Regnesentral – Norwegian Computing Center,
                                  Norway
Saurabh Amin   Massachusetts Institute of Technology, USA
Bo An   Nanyang Technological University, Singapore
Alvaro Cardenas   University of Texas at Dallas, USA
Anil Kumar Chorppath   Technische Universität Dresden, Germany
Sajal Das   Missouri University of Science and Technology, USA
Mark Felegyhazi   Budapest University of Technology and Economics,
                                  Hungary
Andrew Fielder   Imperial College London, UK
Cleotilde Gonzalez   Carnegie Mellon University, USA
Jens Grossklags   Penn State University, USA
Yezekael Hayel   LIA/University of Avignon, France
Karl Henrik Johansson   Royal Institute of Technology, Sweden
Murat Kantarcioglu   University of Texas at Dallas, USA
Christopher Kiekintveld   University of Texas at El Paso, USA
Aron Laszka   University of California, Berkeley, USA
Yee Wei Law   University of South Australia, Australia
Pasquale Malacaria   Queen Mary University of London, UK
Mohammad Hossein   EPFL, Switzerland
  Manshaei
Mehrdad Nojoumian   Florida Atlantic University, USA
Andrew Odlyzko   University of Minnesota, USA
David Pym   University College London, UK
Reza Shokri   Cornell University, USA
Arunesh Sinha   University of Southern California, USA
George Theodorakopoulos   Cardiff University, UK
Pradeep Varakantham   Singapore Management University, Singapore
Athanasios Vasilakos   NTUA, Greece
Yevgeniy Vorobeychik   Vanderbilt University, USA
Nan Zhang   The George Washington University, USA
Jun Zhuang   SUNY Buffalo, USA

Contents

Network Security

Resilience of Routing in Parallel Link Networks

Eitan Altman[1,2], Aniruddha Singhal[2], Corinne Touati[2(✉)], and Jie Li[3]

[1] Université Côte d'Azur, Côte d'Azur, France
[2] Inria, Grenoble, France
{Eitan.Altman,corinne.touati}@inria.fr
[3] Faculty of Engineering, Information and Systems,
University of Tsukuba, Tsukuba, Japan
lijie@cs.tsukuba.ac.jp

Abstract. We revisit in this paper the resilience problem of routing traffic in a parallel link network model with a malicious player using a game theoretic framework. Consider that there are two players in the network: the first player wishes to split its traffic so as to minimize its average delay, which the second player, i.e., the malicious player, tries to maximize. The first player has a demand constraint on the total traffic it routes. The second player controls the link capacities: it can decrease by some amount the capacity of each link under a constraint on the sum of capacity degradation. We first show that the average delay function is convex both in traffic and in capacity degradation over the parallel links and thus does not have a saddle point. We identify best responses strategies of each player and compute both the max-min and the min-max values of the game. We are especially interested in the min max strategy as it guarantees the best performance under worst possible link capacity degradation. It thus allows to obtain routing strategies that are resilient and robust. We compare the results of the min-max to those obtained under the max-min strategies. We provide stable algorithms for computing both max-min and min-max strategies as well as for best responses.

1 Introduction

The current computer networks such as Internet architecture remain remarkably vulnerable to different security attacks and failures which may cause system unaivabilities or performance degradation. It is a great challenge to provide services under such security attacks and failures in computer networks. Resiliency is the ability to provide and maintain an acceptable level of service in the face of faults and challenges to normal operation [1].

In this paper, we study the resilience problem of routing traffic in a parallel link network model with a malicious player using game theory. Although the network model looks simple, it could be taken as a typical one for a computer network with general network configuration in which there are many paths between a source node and a destination node and a path consists of several communications.

© Springer International Publishing AG 2016
Q. Zhu et al. (Eds.): GameSec 2016, LNCS 9996, pp. 3–17, 2016.
DOI: 10.1007/978-3-319-47413-7_1

Although our network is a simple one, the network resilience problem in the network model is not a trivial one. We study the resilience problem of routing traffic in a parallel link network model with a malicious player using a game theoretic framework. Consider that there are two players in the network: the first player wishes to split its traffic so as to minimize its average delay, which the second player, i.e., the malicious player, tries to maximize. The first player has a demand constraint on the total traffic it routes. The second player controls the link capacities: it can decrease some amount of the capacity of each link under a constraint on the sum of capacity degradation. We first show that the average delay function is convex both in traffic and the capacity degradation over the parallel links and thus does not have a saddle point. We identify best responses strategies of each player and compute both the max-min and the min-max value of the game. We are especially interested in the min-max strategy as it guarantees the best performance under worst possible unknown link capacity degradation. It thus allows to obtain routing strategies that are resilient and robust. We compare the results of min-max to those obtained at max-min. We provide numerical algorithms for computing both max-min and min-max values and strategies as well as for best responses.

1.1   Related Work

We restrict in this paper our analysis to the framework of routing in a "parallel link" network. This topology has long been a basic framework for the study of routing, as it is a natural generic framework of load balancing among servers in a network. The study of competitive routing in networks with parallel links using game theory goes back to [2]. They were further studied in [3,4] and many others. The only reference we know that studied adversarial behavior in routing in a model similar to the max-min scenario is [5] but they do not propose an algorithmic solution as we do here. On the other hand, to the best of our knowledge, the min-max setting has not been studied before. While the max-min problem has a water-filling structure, we show that the min-max policy has a form which extends the water filling policy and we call it the "water distribution" policy. We provide an algorithm for computing it.

2   System Model and Problem Formulation

Consider a set $\mathcal{L} = \{1, ..., \mathbf{L}\}$ of parallel links between a common source s and destination d as shown in Fig. 1.

Let the delay density over link $\ell \in \mathcal{L}$ of capacity C_ℓ be given by the following function of the link flow x_l:

$$D(x_\ell, C_\ell) \triangleq \begin{cases} \dfrac{1}{C_\ell - x_\ell} & \text{if } x_\ell < C_\ell, \\ +\infty & \text{otherwise.} \end{cases} \tag{1}$$

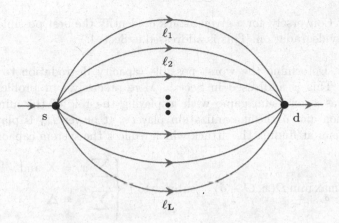

Fig. 1. A system of parallel links

Let x be the flow vector, $x = (x_\ell, 1 \le \ell \le \mathbf{L})$. Define the system delay as the average delay experienced by the flow on the different links:

$$\hat{D}(x, C) \triangleq \sum_{\ell \in \mathcal{L}} x_\ell D_\ell(x_\ell, C_\ell). \tag{2}$$

Such delay system model was already widely used to describe delay in telecommunication network (see, e.g. [2]). In this paper, we address resilience in routing for such networks.

The vector x is controlled so as to minimize the system delay under the demand constraint: $\sum_{\ell \in \mathcal{L}} x_\ell = \mathbf{X}$. Meanwhile, suppose that the capacity C_ℓ of link ℓ is decreased to $C_\ell - \delta_\ell$ where $\delta_\ell \in \mathbb{R}_+$. In this case, the delay of link ℓ becomes

$$D(x_\ell, C_\ell - \delta_\ell) = \begin{cases} \dfrac{1}{C_\ell - \delta_\ell - x_\ell} & \text{if } x_\ell < C_\ell - \delta_\ell, \\ +\infty & \text{otherwise.} \end{cases}$$

Let δ be the degradation vector, $\delta = (\delta_\ell, 1 \le \ell \le \mathbf{L})$. The average system delay is therefore given by

$$\hat{D}(x, C - \delta) \triangleq \sum_{\ell \in \mathcal{L}} x_\ell D_\ell(x_\ell, C_l - \delta_\ell), \tag{3}$$

and the worst degradation is therefore $\max_\delta \hat{D}(x, C - \delta)$ subject to the constraint $\sum_{\ell \in \mathcal{L}} \delta_\ell = \Delta$.

In this paper, we study a routing that would be robust under the worst possible impact of removing an amount Δ of link capacities. Our objectives are as follows:

Objective 1. For a given load vector x, identify the vector δ which is the most harmful. This is addressed in Sect. 3.

Objective 2. Conversely, for a given attack $\boldsymbol{\delta}$, identify the best possible response $\boldsymbol{x}$ to capacity degradation. This is addressed in Sect. 4.

Objective 3. Determine the worst possible capacity degradation to arbitrary flow vector. This is addressed in Sect. 5. More precisely, our problem can be interpreted as a zero-sum game with $\boldsymbol{x}$ playing the role of the minimization player's action and $\boldsymbol{\delta}$ the maximization player's. If player $\Pi_{\boldsymbol{\delta}}$ is playing first, then it will aim at finding the attack which reduces the system capacity most:

$$\max_{\boldsymbol{\delta}} \min_{\boldsymbol{x}} \hat{D}(\boldsymbol{x}, \boldsymbol{C} - \boldsymbol{\delta}) \quad \text{subject to} \quad \begin{cases} \sum_{\ell \in \mathcal{L}} x_\ell = \mathbf{X} \text{ and} \\ \sum_{\ell \in \mathcal{L}} \delta_\ell = \boldsymbol{\Delta}. \end{cases} \tag{4}$$

Objective 4. Determine the flow vector $\boldsymbol{x}^*$ which guaranties the best possible performance under any possible capacity degradation response. This is addressed in Sect. 6. That is, if player $\Pi_{\mathbf{x}}$ is playing first, it will aim at choosing the flow vector $\boldsymbol{x}^*$ that guarantees the best possible performance under the worst possible reaction of attack $\boldsymbol{\delta}$ of player $\Pi_{\boldsymbol{\delta}}$:

$$\min_{\boldsymbol{x}} \max_{\boldsymbol{\delta}} \hat{D}(\boldsymbol{x}, \boldsymbol{C} - \boldsymbol{\delta}) \quad \text{subject to} \quad \begin{cases} \sum_{\ell \in \mathcal{L}} x_\ell = \mathbf{X} \text{ and} \\ \sum_{\ell \in \mathcal{L}} \delta_\ell = \boldsymbol{\Delta}. \end{cases} \tag{5}$$

A crucial question is whether the solutions of the latter two problems coincide. The following result gives a clue:

Proposition 1. *The average delay function $\hat{D}$ is convex both in $\boldsymbol{x}$ and $\boldsymbol{\delta}$.*

The proof is available in the technical report [6].

A very well studied class of games is that of concave-convex games, for which the maximizing player's optimization function is concave while the minimizer player's is convex. These games are known to have a *value*, that is their maximin optimization (4) coincide with their minimax (5). However, in our scenario, the game is convex-convex, and therefore the order at which the players are taking decisions can affect the resulting equilibrium.

In the following, we shall obviously restrict to the case that $\sum_{\ell \in \mathcal{L}} C_\ell > \mathbf{X} + \boldsymbol{\Delta}$ so that there exists a routing strategy with finite cost.

3   Optimal Attack in Response to Link Utilization

In this section, we consider the optimal strategy for player $\Pi_{\boldsymbol{\delta}}$ in response to a given link usage $\boldsymbol{x}$. That is:

$$\boldsymbol{\delta}^*(\boldsymbol{x}) \triangleq \arg\max_{\boldsymbol{\delta} \geq 0} D(\boldsymbol{x}, \boldsymbol{C} - \boldsymbol{\delta}), \text{ s.t } \sum_\ell \delta_\ell = \boldsymbol{\Delta}.$$

The next theorem gives a characterization of the optimal reaction of player Π_δ: it is such that only a single link should be attacked, that is, the one inducing the higher throughput degradation:

Theorem 1 (Optimal attack response). *For any load vector x, there exists a unique optimal reaction of player Π_δ. It is such that:*

$$\delta_\ell^*(x) = \begin{cases} \Delta & \text{if } \ell = \ell^* \\ 0 & \text{otherwise,} \end{cases} \quad \text{with} \quad \ell^* = \arg\max_{\ell \in \mathcal{L}} \frac{x_\ell}{(C_\ell - x_\ell)(C_\ell - \Delta - x_\ell)}. \tag{6}$$

Proof. For a given x vector, note that $\hat{D}$ is convex in δ and defined on the convex polytope P:

$$P \triangleq \{(\delta_\ell)_{\ell \in \mathcal{L}}, \forall \ell, 0 \le \delta_\ell \le \Delta \text{ and } \sum_\ell \delta_\ell = \Delta\}$$

Define e_i the unit vector, i.e. the vector of dimension L with all elements being equal to 0 except for the ith element which is 1. Then P is the convex hull of a set of $L+1$ extreme points: $\{0 \cup \Delta e_i, i \in \mathcal{L}\}$.

Hence, for any point p of P, there exists non-negative $\alpha_0, ..., \alpha_L$ such that $1 = \sum_{\ell \in \mathcal{L}} \alpha_\ell$ and $p = \Delta \sum_\ell \alpha_\ell e_\ell$. Let $\ell^* = \arg\max_\ell \hat{D}(x, C - e_\ell)$.

As $\hat{D}$ is convex, then $\hat{D}(x, C - p) = \hat{D}(x, C - \Delta \sum_\ell \alpha_\ell e_\ell) \le \Delta \sum_\ell \alpha_\ell \hat{D}(x, C - e_\ell) \le \Delta \sum_\ell \alpha_\ell \hat{D}(x, C - e_\ell^*) = \Delta \hat{D}(x, C - e_\ell^*)$ which gives Eq. (6).

The degradation of the delay induced by attacking link ℓ is

$$x_\ell \left(\hat{D}_\ell(\Delta, x_\ell) - \hat{D}_\ell(0, x_\ell) \right) = \frac{x_\ell \Delta}{(C_\ell - \Delta - x_\ell)(C_\ell - x_\ell)} \quad \text{which leads to the}$$

desired result.

Corollary 1. *The degradation induced by player Π_δ on the total delay equals to*

$$\frac{\Delta x_{\ell^*}}{(C_{\ell^*} - x_{\ell^*})(C_{\ell^*} - \Delta - x_{\ell^*})}.$$

Therefore, from Theorem 1, a straightforward algorithm can give the exact optimal attack in $3 \times L$ multiplications and L comparisons.

4   Optimal Link Utilization as Response to Some Attack

We now analyze the optimal link utilization in response to a choice of δ from player Π_δ, which is denoted by $x^*(\delta)$. Then, we seek:

$$x^*(\delta) = \arg\min_{x \ge 0} D(x, C - \delta), \text{ subject to } \sum_\ell x_\ell = X.$$

The next theorem gives a characterization of $x^*(\delta)$:

Theorem 2 (Optimal link usage response). *There exists a unique real value K such that:*

$$x_\ell^*(\boldsymbol{\delta}) = \begin{cases} C_\ell - \delta_\ell - K\sqrt{C_\ell - \delta_\ell} & \text{if } \ell \in \mathcal{X}, \\ 0 & \text{otherwise,} \end{cases} \tag{7}$$

with $\quad \mathcal{X} \triangleq \{\ell, C_\ell - \delta_\ell \geq K^2\} \quad$ *and* $\quad K \triangleq \dfrac{\displaystyle\sum_{\ell \in \mathcal{X}} (C_\ell - \delta_\ell) - \mathbf{X}}{\displaystyle\sum_{\ell \in \mathcal{X}} \sqrt{C_\ell - \delta_\ell}}.$ $\tag{8}$

(Note the fixed point equation between the set of links used at the optimal response $\mathcal{X}$, and the quantity K.)

Proof. For given $\boldsymbol{C}$ and $\boldsymbol{\delta}$, consider the Lagrangian function

$$L(\lambda, \boldsymbol{x}) = \hat{D}(\boldsymbol{x}, \boldsymbol{C} - \boldsymbol{\delta}) - \lambda \left(\sum_{\ell \in \mathcal{L}} x_\ell - \mathbf{X} \right). \tag{9}$$

Then, the optimal link usage response $\boldsymbol{x}^*$ is solution of the optimization problem $\min\limits_{\boldsymbol{x} \geq 0} L(\lambda, \boldsymbol{x})$. Since $x \mapsto L(\lambda, \boldsymbol{x})$ is convex, then

$$\boldsymbol{x}^* = \arg\min_x L(\lambda, \boldsymbol{x}) \Leftrightarrow \frac{\partial L(\lambda, \boldsymbol{x}^*)}{\partial x_\ell^*} \begin{cases} \geq 0 \; \forall \ell \\ = 0 \text{ if } x_\ell^* > 0. \end{cases}$$

Then $\quad \begin{cases} x_\ell^* = 0 \Leftrightarrow \dfrac{\partial \hat{L}}{\partial x_\ell}(\lambda, \boldsymbol{x}|x_\ell = 0) \geq 0 \Leftrightarrow \dfrac{1}{C_\ell - \delta_\ell} \geq \lambda, \\ x_\ell > 0 \Leftrightarrow C_\ell - \delta_\ell - x_\ell = \sqrt{\dfrac{C_\ell \div \delta_\ell}{\lambda}} \end{cases}$ $\tag{10}$

which gives (7) by taking $K = 1/\sqrt{\lambda}$. Then, summing Eq. (7) over $\mathcal{X}$ yields $\mathbf{X} = \sum\limits_{\ell \in \mathcal{X}} \left(C_\ell - \delta_\ell - K\sqrt{C_\ell - \delta_\ell} \right)$, which allows us to express K as Eq. (8).

From this theorem, we can then derive the performance achieved at the optimal link usage response:

Proposition 2 (Performance at the optimal link usage). *At the optimal* $\boldsymbol{x}^*(\boldsymbol{\delta})$, *the total delay on any used link* ℓ *(i.e. such that* $x_\ell^* > 0$*) is given by*

$$x_\ell^*(\boldsymbol{\delta}) D_\ell(x_\ell, C_\ell - \delta_\ell) \triangleq \frac{x_\ell^*(\boldsymbol{\delta})}{C_\ell - \delta_\ell - x_\ell^*(\boldsymbol{\delta})} = \frac{\sqrt{C_\ell - \delta_\ell}}{K} - 1$$

and the total delay is

$$\hat{D}(\boldsymbol{x}^*(\boldsymbol{\delta}), \boldsymbol{C} - \boldsymbol{\delta}) = \frac{\left(\displaystyle\sum_{\ell \in \mathcal{X}} \sqrt{C_\ell - \delta_\ell} \right)^2}{\displaystyle\sum_{\ell \in \mathcal{X}} (C_\ell - \delta_\ell) - \mathbf{X}} - |\mathcal{X}|. \tag{11}$$

Proof. From (7), we have:

$$\frac{x_\ell^*(\delta)}{C_\ell - \delta_\ell - x_\ell^*(\delta)} = \frac{C_\ell - \delta_\ell - K\sqrt{C_l - \delta_\ell}}{K\sqrt{C_\ell - \delta_\ell}} = \frac{\sqrt{C_\ell - \delta_\ell}}{K} - 1.$$

Thus

$$\hat{D}(x^*(\delta), C - \delta) = \sum_{\ell \in \mathcal{X}} \left(\frac{\sqrt{C_\ell - \delta_\ell}}{K} - 1 \right) = \frac{(\sum_{\ell \in \mathcal{X}} \sqrt{C_\ell - \delta_\ell})^2}{\sum_{\ell \in \mathcal{X}} (C_\ell - \delta_\ell) - \mathbf{X}} - |\mathcal{X}|.$$

In order to derive a powerful algorithmic solution, we need the following characterization of the optimal link usage solution:

Proposition 3 (Optimal Usage Characterization). *For each link ℓ, define the normalized delay as*

$$\mathcal{ND}_\ell(x, C - \delta) = \sqrt{C_\ell - \delta_\ell}.D_\ell(x, C - \delta). \tag{12}$$

Then, at the optimal $x^*(\delta)$:

$$\mathcal{ND}_\ell(x^*(\delta), C - \delta) \begin{cases} = K(\mathcal{X}) & if\ C_\ell - \delta_\ell \geq 1/K^2 \\ \geq K(\mathcal{X}) & if\ C_\ell - \delta_\ell \leq 1/K^2 \end{cases} \tag{13}$$

Proof. At the optimal response $x^*(\delta)$, we have, from Eq. (7), for any used link:

$$\mathcal{ND}_\ell(x^*(\delta), C - \delta) = \frac{\sqrt{C_\ell - \delta_\ell}}{C_\ell - \delta_\ell - x_\ell^*(\delta)}$$
$$= \frac{\sqrt{C_\ell - \delta_\ell}}{K\sqrt{C_\ell - \delta_\ell}} = 1/K.$$

For any unused link, we have:

$$\mathcal{ND}_\ell(x^*(\delta), C - \delta) = \sqrt{C_\ell - \delta_\ell}.D_\ell(x^*(\delta), C - \delta)$$
$$= \frac{\sqrt{C_\ell - \delta_\ell}}{C_\ell - \delta_\ell} = \frac{1}{\sqrt{C_\ell - \delta_\ell}}$$

But from Eq. (10), $\frac{1}{C_\ell - \delta_\ell} \geq \lambda$, i.e. $\frac{1}{\sqrt{C_\ell - \delta_\ell}} \geq \sqrt{\lambda} = 1/K$ which concludes the proof.

The proposed water-filling mechanism for the strategy of player $\Pi_\mathbf{x}$ is given in Algorithm 1. The links are initially sorted by decreasing capacity. The mechanism gradually increases the amount of x_ℓ of the various links until reaching $\mathbf{X}$.

More precisely, the algorithm proceeds with initialization of x as zero. At each iteration of the algorithm, the set $\mathcal{X}$ is updated by checking for some potential new candidates.

One can use a direct water-filling algorithm by using ε, a very small quantity, representing the discretization of level increase in the water-filling algorithm.

Algorithm 1. The algorithm of player $\Pi_{\mathbf{x}}$ which defines an optimal strategy in response to some attack.

Input: Vector C of channel capacities, Attack vector $\boldsymbol{\delta}$
Output: Load vector $\boldsymbol{x}$

1   Sort links with decreasing capacity $C_\ell - \delta_\ell$
2   $TA \leftarrow 0$ `// The traffic allocated so far`
3   Link $\leftarrow 1$ `// The link index up to which we inject data`
4   $\varepsilon \leftarrow 0.001$ `// Set it to the desired accuracy`
5   $\boldsymbol{x} \leftarrow \boldsymbol{0}$ `// The traffic vector`
6   **while** $TA < \mathbf{X}$ **do**
7   **while** Link $< \mathbf{L}$ *and* $\mathcal{ND}_1(\boldsymbol{x}, C - \boldsymbol{\delta}) \geq \mathcal{ND}_{\text{Link}+1}(\boldsymbol{x}, C - \boldsymbol{\delta})$ **do**
8   Link$++$
9      $x_1 \leftarrow x_1 + \varepsilon$
10      $K \leftarrow \dfrac{\sqrt{C_1 - \delta_1}}{C_1 - \delta_1 - x_1}$
11   **for** $j = 2$ *to* Link **do**
12         $x_j \leftarrow C_j - \delta_j - \dfrac{\sqrt{C_j - \delta_j}}{K}$
13   Update $TA \leftarrow \sum_{\ell \in \mathcal{L}} x_\ell$

14 **return** $\boldsymbol{x}$

The algorithm would be a direct implementation of Proposition 3, that is, if the current link was filled up to a level (in terms of $\mathcal{ND}$) that is greater or equal than that of next link, then variable Link is incremented so as to start filling the next link. Then, the "for" loop would fill each link j by a small amount η_j which is such that $\mathcal{ND}_j(\boldsymbol{x} + \eta_j \boldsymbol{e}_j, C - \boldsymbol{\delta}) - \mathcal{ND}_j(\boldsymbol{x}, C - \boldsymbol{\delta}) = \varepsilon$ until $\mathbf{X}$ is exhausted.

The performance of such algorithm exhibits average performance though, as the numerical precision errors in the level increases of the different links are summed up over the different iterations and can end up in large inaccuracy if the ratio x_ℓ/η_ℓ turns out to be large. Performance is significantly improved by using one link (for instance that of greater capacity) as the point of reference of the link level and setting up the other links levels accordingly. We propose another variant of the algorithm where ε represents the discretization of x_1. Then, at each iteration of the algorithm, x_1 is increased by ε, then K is updated and then all links in $\mathcal{X}$.

Then, the maximal error is $Err \leq \sum_{\ell \in \mathcal{L}} |x_\ell(x_1) - x_\ell(x_1 + \varepsilon)| = \sum_{\ell \in \mathcal{L}} |\frac{\sqrt{C_j - \delta_j}}{K(x_1)} - \frac{\sqrt{C_j - \delta_j}}{K(x_1 + \varepsilon)}| = \frac{\varepsilon}{\sqrt{C_1 - \delta_1}} \sum_{\ell \in \mathcal{L}} \sqrt{C_\ell - \delta_\ell}$. Since the links are ordered by decreasing capacity, then the error is bounded by $\mathbf{L}\varepsilon$.

5   Optimal Link Degradation Strategy to Unknown Link Usage

Let us now consider that player Π_δ is to choose its attack vector, without knowing the link usage chosen by player Π_x. Then, a natural strategy of player Π_δ is to choose the attack vector that would guarantee the highest value of delay under any action load x. Such strategy of player Π_δ is commonly known as the *maxmin strategy* and is given by Definition 1.

Definition 1. δ^* *is a maxmin strategy if it is solution of*

$$\mathbf{Mm}(\mathbf{C}, \mathbf{X}, \boldsymbol{\Delta}): \qquad \max_{\delta \geq 0} \min_{x \geq 0} D(x, C - \delta),$$
$$s.t \sum_\ell x_\ell = \mathbf{X}, \ \sum_\ell \delta_\ell = \boldsymbol{\Delta}. \tag{14}$$

Note that this is equivalent to a two-player sequential game where player Π_δ plays first, followed by player Π_x, after it observes the action of player Π_δ.

5.1   Existence and Characterization of the Optimal Strategy

Theorem 3 shows that there exists a unique strategy for player Π_δ and provides a characterization of it.

Theorem 3. *There exists a unique real value α such that the optimal strategy for player Π_δ is given by:*

$$\delta_\ell = \begin{cases} C_\ell - \alpha & \text{if } \ell \in \mathcal{D}, \\ 0 & \text{otherwise} \end{cases} \quad \text{with } \alpha = \frac{\sum_{\ell \in \mathcal{D}} C_\ell - \boldsymbol{\Delta}}{|\mathcal{D}|} \quad \text{and } \mathcal{D} = \{\ell \,|\, C_\ell \geq \alpha\}. \tag{15}$$

The proof is given in the technical report [6].

Note that the optimal strategy for player Π_δ is therefore to attack the links of greater capacity in a way so that their remaining capacities $(C_\ell - \delta_\ell)$ are all equal to α. Hence, the optimal strategy for player Π_δ is independent on the weight $\mathbf{X}$ of player Π_x.

5.2   A Decreasing Water Filling Algorithm

Based on Theorem 3, we can derive an algorithm to compute the optimal strategy of player Π_δ, which is given in Algorithm 2.

Similarly to Algorithm 1, at each step of the algorithm, the links 1 to Link are being filled. The algorithm ends whenever all links have been attacked or when the attack level $\boldsymbol{\Delta}$ has been exhausted. More precisely, at any stage of the loop, the links 1 to Link are being filled until either the attack has been exhausted (Line 9–10) or the water-level reaches that of the next link (Line 11–12).

Algorithm 2. The algorithm of player Π_δ which defines an optimal strategy for unknown link usage.

Input: Vector C of channel capacities, of size **L**
Output: Attack vector δ
1 Sort links with decreasing capacity C_ℓ
2 Attack $\leftarrow \Delta$ // Amount of Δ left to be allocated
3 Link $\leftarrow 1$ // The link index up to which we attack
4 Diff $\leftarrow 0$ // Extra capacity of the current link to the next
5 $\eta \leftarrow 0$ // Amount to be allocated in each link
6 $\delta \leftarrow 0$ // The attack vector
7 **while** Link $\leq$ **L** *and* Attack > 0 **do**
8 $\quad$ Diff $= C_{\text{Link}} - C_{\text{Link}+1}$
9 $\quad$ **if** *(*Link $=$ **L** *or* Attack $<$ Link $\times$ Diff*)* **then**
10 $\quad\quad |\ \eta \leftarrow$ Attack $/$ Link
11 $\quad$ **else**
12 $\quad\quad \lfloor\ \eta \leftarrow$ Diff
13 $\quad$ Attack $\leftarrow$ Attack $-\eta.$ Link
14 $\quad$ **for** $j = 1$ *to* Link **do**
15 $\quad\quad \lfloor\ \delta_j \leftarrow \delta_j + \eta$
16 $\quad$ Link $++$
17 **return** δ

Yet, the algorithm differs drastically from Algorithm 1 in its form and complexity. Indeed, from Eq. 15, all links $\ell \in \mathcal{D}$ are such that $C_\ell - \delta_\ell$'s are equal, which amounts to say that for $i, j\ \mathcal{D}$, we have $\delta_i - \delta_j = C_i - C_j$. Hence, the different links are being filled *at the same rate* η, which allows us to simply derive the level of exhaustion of Δ or when the set $\mathcal{D}$ is to be modified. As opposed to Algorithm 1 which computes the solution with arbitrary precision, Algorithm 2 gives the exact solution. Further, the loop runs for at most **L** times and the solution is obtained after at most $\mathcal{O}(\mathbf{L})$ multiplications and $\mathcal{O}(\mathbf{L}^2)$ additions.

Figure 2 shows a typical run of the algorithm, with a set of 5 links. There, the algorithm terminates after 3 loops in the"while" command, as $\sum_\ell \delta_\ell = \Delta$.

6 Optimal Link Usage Strategy with Unknown Degradation Attack

We finally proceed to the case where player $\Pi_\mathbf{x}$ chooses its routing strategy without knowledge of the attack performed by player Π_δ. Then, we consider its strategy to be the one that has the best delay guarantee, i.e. the one such that the delay it will suffer from is the lowest possible one in the worst case scenario (i.e. where player Π_δ has the strongest attack). The problem is referred to as *minmax* and given below:

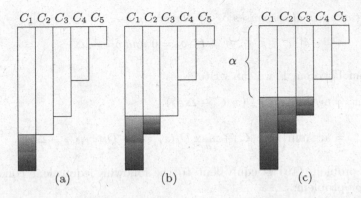

Fig. 2. Typical run of the water-filling algorithm. First, in (a), the channel with highest capacity is filled so as to reach the level of the second. Then in (b) the two channels of largest capacities are filled. Finally, in (c) as the total attack is exhausted before reaching the level of C_4, then channels 1 to 3 are equally filled with the remaining attack.

Definition 2. x^* *is a minmax strategy if it is solution of:*

$$\mathbf{mM}(\mathbf{C}, \mathbf{X}, \mathbf{\Delta}): \qquad \min_{x \geq 0} \max_{\delta \geq 0} D(x, \mathbf{C} - \boldsymbol{\delta}),$$
$$s.t \ \sum_\ell x_\ell = \mathbf{X}, \ \sum_\ell \delta_\ell = \mathbf{\Delta}. \tag{16}$$

Note that the minmax strategy is also the optimal one in a scenario of a two-player sequential game with perfect information where player Π_x plays first followed by player Π_δ.

6.1   Existence and Characterization of the Optimal Strategy

The following theorem states the uniqueness of the solution and gives a characterization:

Theorem 4. *There exists a unique x^* solution of Eq. (16). It is such that there exists a unique α and λ such that*

$$x_\ell^* = \begin{cases} C_\ell - \frac{\Delta}{2} + \frac{\Delta}{2\alpha}\left(1 - \sqrt{4\alpha\frac{C_\ell}{\Delta} + (\alpha - 1)^2}\right) & \text{if } \ell \in C_M, \\ C_\ell - \sqrt{C_\ell/\lambda} & \text{if } \ell \in C_I, \\ 0 & \text{otherwise.} \end{cases} \tag{17}$$

with

$$C_M \triangleq \left\{\ell \in \mathcal{L}, \quad C_\ell \geq \frac{1}{\lambda}\left(\frac{\Delta - \alpha\Delta}{\Delta - \alpha/\lambda}\right)^2\right\}$$
$$C_I \triangleq \left\{\ell \in \mathcal{L}, \quad \frac{1}{\lambda}\left(\frac{\Delta - \alpha\Delta}{\Delta - \alpha/\lambda}\right)^2 \geq C_\ell \geq 1/\lambda\right\} \tag{18}$$

and the set of optimal responses δ^* of player Π_δ are:

$$\{\delta | \exists \ell^* \in C_M, \forall \ell \neq \ell^*, \delta_\ell = 0 \text{ and } \delta_{\ell^*} = \Delta\}.$$

Proof. From Theorem 1, we can write

$$x^* = \arg\min \max_{\ell \in \mathcal{L}} \hat{D}(x, C - \Delta e_\ell)$$

$$= \arg\min \left(\hat{D}(x, C) + \max_{\ell \in \mathcal{L}} D_\ell(x_\ell, C_\ell) - D_\ell(x_\ell, C_\ell - \Delta) \right).$$

Hence, problem (16) is equivalent to the following equivalent constrained optimization problem:

$$\min_{x,\alpha} D(x, C) + \alpha \quad \text{s.t.} \quad \begin{cases} \forall x_\ell, \dfrac{\Delta x_\ell}{(C_\ell - \Delta - x_\ell)(C_\ell - x_\ell)} \leq \alpha, \\ x_\ell \geq 0, \text{ and } \sum_\ell x_\ell = X. \end{cases} \quad (19)$$

The corresponding Lagrangian is

$$L(x, \alpha, \lambda, \mu) = \sum_l \frac{x_\ell}{C_\ell - x_\ell} + \alpha + \lambda \left(X - \sum_\ell x_\ell\right)$$

$$- \sum_\ell \mu_\ell \left(\alpha + \frac{x_\ell}{C_\ell - x_\ell} - \frac{x_\ell}{C_\ell - \Delta - x_\ell} \right)$$

with $\forall \ell, x_\ell \geq 0, \mu_\ell \geq 0$.

Let C^M be the set of links for which $\alpha + \frac{x_\ell}{C_\ell - x_\ell} - \frac{x_\ell}{C_\ell - \Delta - x_\ell} = 0$ and x_ℓ^M the corresponding loads. Then, x_ℓ^M satisfies

$$\frac{x_\ell^M \Delta}{(C_\ell - \Delta - x_\ell^M)(C_\ell - x_\ell^M)} = \alpha$$

$$\text{i.e. } x_\ell^M = C_\ell - \frac{\Delta}{2} + \frac{\Delta}{2\alpha} \left(1 - \sqrt{4\alpha \frac{C_\ell}{\Delta} + (\alpha - 1)^2} \right).$$

If $\ell \notin C^M$, then the Karush Kuhn Tucker conditions give that $\mu_\ell = 0$ and hence the lagrangian reduce to Eq. (9). Then, Eq. (10) leads to Eq. (17).

Finally, $\ell \in C^M$ iff

$$x_\ell^M \leq x_\ell^I \text{ i.e. } \frac{x_\ell^M \Delta}{(C_\ell - \Delta - x_\ell^M)(C_\ell - x_\ell^M)} \geq \alpha$$

But

$$\frac{x_\ell^M \Delta}{(C_\ell - \Delta - x_\ell^M)(C_\ell - x_\ell^M)} = \frac{\lambda \Delta \sqrt{C_\ell} - \Delta \sqrt{\lambda}}{\sqrt{C_\ell} - \Delta \sqrt{\lambda}}.$$

Therefore

$$\ell \in C^M \text{ iff } C_\ell \geq \frac{1}{\lambda} \left(\frac{\Delta - \alpha \Delta}{\Delta - \alpha/\lambda} \right)^2.$$

Note that α represents the degradation induced by player Π_δ. One can readily check that $\alpha = 0 \Leftrightarrow C_M = \emptyset$ which leads to Eq. 8, that is the optimal strategy for Π_δ when there is no capacity degradation.

Algorithm 3. A water-distributed algorithm for optimal strategy of player Π_x with unknown attack

---

1 Sort links with decreasing capacity C_ℓ
2 $\varepsilon \leftarrow 0.01,\ \varepsilon_\alpha \leftarrow 0.1$ `// Set it to the desired accuracy`
3 $TA \leftarrow 0$ `// The traffic to be redistributed`
4 $\ell_M \leftarrow 1$ `// The link index up to which we reduce the flow`
5 $\ell_I \leftarrow 1$ `// The link index up to which we increase the flow`
6 $x \leftarrow$ Solution of Algorithm 1 with no attack ($\delta = 0$)
7 $\alpha \leftarrow \frac{x_1}{C_1 - \Delta - x_1} - \frac{x_1}{C_1 - x_1}$
8 $value \leftarrow \frac{x_1}{C_1 - \Delta - x_1} + \sum_{l=2}^{L} \frac{x_l}{C_l - x_l}$
9 $prec \leftarrow value + 1$
10 **while** $value < prec$ **do**
11     $\alpha \leftarrow \alpha - \varepsilon_\alpha,\ prec \leftarrow value,\ \ell_I \leftarrow \ell_M$
12   **for** $\ell = 1$ **to** ℓ_M **do**   `// Reduce all Links in M`
13         $TA \leftarrow TA + x_\ell$
14         $x_\ell \leftarrow C_\ell - \frac{\Delta}{2} + \frac{\Delta}{2\alpha}\left(1 - \sqrt{4\alpha\frac{C_\ell}{\Delta} + (\alpha - 1)^2}\right)$
15         $TA \leftarrow TA - x_\ell$
16   **while** $TA > 0$ **do**   `// Redistribute TA among the links`
17   **while** $\ell_M < L$ *and* $\alpha \leq D(x_{\ell_M+1}, C_{\ell_M+1} - \Delta) - D(x_{\ell_M+1}, C_{\ell_M+1})$ **do**
18             $\ell_M + +$
19         $\ell_I \leftarrow \ell_M$
20   **while** $\ell_I < L$ *and* $\mathcal{ND}_{\ell_I}(x, C) \geq \mathcal{ND}_{\ell_I+1}(x, C)$ **do**
21             $\ell_I + +$
22   **for** $j = \ell_M + 1$ **to** ℓ_I **do**
23             $\eta \leftarrow \dfrac{\varepsilon(C_j - x_j)^2}{\sqrt{C_j} + \varepsilon(C_j - x_j)}$
24             $x_j \leftarrow x_j + \eta$
25             $TA \leftarrow TA - \eta$
26     $value \leftarrow \frac{x_1}{C_1 - \Delta - x_1} + \sum_{l=2}^{L} \frac{x_l}{C_l - x_l}$

6.2 An Algorithmic Solution

We use the equivalent optimization problem given in Eq. (19). Since it is a convex optimization problem, standard optimization tools (e.g. a projected gradient descent on the Lagrangian) can be used, although they exhibit poor performance, in particular because of the nature of the needed projection and the system's size.

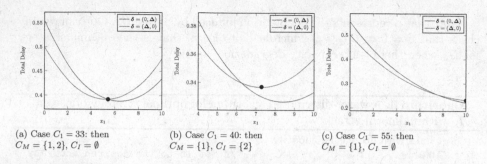

(a) Case $C_1 = 33$: then $C_M = \{1, 2\}$, $C_I = \emptyset$

(b) Case $C_1 = 40$: then $C_M = \{1\}$, $C_I = \{2\}$

(c) Case $C_1 = 55$: then $C_M = \{1\}$, $C_I = \emptyset$

Fig. 3. Different sets C_M and C_I. In the three examples, $C_2 = 30$, $\mathbf{X} = 10$ and $\mathbf{\Delta} = 2$. The min-max solution is represented with the black point. In each plot, the two graphs (red and blue) represent the overall delay that would be experienced by the user if the attack was concentrated on a single link (1 and 2 respectively). (Color figure online)

Therefore, we propose an algorithm, in a similar vein to water-filling algorithms, which we refer to as *water-distributed algorithm*. It can be seen as a water-filling algorithm with a top cap on the water level (represented by α).

The mechanism is given by Algorithm 3. We initialize it by using Algorithm 1 to compute the optimal allocation x if there was no attack. We deduce the initial value of α.

We then iteratively decrease the value of α and compute the corresponding allocation x. The algorithm ends when no gain in the delays is obtained.

Instead of computing the whole allocation at each iteration of the algorithm, we compute the amount of flow which is removed to the links of C_M as a consequence of the decrease of α (lines 21 to 25) and then redistribute this amount to the other links (Line 26 to 36).

6.3   Different C_M and C_I

Note that the set C_M and C_I both depend on the parameter $\mathbf{X}$, $\mathbf{\Delta}$ and the link capacities C_ℓ, $1 \leq \ell \leq \mathbf{L}$.

As long as $\mathbf{\Delta} > 0$, the set C_M is always non-empty (as it should include the link of highest capacity). In contrast, the set C_I can be empty or not. Further, the set $C_M \cup C_I$ may cover all links or not. The different situations are illustrated in the scenario of Fig. 3. The system has a set of two links. In Fig. 3a both of them are in C_M. In this case, the set C_I is empty. Figure 3c shows a scenario where C_I is also empty and C_M consists of only the link of highest capacity. Finally, Fig. 3b shows a case where C_M only contains the link of higher capacity, while the other one is in C_I.

7   Conclusion

We have studied in this paper a game between a router that has a fixed demand to ship and a malicious controller that affects the system capacity by some fixed amount and can decide how to split this among different links. It turned out to be a non standard zero-sum game since the cost is convex for both the minimizer and maximizer and thus does not have a saddle point. We thus focused on computing the max-min and the min-max value and proposed efficient solution algorithms. While the max-min problem is solved using a water-filling algorithm, the solution of the minmax problem requires a more complex algorithm which we call water-distributing algorithm. We plan in the future to extend the problem to several players that try selfishly to minimize their cost in the presence of adversarial capacity degradation controller.

References

1. Smith, P., Hutchison, D., Sterbenz, J., Scholler, M., Fessi, A., Karaliopoulos, M., Lac, C., Plattner, B.: Network resilience: a systematic approach. In: Communications Magazine, vol. 49, no. 7, pp. 88–97. IEEE, July 2011
2. Orda, A., Rom, R., Shimkin, N.: Competitive routing in multiusercommunication networks. IEEE/ACM Trans. Netw. 1(5), 510–521 (1993). http://dx.doi.org/10.1109/90.251910
3. Harks, T.: Stackelberg strategies and collusion in network games with splittable flow. Approximation and Online Algorithms (2009)
4. Koutsoupias, E., Papadimitriou, C.: Worst-case equilibria. In: Meinel, C., Tison, S. (eds.) STACS 1999. LNCS, vol 1563, pp. 404–413. Springer, Heidelberg (1999). doi:10.1007/3-540-49116-3_38
5. Blocq, G., Orda, A.: Worst-case coalitions in routing games. arXiv:1310.3487, August 2014
6. Altman, E., Singhal, A., Touati, C., Li, J.: Resilience of routing in parallel link networks. Hal, Technical report (2015). https://hal.inria.fr/hal-01249188

Deception-Based Game Theoretical Approach to Mitigate DoS Attacks

Hayreddin Çeker[1]($\boxtimes$), Jun Zhuang[1], Shambhu Upadhyaya[1],
Quang Duy La[2], and Boon-Hee Soong[3]

[1] University at Buffalo, Buffalo, NY 14260, USA
{hayreddi,jzhuang,shambhu}@buffalo.edu
[2] Singapore University of Technology and Design, Singapore 487372, Singapore
quang_la@sutd.edu.sg
[3] Nanyang Technological University, Singapore 639798, Singapore
ebhsoong@ntu.edu.sg

Abstract. Denial of Service (DoS) attacks prevent legitimate users from accessing resources by compromising availability of a system. Despite advanced prevention mechanisms, DoS attacks continue to exist, and there is no widely-accepted solution. We propose a deception-based protection mechanism that involves game theory to model the interaction between the defender and the attacker. The defender's challenge is to determine the optimal network configuration to prevent attackers from staging a DoS attack while providing service to legitimate users. In this setting, the defender can employ camouflage by either disguising a normal system as a honeypot, or by disguising a honeypot as a normal system. We use *signaling game* with perfect Bayesian equilibrium (PBE) to explore the strategies and point out the important implications for this type of dynamic games with incomplete information. Our analysis provides insights into the balance between resource and investment, and also shows that defenders can achieve high level of security against DoS attacks with cost-effective solutions through the proposed deception strategy.

Keywords: Game theory · Deception · DoS attacks · Honeypot · Perfect Bayesian equilibrium · Security · Signaling game

1 Introduction

A denial of service (DoS) attack is an attempt to prevent legitimate users from accessing resources. An attacker may target an entire network to cause temporary or permanent unavailability, reduce intended users' bandwidth, or interrupt access to a particular service or a system. The distributed DoS (DDoS) attacks even make it more difficult to prevent and harder to recover. These attacks have already become a major threat to the stability of the Internet [7]. In the survey paper on DDoS attacks, Lau et al. [17] observe that as time has passed, the distributed techniques (e.g., Trinoo, TFN, Stacheldraht, Shaft, and TFN2K) have

© Springer International Publishing AG 2016
Q. Zhu et al. (Eds.): GameSec 2016, LNCS 9996, pp. 18–38, 2016.
DOI: 10.1007/978-3-319-47413-7_2

become more advanced and complicated. Many observers have stated that there is currently no successful defense against a fully distributed DoS attack.

In addition, attackers have the advantage of time and stealth over defenders, since an attacker can obtain information about a defender by pretending to be a legitimate user. Thus, in order to counter this imbalance, *deception* can be utilized to lead an attacker to take actions in the defender's favor by sending fake signals. This way, deception can be used to increase the relative cost of attack, which in turn will delay the attacker because of the uncertainty. In the meantime, the defender can work on solutions to defer and counter the potential attacks. In this setting, although, both the defender and the attacker may spend extra resources to understand the real intention of each other, from the defender's view point, this approach provides a means to mitigate DoS attacks.

Furthermore, the need for protection against DoS attacks extends beyond employing routine intrusion detection system into the domain of survivability. Survivability focuses on the provisioning of essential services in a timely manner without relying on the guarantee that precautionary measures will always succeed against failures, accidents as well as coordinated attacks. It is not an easy task to capture unprecedented DoS attacks while monitoring the entire traffic and providing service to legitimate users. Some resources are to be allocated for attacker detection and advanced tracking tools are to be utilized to protect against patient, strategic and well organized attackers. At the end, it turns out to be an optimization problem from the defender's side about how to allocate the limited resources in a way that the cost will be minimum while the deterrence will be maximum. Similarly, the attacker will try to cause as much damage as possible with limited resources.

In this paper, we propose a game-theoretical approach to model the interaction between the defender and the attacker by deploying honeypots as a means to attract the attacker and retrieve information about the attacker's real intention. A honeypot, unlike a normal system, is a computer system to trap the attacker [3]. Honeypots produce a rich source of information by elaborating the attacker intention and methods used when attackers attempt to compromise a seemingly real server.

In addition to deploying honeypots, we employ deception in our dynamic game in which players (i.e., defender and attacker) take turns choosing their actions. In the scenario under study, the defender moves first by deciding whether to camouflage or not, after which the attacker responds with attack, observe or retreat actions. It is a game of incomplete information because of the attacker's uncertainty of system type. We determine the perfect Bayesian equilibria (PBE) at which both players do not have incentives to deviate from the actions taken.

The contribution of this paper is two-fold: (1) A new defense framework which proactively uses deception as a means to assist in developing effective responses against unconventional, coordinated and complex attacks emanating from adversaries. (2) Determination of the Bayesian equilibrium solutions for this model and analyze the corresponding strategies of the players using a new quantification method for the cost variables.

We also show that deception is an optimal/best response action in some cases where the attacker chooses not to attack a real server because of the confusion caused in the signaling game. Furthermore, we include comprehensive graphics to reflect the possible scenarios that may occur between an attacker and a defender.

The paper continues with the background information and related work on the use of game theory in DoS attacks in Sect. 2. We give the details on the formulation of our model, and specify the assumptions made and notations used in this paper in Sect. 3. In Sect. 4, in case of an attack, the methods for quantifying the damage to a defender and the benefit to an attacker are discussed. Then, we continue with the analysis of PBE and document pooling and separating equilibria in Sect. 5. Section 6 presents the equilibria solutions under various circumstances and find out important implications about the interaction between a defender and an attacker. Section 7 compares our model with real-life systems, and finally Sect. 8 summarizes our findings and gives an insight into how our methodology could be improved further.

Occasionally, the feminine subject she is used to refer to the defender and he to the attacker in the rest of the paper.

2   Background

In this section, we briefly review the basic elements of the game theoretical approach, and relate them to our proposed solution.

2.1   Deception via Honeypots

Game theory has been used in the cyber-security domain ranging from wireless sensor networks [14,28] to DoS attacks [1,18] and information warfare [13] in general. Specifically for DoS attacks, after an attack plan is made by an attacker, even though launching a DoS attack against a victim/defender is always preferred regardless of system type (because of, say, its low cost), the attacker might prefer not to attack if he cannot confirm if a system is of a normal type or a honeypot [6,19].

Defenders can employ deception to increase the effectiveness of their defense system and also to overcome a persistent adversary equipped with sophisticated attack strategies and stealth.

Deception has been used in the military [8,24] and homeland security [29] to protect information critical systems. When attackers cannot determine the type of a system due to deception employed by the defender, they might want to postpone the attack or retreat conditionally. Additional resources might be required to perceive the true system type. In other words, deception hampers the attackers' motivation by increasing the cost. In this paper, we explore the strategies for the defender to estimate the expectations of an attacker and behave accordingly in order to halt him from becoming excessively aggressive and launching DoS attacks. Although we illustrate the solution for DoS attacks, the framework can be used for addressing broader category of attacks in general.

In the context of DoS attacks, a defender can deceive an attacker by deploying several honeypots in her system, and behave as if the attack was successful. However, an intelligent attacker can run simple scripts to understand the real type of the system. For example, an attacker can measure simple I/O time delays or examine unusual and random system calls on the defender server. Similarly, temptingly obvious file names (e.g., "passwords"), and the addition of data in memory as discussed in [10] can disclose the system type obviously [6]. On the other hand, Rowe et al. [24] propose using *fake honeypots* (normal systems that behave as honeypots) to make the job of detecting system type more complicated for the attacker. It is also a form of deception in which the system is camouflaged or disguised to appear as some other types [5].

Similarly, Pibil et al. [21] investigate how a honeypot should be designed and positioned in a network in such a way that it does not disclose valuable information but attracts the attacker for target selection. Also, La et al. [16] analyze a honeypot-enabled network that comprises of IoTs to defend the system against deceptive attacks. In this setting, the attacker might avoid attacking, assuming that the system could be actually a honeypot. As this defensive strategy becomes common knowledge between players, the attacker needs to expend additional resources to determine a system's true type.

Accordingly, a defender can use deception to halt the attacker from executing his contingency plan until she is better prepared, or to effectively recover the system to a secure state that patches all the vulnerabilities exploited by the attacker in the current recovery cycle. The concept of deception is formulated in greater detail in [23,30] as a multi-period game. In this paper, we use a formulation method similar to Zhuang et al. [31] for single period games.

2.2 DoS Attacks from a Game-Theoretical Perspective

The studies that analyze DoS attacks from the game theoretical perspective mostly applied game theory on wireless sensor networks (WSN) considering an intrusion detector as defender and malicious nodes among the sensors as attackers [18,28]. Malicious nodes are those sensors that do not forward incoming packets properly.

Agah and Das [1] formulate the prevention of passive DoS attacks in wireless sensor networks as a repeated game between an intrusion detector and nodes of a sensor network, where some of these nodes act maliciously. In order to prevent DoS, they model the interaction between a normal and a malicious node in forwarding incoming packets, as a non-cooperative N player game.

Lye et al. [18] deal with interactions between an attacker and an administrator of a web server. The game scenario begins with the attacker's attempts to hack the homepage, and the Nash equilibria are computed accordingly. By verifying the usefulness of the approach with network managers, they conclude that the approach can be applied on heterogeneous networks with proper modeling. As for Hamilton et al. [13], they take the security issues from a very general perspective and discuss the role of game theory in information warfare. The paper focuses mostly on areas relevant to tactical analysis and DoS attacks.

Some of the studies that involve game theory about DoS attacks are discussed in a survey by Shen et al. [25]. Here the authors categorize them under non-cooperative, cooperative and repeated game models. However, the use of *signaling game* in DoS attacks is not mentioned under those categories as it is a new area that we explore throughout this paper. Nevertheless, a theoretical analysis of DDoS attacks is proposed using signaling game in [12]. They show the effectiveness and feasibility of a defense strategy based on port and network address hopping compared to packet filtering alone and do not employ any deception. The study is very specific to certain conditions and lacks a comprehensive analysis of possible scenarios that may occur between an attacker and a defender.

The work closest to ours is that of Carroll and Grosu [6] who also use signaling game to investigate the interactions between an attacker and a defender of a computer network. Honeypots are added to the network to enable deception, and they show that camouflage can be an equilibrium strategy for the defender. We extend this study to a broader aspect that includes DoS attacks and we not only find out inequalities that must hold during the game with certain parameters but also propose a parameter valuation mechanism to quantify benefits and damages using existing security evaluations.

Although not directly related to DoS attacks, the authors in [26] study the interactions between a malicious node and a regular node by using PBE to characterize the beliefs the nodes have for each other. Since the best response strategies depend on the current beliefs, the authors apply signaling game to model the process of detecting the malicious nodes in the network.

Despite these studies end up with equilibrium points that represent how a defender and an attacker would act under some conditions, the formulations of the game require all parameters to be known in advance. Also, concrete modeling of DoS attacks requires involving various parameters and valuations of the players to explore equilibria. In this paper, we propose a quantification method with parametric functions under uncertain conditions (incomplete information). This way, the number of all possible scenarios increases and the interactions between players can be reflected in a more comprehensive manner.

3   Model Formulation

We start with a model of incomplete information in which only the defender has private information. In particular, the defender is of a particular type *normal* or *honeypot*. This private information is known by the defender herself but not by the attacker. Although the defender's type is not directly observable by the attacker, the prior probability of her type is assumed to be common knowledge to the attacker. We will let nature make the initial (zero-stage) move, randomly drawing the defender's type from the prior probability distribution.

A defender protects her network by deploying honeypots, which are traps to detect unauthorized access. The defender can disguise normal systems as honeypots and honeypots as normal systems. After the network is created, an attacker then attempts to compromise systems. The attacker can successfully compromise normal systems, but not honeypots. If the attacker attempts to compromise a honeypot, the defender observes the actions and can later improve her defenses. We model this interaction between defender and attacker as a signaling game as described next.

3.1 Assumptions

Although DoS (especially distributed DoS) attacks are launched by a mass (army) of computers, we do restrict our attention to the case of a single centralized attacker where he can submit multiple requests to a server in parallel to cause unavailability (the main purpose of DoS attacks) or temporarily unreachable server error. Thus, we do not address the case of decentralized attackers (such as multiple hacker groups, countries or companies).

During the game, the attacker can update his knowledge about the defender type after observing the signal sent by the defender. However, we do not include any other types of observations (such as spying or probing attacks) for simplicity. Finally, we assume that the players are fully rational, and want to maximize their utilities.

3.2 Signaling Game with Perfect Bayesian Equilibrium

A signaling game is a dynamic game with two players: attacker and defender in our case. The defender has a certain type which is set by nature. The defender observes her own type to take an action during the game, while the attacker does not know the type of the defender. Based on the knowledge of her own type, the defender chooses to send a signal from a set of possible options. Then, the attacker observes the signal sent by the defender and chooses an action from a set of possible actions. At the end, each player gets the payoff based on the defender's type, the signal sent and the action chosen in response.

In our game, the nature decides the defender type to be either normal (N) or honeypot (H). Based on the type, the defender makes truthful disclosure or sends the deception signal. For example, when the defender sends 'H' signal (the apostrophe indicates the message is a signal) for N type, she informs the attacker as if the system is slowing down and the attack is successful. The attacker receives the signal 'H' or 'N' and decides whether to attack (A), observe (O) or retreat (R). Both players will choose the option which yields the maximum utility considering all possibilities.

However, in game theory, sometimes Nash equilibrium results in some implausible equilibria, such as incredible threats from the players. To deal with this type of threats, the concept of PBE which is a strategy profile that consists of sequentially rational decisions is utilized in a game with incomplete information. PBE can be used to refine the solution by excluding theoretically feasible but not probable situations [11].

3.3   Notation and Problem Formulation

We define the notations as follows:

– A and D: Attacker (signal receiver) and defender (signal sender), respectively.
– θ_D is the nature's decision of defender type.
– α^N and α^H are the probabilities of signaling 'N' which originates from a normal type and a honeypot defender, respectively.
– μ refers to the attacker's belief for the probability of receiving the signal 'N' from a normal type defender. Accordingly, $(1 - \mu)$ represents the probability when the signal is 'N' but the defender type is honeypot.
– γ and $(1 - \gamma)$ denote to the attacker's belief for how likely the signal of 'H' might have originated from a normal type defender or a honeypot.
– c_a and c_o are attacker's cost of attacking and observing respectively where $c_a, c_o \geqslant 0$ (we do not incur any charges for retreating in this model).
– b_a and b_o correspond to benefit of attacking and observing where $b_a \geqslant c_a, b_o \geqslant c_o$.
– c_c, c_s, c_h and c_w are defender's costs of compromise, signaling, honeypot and being watched, respectively, where $c_c, c_s, c_h, c_w \geqslant 0$.
– b_{cs} and b_w are customer satisfaction on normal system and benefit of observing the attacker on a honeypot, respectively.
– R_d is the service rate of the defender, and R_a, R_o are the attacking and observing rates of the attacker.
– C is the quantification factor for scaling the rates.

Table 1. Actions and posterior probabilities

$\alpha^N = \Pr(\text{'N'} \mid \text{type N})$	$(1 - \alpha^N) = \Pr(\text{'H'} \mid \text{type N})$
$\alpha^H = \Pr(\text{'N'} \mid \text{type H})$	$(1 - \alpha^H) = \Pr(\text{'H'} \mid \text{type H})$
$\mu = \Pr(\text{type N} \mid \text{'N'})$	$(1 - \mu) = \Pr(\text{type H} \mid \text{'N'})$
$\gamma = \Pr(\text{type N} \mid \text{'H'})$	$(1 - \gamma) = \Pr(\text{type H} \mid \text{'H'})$

3.4   Sequence of Actions in an Extensive Form

Figure 1 illustrates the sequence of deception actions of the signaling game in an extensive form. The nature decides the system type as normal (N) with probability θ_D (top part of the figure) or honeypot (H) with probability $1 - \theta_D$ (bottom shaded part of the figure) and only defender knows it. The defender can choose to disclose her real type by sending message N (top-left branch) or H (bottom-right branch). On the other hand, she can choose to deceive the attacker by signaling 'H' in normal type (top-right branch) or 'N' in honeypot (bottom-left branch). The attacker receives the signal as 'N' or 'H' from the defender, updates his posterior probability and takes an action accordingly.

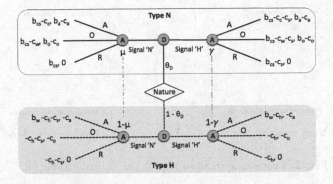

Fig. 1. Signaling game in extensive form

4   Quantification of Damage

We consider a game scenario in which the attacker is uncertain about the defender's asset valuation and the cost. In this section, we first quantify the cost of a DoS attack to the defender and to the attacker, then solve the perfect Bayesian equilibrium (PBE) using sequential rationality.

Basagiannis et al. [4] propose a probabilistic model to quantify the cost of a DoS attack to the defender and to the attacker using Meadows' framework [20]. Although the model makes the cost calculation by including a security protocol, the costs for both parties generically reflect the level of resource expenditure (memory, capacity, bandwidth) for the related actions. As the security level increases, the cost of providing security on the defender's side and the cost of breaking security on the attacker's side increase too. In [4], there is an analysis of how defender/attacker costs change with respect to security level. We refer to the high and low security level cases in Fig. 2a and b respectively. The security level referred in [4] is determined by the complexity of a puzzle that the attacker tries to solve by intercepting messages. In comparison of the processing costs at high security level with low security level, the relative costs to the defender and to the attacker can be approximated by the values specified in the figures for the quantification of equilibrium points. For example, the processing costs at high security level for 100 messages can be used to determine the cost of compromise (c_c) and cost of attacking (c_a), e.g., $c_c = 4000$ units and $c_a = 600$ units. Similarly considering the relative costs, the rewards at low level security can be used to quantify the costs when the defender chooses to disclose her own type and the attacker chooses to observe.

Moving forward with that analogy, the cost variables introduced in the extensive form of the game turn out to be: $c_c = 4000$ units, $c_a = 600$ units, $c_w = 80$ units and $c_o = 30$ units. We fit these values to estimate the service rate of the defender so that our analysis can explore the degradation as a result of the attacker's strategies. We use the formula derived in [15] to measure the

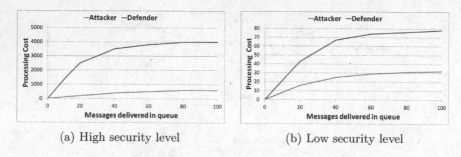

(a) High security level   (b) Low security level

Fig. 2. Processing costs in high & low security levels [4]

satisfaction rate of customers (R) with respect to effective service rate:

$$U(R) = 0.16 + 0.8\,ln(R - 3) \tag{1}$$

Equation (1) quantifies the satisfaction of customers who visit the defender's resources (e.g., website) when she serves at a rate of R. The rate R can be disrupted by an attack as the attacker's aim is to cause unavailability for intended customers. Using this equation helps us reflect the degradation when there is an attack against the defender.

In [15], the maximum satisfaction is rated out of 5, we accept that value as normal case for which $R_d = 427.11$. We assume that the decrease in service rate will reduce the satisfaction of the customers, and eventually it will turn out to be a negative cost for the defender. This way, the satisfaction rate can be referred as the difference between the service rate of the defender and the degradation caused by the attacker. However, since the costs referred in [4] are of large magnitudes, to be able to properly measure the satisfaction rate, we scale it with a quantification factor, C.

We refer to the cost of defender as the degradation in the satisfaction, which corresponds to the difference between the satisfaction in normal case, $C \cdot U(R_d)$ and attack case, $C \cdot U(R_d - R_a)$ or observe case, $C \cdot U(R_d - R_o)$. It can be summarized as follows:

$$
\begin{aligned}
C \cdot U(R_d) - C \cdot U(R_d - R_a) &= C \cdot 0.8 \cdot ln(\frac{R - 3}{R - R_a - 3}) = c_c \\
C \cdot U(R_d) - C \cdot U(R_d - R_o) &= C \cdot 0.8 \cdot ln(\frac{R - 3}{R - R_o - 3}) = c_w
\end{aligned}
\tag{2}
$$

Also, we assume that the cost of attacker is proportional to the rate that they send traffic to cause DoS attack: $\dfrac{R_a}{R_o} = \dfrac{c_a}{c_o} = \dfrac{600}{30} = 20$.

Solving these equations, we end up with $R_a = 389.31$ and $R_o = 19.46$. Having real values for the players' rates helps us estimate the constants in the cost table and make reasonable assumptions accordingly. Substituting the numeric values, we set $C = 1600$ and $c_s = 50$.

As a result, we represent the players' utilities for every action more accurately. Figure 3 updates the notation used in extensive form as a function of service/attack rate so that cost and benefit for each strategy are reflected to both players in a precise manner. New constant values such as v_a, v_1 and v_2 are introduced to reflect the conditional variables that arise based on the strategies taken by the players, e.g., the original service rate $(v_1 \cdot R_d)$ reduces to $v_2 \cdot R_d$ when signaling 'H' in normal type.

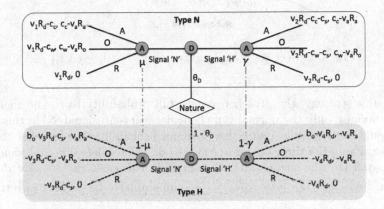

Fig. 3. Updated signaling game

5 Analysis of Perfect Bayesian Equilibrium

In a game with incomplete information, players might update their beliefs through observations about types of the opponent. This belief updating process must satisfy Bayes' rule in which posterior probability is determined by the priori and the likelihood of each type.

5.1 Separating Equilibria

In this section, we provide the steps to find out if there is a *separating equilibrium* where defenders' signals are different. We first examine the game depicted in Fig. 4 where both defender types are truthful, i.e., normal type's signal is 'N' and honeypot's signal is 'H'.

Based on the scenario, when the attacker receives the signal 'N' (on the left side), he updates the posterior probability, μ:

$$\mu = \frac{\theta_D \cdot \alpha^N}{\theta_D \cdot \alpha^N + (1 - \theta_D) \cdot \alpha^H} = \frac{\theta_D \cdot 1}{\theta_D \cdot 1 + (1 - \theta_D) \cdot 0} = 1$$

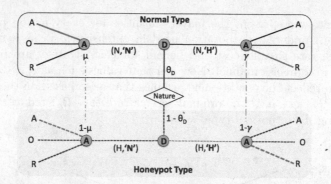

Fig. 4. Perfect Bayesian equilibrium of ('N','H') - (A,R)

With this strategy, the attacker assigns full probability to the normal type defender because only the normal type defender can send signal 'N' in this separating equilibrium setting. Using this posterior probability, the attacker chooses a response among the three options (A, O, R) on the top-left side that yields the highest payoff (by the sequential rationality). In this case, the attacker decides to attack if $c_c - v_a \cdot R_a \geq 0 \Rightarrow R_a \leq \dfrac{c_c}{v_a}$, and similarly he decides to retreat if $R_a > \dfrac{c_c}{v_a}$.

We omit comparison of attack (A) with observe (O) option since $c_c > c_w$ by Eq. 2. When the attacker chooses to attack, we mark the corresponding branches for both defender types on the top and bottom left side. Similarly, if the attacker receives 'H', he believes that the signal comes from a honeypot by the posteriori calculation:

$$\gamma = \frac{\theta_D \cdot (1 - \alpha^N)}{\theta_D \cdot (1 - \alpha^N) + (1 - \theta_D) \cdot (1 - \alpha^H)} = \frac{\theta_D \cdot 0}{\theta_D \cdot 0 + (1 - \theta_D) \cdot 1} = 0$$

In this case, the attacker chooses to retreat (R) with 0 payoff among the three options because A and O have negative values in Fig. 3. Accordingly, the branches are shaded (top and bottom right) for both defender types in Fig. 4.

Once the actions are taken for both players, PBE seeks for any deviations from the players' decisions. In other words, if a player has incentive to change the decision among the shaded branches, we say that PBE does not exist for such a case. We first consider the scenario in which the attacher decides to attack against receiving signal 'N' (shaded branches). The normal type defender compares the utility of signaling 'N' ($v_1 \cdot R_d - c_c$) with signaling 'H' ($v_2 \cdot R_d - c_s$). She does not deviate from the decision as long as:

$$v_1 \cdot R_d - c_c \geq v_2 \cdot R_d - c_s \Rightarrow R_d \geq \frac{c_c - c_s}{(v_1 - v_2)}$$

Similarly, the honeypot type compares the shaded branches and does not deviate if and only if:

$$-v_4 \cdot R_d \geq b_o - v_3 \cdot R_d - cs \Rightarrow R_d \geq \frac{b_o - c_s}{v_3 - v_4}$$

Consequently, this separating equilibrium strategy (the defender plays ('N','H') and the attacker plays (A,R) represents a PBE of this incomplete information game, if and only if:

$$R_d \geq \frac{c_c - c_s}{(v_1 - v_2)}, R_d \geq \frac{b_o - c_s}{v_3 - v_4}, R_a \leq \frac{c_c}{v_a}$$

Now we consider the scenario in which the attacker decides to retreat against receiving signal 'N'. In a similar way, both defender types seek for incentives to deviate from current strategy by comparing the utility of signaling 'N' with that of 'H'. After substituting the payoffs, we conclude that she does not deviate if:

$$R_d \leq \frac{c_s}{(v_2 - v_1)}, R_d \leq \frac{c_s}{v_4 - v_3}, R_a \geq \frac{c_c}{v_a}.$$

For illustration purposes, we show the exhaustive analysis of the strategy in which the defenders signal ('N','H') and the attacker responds by (A,R). All separating equilibria (including the above solution) that satisfy PBE and the corresponding conditions are listed in Table 2.

Table 2. Perfect Bayesian equilibrium for separating equilibria

	$(s_1, s_2), (a_1, a_2)$	Conditions	μ, γ
E1	('N','H') - (A,R)	$R_d \geq \dfrac{c_c - c_s}{(v_1 - v_2)}, R_d \geq \dfrac{b_o - c_s}{v_3 - v_4}, R_a \leq \dfrac{c_c}{v_a}$	1, 0
E2	('N','H') - (R,R)	$R_d \leq \dfrac{c_s}{(v_2 - v_1)}, R_d \leq \dfrac{c_s}{v_4 - v_3}, R_a > \dfrac{c_c}{v_a}$	1, 0
E3	('H','N') - (A,R)	$R_d > \dfrac{c_c + c_s}{(v_2 - v_1)}, R_d > \dfrac{b_o + c_s}{v_4 - v_3}, R_a \leq \dfrac{c_c}{v_a}$	0, 1
E4	('H','N') - (R,R)	$R_d > \dfrac{c_s}{(v_2 - v_1)}, R_d > \dfrac{c_s}{v_4 - v_3}, R_a > \dfrac{c_c}{v_a}$	0, 1

s_1 and s_2 represent the signals sent by normal type and honeypot defenders. a_1 and a_2 represent attacker's responses against normal type and honeypot defenders.

5.2 Pooling Equilibria

In this section, we provide the steps to find out potential PBEs where both defender types send the same signal. We examine the scenario shaded on the left half of Fig. 5 when both defender types send the signal 'N'. The attacker updates the posterior probability, μ in a similar way for which $\alpha^N = 1$ and $\alpha^H = 1$ based on the definition in Table 1.

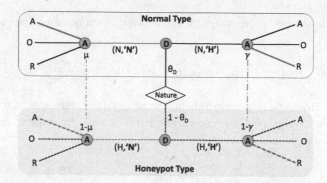

Fig. 5. Perfect Bayesian equilibrium of ('N','N') - (A,R)

$$\mu = \frac{\theta_D \cdot \alpha^N}{\theta_D \cdot \alpha^N + (1 - \theta_D) \cdot \alpha^H} = \frac{\theta_D \cdot 1}{\theta_D \cdot 1 + (1 - \theta_D) \cdot 1} = \theta_D$$

With this strategy, the attacker cannot distinguish between the defender types, hence the announcements from the defenders are uninformative. In contrast to strategies in separating equilibria, the attacker cannot assign a full probability to a certain type, and must consider the nature's probability (priori) θ_D as a result of the updated μ value. In this scenario, the posteriori coincides with the prior probability which is a common case in pooling equilibria [9].

After μ is updated, the attacker chooses between the actions. He chooses A, if these conditions hold from Fig. 3:

$$\theta_D(c_c - v_a \cdot R_a) + (1 - \theta_D)(-v_a \cdot R_a) \geq \theta_D(c_w - v_a \cdot R_o) + (1 - \theta_D)(-v_a \cdot R_o)$$

$$\theta_D(c_c - v_a \cdot R_a) + (1 - \theta_D)(-v_a \cdot R_a) \geq 0$$

which holds for

$$\theta_D \geq \frac{v_a \cdot (R_a - R_o)}{c_c - c_w} \text{ and } \theta_D \geq \frac{R_a \cdot v_a}{c_c}$$

On the other hand, the attacker decides to observe (O) if:

$$\theta_D < \frac{v_a \cdot (R_a - R_o)}{c_c - c_w} \theta_D \geq \frac{R_o \cdot v_a}{c_w}$$

and finally he decides to retreat (R) if:

$$\theta_D < \frac{R_a \cdot v_a}{c_c} \text{ and } \theta_D < \frac{R_o \cdot v_a}{c_w}$$

Despite the probability of signaling '**H**' is 0 for both defenders, the attacker must still update γ to finish the game:

$$\gamma = \frac{\theta_D \cdot (1 - \alpha^N)}{\theta_D \cdot (1 - \alpha^N) + (1 - \theta_D) \cdot (1 - \alpha^H)} = \frac{\theta_D \cdot 0}{\theta_D \cdot 0 + (1 - \theta_D) \cdot 0} = \frac{0}{0}$$

which is a special case where γ can have an arbitrary value ($\gamma \in [0,1]$) because the player is at a state which should not be reached in equilibrium [2]. To handle such cases, we first set restrictions on the range of γ based on the attacker's decisions, then check whether there is a deviation in any of the defenders. For example, let us assume that the attacker chooses to retreat when he receives the signal '**H**' on the right half of Fig. 5. Then, this equations must hold for the retreat option to be optimal:

$$\gamma \cdot 0 + (1 - \gamma) \cdot 0 > \gamma(c_c - v_a \cdot R_a) + (1 - \gamma)(-v_a \cdot R_a) \Rightarrow \gamma < \frac{v_a \cdot R_a}{c_c}$$

$$\gamma \cdot 0 + (1 - \gamma) \cdot 0 > \gamma(c_w - v_a \cdot R_o) + (1 - \gamma)(-v_a \cdot R_o) \Rightarrow \gamma < \frac{v_a \cdot R_o}{c_w}$$

After setting the restrictions and assuming that the attacker has chosen to attack against normal type defender (A in the first computation), we check if there is deviation by the defender types by comparing the marked selections in Fig. 5. Then, we can conclude that the PBE can be sustained with this scenario, if these conditions hold:

$$v_1 \cdot R_d - c_c > v_2 \cdot R_d - c_s \Rightarrow R_d > \frac{c_c \quad c_s}{v_1 - v_2}$$

$$b_o - v_3 \cdot R_d - c_s \geq -v_4 \cdot R_d \Rightarrow R_d \leq \frac{b_o - c_s}{v_3 - v_4}$$

The remaining pooling equilibrium scenarios that satisfy PBE in the exhaustive analysis are all listed in Table 3 with respective conditions.

6 Results

Using the valuations of players (e.g., cost variables, service rate), we explore the Nash equilibria by finding out steady states where neither player has incentives to deviate from the actions taken. We take all possibilities into account for both defender types (normal and honeypot) and one attacker including the nature's decision, our results in Fig. 6 show that the equilibrium can be at 4 different settings based on the valuation of the players. In particular, when normal defender's service rate is very high compared to the attacker (the square □ and triangle ▽ area), the defender does not anticipate the use of deception because the overhead caused by signaling is more than the damage the attacker may cause. In response, when the attacker's rate is comparable to defender's service rate (triangle ▽ area), he wants to attack in normal type and retreat in honeypot; whereas if the defender's service rate is extremely high (the square □), then the attacker chooses to retreat with certainty. That is, the attacker's utility which takes into account the prior belief (θ_D), the signal (s) and the posterior probability (μ), makes it infeasible to attack. However, in the former case (triangle ▽ area), since the attack rate is relatively close to the defender's service rate, the attacker finds initiatives to attack in the case where he receives the signal 'N'. In other words, the potential damage he may cause (if the target is normal) is larger than the cost incurred on the attacker (in case he fails to attack a normal server).

In pooling equilibria where defenders with different types all choose the same signal to be sent (the diamond ◊ and circle ◯ area), we see that if the attacker's rate is very close to the defender's service rate (circle ◯ area), the attacker chooses to attack with certainty. If the attack succeeds, the damage to the defender is huge and a permanent unavailability can occur on the server side. However, even if the attacker's rate is high enough in some cases (the diamond ◊), the attacker may prefer retreating because of the likelihood that the defender's

Table 3. Perfect Bayesian equilibrium for pooling equilibria

	(s_1, s_2) - (a_1, a_2)	Conditions	Prior & Posterior*
E5	('N','N') - (A,A)	$\dfrac{c_s}{v_2-v_1} \geq R_d \geq \dfrac{c_s}{v_4-v_3}$	$\theta_D \geq \dfrac{v_a\cdot(R_a-R_o)}{c_c-c_w}, \theta_D \geq \dfrac{R_a\cdot v_a}{c_c}$,
E6	('H','H') - (A,A)	$\dfrac{c_s}{v_4-v_3} > R_d > \dfrac{c_s}{v_2-v_1}$	$\gamma \geq \dfrac{v_a\cdot(R_a-R_o)}{c_c-c_w}, \gamma \geq \dfrac{v_a\cdot R_a}{c_c}$
E7	('N','N') - (A,O)	$\dfrac{c_s+c_w-c_c}{v_2-v_1} \geq R_d \geq \dfrac{c_s-b_o}{v_4-v_3}$	$\theta_D \geq \dfrac{v_a\cdot(R_a-R_o)}{c_c-c_w}, \theta_D \geq \dfrac{R_a\cdot v_a}{c_c}$,
E8	('H','H') - (A,O)	$\dfrac{c_s-b_o}{v_4-v_3} > R_d > \dfrac{c_s+c_w-c_c}{v_2-v_1}$	$\gamma < \dfrac{v_a\cdot(R_a-R_o)}{c_c-c_w}, \gamma \geq \dfrac{v_a\cdot R_o}{c_w}$
E9	('N','N') - (A,R)	$\dfrac{b_o-c_s}{v_3-v_4} \geq R_d \geq \dfrac{c_c-c_s}{v_1-v_2}$	$\theta_D \geq \dfrac{v_a\cdot(R_a-R_o)}{c_c-c_w}, \theta_D \geq \dfrac{R_a\cdot v_a}{c_c}$,
E10	('H','H') - (A,R)	$\dfrac{c_c-c_s}{v_1-v_2} > R_d > \dfrac{b_o-c_s}{v_3-v_4}$	$\gamma < \dfrac{v_a\cdot R_a}{c_c}, \gamma < \dfrac{v_a\cdot R_o}{c_w}$
E11	('N','N') - (O,A)	$\dfrac{c_s+c_c-c_w}{v_2-v_1} \geq R_d \geq \dfrac{c_s+b_o}{v_4-v_3}$	$\theta_D < \dfrac{v_a\cdot(R_a-R_o)}{c_c-c_w}, \theta_D \geq \dfrac{R_o\cdot v_a}{c_w}$
E12	('H','H') - (O,A)	$\dfrac{c_s+b_o}{v_4-v_3} > R_d > \dfrac{c_s+c_c-c_w}{v_2-v_1}$	$\gamma \geq \dfrac{v_a\cdot(R_a-R_o)}{c_c-c_w}, \gamma \geq \dfrac{v_a\cdot R_a}{c_c}$
E13	('N','N') - (O,O)	$\dfrac{c_s}{v_2-v_1} \geq R_d \geq \dfrac{c_s}{v_4-v_3}$	$\theta_D < \dfrac{v_a\cdot(R_a-R_o)}{c_c-c_w}, \theta_D \geq \dfrac{R_o\cdot v_a}{c_w}$
E14	('H','H') - (O,O)	$\dfrac{c_s}{v_4-v_3} > R_d > \dfrac{c_s}{v_2-v_1}$	$\gamma < \dfrac{v_a\cdot(R_a-R_o)}{c_c-c_w}, \gamma \geq \dfrac{v_a\cdot R_o}{c_w}$
E15	('N','N') - (O,R)	$\dfrac{c_w}{v_1-v_2} \geq R_d \geq \dfrac{c_s}{v_4-v_3}$	$\theta_D < \dfrac{v_a\cdot(R_a-R_o)}{c_c-c_w}, \theta_D \geq \dfrac{R_o\cdot v_a}{c_w}$
E16	('H','H') - (O,R)	$\dfrac{c_s}{v_4-v_3} > R_d > \dfrac{c_w}{v_1-v_2}$	$\gamma < \dfrac{v_a\cdot R_a}{c_c}, \gamma < \dfrac{v_a\cdot R_o}{c_w}$
E17	('N','N') - (R,A)	$\dfrac{c_s+c_c}{v_2-v_1} \geq R_d \geq \dfrac{c_s+b_o}{v_4-v_3}$	$\theta_D < \dfrac{R_a\cdot v_a}{c_c}, \theta_D < \dfrac{R_o\cdot v_a}{c_w}$
E18	('H','H') - (R,A)	$\dfrac{c_s+b_o}{v_4-v_3} > R_d > \dfrac{c_s+c_c}{v_2-v_1}$	$\gamma \geq \dfrac{v_a\cdot(R_a-R_o)}{c_c-c_w}, \gamma \geq \dfrac{v_a\cdot R_a}{c_c}$
E19	('N','N') - (R,O)	$\dfrac{c_s+c_w}{v_2-v_1} \geq R_d \geq \dfrac{c_s}{v_4-v_3}$	$\theta_D < \dfrac{R_a\cdot v_a}{c_c}, \theta_D < \dfrac{R_o\cdot v_a}{c_w}$
E20	('H','H') - (R,O)	$\dfrac{c_s}{v_4-v_3} > R_d > \dfrac{c_s+c_w}{v_2-v_1}$	$\gamma < \dfrac{v_a\cdot(R_a-R_o)}{c_c-c_w}, \gamma \geq \dfrac{v_a\cdot R_o}{c_w}$
E21	('N','N') - (R,R)	$\dfrac{c_s}{v_2-v_1} \geq R_d \geq \dfrac{c_s}{v_4-v_3}$	$\theta_D < \dfrac{R_a\cdot v_a}{c_c}, \theta_D < \dfrac{R_o\cdot v_a}{c_w}$
E22	('H','H') - (R,R)	$\dfrac{c_s}{v_4-v_3} > R_d > \dfrac{c_s}{v_2-v_1}$	$\gamma < \dfrac{v_a\cdot R_a}{c_c}, \gamma < \dfrac{v_a\cdot R_o}{c_w}$

s_1 and s_2 represent the signals sent by normal type and honeypot defenders. a_1 and a_2 represent attacker's responses against normal type and honeypot defenders.
* γ becomes μ in the equation when defenders' signals are ('H','H')

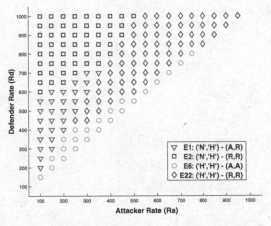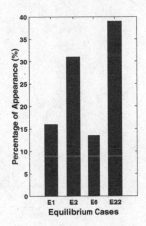

(The legend icons consist of normal type and honeypot defenders' strategies ('s1','s2'), and attacker's strategy against each defender type (a1,a2). In circle ◯ deception, the normal type defender signals 'H', the honeypot defender sends message H, and the attacker chooses to attack against both types)

Fig. 6. Nash equilibria by attacker/defender rate & histogram

type is honeypot. In this case, the attacker not only fails to attack and consume his resources but also is observed by the defender. In other words, the defender takes advantage of the confusion caused on the attacker's side by sending the same signals to the attacker.

In Fig. 6, we remove the cases where the attacker's rate exceeds the defender's service rate. Since those cases signify that the attack is already successful and the defender cannot serve her customers, we do not include the equilibrium analysis for the bottom right side of the figure.

Another interesting inference that can be drawn from Fig. 6 is that the defender doesn't anticipate signaling while the attacker's rate is approximately less than 50 % of the defender's rate (the square □ and triangle ▽ area). This result might depend on our game setting and the nature's probability of choosing defender type. Nevertheless, it is of paramount importance to point out that the defender might not need to deploy honeypots if she believes the attacker's rate is below a certain threshold. That is, on the defender side, she can come up with a tolerance rate that the attacker can consume up to without a major degradation on customer satisfaction.

Now that we observe the interaction between the players, we can focus on specific equilibrium cases and examine how they behave under different circumstances. Figure 7a and b show how the equilibria change when we modify the nature's probability of deciding if a system is of a normal type. We pick two extreme cases where $\theta_D = 0.1$ and $\theta_D = 0.9$. In Fig. 7a, since the probability of a system being normal type is very low ($\theta_D = 0.1$), the server that the attacker

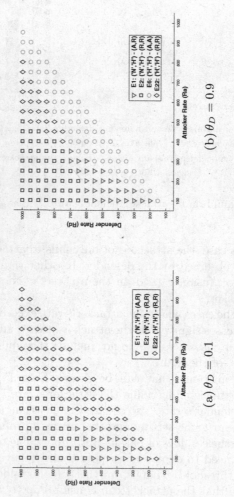

(a) $\theta_D = 0.1$ (b) $\theta_D = 0.9$

Fig. 7. Nash equilibria with different θ_D settings

targets is more likely to be a honeypot. Accordingly, we see that the attacker is less likely to choose attacking, and all circles in Fig. 6 (○) turn into diamonds (◊). Whereas, in Fig. 7b, the circles expand more and constitute a larger area as the likelihood of a system being normal type is set high. The attacker anticipates attacking whichever the signal he receives since the server that he will attack is more likely to be a normal type. In other words, the overall benefit of attacking (despite it can rarely be a honeypot) becomes an always-advantageous option for the attacker, when the nature decides the probability of being a normal server to be high ($\theta_D = 0.9$).

Figure 8 shows how the equilibria change when we vary the signaling cost by the defender rate (keeping $R_a = 500$ constant). The changes in equilibrium points indicate important implications about how the players switch strategies with respect to the parameters. The equilibria line where $R_d = 500$ begins with honeypot defender's deception strategy (plus sign +), but she switches to truthful disclosure (diamond ◊) as the cost of signaling increases. From the attacker's perspective, as we increase the defender's rate (R_d) while keeping the cost of signaling low ($c_s = 0$ or $c_s = 50$), the attacker's choices switch first from fully attacking (A,A) (plus sign +) to (A,R) (square □) and then to (R,R) (cross × or circle ○) because the attacker's degradation incurred on the customer satisfaction becomes relatively smaller. Similarly, after a certain R_d value ($R_d \geq 1000$), the defenders do not involve (do not need) any deceptions since the attacker retreats in both options because of the high defender rate.

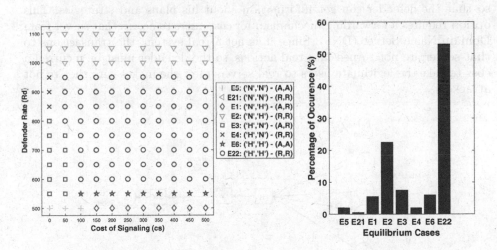

Fig. 8. Nash equilibria by cost of signaling & histogram

When we examine the strategies taken by both players in this work, we see strategy O is never the best response. Since the game is single period and the attacker takes action only once, naturally he never finds *observing* more advantageous than other strategies. In game theory, this fact can be easily proven by the dominant strategy notion in which for every possible play of strategy O, keeping everything else constant, there is at least one other strategy that yields higher payoff. It can be seen in Fig. 3 that A is always more beneficial than O for normal type defender after substituting the corresponding formulas and constant values. Similarly, strategy R is the dominant strategy for the attacker in response to the honeypot defender type. Therefore, O is a dominated strategy.

7   Discussion

For ease of calculation, the utility functions of the defender and the attacker are kept simple in this study. However, the defender's utility should be based not only on the effect of the attacks but also the satisfaction of the customers she is providing service to. In real systems (Fig. 9), attackers might not be able to drop the entire traffic but only a part of it. Similarly, when the defender blocks certain attackers, she may be preventing some legitimate users from accessing to servers, too. Therefore, it is desirable to come up with an advanced and more capable model which involves the satisfaction rate of customers and the degradation caused by the attackers [22].

In our game, when the defender sends honeypot (H) signal for the normal (N) type, she basically informs the attacker as if the system is slowing down and the attack is successful. However, the system might send the same signal to legitimate users and degrade their experience. Similarly, the defender can send 'N' signal even if the type is H to attract the attacker and have him attack so that the defender can get information about his plans and strategies. This option requires a forwarding mechanism for customers to access that server from Domain Name Server (DNS). Since it is not a real system, the transactions to that server are not turned into real actions, so the defender must be incurred a cost to take the legitimate users to real servers after she makes sure they're not attackers.

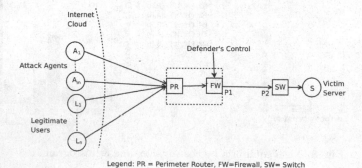

Legend: PR = Perimeter Router, FW=Firewall, SW= Switch

Fig. 9. A generic network topology in DoS attack [27]

Similarly, the constant costs that set in our experiments, e.g., c_s, v_a can be converted into a function that may reflect more realistic effect on the equilibria.

8   Conclusion and Future Work

We propose a new defense framework by proactively using deception as a means to assist in developing effective responses to DoS-type attacks and threats emanating from adversaries who may employ unconventional multi-stage stealth. Furthermore, our methodology can be generalized to be used through a game-theoretic formulation and simulation of any kind of attacks. We use game theory-based approach to gain insights and recommendations so as to increase the probability of surviving advanced and complicated attacks. The framework itself is concrete with quantification of the cost variables and can be generalized to protect critical enterprise systems such as data centers and database servers, and military fault-tolerant mission-critical systems from a persistent adversary.

In this paper, for simplicity, we examine a single target/one period game between an attacker and a defender. Investigation of multiple players (e.g., decentralized attacks by various agents and bots) in multi-period (taking turns) games is of paramount importance to explore the real-life scenarios taking place during a distributed DoS attack. Employing an advanced network configuration and a real-world DoS attack scenario for our model is also left as future work to involve the satisfaction rate of customers and reflect effects of attacks on the defender.

References

1. Agah, A., Das, S.K.: Preventing DoS attacks in wireless sensor networks: a repeated game theory approach. IJ Netw. Secur. **5**(2), 145–153 (2007)
2. Bagwell, K., Ramey, G.: Advertising and pricing to deter or accommodate entry when demand is unknown. Int. J. Indus. Organ. **8**(1), 93–113 (1990)
3. Balas, E.: Know Your Enemy: Learning About Security Threats. Addison Wesley, Boston (2004)
4. Basagiannis, S., Katsaros, P., Pombortsis, A., Alexiou, N.: Probabilistic model checking for the quantification of DoS security threats. Comput. Secur. **28**(6), 450–465 (2009)
5. Bell, J.B., Whaley, B.: Cheating and Deception. Transaction Publishers, Brunswick (1991)
6. Carroll, T.E., Grosu, D.: A game theoretic investigation of deception in network security. Secur. Commun. Netw. **4**, 1162–1172 (2011)
7. Center, C.C.: Results of the distributed-systems intruder tools workshop. Software Engineering Institute (1999)
8. Cohen, F., Koike, D.: Misleading attackers with deception. In: Proceedings from the Fifth Annual IEEE SMC Information Assurance Workshop, pp. 30–37. IEEE (2004)
9. Fong, Y.: Private information of nonpaternalistic altruism: exaggeration and reciprocation of generosity. Adv. Theor. Econ. **9**(1), 1 (2009)
10. Fu, X., Yu, W., Cheng, D., Tan, X., Streff, K., Graham, S.: On recognizing virtual honeypots and countermeasures. In: 2nd IEEE International Symposium on Dependable, Autonomic and Secure Computing, pp. 211–218. IEEE (2006)
11. Fudenberg, D., Tirole, J.: Perfect Bayesian equilibrium and sequential equilibrium. J. Econ. Theor. **53**(2), 236–260 (1991)

12. Gao, X., Zhu, Y.-F.: DDoS defense mechanism analysis based on signaling game model. In: 2013 5th International Conference on Intelligent Human-Machine Systems and Cybernetics, pp. 414–417 (2013)
13. Hamilton, S.N., Miller, W.L., Ott, A., Saydjari, O.S.: The role of game theory in information warfare. In: 4th Information Survivability Workshop (ISW-2001/2002), Vancouver, Canada (2002)
14. Heitzenrater, C., Taylor, G., Simpson, A.: When the winning move is not to play: games of deterrence in cyber security. In: Khouzani, M.H.R., Panaousis, E., Theodorakopoulos, G. (eds.) Decision and Game Theory for Security, pp. 250–269. Springer, Heidelberg (2015)
15. Jiang, Z., Ge, Y., Li, Y.: Max-utility wireless resource management for best-effort traffic. IEEE Trans. Wirel. Commun. 4(1), 100–111 (2005)
16. La, Q.D., Quek, T., Lee, J., Jin, S., Zhu, H.: Deceptive attack and defense game in honeypot-enabled networks for the internet of things. IEEE Internet Things J. PP(99), 1 (2016)
17. Lau, F., Rubin, S.H., Smith, M.H., Trajkovic, L.: Distributed denial of service attacks. In: 2000 IEEE International Conference on Systems, Man, and Cybernetics, vol. 3, pp. 2275–2280. IEEE (2000)
18. Lye, K.W., Wing, J.M.: Game strategies in network security. Int. J. Inf. Secur. 4(1–2), 71–86 (2005)
19. McCarty, B.: The honeynet arms race. IEEE Secur. Priv. 1(6), 79–82 (2003)
20. Meadows, C.: A cost-based framework for analysis of denial of service in networks. J. Comput. Secur. 9(1), 143–164 (2001)
21. Píbil, R., Lisý, V., Kiekintveld, C., Bošanský, B., Pěchouček, M.: Game theoretic model of strategic honeypot selection in computer networks. In: Decision and Game Theory for, Security, pp. 201–220 (2012)
22. Rasouli, M., Miehling, E., Teneketzis, D.: A supervisory control approach to dynamic cyber-security. In: Poovendran, R., Saad, W. (eds.) Decision and Game Theory for Security, pp. 99–117. Springer, Heidelberg (2014)
23. Rass, S., Rainer, B.: Numerical computation of multi-goal security strategies. In: Poovendran, R., Saad, W. (eds.) Decision and Game Theory for Security, pp. 118–133. Springer, Heidelberg (2014)
24. Rowe, N.C., Custy, E.J., Duong, B.T.: Defending cyberspace with fake honeypots. J. Comput. 2(2), 25–36 (2007)
25. Shen, S., Yue, G., Cao, Q., Yu, F.: A survey of game theory in wireless sensor networks security. J. Netw. 6(3), 521–532 (2011)
26. Wang, W., Chatterjee, M., Kwiat, K.: Coexistence with malicious nodes: a game theoretic approach. In: International Conference on Game Theory for Networks, GameNets 2009, pp. 277–286. IEEE (2009)
27. Wu, Q., Shiva, S., Roy, S., Ellis, C., Datla, V.: On modeling and simulation of game theory-based defense mechanisms against DoS and DDoS attacks. In: Proceedings of the 2010 Spring Simulation Multiconference, p. 159. Society for Computer Simulation International (2010)
28. Yang, L., Mu, D., Cai, X.: Preventing dropping packets attack in sensor networks: a game theory approach. Wuhan Univ. J. Nat. Sci. 13(5), 631–635 (2008)
29. Zhuang, J., Bier, V.M.: Reasons for secrecy and deception in homeland-security resource allocation. Risk Anal. 30(12), 1737–1743 (2010)
30. Zhuang, J., Bier, V.M.: Secrecy and deception at equilibrium, with applications to anti-terrorism resource allocation. Defence Peace Econ. 22(1), 43–61 (2011)
31. Zhuang, J., Bier, V.M., Alagoz, O.: Modeling secrecy and deception in a multiple-period attacker-defender signaling game. Eur. J. Oper. Res. 203(2), 409–418 (2010)

Data Exfiltration Detection and Prevention: Virtually Distributed POMDPs for Practically Safer Networks

Sara Marie Mc Carthy[1(✉)], Arunesh Sinha[1], Milind Tambe[1],
and Pratyusa Manadhata[2]

[1] University of Southern California, Los Angeles, USA
{saramarm,tambe}@usc.edu, aruneshsinha@gmail.com
[2] Hewlett Packard Labs, Princeton, USA
pratyusa.k.manadhata@hpe.com

Abstract. We address the challenge of detecting and addressing advanced persistent threats (APTs) in a computer network, focusing in particular on the challenge of detecting data exfiltration over Domain Name System (DNS) queries, where existing detection sensors are imperfect and lead to noisy observations about the network's security state. Data exfiltration over DNS queries involves unauthorized transfer of sensitive data from an organization to a remote adversary through a DNS data tunnel to a malicious web domain. Given the noisy sensors, previous work has illustrated that standard approaches fail to satisfactorily rise to the challenge of detecting exfiltration attempts. Instead, we propose a decision-theoretic technique that sequentially plans to accumulate evidence under uncertainty while taking into account the cost of deploying such sensors. More specifically, we provide a fast scalable POMDP formulation to address the challenge, where the efficiency of the formulation is based on two key contributions: (i) we use a *virtually distributed POMDP* (VD-POMDP) formulation, motivated by previous work in distributed POMDPs with sparse interactions, where individual policies for different sub-POMDPs are planned separately but their sparse interactions are only resolved at execution time to determine the joint actions to perform; (ii) we allow for abstraction in planning for speedups, and then use a fast MILP to implement the abstraction while resolving any interactions. This allows us to determine optimal sensing strategies, leveraging information from many noisy detectors, and subject to constraints imposed by network topology, forwarding rules and performance costs on the frequency, scope and efficiency of sensing we can perform.

1 Introduction

Advanced persistent threats can be one of the most harmful attacks for any organization with a cyber presence, as well as one of the most difficult attacks to defend against. While the end goal of such attacks may be diverse, it is often the case that intent of an attack is the theft of sensitive data, threatening

© Springer International Publishing AG 2016
Q. Zhu et al. (Eds.): GameSec 2016, LNCS 9996, pp. 39–61, 2016.
DOI: 10.1007/978-3-319-47413-7_3

the loss of competitive advantage and trade secrets as well as the leaking of confidential documents, and endangerment of national security [4,13]. These attacks are sophisticated in nature and often targeted to the vulnerabilities of a particular system. They operate quietly, over long periods of time and actively attempt to cover their tracks and remain undetected. A recent trend in these attacks has relied on *exploiting Domain Name System (DNS) queries* in order to provide channels through which exfiltration can occur [8,20]. These DNS based exfiltration techniques have been used in well-known families of malware; e.g., FrameworkPOS, which was used in the Home Depot data breach involving 56 million credit and debit card information [2].

At a high level, DNS exfiltration involves an attacker-controlled malware inside an organization's network, an external malicious domain controlled by the attacker, and a DNS server authoritative for the domain that is also controlled by the same attacker. The malware leaks sensitive data by transmitting the data via DNS queries for the domain; these queries traverse the DNS hierarchy to reach the attacker controlled DNS server. Attackers can discretely transfer small amounts of data over long periods of time disguised as legitimate user generated DNS queries. Detecting and protecting against such an attack is extremely difficult as the exfiltration attempts are often lost in the high volume of DNS query traffic and any suspicious activity will not be immediately obvious. In both academia and industry, multiple detectors have been proposed to detect DNS exfiltration. However, because of the sophisticated and covert nature of these attacks, detectors designed to protect against these kinds of attacks either often *miss attacks* or are plagued by *high false positive rates, misclassifying legitimate traffic* as suspicious, and potentially *overwhelming a network administrator* with suspicious activity alerts; these issues have been identified with machine learning based detectors [23], pattern matching based detectors [1] and information content measuring detector [19].

We focus on the problem of *rapidly determining malicious domains* that could be potentially exfiltrating data, and then *deciding whether to block traffic or not*. In our problem, the defender observes a stream of suspicious DNS based exfiltration alerts (or absence of alerts), and is tasked with inferring which of the domains being queried are malicious, and determining the best response (block traffic or not) policy. Unfortunately, as stated earlier, detectors are inherently *noisy* and each single alert does not provide a high confidence estimate about the security state. Thus, the defender needs to come up with a sequential plan of actions while dealing with uncertainty in the network and in the alerts, and must weight the cost of deploying detectors to increase their knowledge about malicious domains with the potential loss due to successful attacks as well as the cost of misclassifying legitimate network use. This problem of active sensing is common to a number of cyber security problems; here we focus on the challenge of data exfiltration over DNS queries.

There has been a large amount of work on how to deal with and make decision under uncertainty. Problems such as ours can be well modeled using Partially Observable Markov Decision Process (POMDP) to capture the dynamics of real-world sequential decision making processes, and allow us to reason about uncertainty and compute optimal policies in these types of environments. However a major drawback to these models is that they are *unable to scale* to solve any problem instances of reasonable size. In order to be successful in the cyber domain, such a models needs to be able to handle extremely large problem instances, as networks are often extremely complex, with lots of moving parts. Additionally, due to the salient nature of network states, we need to be able to make *decisions in real time* in order to observe and quickly react to a potential threat.

To address this challenge we make the following key contributions: (1) We provide a formal model of the DNS data exfiltration problem. We propose a new decision making framework using Partially Observable Markov Decision Processes (POMDPs). (2) We address the scalability issued faced when dealing with large networks by proposing a series of abstractions of the original POMDP. These include using abstract action and observation space. (3) Another step in the abstraction is a *new paradigm* for solving these models by factoring the POMDP into several sub-POMDPs and solving each individual sub-POMDP separately offline; this is motivated by previous work in distributed POMDPs with sparse interactions. We provide techniques for policy aggregation to be performed at runtime in order to combine the abstract optimal actions from each sub-POMDP to determine the final joint action. We denote this model as a *virtually distributed POMDP* (VD-POMDP). We provide conditions under which our methods are guaranteed to result in the optimal joint policy, and provide empirical evidence to show that the final policy still performs well when these conditions do not hold. (4) Finally we provide experimental evaluation of our model in a real network testbed, where we demonstrate the ability to correctly identify real attacks.

2   Background and Related Work

We split our discussion of the required background and related work for this paper along two broad categories that are covered in the two sub-sections below.

2.1   DNS Exfiltration

Sensitive data exfiltration from corporations, governments, and individuals is on the rise and has led to loss of money, reputation, privacy, and national security. For example, attackers stole 100 million credit card and debit card information via breaches at Target and Home Depot [11]; a cluster of breaches at LinkedIn, Tumblr, and other popular web services led to 642 million stolen passwords [5]; and the United States Office of Personnel Management (OPM) data breach resulted in 21.5 million records, including security clearance and fingerprint information, being stolen [27].

In the early days, exfiltration happened over well known data transfer protocols such as email, File Transfer Protocol (FTP), and Hypertext Transfer

Protocol (HTTP) [13]. The seriousness of the problem has led to several "data loss prevention (DLP)" products from the security industry [15,24] as well as academic research for monitoring these protocols [7,10]. These solutions monitor email, FTP, and other well known protocols for sensitive data transmission by using keyword matching, regular expression matching, and supervised learning.

The increased monitoring of the protocols has forced attackers to come up with ingenious ways of data exfiltration. One such technique used very successfully in recent years is exfiltration over DNS queries [1,20]. Since DNS is fundamental to all Internet communication, even the most security conscious organizations allow DNS queries through their firewall. As illustrated in Fig. 1, an adversary establishes a malicious domain, *evil.com*, and infects a client in an organization with malware. To exfiltrate a data file, the malware breaks the file into blocks, $b_1, b_2, \cdots, b_n$, and issues a sequence of DNS queries, $b_1.evil.com$, $b_2.evil.com$, $\cdots$, $b_n.evil.com$. If their responses are not cached, the organization's DNS server will forward them to the nameserver authoritative for *evil.com*; at this point, the adversary controlling the authoritative nameserver can reconstruct the data file from the sequence of blocks.

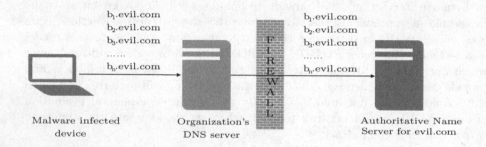

Fig. 1. Data exfiltration over DNS.

The data transmission is covert and can be accomplished by various means such as a particular sub-domain query meaning bit 1 and another sub-domain query meaning bit 0, or even the timing between queries can leak information. By compressing the data at the client, and by varying query lengths and the time interval between successive queries an adversary can adjust the bandwidth of the communication channel. The adversary could choose to transfer data as quickly as possible (long and rapid domain queries) or slowly (short queries spaced apart in time), depending on the intent behind the attack. To further hide exfiltration activity, the data blocks can be encrypted by the client before the queries are issued, and decrypted by the adversary. Further, the adversary can encode instructions within its responses to establish a two-way communication tunnel.

Hence building a reliable DNS exfiltration detector is extremely challenging. A recent work on building a detector for DNS exfiltration using measurement of information content of a DNS channel provides techniques that we use to build the low level detector in our problem setting [19]. Apart from this work in academia, there has been some work in the industry that use various heuristics to build low level detectors for DNS exfiltration [1]; examples of such heuristics are lengths of DNS queries and responses, sizes of query and response packets, entropy of DNS queries, total volume of DNS queries from a device, and total volume of DNS traffic to a domain. As far as we know, we are the first to build a cost based sequential planning tool that uses the imperfect low level detectors to determine if a domain is involved in exfiltrating data over DNS.

2.2 POMDP

There has been a large amount of work on how to deal with and make decisions in stochastic environments under uncertainty using POMDPs. However, it is known that offline POMDP solving is intractable for large problems [9,14,18] and given our fast response requirements an online POMDP solver is also not feasible [21]. We show empirically that our original POMDP is simply impractical to solve for even a small network of 3–4 computers. Thus, in this paper, we focus on speeding up the offline POMDP solving by performing a series of abstractions of the original POMDP. Our technique of solving the POMDP is inspired by conflict resolution techniques in solving distributed POMDP [12,17] and distributed POMDPs with sparse interactions [25,26]. While our VD-POMDP does not have an inherent distributed structure, we break up the original POMDP into multiple domain specific POMDPs to build a virtually distributed POMDP; this allows for scalability. However, instead of resolving conflicts at runtime or using sparse interaction (which does not exist in our split POMDPs), we modify our action space so that the actions output by each domain specific POMDP is at a higher level of abstraction. With these abstract actions from each POMDP we provide a fast technique to come up with the joint action by combining the abstract actions at execution time. The main difference with existing work on POMDP and distributed POMDP is that we reason about policies at execution time, allowing efficient "groundlevel" implementation of abstract policy recommendation from multiple virtual POMDPs. Further, the abstraction of actions is possible in our problem due to the special relation between detecting malicious domains and sensing of traffic on a network node (see Model section for details).

While, the main step in VD-POMDP compaction is based on similar ideas of factored POMDPs used in past literature [16], our approach critically differs as we do not just factor the belief state of the POMDP, but split it into multiple POMDPs per domain. Also, distinct from general distributed POMDP [6] the multiple agents (for each domain) do not share a common reward function and neither is the reward function a sum of each of their rewards (Table 1).

3 Model

The local computer network can represented as a graph $G(N, E)$, where the nodes N correspond to the set of hosts in the network, with edges E if communication between the hosts is allowed. Each node n has a particular value v_n

Table 1. Notation

G(N, E)	Graph representing network
v_n	Value of data at the nth node
$v_{[d]}$	Average value of the set of channels to the dth domain
w_n	Volume of traffic at the nth node
$w_{[d]}$	Total volume of the set of channels to the dth domain
d	The dth domain
X_d	True $\{0, 1\}$ state of the dth domain
M_d	Estimated $\{0, 1\}$ state of the dth domain
X	Set of all X_d random variables
$c_k = <n, \ldots d>$	k^{th} channel from node n to domain d
$C_{[d]}$	Subset of channels ending with the dth domain
C	Set of all channels
τ_k	Threshold set for channel c_k
a_n	Binary variable indicating if node n is sensed or not
z_k	Binary variable indicating if channel c_k is sensed or not
Ω_k	$\{0, 1\}$ observation on k^{th} channel
$\Omega_{[d]}$	Subset of observations for channels ending with the dth domain

corresponding to the value of data stored at that computer. At any point in time t each node has a variable traffic volume of requests w_n^t passing through it. We assume there are D domains, where for tractability we assume D is the number of domains that have ever been queried for the given computer network. DNS queries made from internal nodes in the network are forwarded to special nodes, either access points or internal DNS servers, and then forwarded to external servers from these points. A channel c_k over which exfiltration can occur is then a path, starting at source node, where the query originated, traveling through several nodes in the network and finishing at a target domain d. The total number of channels is K. We use $n \in c_k$ to denote any node in the path specified by c_k.

Let X_d be a random binary variable denoting whether domain d is malicious or legitimate. We assume that a malicious domain will always be malicious, and that legitimate domains will not become compromised; this means that the state X_d does not change with time. Even though legitimate domains get compromised in practice, attackers often use new malicious domains for DNS exfiltration since an attacker needs to control both a domain and the authoritative name server for a domain to successfully carry out exfiltration. In other words, legitimate domains that get compromised are rarely used in DNS exfiltration. Hence in our model, it is reasonable to assume that domains don't change their states. We call the active sensing problem, the challenge of determining the values of X_d. In order to do this we may place detectors at nodes in the network; the

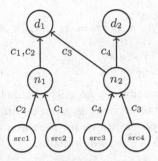

Fig. 2. Example of a network with two domains, 4 source hosts and 4 channels. Channels $c1$, $c2$, $c3$ go from sources 1, 2, and 3 to domain d_1 while channel c_4 goes from source 4 to domain d_2. We may consider the situation where we can only turn on one detector at any time step, either at node n_1 or n_2, and choose to sense on channels $\{c_1, c_2\}$ or $\{c_3, c_4\}$. We can additionally chose thresholds τ_j for each channel. Each source host has a value v_n and each node n has traffic volume w_n.

state of a detector (off/on) at any node in the network is $a_n \in \{0, 1\}$. Each detector monitors all the channels passing through that particular node, i.e., all $c_k : n \in c_k$. We use the binary variable z_k to indicate if channel c_k is monitored. We can set discrete thresholds individually for each channel; lower thresholds correspond to higher sensitivity to information flow out of any particular channel. Because each channel is associated with a domain, we set a threshold τ_k for each channel. We use $|\tau|$ to denote the number of discrete threshold choices available. We then get observations in the form of alerts for each channel $\Omega_k \in \{0, 1\}$. The probability of receiving an alert for any channel is characterized by some function $\alpha(\tau_k)$ if the channel is malicious and $\beta(\tau_k)$ if the channel is legitimate. Finally, the defender classifies the state of domain d as malicious or legitimate, indicated by M_d.

3.1   The POMDP Model

Our POMDP model is a tuple (S, A, T, Ω, O, R) with state space S, action space A, state transition function T, observation space Ω, observation probabilities O and reward function R. Additionally define the average value of the channels to domain d as $v_{[d]} = \sum\limits_{n:n \in C_{[d]}} \dfrac{v_n}{|C_{[d]}|}$. Below we list the details of components of POMDP model. The state captures the true security state of every domain and the actions specify the thresholds for monitoring each channel, the nodes to be monitored and the decision about which domains are classified as malicious. As we assume the security state of the system does not change, the transition function is straightforward.

States $S = \langle X_1, \ldots X_D \rangle$

Actions $A = A_c \times A_n \times A_d$ where $\langle \tau_1, \ldots \tau_K \rangle \in A_c$, $\langle a_1 \ldots a_N \rangle \in A_n$
                 and $\langle M_1 \ldots M_D \rangle \in A_d$

Transision $T(s', s) = \begin{cases} 1 & \text{iff } s' = s \\ 0 & \text{else} \end{cases}$

Next, we obtain an observation Ω_k for each channel, and as stated earlier for each channel the probability of an alert is given by functions α and β. We state the probability first for the observations for each domains, and then for all the observations using independence across domains.

Observations $\Omega = \langle \Omega_1 \ldots \Omega_K \rangle$

Observation Prob $O(\Omega_{[d]} | X_d, A) = \begin{cases} \displaystyle\prod_{k:k \in C_{[d]} \wedge z_k = 1} \alpha(\tau_k)^{\Omega_k} (1 - \alpha(\tau_k))^{1 - \Omega_k} & \text{if } X_d = 1 \\ \displaystyle\prod_{k:k \in C_{[d]} \wedge z_k = 1} \beta(\tau_k)^{\Omega_k} (1 - \beta(\tau_k))^{1 - \Omega_k} & \text{if } X_d = 0 \\ 0 & \text{else} \end{cases}$

$$O(\Omega | X, A) = \prod_d O(\Omega_{[d]} | X_d, A_{[d]})$$

Finally, the reward for the POMDP is given by the following equation:

$$R(S, A) = -\left(\sum_d \left(X_d(1 - M_d)v_{[d]} + (1 - X_d)M_d w_{[d]} \right) + \sum_n^N a_n w_n \right)$$

The reward contains two cost components: the first component has two terms for each domain that specify the penalty for mislabeling a domain and the second component is the cost of sensing over the nodes. When a malicious domain d is labeled safe then the defender pays a cost $v_{[d]}$, i.e., the average value of channels going to domain d; in the opposite mislabeling the defender pays a cost $w_{[d]}$, i.e., a cost specified by loss of all traffic going to domain d. While this POMDP model captures all relevant elements of the problem, it is not at all tractable. Consider the input variables to this model, the number of domains D, the number of nodes N and the number of channels k. The state space grows as $O(2^D)$, the action space is $O(2^N |\tau|^K 2^D)$ and the observation space is $O(2^K)$. This full formulation is exponential in all the input variables and cannot scale to larger, realistic network instances (we also show this experimentally in the Evaluation Section). In order to reduce the combinatorial nature of the observation space, action space and state space, we introduce a compact representation for the observation and action space and a factored representation for the state space that results in splitting the POMDP into multiple POMDPs.

4   POMDP Abstraction

We represent the POMDP compactly by using three transformations: (1) we use the same threshold for very channel going to the same domain and change the action space from sensing on nodes to sensing on channels, (2) reduce the observation space by noting that only the number of alerts for each domain are required and not which of the channels generated these alerts and (3) factoring the whole POMDP by domains, then solve a POMDP per domain and combine the solutions at the end. Next, we describe these transformations in details.

Abstract Actions. We can reduce the action space by (1) enforcing that the same threshold is set for all channels going to the same domain and (2) by reasoning about which channels to sense over instead of which nodes to sense on. The first change reduces the action space from a $|\tau|^K$ dependance to $|\tau|^D$, where $|\tau|$ is the discretization size of the threshold for the detector. The new set of threshold actions is then $A_c = \langle \tau_1, \ldots \tau_D \rangle$. The second change replaces the set of actions on nodes A_n with a set of actions on channels $A_k = \langle s_{k_{[1]}} \ldots s_{k_{[D]}} \rangle$, where $s_{k_{[d]}}$ is the number of channels to be sensed out of the $|C_{[d]}|$ channels that end in domain d. This changes the action space complexity from 2^N to $|C_{[d]}|^D$. Then the action space is given by

Actions $A = \Lambda_c \times \Lambda_k \times \Lambda_d$

where $\langle \tau_1, \ldots \tau_D \rangle \in A_c, \langle s_{k_{[1]}} \ldots s_{k_{[D]}} \rangle \in A_k$ and $\langle M_1 \ldots M_D \rangle \in A_d$.

In order to properly compute the reward we need to compute the cost of any action in A_k. To do this we need to build a lookup table mapping each action in A_k to an action in A_n, and hence obtain the cost of actions in A_k. Because we will always choose the lowest cost way to sense on a number of channels, the action of sensing a specified number of channels can be mapped to the set of nodes that minimizes the cost of sensing the specified number of channels. We can compute this using the following mixed integer linear program (MILP) mincost($\langle s_{k_{[1]}} \ldots s_{k_{[D]}} \rangle$).

$$\min_{z_k, a_n} \sum_n a_n w_n \tag{1}$$

$$z_k \leq \sum_{n \in c_k} a_n \quad \forall k \in \{1, \ldots, K\} \tag{2}$$

$$\sum_{c_k \in C_{[d]}} z_k \geq s_{k_{[d]}} \quad \forall d \in \{1, \ldots, D\} \tag{3}$$

$$z_k \in \{0, 1\} \quad a_n \in \{0, 1\} \tag{4}$$

The mincost($\langle s_{k_{[1]}} \ldots s_{k_{[D]}} \rangle$) MILP needs to be solved for every action $\langle s_{k_{[1]}} \ldots s_{k_{[D]}} \rangle \in A_k$, i.e., we need to fill in a table with $O(|C_{[d]}|^D)$ entries. If we take the example network in Fig. 2, the old action space is $A_n = \{\{\emptyset\}, \{n_1\}, \{n_2\}, \{n_1, n_2\}\}$, and the new action space is $A_k = \{\{0,0\}, \{1,0\}, \{2,0\}, \{3,0\}, \{0,1\}, \{1,1\}, \{2,1\}, \{3,1\}\}$. In order to map back to the representation using nodes, we build the mapping: $\{0,0\} \to \emptyset, \{1,0\} \to \{n_1\}, \{2,0\} \to \{n_1\}, \{3,0\} \to \{n_1\}, \{0,1\} \to \{n_2\}, \{1,1\} \to \{n_1, n_2\}, \{2,1\} \to \{n_1, n_2\}, \{3,1\} \to \{n_1, n_2\}$. However, the problem of converting from number of channels to nodes (stated as mincost($\langle s_{k_{[1]}} \ldots s_{k_{[D]}} \rangle$) above) is not easy as the following theorem shows:

Theorem 1. *The problem of converting from number of channels to nodes is NP hard to approximate to any factor better than $\ln |N|$.*

Proof. We perform a strict approximation preserving reduction from the set cover problem. Consider a set cover problem. We are given a universe of m

elements E and u subsets of E: U. Form a node n_u for each subset $u \in U$ and a domain d_e for each element $e \in E$. For any particular element e and any node containing that element, connect it to the domain d_e. Then, these connections, say from l nodes, defines l channels each starting from a node and ending in d_m. For any domain d choose $s_{k_{[d]}} = 1$, i.e., at least one channel needs to be sensed. It can be easily seen that for any channel c_k in this network there is a unique node it passes through: call it $n(k)$. Choose $w_n = 1$. Then, the optimization problem to be solved is the following:

$$\min_{z_k, a_n} \quad \sum_n a_n \tag{5}$$

$$z_k \leq a_{n(k)} \quad \forall k \in \{1, \ldots, K\} \tag{6}$$

$$\sum_{c_k \in C_{[d]}} z_k \geq 1 \quad \forall d \in \{1, \ldots, D\} \tag{7}$$

$$z_k \in \{0, 1\} \quad a_n \in \{0, 1\} \tag{8}$$

First, we prove that the constraints of this optimization specify a choice of subsets (nodes) the union of which equals the set E. Since all channels going to domain d corresponds to a unique element e and at least one channel going to d is chosen (Eq. 7), this implies at least one node containing e is selected (Eq. 6). Thus, the set of nodes (hence subsets) contains all elements e.

Given, the feasible space is given by a set of subsets (nodes) the union of which produces E, the objective clearly produces the minimum number of such sets. Also, any approximate solution with guarantee α maps to an α approximate solution of the set cover problem. The theorem follows from the lack of better than $\ln n$ approximatability of set cover. $\square$

Abstract Observations. As we only reason about the state of each domain in the network, not each individual channel, we can aggregate the observation in order to reduce the observation space. Thus, instead of recording which channel generated an alert, we only record total number of alerts per domain. Given there are $|C_{[d]}|$ channels going to domain d then the observations for each domain lie in $\{0 \ldots |C_{[d]}|\}$. This observation space for each domain is then linear in the number of channels $O(|C_{[d]}|)$. The full joint observation space is exponential in the number of domains $O(|C_{[d]}|^D)$.

The set of observations is then $\Omega = \langle \Omega_1, \ldots, \Omega_D \rangle$ where $\Omega_d \in \{0 \ldots |C_{[d]}|\}$ corresponding to the number of alerts from all $|C_{[d]}|$ channels going to domain d. Because there is now multiple way for us to get this single observation, the observation probability function for each domain also needs to be modified.

$$O(\Omega_d | X_d, A) = \begin{cases} \binom{s_{k_{[d]}}}{\Omega_d} \alpha(\tau_d)^{\Omega_d} (1 - \alpha(\tau_d))^{s_{k_{[d]}} - \Omega_d} & \text{if } X_d = 1 \\ \binom{s_{k_{[d]}}}{\Omega_d} \beta(\tau_d)^{\Omega_d} (1 - \beta(\tau_d))^{s_{k_{[d]}} - \Omega_d} & \text{if } X_d = 0 \\ 0 & \text{else} \end{cases}$$

VD-POMDP Factored Representation. Looking at both the observation probability function as well as the belief update, we can consider a factored representation of this POMDP, by factoring these by domains. If we then separate out these factored components and create a new sub-agent for each factor, so that we now have a total of D POMDP's, we can greatly reduce the state space, observation space and action space for each individual sub agent. The model for each of these individual POMDP is then given as follows.

States $\qquad\qquad$ $S = X_d$

Actions $\qquad\qquad$ $A = \tau_d \times \{0,\ldots,|C_{[d]}|\} \times M_d$

Transition $\qquad$ $T(s', s) = \begin{cases} 1 & \text{iff } s' = s \\ 0 & \text{else} \end{cases}$

Observations $\qquad$ $\Omega = \langle \Omega_1,\ldots,\Omega_D \rangle$ where $\Omega_d \in \{0,\ldots,|C_{[d]}|\}$

$$O(\Omega_d|X_d, A) = \begin{cases} \binom{s_{k_{[d]}}}{\Omega_d} \alpha(\tau_d)^{\Omega_d}(1 - \alpha(\tau_d))^{s_{k_{[d]}} - \Omega_d} & \text{if } X_d = 1 \\ \binom{s_{k_{[d]}}}{\Omega_d} \beta(\tau_d)^{\Omega_d}(1 - \beta(\tau_d))^{s_{k_{[d]}} - \Omega_d} & \text{if } X_d = 0 \\ 0 & \text{else} \end{cases}$$

Reward $\qquad$ $R(S, A) = -\left(X_d(1 - M_d)v_{[d]} + (1 - X_d)M_d w_{[d]} + \text{mincost}(s_{k_{[d]}})\right)$

The complexity of the state space is reduced to $O(1)$, the action space is $O(|\tau||C_{[d]}|)$ and the observation space is $O(|C_{[d]}|)$. Table 2 shows the comparative complexities of the original POMDP model and the VD-POMDP model. As we use channels as actions for each domain specific POMDP, we still need to construct the lookup table to map channels as actions to nodes as actions in order to obtain the cost of each action on channels. Factoring the model in the way described above also simplifies the construction of this lookup table from actions on channels to actions on nodes, and hence computing $\text{mincost}(s_{k_{[d]}})$ can be done in a much simpler way for the VD-POMDPs. We solve a similar (MILP) as in (2)–(4) but for each VD-POMDP for domain d; thus, we only need to fill in a table with $O(|C|)$ entries, one for each of the $s_{k_{[d]}}$ actions for each domain d. The new MILP formulation is given in Eqs. 10–12. Observe that unlike the MILP (2)–(4) used to build the lookup table for the original POMDP, this MILP is solved for a fixed domain d.

Table 2. Complexities of full and VD-POMDP models with original and compact representations.

	Full POMDP		VD-POMDP
	Original	Abstract	
State	$O(2^D)$	$O(2^D)$	$O(1)$
Action	$O(2^N|\tau|^K 2^D)$	$O(|C_{[d]}|^D|\tau|^D 2^D)$	$O(2^{|C_{[d]}|}|\tau|)$
Observation	$O(2^K)$	$O(|C_{[d]}|^D)$	$O(|C_{[d]}|)$

$$\min_{z_k,a_n} \quad \sum_n a_n w_n \tag{9}$$

$$z_k \le \sum_{n \in C_k} a_n \tag{10}$$

$$\sum_{c_k \in C_{[d]}} z_k \ge s_{k_{[d]}} \tag{11}$$

$$z_k \in \{0,1\} \quad a_n \in \{0,1\} \tag{12}$$

While the above optimization is much more simpler than the corresponding optimization for the original POMDP, it is still a hard problem:

Theorem 2. *The problem of coverting the number of channels to nodes for each VD-POMDP is NP Hard.*

Proof. We reduce from the min knapsack problem. The min knapsack problem is one where the objective is to minimize the value of chosen items subject to a minimum weight W being achieved, which is a well known hard problem. Also, wlog, we can assume weights of items and W to be integers. Given a min knapsack with n items of weights w_i' and value v_i and min weight bound W form an instance of our problem with n nodes (mapped to items) and each node i having w_i' channels going directly to domain d. It can be easily seen that for any channel c_k in this network there is a unique node it passes through: call it $n(k)$. Each node i also has traffic $w_i = v_i$. Also, $s_{k_{[d]}} = W$. Then, the optimization problem being solved is

$$\min_{z_k,a_n} \quad \sum_n a_n v_n \text{ subject to } z_k \le a_{n(k)}, \sum_{c_k \in C_{[d]}} z_k \ge W, z_k \in \{0,1\}, a_n \in \{0,1\}$$

Note that in the constraints, whenever a node is selected $a_{n(k)} = 1$ then making all w_i channels in it one makes the weight constraints less tight. Thus, any values of a_n, z_k satisfying the constraints specify a set of nodes such that the sum of its weights is $\ge W$. Coupled with the fact that the objective minimizes the values of selected nodes, the solution to this optimization is a solution for the min knapsack problem. □

Policy Execution: The solutions to each of these VD-POMDP's give us an action $\langle M_d^*, \tau_d^*, s_{k_{[d]}}^* \rangle$ corresponding to a labeling of malicious or legitimate for that particular domain d, the threshold, and the desired number of channels to sense over. However, at execution time we need to turn on detectors on nodes. Thus, in order to aggregate these factored actions to determine the joint action to take at execution, we need to map the output from each POMDP back to a set of sensing actions on nodes. This can be easily accomplished by solving a single instance of the larger MILP (2)–(4) with the $s_{k_{[d]}}$ values set to $s_{k_{[d]}}^*$ (Fig. 3).

We *emphasize* here the importance of using the abstract channels as actions instead of nodes. The possibly alternate approach with nodes as action for each sub-POMDP and just taking union of the nodes output by each domain specific

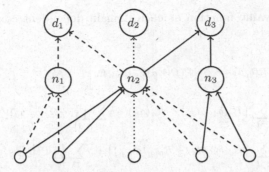

Fig. 3. Sample network with 3 domains, 3 nodes and 5 sources. The dashed lines are channels to domain d_1 the dotted line is the channel to domain d_2 and the solid lines are channels to d_3.

POMDP, when the channels are not disjoint, can result in over sensing. Consider the example below, where there are 4 channels going to domains d_1 and d_3 and one to d_2 and let us currently be in a belief state where the optimal action for domain d_1 and d_3 would be to sense on 2 channels out of the 4 going to each domain and the optimal action for d_2 is to sense on the one channel. Working in an action space of nodes, the VD-POMDP for d_1 would choose to sense on node n_1, the VD-POMDP for d_3 would choose to sense on n_3 as it has the lowest amount of traffic for 2 channels and the one for d_2 would choose to sense on n_2 as it is the only option. Taking the union of these would result in all the sensors being turned on. However, we can see that choosing only to sense on node n_2 satisfies the sensing requirements of all three separate VD-POMDPs.

Next, we identify a condition under which the solution from the larger MILP is optimal. In the next section, we show empirically that even when this condition is not met our approach is close to optimal.

Theorem 3. *The above described technique of aggregating solutions of the VD-POMDPs is optimal for the original POMDP iff the solution to the MILP (2)–(4) for any VD-POMPD policy action results in an equality for the constraint (3).*

Proof. First, with the standard representation the global value function given in Eqs. 13 and 14, cannot generally be fully decomposed by domain due to the $R_n(a_n)$ term which couples the actions for each domain through the sending cost. The decomposition is only possible in special instances of the problem, such as if network of channels were completely disconnected. The action of selecting nodes can be partitioned by domains as $a_{n_{[d]}}$. Then, the cost associated with sensing on the nodes could be written as a sum of domain dependent terms $R_n(a_n) = \sum_d R_{n_{[d]}}(a_{n_{[d]}})$. Also, all actions (threshold, choice of nodes and decision about each domain) are now partitioned by domain, thus any action a is a combination of actions per domain a_d. Let b_d denote the belief state for domain d. The choice of nodes in this case should just be a union of the nodes chosen by each POMDP as seen from the value function as each domain dependent component can be optimized separately.

$$V^* = \max_a \left[R(b,a) + \gamma \sum_{\Omega} P(\Omega|b,a)V^*(b,a,\Omega) \right] \tag{13}$$

$$= \max_a \left[\sum_d \Big(R_d(b_d, M_d) \Big) + R_n(a_n) + \gamma \sum_{\Omega} \prod_d P(\Omega_d|b_d, \tau_d)V^*(b,a,\Omega) \right] \tag{14}$$

$$= \max_a \left[\sum_d \Big(R_d(b_d, M_d) + R_{n_{[d]}}(a_{n_{[d]}}) \Big) + \gamma \sum_{\Omega} \prod_d P(\Omega_d|b_d, \tau_d)V^*(b,a,\Omega) \right] \tag{15}$$

$$= \max_a \left[\sum_d \Big(R_d(b_d, a_d) \Big) + \gamma \sum_{d,\Omega_d} P(\Omega_d|b_d, a_d)V_d^*(b_d, a_d, \Omega_d) \right] = \sum_d V_d^* \tag{16}$$

$$\text{where } V_d^* = \max_{a_d} \left[R_d(b_d, a_d) + \gamma \sum_{\Omega_d} P(\Omega_d|b_d, a_d)V_d^*(b_d, a_d, \Omega_d) \right] \tag{17}$$

If we instead use the compact representation of the action space, and let the actions simply be the number of channels to be sensed on and equality is obtained for the constraint (3) for any action from each domain specific POMDP (i.e., each $s_{k_{[d]}}$ can be implemented), then the value function can be decomposed by domain, because the term $R_n(a_n)$ is replaced by the $\sum_d \text{mincost}(s_{k_{[d]}})$, which can be factored by domain and does not couple the VD-POMDP's. Reconstructing the joint policy then just amounts to finding the best action from each POMDP and taking a union of these individual actions. We then just need to map back to the representation of actions on nodes, by solving the MILP (2)–(4).

$$V^* = \max_{a_d} \left[\sum_d \Big(R_d(b_d, a_d) + \text{mincost}(s_{k_{[d]}}) \Big) + \gamma \sum_{d,\Omega_d} P(\Omega_d|b_d, a_d)V_d^*(b_d, a_d, \Omega_d) \right] = \sum_d V_d^*$$

$$\square$$

5   VD-POMDP Framework

Here we explain at a high level, the VD-POMDP framework as applied to the data exfiltration problem, and how it is implemented. With the VD-POMDP, entire planning model is broken up into two parts as depicted in Fig. 4. The first is the offline factoring and planning, where the POMDP is factored into several sub-agents, and each solved individually. Second is the online policy aggregation and execution, where the policies of each sub-agent are aggregated as each of them choose actions to perform.

In order to build the VD-POMDP for data exfiltration problem, we first construct the network graph, based on the topology of the actual computer network we are modeling as well as the set of domains under consideration, shown at point (a) in Fig. 4. Then at (b), for each domain in our network, we construct a separate POMDP sub-agent. In order to do this we solve the MILP mincost(s_{k_d}) for each agent, in order to abstract away from the network layer and construct the compact representation of the network. At point (c) each individual POMDP agent is solved, offline, ignoring the presence of the other agents to obtaining a policy for each domain.

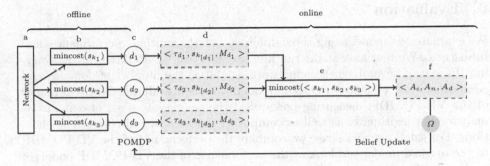

Fig. 4. Flowchart for the data exfiltration VD-POMDP

The policies are then aggregated in an online fashion, shown at point (d) in Fig. 4 to obtain a joint action (f). At each time step the agents receive observations from the network and update their beliefs individually. Each agent then presents the individual action to be performed consisting of a number of channels to be sensed on, a threshold for sensing and a classification of malicious or legitimate for their respective domain. The required number of channels for each agent is then fed into the MILP mincost($\langle s_{k_{[i]}} \cdots s_{k_{[D]}} \rangle$) to determine the set of nodes to be sensed on. The agents then again receive observations from the resulting set of detectors and iterate through this process again.

Policy aggregation is performed online as it would be infeasible to do offline policy aggregation for all but the smallest policies. If aggregation were to be performed offline, we would need to consider every possible combination of actions from each policy and then solve the MILP mincost($\langle s_{k_{[i]}} \cdots s_{k_{[D]}} \rangle$) for each of these, in order to compute the combinatorially large joint policy. Because the MILP is fast to solve, it does not result in much overhead when these joint actions are computed in an online fashion.

It is important to note here that the policy we compute is not an optimal sequence of actions, but rather a mapping of belief state to actions. This distinction is important, as it may be the case that, upon policy aggregation, there is no feasible implementation of the individual action. In such a scenario, an agent may choose an action to sense on a subset of k channels; however, given the sensing requirements of the other agents, the agent in question may actually get to sense on more channels than they had initially wanted. The agent may then end up in a belief state that they had not originally planned for, but because we are solving for the entire belief space, we still know how to behave optimally. Additionally, from Theorem 3, we know that the joint action will only be optimal if we can exactly implement each individual action, and no agent get to sense on more channels than it requests. Our policy aggregation may then result in a suboptimal joint action being taken, however, we show later in Sect. 6, that even when the optimality condition does not hold, we can still achieve good performance.

6 Evaluation

We evaluate our model using three different metrics: runtime, performance, and robustness. We first look at the runtime scalability of VD-POMDP model, varying the size of several synthetic network as well as the number of domains and compare to the standard POMDP model. We then evaluate the performance of the VD-POMDP, measuring how quickly it can classify a set of domains as malicious or legitimate, as well as computing the accuracy of correct classifications. For small network sizes, we compare the performance of the VD-POMDP to the original model and look at the performance of the VD-POMDP on larger synthetic networks.

Synthetic Networks. In order to test a variety of network sizes we created synthetic networks using a tree topology. Leaf nodes in the tree network correspond to source computers. Channels travel upwards from these nodes to the root of the tree; for each domain we create one such channel on each source computer. The size of the network is varied by varying the depth and branching factor of the tree.

6.1 DETER Testbed Simulation

We also evaluated the performance of our model using a real network, by running simulations on the DETER testbed. The DETER testbed provides capabilities of simulating a real computer network with virtual machines and simulating agents that perform tasks on each computer. Every agent is specified in a custom scripting language, and allows simulating attackers, defender and benign users. For our simulation we simulated legitimate DNS queries as well as launched real attacks. We performed a simple attack, by attempting to exfiltrate data from a file to a chosen malicious domain by embedding data from the file into the DNS queries. We conducted the attack using the free software Iodine [3] which allows for the creation of IP tunnels over the DNS protocol in order to generate these DNS queries. We were provided with 10 virtual machines, from which we formed a tree topology with 7 of them as host computers and sources of traffic. We then built and implemented a real time data exfiltration detector based off of the techniques proposed in [19]. The detector uses off the shelf compression algorithms like *gzip* in order to measure the information content of any channel in the network. We then set a cut off threshold for the level of allowable information content in any channel. Channels exceeding this threshold are flagged as malicious. While we chose to use this specific detector to generate observations for our model, it is important to note that any other methods for detecting exfiltration would have been equally viable.

6.2   Runtime

We first look at runtimes needed to solve the original model compared to our VD-POMDP model with increasing number of domains. Unless otherwise stated, all test used a threshold discretization of 2. We used an offline POMDP solver ZMPD [22] to compute policies; however, any solver which computes policies for the entire belief space may be used. The largest network we were able to solve for with the original model was one of only 3 nodes. For larger than 2 domains with discount factors $\gamma = -0.2$ and all cases with $\gamma = -0.4$ and $\gamma = -0.8$ the original POMDP did not finish solving in 24 h and is shown cut off at the 24 h mark in Fig. 5a. Consistent with the complexities in Table 2, in Fig. 5a we see the runtimes on the y-axis increase exponentially with the number of domains on the x-axis, for the original POMDP. If the VD-POMDP models are solved in parallel, runtimes do not vary with increasing domain. If the models are solved sequentially, then we would see only a linear increase in runtime. However in the case where networks have the channels uniformly distributed among all hosts, i.e. there exists one channel from every host to every domain, then the models become identical, and it becomes only necessary to solve one of them.

We show the scalability of computing policies for the VD-POMDP in Fig. 5a. On the y-axis, in log scale we have the runtime in seconds, and on the x-axis, also in log scale we have the number of nodes in the network, achieved by varying both the depth and branching factor of our tree network structure. We can see that there appears to be a linear scaling with the size of the network. We also show in Fig. 5d the time it takes to build the network, the lookup table of costs computed by repeatedly solving (10)–(12) and pomdp files to be fed to the solver. This time corresponds to the steps (a) and (b) in Fig. 4. On the y-axis we have again the runtime and on the x-axis the number of nodes in the network.

Figure 5c shows the runtime for computing the policy of a single factored agent with increasing action space. The runtime is shown on the y-axis in seconds, while the increasing action space is measured by the threshold discretization on the x-axis. We first divided the space of possible true positive and true negative rates into a number of segments equal to the discretization number. For each discretization level, we then combinatorially formed all the true positive and true negative pairs possible within that discretization number and averaged over the runtimes, in order to ensure that we were not only testing easy cases, where one choice threshold was dominant over another.

6.3   Performance

We evaluate the performance of the model by looking at the reward, the number of time steps taken to classify all domains and the accuracy of the classifications. For each test, we averaged the values over 100 simulations. Table 3 compares the performance of the original POMDP model with the VD-POMDP model.

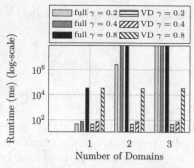

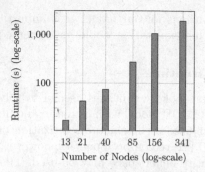

(a) Comparing runtimes of the POMDP to the VD-POMDP for a small network of 3 nodes and varying discount factor γ. On the y-axis, we have runtime in log scale and on the x-axis, we have increasing number of domains.

(b) Log-Log graph of runtimes for a single VD-POMDP for networks of increasing size. Both axes are in log scale, with runtimes on the y-axis in seconds and number of nodes on the x-axis, and showing a linear increase with network size.

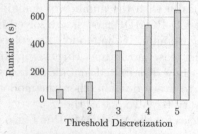

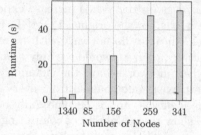

(c) Runtime for a single VD-POMDP with increasing action space. Test were run on a tree network with branching factor 4 and depth of 3 for a total of 85 nodes.

(d) Runtime required to build network for a single VD-POMDP. Time in seconds is shown on the y-axis and size of network in number of nodes is shown on the x-axis.

Fig. 5. Runtime results

The largest tests we could run using the full POMDP were on a network of 3 nodes with a total of two domains, while solving the model with a discount factor of $\gamma = 0.2$. The VD-POMDP model performs as well in terms of accuracy and time compared to the full POMDP model. We show a normalized average reward, computed by dividing the total reward by the number of time steps taken to classify the domains to better compare the models. Since we stop the simulation after all the domains have been classified, the total reward is not the expected infinite horizon reward, so simulations which run for different amounts of time will have had the chance to accumulate different amounts of reward. The normalized reward is meant to give a better indication of what the long term average reward would be, which would be a much fairer comparison. We also looked at the VD-POMDP solved with a discount factor of $\gamma = 0.8$, where we can clearly see the benefit of longer term planning. Although this VD-POMDP takes longer to classify both domains, it has a perfect accuracy and lower normalized reward than the other two models. This shows that the model is placing more value on potential future information, by preferring to wait and collect more alerts before making a final decision. This is clearly the better choice as we see a much better accuracy. It is clear that it is necessary to be able to plan for the future to perform well in this kind of domain; it is therefore necessary to be able to solve for large planning horizons, something that we cannot do using just the original POMDP model. This demonstrates the merit of the VD-POMDP framework, as solving this problem with a simple POMDP framework is clearly infeasible.

Table 3. Comparing performance of full POMDP model to factored model on a small test network of 3 nodes, with 2 domains. One domain is malicious and the other domain is legitimate.

Model	Timesteps to classify	Attack traffic accuracy	User traffic accuracy	Normalized reward
Full POMDP $\gamma = 0.2$	11.814	0.948	0.979	−470.594
VD-POMDP $\gamma = 0.2$	11.144	0.944	0.961	−675.100
VD-POMDP $\gamma = 0.8$	29.044	1.0	1.0	−386.982

Looking at just the VD-POMDP we test performance on a variety of larger networks in Table 4. Each of the synthetic networks are tested with 50 domains, averaged over 30 trials. The DETER network is tested with 100 domains averaged over 30 trials. For the DETER network, we used two thresholds, and determined the true and false positive rates of our detector by letting it monitor traffic at each threshold setting and observing the number of alerts obtained for each channel. We found our simple implementation of the detector had true positive rates of $\alpha(\tau 1) \sim 0.35$, $\alpha(\tau 2) \simeq 0.45$ and true negative rates of $\beta(\tau 1) \simeq 0.8$, $\alpha(\tau 2) \simeq 0.7$, and these were the parameters used in the model for this experiment as well as all the synthetic ones. We can see that, although the synthetic simulations all perform extremely well, and have a perfect accuracy, the deter simulation occasionally misclassifies legitimate traffic. This is due to the uncertainty in the characterization of the detector, as network traffic is variable and may not always follow a static distribution. Observations for the synthetic experiments were drawn from the distributions that the VD-POMDP had planned for, while in the DETER experiments, traffic did not always follow the input distributions. However, even with this uncertainty, the model still performs well in this realistic network setting. A more sophisticated implementation of this detector along with a more intensive characterization would even further boost the performance.

Table 4. Performance of the factored model on larger networks.

Network	Timesteps to classify	Attack traffic accuracy	User traffic accuracy	Normalized reward
Synthetic 40 nodes	4.079	1.0	1.0	−13523275.239
Synthetic 85 nodes	3.252	1.0	1.0	−15514333.580
Synthetic 156 nodes	3.235	1.0	1.0	−22204095.194
Synthetic 341 nodes	3.162	1.0	1.0	−21252069.929
DETER	5.3076	1.0	0.995	−6835.588

We also show an example of the diversity of actions chosen by the VD-POMDP. In Fig. 6 we show a trace of the actions taken by a single agent planning for a single domain. We show the number of channels chosen to sense on, the choice of threshold, along with the classification of the domain. We also show the observations, which the number of channels that triggered alerts. The simulation ends when no more channels are to be sensed on. We can see the agent varying the number of channels as well as the threshold of the detector, as they become more and more sure that they domain is legitimate.

Time	Action			Observations
	# Channels	τ	M_d	
1	64	1	0	23
2	62	1	0	23
3	3	0	0	0
4	5	0	0	2
5	8	0	0	2
6	18	0	0	4
7	0	0	0	0

Fig. 6. Trace of a run on a network of 85 nodes of a single legitimate domain.

6.4   Robustness

Lastly, we looked at the robustness of our model to errors in the input parameter. As evidenced with the DETER experiment, the model requires known false positive and true positive rates for the detector. While it may be reasonable to assume that with enough monitoring, it is possible to get accurate measures of false positive rates in a network by simply running the detector on known legitimate traffic for long periods of time, it is more difficult to characterize the true positive rates, as attacks can take many forms and exfiltration can occur over varying rates. In order to test the robustness of our model, we solved for the policy using one set of rates and then tested the model in simulation against a variety of actual rates. For our tests, the model was solved with a true negative rate of 0.8 and true positive rate of 0.55. We then drew alerts from a range of distributions for the true positive and negative rates as shown in Fig. 7 on the y-axis and x-axis respectively. The metrics used to measure robustness are shown as a heat-map for each true positive, true negative pair.

In Fig. 7a performance of the model was tested by looking at the percent of incorrect legitimate domain classifications i.e. the percent of legitimate domains flagged as malicious. In all cases except for one, all legitimate domains were correctly flagged as non-malicious, and in one case legitimate domains were mis-classified in only 0.4 % of the trials. In Fig. 7b the percent of correct malicious domain classifications is shown, where in all but two cases, the correct domain was always identified. Figure 7c shows the number of time steps taken to classify all the domains, while Fig. 7d shows the average reward (in this case a penalty) for the simulations. We can see that the model is robust to mischaracterization of the detectors, where the only dips in performance occur when either the detector has a low true negative rate and when the error in both the true positive and negative rates are large.

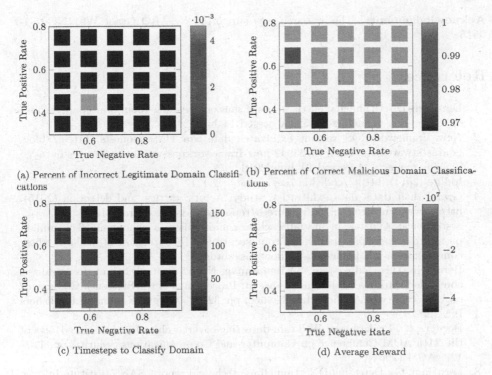

(a) Percent of Incorrect Legitimate Domain Classifications

(b) Percent of Correct Malicious Domain Classifications

(c) Timesteps to Classify Domain

(d) Average Reward

Fig. 7. Testing the robustness with respect to error in the planned true positive and true negative rate.

7 Conclusion and Future Work

We demonstrated the effectiveness of POMDP based planning tool in making intelligent decisions to tackle the problem of DNS based data exfiltration. These decisions were made by aggregating information from multiple noisy detectors and using sequential planning under uncertainty based reasoning. In doing so, we also proposed a new class of POMDPs called VD-POMDP that uses domain characteristics to split the POMDP into sub-POMDPs and allows for abstract actions in each sub-POMDP that can then be easily converted to a full joint action at execution time. VD-POMDP allows scaling up our approach to real world sized networks. The approach also detects attacks in near real time, thereby providing options to minimize the damage from such attacks. More generally, we believe that our approach applies to other security detection and response problems such as exfiltration over other protocols like HTTP and intrusion detection. While this work is an important step in addressing the problem of APT's in a realistic and scalable manner, we recognize that having a non-adaptive adversary is a simplification of the potentially complex interaction between attacker and defender in this environment. Building an appropriate adversary model, and considering the underlying game in this domain is a key avenue for future work in this area.

Acknowledgements. This research was supported by ARO Grant W911NF-15-1-0515.

References

1. Detecting DNS Tunneling. https://www.sans.org/reading-room/whitepapers/dns/detecting-dns-tunneling-34152. Accessed 14 June 2016
2. New FrameworkPOS variant exfiltrates data via DNS requests. https://blog.gdatasoftware.com/2014/10/23942-new-frameworkpos-variant-exfiltrates-data-via-dns-requests. Accessed 14 June 2016
3. Iodine (2014). http://code.kryo.se/iodine/
4. Grand theft data, data exfiltration study: Actors, tactics, and detection (2015). http://www.mcafee.com/us/resources/reports/rp-data-exfiltration.pdf
5. arstechnica: Cluster of megabreaches compromises a whopping 642 million passwords. http://arstechnica.com/security/2016/05/cluster-of-megabreaches-compromise-a-whopping-642-million-passwords/
6. Bernstein, D.S., Zilberstein, S., Immerman, N.: The complexity of decentralized control of Markov decision processes. In: Proceedings of the Sixteenth Conference on Uncertainty in Artificial Intelligence, pp. 32–37. Morgan Kaufmann Publishers Inc. (2000)
7. Borders, K., Prakash, A.: Web tap: detecting covert web traffic. In: Proceedings of the 11th ACM Conference on Computer and Communications Security, pp. 110–120. ACM (2004)
8. FarnHam, G.: Detecting DNS tunneling. Technical report, SANS Institute InfoSec Reading Room, Februrary 2013
9. Gerkey, B.P., Mataric, M.J.: Multi-robot task allocation: analyzing the complexity and optimality of key architectures. In: IEEE International Conference on Robotics and Automation Proceedings, ICRA 2003, vol. 3, pp. 3862–3868. IEEE (2003)
10. Hart, M., Manadhata, P., Johnson, R.: Text classification for data loss prevention. In: Proceedings of the 11th International Conference on Privacy Enhancing Technologies, PETS 2011 (2011)
11. Journal, T.W.S.: Home depot's 56 million card breach bigger than target's. http://www.wsj.com/articles/home-depot-breach-bigger-than-targets-1411073571
12. Jung, H., Tambe, M.: Performance models for large scale multiagent systems: using distributed POMDP building blocks. In: Proceedings of the Second International Joint Conference on Autonomous Agents and Multiagent Systems, pp. 297–304. ACM (2003)
13. Labs, T.: Data exfiltration: How do threat actors steal yourdata? (2013). http://about-threats.trendmicro.com/cloud-content/us/ent-primers/pdf/how_do_threat_actors_steal_your_data.pdf
14. Madani, O., Hanks, S., Condon, A.: On the undecidability of probabilistic planning and infinite-horizon partially observable Markov decision problems. In: Proceedings of the Sixteenth National Conference on Artificial Intelligence and the Eleventh Innovative Applications of Artificial Intelligence Conference Innovative Applications of Artificial Intelligence, AAAI 1999/IAAI 1999, pp. 541–548. American Association for Artificial Intelligence, Menlo Park (1999)
15. McAfee: Data loss prevention. http://www.mcafee.com/us/products/total-protection-for-data-loss-prevention.aspx

16. McAllester, D.A., Singh, S.: Approximate planning for factored POMDPS using belief state simplification. In: Proceedings of the Fifteenth Conference on Uncertainty in Artificial Intelligence, pp. 409–416. Morgan Kaufmann Publishers Inc. (1999)

17. Nair, R., Varakantham, P., Tambe, M., Yokoo, M.: Networked distributed POMDPS: a synthesis of distributed constraint optimization and POMDPS. AAAI 5, 133–139 (2005)

18. Papadimitriou, C.H., Tsitsiklis, J.N.: The complexity of Markov decision processes. Math. Oper. Res. 12(3), 441–450 (1987)

19. Paxson, V., Christodorescu, M., Javed, M., Rao, J., Sailer, R., Schales, D., Stoecklin, M.P., Thomas, K., Venema, W., Weaver, N.: Practical comprehensive bounds on surreptitious communication over DNS. In: Proceedings of the 22nd USENIX Conference on Security, SEC 2013, pp. 17–32. USENIX Association, Berkeley (2013). http://dl.acm.org/citation.cfm?id=2534766.2534769

20. Bromberger, S.: Co-Principal Investigator, NESCOCo-Principal Investigator, N.: DNS as a covert channel within protected networks. Technical Report WP2011-01-01, National Electric Sector Cyber Security Organization, January 2011

21. Silver, D., Veness, J.: Monte-carlo planning in large POMDPS. In: Advances in Neural Information Processing Systems, pp. 2164–2172 (2010)

22. Smith, T.: Probabilistic Planning for Robotic Exploration. Ph.D. thesis. The Robotics Institute, Carnegie Mellon University, Pittsburgh, PA, July 2007

23. Sommer, R., Paxson, V.: Outside the closed world: on using machine learning for network intrusion detection. In: 2010 IEEE Symposium on Security and Privacy (SP), pp. 305–316. IEEE (2010)

24. Symantec: Data Loss Prevention and Protection. https://www.symantec.com/products/information-protection/data-loss-prevention

25. Varakantham, P., young Kwak, J., Taylor, M., Marecki, J., Scerri, P., Tambe, M.: Exploiting coordination locales in distributed POMDPS via social modelshaping (2009). http://aaai.org/ocs/index.php/ICAPS/ICAPS09/paper/view/733/1128

26. Velagapudi, P., Varakantham, P., Sycara, K., Scerri, P.: Distributed model shaping for scaling to decentralized POMDPS with hundreds of agents. In: The 10th International Conference on Autonomous Agents and Multiagent Systems, vol. 3, pp. 955–962. International Foundation for Autonomous Agents and Multiagent Systems (2011)

27. Wikipedia: Office of personnel management data breach. https://en.wikipedia.org/wiki/Office_of_Personnel_Management_data_breach

On the Mitigation of Interference Imposed by Intruders in Passive RFID Networks

Eirini Eleni Tsiropoulou[1(✉)], John S. Baras[1], Symeon Papavassiliou[2], and Gang Qu[1]

[1] Department of Electrical and Computer Engineering,
Institute for Systems Research, University of Maryland,
College Park, MD 20742, USA
{eetsirop, baras, gangqu}@umd.edu

[2] School of Electrical and Computer Engineering,
National Technical University of Athens, 15773 Zografou, Athens, Greece
papavass@mail.ntua.gr

Abstract. RFID networks are becoming an integral part of the emerging Internet of Things (IoT) era. Within this paradigm passive RFID networks have emerged as low cost energy-efficient alternatives that find applicability in a wide range of applications. However, such RFID networks and devices, due to their limited capabilities, can easily become vulnerable to several intrusive actions. In this paper, the problem of proactively protecting a passive RFID network from security threats imposed by intruders that introduce high interference to the system resulting in the possible disruption of the network's proper operation is investigated. Passive RFID tags are associated with a well-designed utility function reflecting on one hand their goal to have their signal properly demodulated by the reader, and on the other hand their risk level of participating in the network, stemming from their hardware characteristics among others, thus characterizing them as normal or intruder tags. An interference mitigation risk aware (IMRA) problem is introduced aiming at maximizing each tag's utility function, thus implicitly enforcing tags to conform to a more social behavior. Due to its nature, the proposed problem is formulated as a non-cooperative game among all the tags (normal and intruders) and its Nash equilibrium point is determined via adopting the theory of supermodular games. Convergence of the game to its Nash equilibrium is also shown. A distributed iterative and low-complexity algorithm is proposed in order to obtain the Nash equilibrium point and the operational effectiveness of the proposed approach is evaluated through modeling and simulation.

Keywords: Intruders · Interference mitigation · Passive RFID networks · Risk · Game theory

1 Introduction

Radio Frequency Identification (RFID) technology aims at tagging and identifying an object. The concept of RFID is envisioned as part of the Internet of Things and has been recently used in numerous applications from asset tracking to supply chain

© Springer International Publishing AG 2016
Q. Zhu et al. (Eds.): GameSec 2016, LNCS 9996, pp. 62–80, 2016.
DOI: 10.1007/978-3-319-47413-7_4

management and from medication compliance and home navigation for the elderly and cognitively impaired to military troop movements monitoring. RFID networks are exposed to a broader attack surface given their IoT nature, thus it is of great interest not only to develop security mechanisms that can protect critical data from harm (e.g. data encryption techniques), but also the application of intelligent control mechanisms that will enable an RFID network to work properly and in a reliable manner with minimum intervention [1].

An RFID basic characteristic is the conversion of a set of objects into a mobile network of nodes, which is of dense and ad-hoc nature and it is mainly utilized for objects tracking, environmental monitoring and events triggering [2]. The fundamental components of an RFID network are: (a) the RFID reader/interrogator and (b) the RFID tag, which can be either active or passive or semi-passive. The RFID reader communicates with the RFID tags via emitting radio waves and receiving signals back from the tags. The active RFID tags and semi-passive RFID tags embed a radio signal transceiver and an internal power source. The main advantages of active RFID tags are that they can activate themselves regardless of the presence of a reader in proximity, while providing greater operating range and supporting advanced functionalities compared to passive RFID tags. On the other hand, their main disadvantages are their high cost and significant environmental limitations due to the presence of the battery, i.e., large size, and their high transmission power [3]. Therefore, passive RFID tags emerge as the most energy-efficient, inexpensive solution to build an RFID network. Their low transmission power backscatter commands and low cost make them suitable for a wide range of IoT applications.

1.1 Motivation

A passive RFID network consists of a number of RFID readers and a number of passive RFID tags. The RFID tags have no on-board power source and derive their reflection power from the signal of an interrogating reader. A passive RFID tag is activated by the reader's forward/transmission power, which is much more powerful than the reverse/reflection power sent back by the tag to the reader. Each tag must be able to reflect sufficient amount of power to the reader, which is mapped to a targeted signal-to-interference-plus-noise ratio (SINR), in order for its signal to be demodulated by the reader. The reflection power of all passive RFID tags within the RFID network contribute to the overall existing interference, which consequently drives the tags to reflect with even more power (while their maximum reflection power is limited) to ensure the demodulation of their signal at the reader.

Within such a passive RFID network, a security threat with respect to the reliable operation of the system is the presence of one or more intruding passive RFID tags that could act as interferers. In other words, such "attacker/intruder tags" can take advantages of their position in the network and their hardware characteristics may simply introduce strong interference in the rest of the passive RFID tags' reflections rendering their signals hard or almost impossible to be demodulated at the RFID reader side. Taking into account the difficulty in identifying those intruder-tags and eventually removing them from the network an alternative strategy in dealing with this problem is to reduce the potential harm that they can impose on the system. This can be achieved by enforcing the tags to conform to a more social behavior with respect to their reflection behavior (for example, not using unnecessarily high reflection power), thus limiting the potential risks. The latter may range from simply wasting unnecessarily power to completely disturbing the proper operation of the system by making some objects impossible to be tracked.

Passive RFID tags share the same system bandwidth towards reflecting back their signal to the reader. Thus, increased level of interference caused by the rest of the tags will enforce a tag to increase also its reflection power in order to achieve a desired power level (which is translated to a target SINR value) that eventually will enable the demodulation of its signal by the reader. Therefore, passive RFID tags compete with each other to determine their optimal reflection powers that enable their signal demodulation. Masked or disguised intruder-tags pretending to act as normal passive RFID tags, tend to introduce high interference level to the passive RFID network, thus disrupting or even causing failure of its proper operation. Furthermore, due to the distributed nature of passive RFID networks and the absence of a single administrative entity to control tags' reflection powers, while considering the potential risk level associated with the operation of each tag, distributed solutions should be devised in order to secure the reliable operation of the RFID networks and impose on participating entities to adhere to proper operation rules and behavior.

1.2   Contributions and Outline

In this paper, the problem of risk-aware mitigation of interference imposed by intruders in passive RFID networks is studied and treated via a game theoretic approach. Envisioning the Internet of Things (IoT) and battery-free wireless networks as key part of the emerging 5G era, the system model of a passive RFID network is initially introduced (Sect. 2.1). A utility-based framework is adopted towards representing passive RFID tag's goal to have its signal being properly demodulated by the reader, while simultaneously considering its reflection power and its corresponding risk level – the latter being mapped to tag's hardware related characteristics (Sect. 2.2). Due to the distributed nature of the proposed interference mitigation risk aware (IMRA) problem, it has been formulated as a non-cooperative game among passive RFID tags, which can be either normal or intruder-tags (Sect. 3.1) and IMRA game's Nash equilibrium point is determined (Sect. 3.2). The convergence of the IMRA game to the Nash equilibrium is shown (Sect. 4), while a non-cooperative distributed low-complexity and iterative algorithm is presented to determine the Nash equilibrium of the IMRA game (Sect. 5). The performance of the proposed approach is evaluated in detail through modeling and simulation (Sect. 6), while related research work from the recent literature is presented in Sect. 7. Finally, Sect. 8 concludes the paper.

2 System Model

2.1 Passive RFID Networks

Figure 1 presents the considered topology of a passive RFID network. An RFID reader is assumed to activate the $N = N_n + N_{in}$ passive RFID tags, which reflect back their information in order for their signal to be demodulated by the reader. The number of normal passive RFID tags is denoted by N_n, while the number of intruder-tags is N_{in}. Respectively, the set of normal RFID tags is denoted by S_n and the corresponding set of intruder-tags by S_{in}. The overall set of passive RFID tags within the network is $S = S_n \cup S_{in}$. Representative real life examples of this assumed topology include: (a) monitoring stock availability on retail shelves, (b) identifying books in shelves of library systems, and (c) monitoring the military equipment supply chain.

RFID reader's transmission power is assumed to be fixed, i.e. P_R, depending on its technical characteristics. In the examined topology, a simplified RFID network has been considered, consisting of one RFID reader and multiple passive RFID tags, which can be either normal or intruder-tags. The proposed framework can be easily extended to multiple RFID readers and multiple tags, while each one of the tags will be associated to its nearest RFID reader. Let P_i, $i = 1, 2, \ldots N$ denote the reflection power of the ith, $i \in S = S_n \cup S_{in}$ passive RFID tag, where $P_i \in A_i$, $A_i = \left[0, P_i^{Max}\right]$. The maximum feasible reflection power P_i^{Max} of each tag depends on: (a) the characteristics of the topology (e.g. distance d_i between the RFID reader and the tag) and (b) tag's hardware characteristics. Assuming single hop communication among the reader and the tag, the upper bound of passive RFID tag's reflection power is:

$$P_i^{Max} = P_R \cdot G_R \cdot G_i . K_i \left(\frac{\lambda}{4\pi d_i}\right)^2 \tag{1}$$

where P_R is the transmission power of the RFID reader R communicating directly with the ith passive RFID tag, G_R and G_i are the RFID reader's and passive RFID tag's directional antenna's gain, respectively, K_i is the backscatter gain of the ith tag and the factor $\left(\frac{\lambda}{4\pi d_i}\right)^2$ describes the free space path loss.

In a backscatter communication system, i.e. communication from the N passive RFID tags to the reader, the signal-to-interference-plus-noise ratio (SINR), γ_i, must meet a required threshold γ_i^{target} for the tag's signal to be able to be demodulated by the reader. The SINR at the RFID reader R for each passive RFID tag $i, i \in S = S_n \cup S_{in}$ is given by [12]:

$$\gamma_i = \frac{h_i P_i}{\sum_{j \neq i} h_j P_j + n} \tag{2}$$

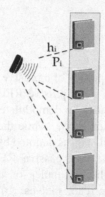

Fig. 1. Passive RFID network – library system example.

where h_i represents the channel loss from the ith tag to the reader and n contains the background noise. The term $\sum_{j \neq i} h_j P_j$ denotes the RFID network interference at the RFID reader when receiving data from the ith tag.

2.2 Utility Function

Towards formulating passive RFID tag's behavior under a common optimization framework, the concept of utility function is adopted. Each passive RFID tag (either normal or intruder) is associated with a utility function, which consists of two parts: (a) the pure utility function and (b) the risk function. The pure utility function represents the tag's degree of satisfaction in relation to the achievement of the targeted SINR γ_i^{target} and the corresponding power consumption. The risk function represents the risk level (with respect to its impact and potential harm to the system) of each passive RFID tag considering its reflection power and its hardware characteristics, i.e., directional antenna's gain G_i and backscatter gain K_i. It is noted that a passive RFID tag is considered as a potential attacker/intruder of the overall RFID network if it introduces high level of interference due to its hardware characteristics, thus it should be penalized for its malicious and non-social behavior. The latter could result in increased reflection power P_i from the rest of the tags. Considering that P_i^{Max} is limited it could be the case that the tags cannot achieve their targeted SINR and consequently the reader will be unable to demodulate their signal. Therefore, the risk function provides the means to enforce the tags to conform to a more social behavior and limiting the potential impact of an intruder. Also note that an intruder will be masking its presence and behavior, and other than trying to impose high interference in the rest of the tags and therefore disrupt the normal system operation, its behavior will look normal to avoid being detected.

Based on the above discussion, each passive RFID tag's utility function can be formulated as follows:

$$U_i(P_i, \boldsymbol{P}_{-i}) = U_{pure}(P_i, \boldsymbol{P}_{-i}) - R(G_i, K_i, P_i) \tag{3}$$

where $U_{pure}(\cdot)$ denotes passive RFID tag's pure utility function and $R(\cdot)$ its risk function. As it was discussed above, $U_{pure}(\cdot)$ reflects the tradeoff between achieving the target SINR and the necessary corresponding reflection power, while considering the imposed interference by the rest of the tags. The risk function is introduced as a cost function penalizing the tags, which present non-social/malicious behavior and tend to damage/attack the RFID network via introducing high interference level due to their increased reflection power. Thus, the penalty increases for the tags that try to reflect with high power and have privilege against other tags due to their hardware characteristics.

Throughout the rest of the paper, without loss of generality and for presentation purposes, we consider the following passive RFID tag's utility function:

$$U(P_i, \boldsymbol{P}_{-i}) = \frac{f_i(\gamma_i)}{P_i} - G_i \cdot K_i \cdot P_i \tag{4}$$

where $f_i(\gamma_i)$ is a sigmoidal-like function with respect to γ_i, where the inflection point is mapped to the target SINR γ_i^{target} of the $i, i \in S$ tag. For presentation purposes, we set $f_i(\gamma_i) = (1 - e^{-A\gamma_i})^M$, where A, M are real valued parameters controlling the slope of the sigmoidal-like function.

3 Interference Mitigation Risk Aware (IMRA) Game

3.1 Problem Formulation

Let $G_{IMRA} = [S, \{A_i\}, \{U_i(\cdot)\}]$ denote the corresponding non-cooperative interference mitigation risk aware game, where $S = \{1, 2, \ldots, N\}$ is the index set of the passive RFID tags, $A_i = (0, P_i^{Max}] \subseteq R^N$ is the strategy set of the ith passive RFID tag and $U_i(\cdot)$ is its utility function, as defined before. Each passive RFID tag aims at maximizing its utility via determining its reflection power P_i in a non-cooperative manner. Thus, the Interference Mitigation Risk Aware (IMRA) game can be expressed as the following maximization problem:

$$(IMRA\ game) \quad \begin{aligned} \max_{P_i \in A_i} U_i = \max_{P_i \in A_i} U_i(P_i, \boldsymbol{P}_{-i}), \ \forall i \in S \\ s.t.\ 0 < P_i \leq P_i^{Max} \end{aligned} \tag{5}$$

The solution of the IMRA game determines the optimal equilibrium for the RFID system, consisting of the individual decisions of each passive RFID tag (either normal or intruder-tag), given the decisions made by the rest of the tags in the passive RFID network. The solution of the IMRA game is a vector of passive RFID tags' reflection powers $P^* = (P_1^*, P_2^*, \ldots, P_N^*) \in A, A = \cup A_i, i \in S = S_n \cup S_{in}$, where P_i^* is the reflection power of tag i. The Nash equilibrium approach is adopted towards seeking analytically the solution of the non-cooperative IMRA game. Based on this approach, which is most widely used for game theoretic problems, we have the following definition.

Definition 1. The power vector $P^* = (P_1^*, P_2^*, \ldots, P_N^*) \in A, A = \cup A_i, i \in S = S_n \cup S_{in}$, is a Nash equilibrium of the IMRA game, if for every $i \in S = S_n \cup S_{in}$ $U_i(P_i^*, P_{-i}^*) \geq U_i(P_i, P_{-i}^*)$ for all $P_i \in A_i$.

The interpretation of the above definition of Nash equilibrium point is that no passive RFID tag, either normal or intruder-tag, has the incentive to change its strategy (i.e., reflection power), due to the fact that it cannot unilaterally improve its perceived utility by making any change to its own strategy, given the strategies of the rest of the tags. Moreover, it is concluded that the existence of a Nash equilibrium point guarantees a stable outcome of the IMRA game, while on the contrary the non-existence of such an equilibrium point is translated to an unstable and unsteady situation of the RFID system, stemming from high risk and interference levels imposed by the intruder-tags.

Furthermore, note that the utility function introduced in Eqs. (3) and (4) is generic enough to capture both normal and intruder-tags behavior, however it is not characterized by desirable properties, e.g., quasi-concavity. Therefore, alternative techniques from the field of game theory should be adopted in order to prove the existence of Nash equilibrium for the IMRA game.

3.2 Towards Determining the Nash Equilibrium

Towards proving the existence of at least one Nash equilibrium of the IMRA game, the theory of supermodular games is adopted. Supermodular games are of great interest as an optimization and decision making tool, due to the fact that they encompass many applied models, they tend to be analytically appealing since they have Nash equilibria and they have the outstanding property that many solutions yield the same predictions [13]. Moreover, supermodular games comply very well with intruder-tags' behavior in the IMRA game, due to the fact that they are characterized by strategic complementarities, i.e., when one intruder-tag takes a more competitive and aggressive action (i.e., increase its reflection power), then the rest of the tags want to follow the same behavior, causing the RFID system to be led to borderline operation.

Considering the Interference Mitigation Risk Aware (IMRA) problem studied in this paper, we examine a single-variable supermodular game, which is defined as follows:

Definition 2. A game $G = [S, \{A_i\}, \{U_i(\cdot)\}]$ with strategy spaces $A_i \subset \Re, \forall i \in S = S_n \cup S_{in}$ is supermodular if for each $i, i \in S$, the utility function $U_i(P_i, P_{-i})$ has non-decreasing differences (NDD) in (P_i, P_{-i}) [13].

The property of non-decreasing differences (NDD) for the objective function $U_i(P_i, P_{-i})$ is formally defined as follows.

Definition 3. The objective function $U_i(P_i, P_{-i})$ has non-decreasing differences (NDD) if for all $P_{-i} \geq P_{-i}'$, the difference $U_i(P_i, P_{-i}) - U_i(P_i, P_{-i}')$ is non-decreasing in P_i. Moreover, if the objective function $U_i(P_i, P_{-i})$ is smooth (i.e., it has derivatives of all orders), then it has non-decreasing differences in (P_i, P_{-i}) if and only if

$$\frac{\partial^2 U_i(P)}{\partial P_i \partial P_j} \geq 0, j \neq i,\ j, i \in S \tag{6}$$

Examining the IMRA game as it has been formulated in relation (5), it is observed that it is not a supermodular game according to Definition 3, due to the exogenous risk factors G_i, K_i included in the objective function. Therefore, the strategy space of each passive RFID tag should be slightly modified, in order to show that condition (6) holds true, so that the resulting game is supermodular.

Theorem 1. The IMRA game's utility function $U_i(P_i, \boldsymbol{P}_{-i})$ as defined in (4) has non-decreasing differences (NDD) in $(P_i, \boldsymbol{P}_{-i})$, i.e. $\frac{\partial^2 U_i(P)}{\partial P_i \partial P_j} \geq 0, j \neq i,\ j, i \in S$, if and only if

$$\gamma_i \in \left[\frac{\ln M}{A}, +\infty\right) \tag{7}$$

Proof. Towards showing that the IMRA game's utility function has non-decreasing differences (NDD) in $(P_i, \boldsymbol{P}_{-i})$, the sign of the second order partial derivative, i.e. $\frac{\partial^2 U_i(P)}{\partial P_i \partial P_j}$, is examined as follows:

$$\frac{\partial^2 U_i(P)}{\partial P_i \partial P_j} = \frac{AM}{P_i^2} \frac{h_i}{\sum\limits_{j \neq i} h_j P_j + n} \gamma_i^2 e^{-A\gamma_i} \left(1 - e^{-A\gamma_i}\right)^{M-2} \left(1 - M e^{-A\gamma_i}\right)$$

It is noted that the term $\frac{AM}{P_i^2} \frac{h_i}{\sum\limits_{j \neq i} h_j P_j + n} \gamma_i^2 e^{-A\gamma_i}$ is non-negative for all $\gamma_i \geq 0$. Moreover, considering the term $\left(1 - e^{-A\gamma_i}\right)^{M-2}$, we have: $\left(1 - e^{-A\gamma_i}\right)^{M-2} \geq 0 \Leftrightarrow \gamma_i \geq 0$. Furthermore, considering the sign of the term $\left(1 - M e^{-A\gamma_i}\right)$, we have: $\left(1 - M e^{-A\gamma_i}\right) \geq 0 \Leftrightarrow \gamma_i \geq \frac{\ln M}{A}$.

Based on the above, it is concluded that the IMRA game's utility function $U(P_i, \boldsymbol{P}_{-i}) = \frac{f_i(\gamma_i)}{P_i} - G_i \cdot K_i \cdot P_i$ has non-decreasing differences in $(P_i, \boldsymbol{P}_{-i})$, if $\gamma_i \geq \frac{\ln M}{A}$. ∎

Based on Definitions 2 and 3 and Theorem 1, we easily conclude the following.

Theorem 2. The IMRA game $G_{IMRA} = [S, \{A_i\}, \{U_i(\cdot)\}]$ is supermodular in a modified strategy space $A_i' = \left[P_i^{Min}, P_i^{Max}\right] \subset A_i$, where P_i^{Min} is derived from $\gamma_i \geq \frac{\ln M}{A}$.

At this point, it should be noted that the constraint $\gamma_i \geq \frac{\ln M}{A}$ is not an additional constraint to the initial formulation of the IMRA game, due to the fact that the target SINR value γ_i^{target} introduced in Sect. 2 is equivalent to the value $\gamma_i^{target} = \frac{\ln M}{A}$. Specifically, it has already been explained in Sect. 2 that γ_i^{target} is mapped to the inflection point of $f_i(\gamma_i)$. Thus, we have: $\frac{\partial^2 f_i(\gamma_i)}{\partial \gamma_i^2} = 0 \Leftrightarrow \gamma_i^{target} = \frac{\ln M}{A}$. The meaning of the above description is that the passive RFID tag should have sufficient reflection power $P_i \in (0, P_i^{Max}]$ such that $\gamma_i \geq \gamma_i^{target}$ is ensured in order for its signal to be demodulated by the reader. Thus, assuming an ideal scenario where we do not have intruder-tags and the topology is favorable (i.e., not relatively extremely large distances for an RFID network) so as tag's available power $P_i \in (0, P_i^{Max}]$ is sufficient in order to be read by the reader, then each tag's goal is to achieve an SINR value greater or at least equal to the target one, i.e., $\gamma_i \geq \gamma_i^{target}$. Therefore, in the case that intruder-tags introduce high interference resulting in violation of the condition $\gamma_i \geq \gamma_i^{target} = \frac{\ln M}{A}$, this is essentially translated to no guarantee of Nash equilibrium existence (i.e., unstable situation of the RFID system), thus some or even all tags will not achieve γ_i^{target} and consequently their signal will not be demodulated, and as a consequence the reader's objective will not be fulfilled.

Theorem 2, i.e., proving that the IMRA game is supermodular in the modified strategy space $A_i' \subset A_i, \forall i \in S = S_n \cup S_{in}$, guarantees the existence of a non-empty set of Nash equilibria [13]. Therefore, the following holds true:

Theorem 3. The modified IMRA game $G_{IMRA}' = [S, \{A_i'\}, \{U_i(\cdot)\}]$ has at least one Nash equilibrium, which is defined as follows:

$$P_i^* = \underset{P_i \in A_i'}{\arg \max}\, U_i(P_i, \boldsymbol{P}_{-i}) \tag{8}$$

It should be noted that Theorem 3 guarantees the existence of at least one Nash equilibrium, while this point is not necessarily unique. Practically, the best response in (8) can be solved via single variable calculus utilizing the Extreme Value Theorem [14], and the most energy-efficient Nash equilibrium (i.e. the Nash equilibrium characterized by less reflection power P_i, while guaranteeing the target SINR γ_i^{target}) is adopted by each passive RFID tag.

4   Convergence of the IMRA Game

In this section, we prove the convergence of the interference mitigation risk aware (IMRA) game to a Nash equilibrium point, as this is determined by relation (8). Towards this direction, the best response strategy of each passive RFID tag $i, i \in S = S_n \cup S_{in}$ is denoted by BR_i and is given as follows:

$$BR_i(P_i) = \underset{P_i \in A_i'}{\arg \max}\, U_i(P_i, \boldsymbol{P}_{-i}) = P_i^* \tag{9}$$

As shown in [15], the fundamental step for showing the convergence of the IMRA game to a Nash equilibrium, as obtained by Eq. (8), is to show that the best response function $\boldsymbol{BR}(\boldsymbol{P})$ is standard. In general, a function is characterized as standard if for all $\boldsymbol{P} > 0$, where $\boldsymbol{P} = (P_1, P_2, \ldots, P_N)$, the following conditions/properties hold true:

(i)   Positivity: $BR(P) > 0$;
(ii) Monotonicity: if $P' \geq P$ then $BR(P') \geq BR(P)$;
(iii) Scalability: for all $\alpha > 1$, $\alpha BR(P) \geq BR(\alpha P)$.

Theorem 4. The modified IMRA game $G'_{IMRA} = \left[S, \{A'_i\}, \{U_i(\cdot)\} \right]$ converges to a Nash equilibrium, as expressed in (8).

Proof. As presented in Eq. (9) each passive RFID tag's best response strategy is the argument of the maximum of the tag's utility function with respect to the reflection power $P_i \in A'_i$. Considering all the passive RFID tags participating in the IMRA game, we have $BR(P) = (BR_1(P_1), BR_2(P_2), \ldots, BR_N(P_N)) = (P_1^*, P_2^*, \ldots, P_N^*)$. Towards proving that the best response function $BR(P)$ is standard, the corresponding afore-mentioned properties can be easily shown:

(i) $P = (P_1, P_2, \ldots, P_N) > 0$, thus $BR(P) > 0$;
(ii) if $P' \geq P$ then via Eq. (9), i.e., $BR_i(P_i) = P_i^*$ we conclude that $BR(P') \geq BR(P)$;
(iii) for all $\alpha > 1$, then via Eq. (9), i.e., $BR_i(P_i) = P_i^*$ we conclude that $BR(P') \geq BR(P')$, where the equality holds true.   ∎

Based on Theorem 4, it is guaranteed that the IMRA game converges to a stable situation, i.e. to a Nash equilibrium point. Detailed numerical results with respect to the convergence of the proposed IMRA game to a Nash equilibrium are presented in Sect. 6.

5   The IMRA Algorithm

Passive RFID networks, as part of the Internet of Things, are characterized by their distributed nature and the absence of any central entity that can take decisions about the actions of the passive RFID tags on their behalf. Thus, each RFID tag should determine in a distributed manner its equilibrium reflection power after being activated by the reader. Except for its hardware characteristics and its channel loss, which is customized/personal information already known by each tag, the only supplementary necessary information, towards determining the equilibrium powers, is the overall network interference which is broadcasted by the reader to the tags. Therefore, in this section we propose a distributed iterative and low complexity algorithm in order to determine the Nash equilibrium point(s) of the IMRA game. The proposed IMRA algorithm runs every time the RFID reader activates the passive RFID tags in order to collect their information.

IMRA Algorithm

Step 1:   Each tag reflects with a randomly selected
           feasible reflection power $P_i^{(ite=0)}$, where
           $P_i^{Min} \leq P_i^{(ite=0)} \leq P_i^{Max}$, $\forall i, i \in S = S_n \cup S_{in}$. Set $ite=0$, where
           ite denotes the number of iterations of the IMRA
           algorithm.
Step 2:   The RFID reader broadcasts the overall sensed
           interference as a global information in the RFID
           network, i.e., $\sum_{i \in S} h_i P_i$, each tag determines its
           sensed interference, i.e., $\sum_{j \neq i} h_j P_j$, and
           determines its best response strategy, i.e.,
           $BR_i(P_i) = \arg\max_{P_i \in A_i'} U_i(P_i, \mathbf{P}_{-i})$. Each passive RFID tag
           assigns its reflection power $P_i^{(ite)} = BR_i(P_i)$.
Step 3:   If the reflection powers of all tags converge,
           i.e., $\left| P_i^{(ite+1)} - P_i^{(ite)} \right| \leq \varepsilon$, where ε is a very small
           value (e.g., $\varepsilon = 10^{-6}$), this means that the RFID
           reader has read the information from all tags,
           therefore it stops activating them and the
           algorithm stops. Otherwise, set $ite=ite+1$, the
           reader transmits with P_R and return to step 2.

6   Numerical Results and Discussions

In this section, we provide some numerical results illustrating the operation, features and benefits of the proposed overall framework and in particular the IMRA algorithm. Furthermore, the efficiency and effectiveness of the proposed approach is demonstrated via representative comparative scenarios.

Specifically, in Sect. 6.1 we initially demonstrate the convergence of the proposed Interference Mitigation Risk Aware (IMRA) algorithm. Moreover, the convergence time of the algorithm in terms of required iterations is studied and indicative real time-values are provided in order to show its applicability in realistic passive RFID scenarios. Then, in Sect. 6.2, the advantages of adopting the IMRA framework, in terms of controlling intruder-tags reflection power, are presented. The results obtained by the proposed IMRA approach are compared against two alternatives, namely: (a) the case where passive RFID tags reflect with their maximum available reflection power without considering any interference mitigation and/or power control scheme (in the following referred to as Max Reflection Scenario), and (b) the case where the IMRA adopts a more strict risk aware policy by the tags (e.g., convex risk function with respect to tag's reflection power) enforcing intruders in a more strict manner, compared to a linear risk aware policy, to adopt a social behavior (in the following referred to as IMRA - Convex Risk Scenario). Finally, in Sect. 6.3, an evaluation of intruders' impact on system's reliability and effectiveness is provided for the IMRA framework and the results are compared to the corresponding outcome from the Max Reflection Scenario, described above.

Throughout our study, we consider a passive RFID network consisting of one RFID reader and $N = N_n + N_{in}$ passive RFID tags. RFID reader's transmission power is fixed, i.e., $P_R = 2W$ and also the gain of its antenna is considered to be $G_R = 6\,dBi$. The minimum received power by the RFID reader, in order to demodulate the received signal from the tags is assumed $P_{TH} = -15\,dBm$ and corresponds to the passive RFID tag's target SINR γ_i^{target}. The passive RFID network operates at $f = 915\,MHz$. The channel loss from the ith tag to the reader is formulated using the simple path loss model, $h_i = c_i/d_i^a$, where d_i is the distance of tag i from the reader, a is the distance loss exponent (e.g. $a = 4$) and c_i is a log-normal distributed random variable with mean 0 and variance $\sigma^2 = 8(dB)$ [12]. The normal passive RFID tags are characterized by their backscatter gain $K_{i,n} = 60\%$ and the gain of their directional antenna is $G_{i,n} = 12\,dBi$, while the corresponding values for the intruder-tags are: $K_{i,in} = 90\%$ and $G_{i,in} = 16\,dBi$. The topology that has been considered in Sects. 6.1 and 6.2, corresponds to a shelve of a library (equivalently it could be a part of any linear supply chain) containing $N = 100$ passive RFID tags and the distance d_i among the reader and each tag ranges in the interval [0.2 m, 1.5 m].

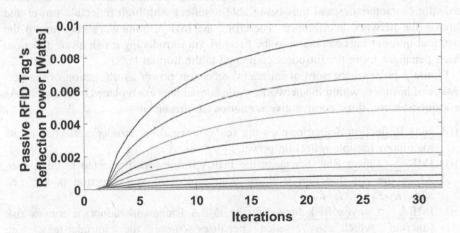

Fig. 2. IMRA algorithm's convergence (10 selected tags presented in the graph).

6.1 Convergence Evaluation of the IMRA Algorithm

We assume that the RFID network consists of $N_n = 100 = N$ passive RFID tags while for demonstration purposes only in the following we present the behavior of 10 tags that are placed in increasing distance from the RFID reader. Figure 2 illustrates tags' reflection powers' evolution as a function of the iterations required for the IMRA algorithm to converge at game's G'_{IMRA} Nash equilibrium point. It should be noted that the same results hold true in terms of necessary iterations for convergence, if intruder-tags were residing in the network, while the absolute values of their reflection powers would be different.

The corresponding results reveal that the convergence of the proposed IMRA algorithm is very fast since less than thirty-five iterations are required in order to reach the equilibrium for all tags, starting from randomly selected feasible initial reflection powers. Moreover, for all practical purposes we notice that in less than twenty five iterations the values of the reflection powers have approximately reached their corresponding equilibrium values. The IMRA algorithm was tested and evaluated in an Intel (R) Core (TM) 2 DUO CPU T7500 @ 2.20 GHz laptop with 2.00 GB available RAM and its runtime was less than 0.5 ms, thus it can be easily adopted in a realistic scenario. Furthermore, given the distributed nature of the IMRA algorithm, i.e., the calculations are made by each RFID tag, its runtime does not depend on the number of passive RFID tags residing in the RFID network, therefore it is quite a scalable approach in single hop communication passive RFID networks.

6.2 Improving System Operational Efficiency Through Interference Mitigation

As it has been presented and discussed in detail in this paper, one of the main reasons that can disturb the proper operation of an RFID network (in terms of properly reading the passive RFID tags) is the presence of intruder-tags that are enabled with favorable hardware characteristics and thus being able to reflect with high reflection power and increase the network interference. Therefore, the IMRA framework can control the harm that intruder-tags can cause to the network via introducing a risk aware function, which penalizes more the intruders compared to the normal tags.

Figure 3 presents the sum of intruders' reflection power as a function of the percentage of intruders within the network, while normal tags are replaced by intruders. As mentioned before, three comparative scenarios are presented:

(i)   Max Reflection Scenario: each tag (either normal or intruder) reflects with its maximum feasible reflection power.
(ii) IMRA – Linear Risk Scenario: the IMRA framework presented in this paper, where the risk function is linear with respect to the reflection power, i.e., $R(G_i, K_i, P_i) = G_i \cdot K_i \cdot P_i$.
(iii) IMRA – Convex Risk Scenario: the IMRA framework adopts a convex risk function which in essence penalizes more the intruder-tags, i.e., $R(G_i, K_i, P_i) = G_i \cdot K_i \cdot e^{P_i}$.

Based on the results of Fig. 3, it is clearly observed that the IMRA framework decreases considerably the impact of the intruder-tags on the network via keeping their reflection powers at low levels, thus mitigating the interference caused by them. Moreover, it is observed that as the risk function becomes more strict, thus imposing an even more social behavior to the intruders, the sum of intruders' reflection powers can

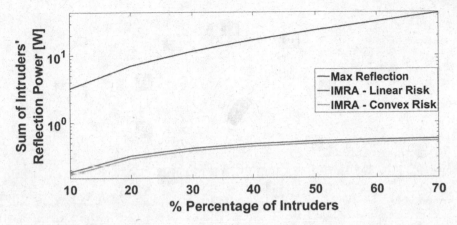

Fig. 3. Sum of Intruders' reflection power as a function of the percentage of intruders.

be further decreased. Therefore, based on the potential threat that an RFID network is expected to confront, different risk functions can be adopted, resulting in better level of protection.

6.3   Evaluation of Intruders' Impact on System Reliability and Effectiveness

Towards studying the impact of intruders on system's reliability and effectiveness, a detailed comparative study between the Max Reflection Scenario and the IMRA – Linear Risk Scenario is presented. A simplified topology has been considered as presented in Fig. 4 towards keeping most of the parameters the same among the passive RFID tags (e.g., distance from the reader), thus observing the impact of replacing normal RFID tags with intruders. The tags with x symbol refer to those tags that do not achieve their target SINR, while the tags with $\sqrt{}$ symbol are those that can be read by the reader. The star-tag depicts the intruder.

In Fig. 5, the results reveal that in the Max Reflection Scenario, the intruder-tag that replaces a normal tag, dominates the rest of the tags and achieves to be read by the RFID reader, due to its comparatively larger reflection power. In parallel, it causes high interference to the network, thus normal RFID tags cannot be read, due to the fact that their maximum available reflection power is not sufficient to overcome the imposed interference. Observing the multiple examples in Fig. 5 for different number of intruders in the Max Reflection Scenario, we conclude that the intruder-tags achieve to be read, while the normal tags fail. However, this is completely undesirable due to the fact that an intruder-tag may reflect erroneous or misleading data, or alternatively few intruder-tags suffice to cause the non-reading of many normal tags.

On the other hand, the IMRA – Linear Risk Scenario achieves to isolate the intruder-tags and not read them, while it enables the RFID reader to properly read the normal tags. This observation stems from the fact that intruder tags are penalized via the linear risk function towards reducing their reflection power, which becomes quite

Fig. 4. Circle topology with $N = 10$ passive RFID tags and $d = 0.4$ m.

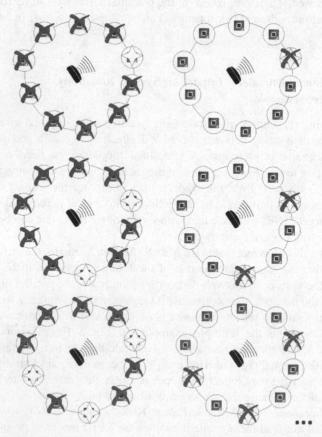

Fig. 5. Read ($\sqrt{}$) and non-read tags (x) for different numbers of intruders ($\diamondsuit$).

low so that it is not sufficient to enable the intruder-tag to be read by the reader. This outcome is of great practical importance because it can be adopted as a methodology to isolate intruder-tags and support RFID network's proper operation.

7 Related Work

Towards guaranteeing the non-disruptive reliable operation of a passive RFID network two critical dimensions should be considered: (a) energy-efficiency and (b) risk level of RFID devices, mainly for the following reasons:

(i)   The maximum RFID reader's transmission power is limited by regulations [4] and it is the only source power enabling the RFID network's operation, thus it should be utilized/spent in a sophisticated manner.

(ii) RFID readers' and tags' emissions and reflections, respectively, can cause interference in the passive RFID network (resulting in limited read range and inaccurate reads) and in the neighboring systems.

(iii) The optimization of readers'/tags' transmission/reflection power contributes to readers' energy saving, prolonging passive RFID network's lifetime, building an energy-efficient network and extending passive RFID tags' reflection range.

(iv) Malicious passive RFID tags characterized by high risk level can cause great interference levels in the passive RFID network, thus threatening its proper operation.

Several frameworks have been proposed in the recent literature in order to deal with energy-efficiency and/or secure and reliable operation mainly in active RFID networks (i.e. including active or semi-passive RFID tags). In [5], a security solution for RFID supply chain systems has been proposed via classifying the supply chain environments in two categories, i.e. weak and strong security mode. A set of RFID protocols, e.g., tag reading, security mode switching, secret updating protocols, are introduced to enable the dual security modes. The authors in [6], propose a key management protocol to ensure the privacy of the RFID reader and tags in the communication channel among tags, reader and backend server. The European research project BRIDGE [7] has focused its efforts in providing security technology that supports RFIDs' potential in mitigating existing business and security process risks. In [8], a trusted platform module is introduced only for the RFID readers, which constitute the core root of trust measurement for the overall framework.

Additional research works have targeted their efforts mainly to the power control and energy-efficiency improvement problem. In [9], a power control mechanism of RFID reader's transmission power considering the proximity and motion sensors towards detecting an RFID tag in reader's range is presented. In [10], an energy-efficient RFID tags inventory algorithm is proposed towards adjusting RFID reader's transmission power via automatically estimating the number of tags in its coverage area. In [11], a dynamic power algorithm is introduced, where a Received Signal Strength Indication (RSSI) receiver is employed at RFID reader's side to measure the strength of the received signal and adapt RFID reader's transmission power accordingly. In [12], two heuristic power control algorithms are presented considering the interference measured at each RFID reader or its achieved signal-to-interference ratio (SIR), respectively, as local feedback parameters in order to adapt RFID readers' transmission power.

The proposed framework in this paper differs from the aforementioned approaches associated with the secure and reliable operation of an RFID network in the sense that the IMRA framework capitalizes on power control and interference management techniques in order to mitigate potential risks introduced by intruding passive RFID tags. Its main novelty is that the IMRA approach proactively protects the RFID network from malicious behaviors of passive RFID tags, thus supporting its proper and non-disturbed operation. Based on an interference mitigation risk aware technique, masked or disguised intruder-tags pretending to act as normal within the RFID network are enforced to conform to a social-behavior, otherwise their existence can be a priori identified due to their increased reflection power levels. As such, the IMRA framework is able to contribute towards securing the proper and reliable operation of the RFID network reducing the threat and harm stemming from intruder-tags.

8 Concluding Remarks and Future Work

In this paper, the problem of mitigating the interference imposed by the intruders towards protecting the proper operation of passive RFID networks has been studied. Passive RFID networks are characterized by limited available power, thus they can become vulnerable to intruder-tags, which cause high interference to the network, resulting in inability of reading passive RFID tags. Passive RFID tags are characterized as normal or intruders and all of them adopt a well-designed utility function, which reflects their goal of being read by the reader, while it also captures their risk level depending on their hardware characteristics. An Interference Mitigation Risk Aware (IMRA) problem is formulated as a maximization problem of each tag's utility function and solved based on a game theoretic approach, i.e., supermodular games. The Nash equilibrium of the IMRA game (i.e., vector of passive RFID tags' reflection powers) is determined and a distributed algorithm towards calculating it is introduced. Indicative numerical results show the superiority of the proposed framework and more specifically its important attribute to identify and isolate the intruder-tags from the network.

Though in the current work as a proof of concept we focused on simple topologies where for example only one reader exists in the network, as part of our current research work we are performing additional extensive simulations in order to evaluate the performance of the proposed approach under more complex topologies, including additional variable (mobile) readers. The additional power overhead imposed to the tags by introducing the risk function can be further investigated and quantified. To further validate the applicability of our proposed interference mitigation risk aware framework, this framework should be also tested either in experimental IoT infrastructures or realistic large scale passive RFID networks, e.g., library systems, warehouses, etc. Furthermore, the IMRA framework can be extended in multi-hop (tag-to-tag communication) passive RFID networks, where the constraints of tags' maximum reflection powers and the appropriate communication path/route should be considered and investigated. Moreover, the utility-based framework that has been proposed in this work can be utilized towards implementing a utility-based risk-aware/secure routing protocol in passive tag-to-tag RFID networks. In addition, different forms and/or expressions of the utility functions should be investigated in order to better represent scenarios where different RFID tags with different criticality and priority are included in the system or alternatively to express intruders' utilities with differentiated forms compared to those of normal tags. Finally, part of our current and future research work in this area, considers additional game theoretic analysis where a team of intruders is strategically placing themselves and acting so as to induce maximum damage in the network, while the proposed network control and management framework attempts to react against such malicious attempts, by minimizing if not totally eliminating the potential damage. Given the distributed nature of the emerging IoT paradigm, additional types of attacks may be considered including localized ones that mainly aim at damaging a subset of RFIDs only.

Acknowledgement. The research of Eirini Eleni Tsiropoulou and John S. Baras was partially supported by NSF grant CNS-1035655 and by AFOSR MURI grant FA9550-10-1-0573.

References

1. Juels, A.: RFID security and privacy: a research survey. IEEE J. Sel. Areas Commun. **24**(2), 381–394 (2006)
2. Ngai, E.W.T., Moon, K.K.L., Riggins, F.J., Yi, C.Y.: RFID research: an academic literature review (1995–2005) and future research directions. Int. J. Prod. Econ. **112**(2), 510–520 (2008)
3. Finkenzeller, K.: RFID Handbook Fundamentals and Applications in Contactless Smart Cards, Radio Frequency Identification and Near-Field Communication. Wiley, New York (2003)
4. http://www.gs1.org/docs/epc/UHF_Regulations.pdf
5. Shaoying, C., Yingjiu, L., Tieyan, L., Deng, R.H., Haixia, Y.: Achieving high security and efficiency in RFID-tagged supply chains. Int. J. Appl. Crypt. **2**(1) (2010). http://dx.doi.org/10.1504/IJACT.2010.033794
6. Bai, E., Ge, H., Wu, K., Zhang, W.: A trust-third-party based key management protocol for secure mobile RFID service. In: International Conference on Wireless Communications, Networking and Mobile Computing, pp. 1–5 (2009)
7. Aigner, M., Burbridge, T., Ilic, A., Lyon, D., Soppera, A., Lehtonen, M.: BRIDGE: RFID Security, White Paper. http://www.bridge-project.eu/data/File/BridgesecuritypaperDL_9.pdf
8. Sun, Y., Yin, L., Liu, L.: Towards a trusted mobile RFID network framework. In: International Conference on Cyber-Enabled Distributed Computing and Knowledge Discovery, pp. 53–58 (2013)
9. Chang, T.-H., Keppeler, K.E., Rinkes, C.: Patent: methods and systems for RFID reader power management, US20090309704 A1 (2009)
10. Xu, X., Gu, L., Wang, J., Xing, G.: Negotiate power and performance in the reality of RFID systems. In: IEEE International Conference on Pervasive Computing and Communications, pp. 88–97 (2010)

11. Boaventura, A.S., Carvalho, N.B.: A proposal for dynamic power control in RFID and passive sensor systems based on RSSI. In: European Conference on Antennas and Propagation, pp. 3473–3475 (2012)
12. Cha, K., Ramachandran, A., Jagannathan, S.: Adaptive and probabilistic power control algorithms for RFID reader networks. Int. J. Distrib. Sens. Netw. **4**(4), 347–368 (2008)
13. Fudenberg, D., Tirole, J.: Game Theory. MIT Press, Cambridge (1991)
14. Apostol, T.M.: Calculus. In: One-Variable Calculus, with an Introduction to Linear Algebra, 2nd edn., vol. 1. Blaisdell, Waltham (1967)
15. Yates, R.D.: A framework for uplink power control in cellular radio systems. IEEE J. Sel. Areas Commun. **13**, 1341–1347 (1995)

Security Risks and Investments

Risk Averse Stackelberg Security Games
with Quantal Response

Renaud Chicoisne[1(✉)] and Fernando Ordóñez[2]

[1] University of Colorado Denver, 1201 Larimer Street, Denver, CO 80204, USA
renaud.chicoisne@ucdenver.edu
[2] Universidad de Chile, Republica 701, Santiago, Chile
fordon@dii.uchile.cl

Abstract. In this paper, we consider a Stackelberg security game (SSG) where a defender can simultaneously protect m out of n targets with $n > m$ from an adversary that uses a quantal response (QR) to decide which target to attack. The main contribution consists in introducing risk aversion in the defender's behavior by using an entropic risk measure. Our work extends the work in [20] to a model that considers a risk averse defender. In addition we improve the algorithms used in [20] by reducing the number of integer variables, outlining how this adapts to arbitrary linear constraints. Computational results are presented on large scale artificial instances, showing the qualitative advantages of using a risk measure rather than the expected value.

Keywords: Stackelberg security games · Risk averse optimization · Entropic risk measure · Quantal response

1 Introduction

In this paper, we introduce risk aversion in a special class of Stackelberg games [16]. In airport security or coast guard patrol, security forces - the *leader* or *defender* - has limited capacity to defend a finite set of targets against human adversaries - the *followers* or *attackers*. A Stackelberg game is defined as a game where the leader decides a mixed strategy to maximize its utility, taking into account that the follower will observe this strategy and in turn decide its action to maximize its utility. In this situation, it is crucial to use resources wisely to minimize the damage done to the targets. Hence, an accurate knowledge of the attackers' behavior is key. Standard models assume a perfectly rational attacker that maximizes its utility knowing the defense strategy [7,11], or that can deviate slightly from the optimal attack [12]. Nevertheless, it is commonly accepted that humans take decisions that are in general different from the policy that optimizes a given reward function [1]. Consequently, assuming a highly intelligent adversary can lead to weak defense strategies, that fail to take advantage of knowledge of the attacker. The work presented in [9] assumes that human adversaries do not behave rationally, sometimes selecting actions that do not maximize

© Springer International Publishing AG 2016
Q. Zhu et al. (Eds.): GameSec 2016, LNCS 9996, pp. 83–100, 2016.
DOI: 10.1007/978-3-319-47413-7_5

their utility. The model in [9] assumes attackers follow a quantal response (QR). This idea models the decision probability of an attacker with a logit distribution derived from discrete choice theory. This model is parametrized with a *degree of rationality* and in fact, considers perfect rationality or indifference as special cases. Furthermore, it is strongly backed in the literature and in practice by its superior ability to model human behavior [5,14,18,19]. A polynomial time algorithm is presented in [20] to solve the problem of finding an optimal - in expectation - defense strategy against QR adversaries in SSG solving a polynomial number of continuous convex optimization problems. In this work, we present a natural extension of this expected utility maximization approach, by including risk aversion in the objective of the defender.

In a security domain, worst case outcomes can require more attention than average, more common events. For instance most smuggling activities could involve commercial goods or drugs, while a worst case outcome could arise from a terrorist group smuggling materials or people for a large scale attack. Focusing on expected outcomes can divert resources from potential catastrophic events to addressing more common threats. Admittedly, only considering a catastrophic event could also be inefficient, as the resources would be concentrated on events that might never occur. It becomes therefore important to balance likely outcomes with rare but catastrophic ones. The literature on risk measures provides various models of incorporating risk in decision models, usually in a parametrizable form to gauge the tradeoffs of considering expected outcomes or catastrophic ones. Toward this objective, we use an Entropic risk measure [13] that amplifies the importance of bad outcomes that are under a determined threshold. The entropic risk measure of parameter $\alpha > 0$ of a random variable X is defined by $\alpha \ln \mathbb{E}\left[e^{\frac{X}{\alpha}}\right]$. Scenarios whose corresponding payoff are larger than the value α contribute more to this risk measure. Therefore, the parameter α corresponds to a payoff value of risky outcomes and must be chosen carefully to tune the risk aversion level of the decision maker. Using an Entropic risk measure gives a solution that reduces the possible bad outcomes, thus reducing the variance that the solution observes over the possible outcomes. For example, consider Table 1, where we compare the solutions obtained for an example that will be explained in Sect. 3. Here, the solution that optimizes the Entropic risk measure, $\tilde{x}$ has a slightly worse expected value but has a much better variability and worst case than the solution that optimizes the expected value, x^*. We show in this work that the best defense strategy for a defender that optimizes an Entropic risk measure against a QR adversary can be found in polynomial time using the change of variables introduced in [20]. Further, we present a computationally fast algorithm to solve the resulting sequence of convex optimization problems.

We structured the rest of the paper as follows: in the next section we present the results of [20] to solve the SSG with a risk neutral defender and an adversary that uses a QR strategy. In Sect. 3 we present a polynomial time algorithm for the problem when introducing risk aversion. Section 4 describes an algorithm that solves a generalization of the original problem by solving $O\left(\ln\frac{1}{\epsilon}\right)$ convex minimization problems when additional linear constraints with positive coefficients

Table 1. Comparison of x^* vs. $\tilde{x}$

	$\mathbb{E}$	$\mathbb{V}$	Worst case $\mathbb{P}$
x^*	0.245	4.980	0.192
$\tilde{x}$	0.233	4.546	0.159
Difference	-4.9%	-8.7%	-16.9%

are involved. In the same section, we introduce a further generalization of the original problem when additional linear constraints with arbitrary coefficients are involved, and propose a solution framework solving a succession of $O\left(\ln\frac{1}{\epsilon}\right)$ Mixed Integer Linear Programming problems. We show experimental results in Sect. 5 on large scale artificial instances and compare the performance of state of the art algorithms with the methods presented in this work. We present our conclusions in Sect. 6.

2 Quantal Response Equilibria in Security Games

We first consider a SSG with a single leader (defender) maximizing its expected utility and a single attacker following a QR as was considered in [20]. If the follower attacks target $i \in \{1, \ldots, n\}$ and the defender blocks the attack, then the reward of the defender is $\bar{R}_i \geqslant 0$ and the penalty of the attacker is $P_i \leqslant 0$. On the other hand, if there is an attack on an undefended target $i \in \{1, \ldots, n\}$, the defender receives a penalty $\bar{P}_i \leqslant 0$ but the attacker obtains a reward $R_i \geqslant 0$. Taking the role of the defender we want to know how to maximize our utility using a total of $m < n$ resources to cover the n targets.

2.1 Problem Formulation

Let $x_i \in [0, 1]$ be the frequency of protecting target i. It follows that the expected utility of the defender and the attacker when the target i is attacked are respectively $\bar{U}_i(x_i) = x_i\bar{R}_i + (1 - x_i)\bar{P}_i$ and $U_i(x_i) = x_iP_i + (1 - x_i)R_i$. Assuming that the attacker is not perfectly rational and follows a QR of rationality factor $\lambda > 0$ [9], the probability that target i is attacked is given by:

$$y_i(x) = \frac{e^{\lambda U_i(x_i)}}{\sum\limits_{j=1}^{n} e^{\lambda U_j(x_j)}}. \tag{1}$$

Perfect rationality ($\lambda \longmapsto +\infty$) and indifference ($\lambda \longmapsto 0$) of the adversary can be represented as limiting cases of the QR model in Eq. (1). We will see later that for theoretical complexity and computational tractability purposes, it is better to use the following alternative form by dividing by $e^{\lambda R}$ both numerator and denominator in (1):

$$y_i(x) = \frac{e^{\lambda(U_i(x_i)-R)}}{\sum\limits_{j=1}^{n} e^{\lambda(U_j(x_j)-R)}}$$

where $R := \max\limits_{i \in \{1,\dots,n\}} R_i$. The defender solves the following nonlinear and non-convex optimization problem to maximize its expected utility:

$$\max_{x \in [0,1]^n} \left\{ \sum_{i=1}^{n} y_i(x)\bar{U}_i(x_i) : \sum_{i=1}^{n} x_i \leqslant m \right\}.$$

In other words, the problem solved by the defender is as follows:

$$\max_{x \in [0,1]^n} \left\{ \frac{\sum\limits_{i=1}^{n} \beta_i e^{-\gamma_i x_i} \left(\bar{P}_i + \delta_i x_i \right)}{\sum\limits_{i=1}^{n} \beta_i e^{-\gamma_i x_i}} : \sum_{i=1}^{n} x_i \leqslant m \right\} \tag{2}$$

where $\beta_i := e^{\lambda(R_i-R)} \geqslant 0$, $\gamma_i := \lambda(R_i - P_i) \geqslant 0$ and $\delta_i := \bar{R}_i - \bar{P}_i \geqslant 0$.

2.2 Solution Approach

The authors in [20] present the following polynomial time algorithm to solve (2). First, given two functions $N : X \subseteq \mathbb{R}^n \longmapsto \mathbb{R}$ and $D : X \subseteq \mathbb{R}^n \longmapsto \mathbb{R}^+\backslash\{0\}$, they establish that for any $r \in \mathbb{R}$ we have:

$$w^* := \max_{x \in X} \left\{ w(x) := \frac{N(x)}{D(x)} \right\} \leqslant r \quad \Leftrightarrow \quad \forall x \in X : N(x) - rD(x) \leqslant 0 \tag{3}$$

The equivalence (3) suggests the following scheme to solve approximately the optimization problem (2): Given a lower bound L and an upper bound U of w^*, we can find an ϵ-optimal solution of (2) by successively solving

$$\max_{x \in X} \left\{ N(x) - rD(x) \right\}$$

with at most $\log_2 \frac{U-L}{\epsilon}$ different values of r using a binary search. At each step of the binary search, the following problem has to be solved:

$$\max_x \sum_{i=1}^{n} \beta_i e^{-\gamma_i x_i} \left(\bar{P}_i + \delta_i x_i \right) - r \sum_{i=1}^{n} \beta_i e^{-\gamma_i x_i}$$

$$\text{s.t.:} \qquad \sum_{i=1}^{n} x_i \leqslant m$$

$$x_i \in [0,1], \forall i \in \{1,\dots,n\}$$

Using the invertible change of variables: $z_i := e^{-\gamma_i x_i}$ (i.e., $x_i := -\frac{1}{\gamma_i} \ln z_i$) the problem to solve can be rewritten as the following convex optimization problem:

$$\max_z - \sum_{i=1}^n \frac{\delta_i \beta_i}{\gamma_i} z_i \ln z_i + \sum_{i=1}^n (\bar{P}_i - r) \beta_i z_i$$

$$\text{s.t.:} \qquad - \sum_{i=1}^n \frac{1}{\gamma_i} \ln z_i \leqslant m$$

$$z_i \in [e^{-\gamma_i}, 1], \forall i \in \{1, \ldots, n\}$$

Proposition 1. *We can approximately solve Problem (2) at ϵ precision in $O\left(\ln \frac{\bar{R} - \bar{P}}{\epsilon}\right)$ binary search iterations, where*

$$\bar{R} := \max_{i \in \{1, \ldots, n\}} \bar{R}_i \text{ and } \bar{P} := \min_{i \in \{1, \ldots, n\}} \bar{P}_i.$$

Proof. First, $L_r := \dfrac{\sum_{i=1}^n \beta_i e^{-\gamma_i \frac{m}{n}} \left(\bar{P}_i + \delta_i \frac{m}{n}\right)}{\sum_{i=1}^n \beta_i e^{-\gamma_i \frac{m}{n}}}$ and $U_r := \bar{R}$ are respectively lower and

upper bounds for the optimal value of problem (2). Because we are maximizing, the expected value of any feasible solution provides a lower bound. In particular, we obtain L_r evaluating the uniform strategy $x_i = \frac{m}{n}$. We obtain U_r noticing that $U_i(x_i) \leqslant \bar{R}_i \leqslant \bar{R}$ and the $y_i(x)$ sum one. In consequence, the binary search in r reaches an ϵ-optimal solution in at most $O\left(\ln \frac{U_r - L_r}{\epsilon}\right)$ iterations. Finally, given that $U_i(x_i) \geqslant \bar{P}_i \geqslant \bar{P}$ and the $y_i(x)$ sum one, we know that $L_r \geqslant \bar{P}$, obtaining the complexity bound.

We will refer to this problem as the basic model with only the resource constraint for Expected value maximization (EXP-B). We will now show that we can use this methodology to solve a generalization of (2) where the defender is risk averse.

3   Risk Averse Defender

A natural extension of the last model is to assume that the defender is risk averse with respect to the attacker's actions, i.e.: the defender prefers to minimize the risk associated to have bad outcomes even if it can imply a lower expected payoff.

3.1   Motivation

We motivate the risk averse model with the example that gave the results presented in the Introduction. Lets consider the case where we have $n = 2$ targets, a single resource $m = 1$ and the attacker has a rationality factor $\lambda = 0.25$. We describe the payoffs in Table 2. Putting aside the QR strategy of the attacker,

Table 2. Example's payoffs

	$\bar{R}_i = R_i$	$\bar{P}_i = P_i$
$i = 1$	3	-1
$i = 2$	1	-3

we can remark that for both players, the average payoff of the first target is higher than the second one ($1 > -1$). We can write the formulation (2) of this simple example as the following optimization problem:

$$\max_{0 \leqslant x_1, x_2 \leqslant 1} \frac{e^{x_1 - 0.75}(4x_1 - 1) + e^{x_2 - 0.25}(4x_2 - 3)}{e^{x_1 - 0.75} + e^{x_2 - 0.25}}$$

The objective function attains its maximum value at $x^* = (0.505, 0.495)$. Notice that the worst case scenario occurs when target 2 is attacked. Now let us compare the properties of this optimal solution against the following solution $\tilde{x} = (0.450, 0.550)$ as shown in Table 1. We can see that using the defense strategy $\tilde{x}$ instead of x^*, we can improve the worst case scenario's probability to occur by 16.9 % at the cost of losing 4.9 % on the average payoff. Moreover, the variance of the payoffs is reduced by 8.7 %, meaning that the payoffs have less variability when using strategy $\tilde{x}$. In consequence, there are other solutions that might provide a better tradeoff between risky and expected outcomes. Depending on the risk aversion of the defender, maximizing an expected utility might not be a reasonable approach.

3.2 Problem Formulation

In the following, we assume that the leader is risk averse and wants to minimize an entropic risk measure. The entropic risk measure of parameter $\alpha > 0$ of a random variable X is defined by $\alpha \ln \mathbb{E}\left[e^{\frac{X}{\alpha}}\right]$. In our case, the uncertainty comes from the mixed strategies of both defender and attacker: x_i and $y_i(x)$ are respectively interpreted as the probability of target i being defended and the probability of target i being attacked. In other words, given a defense mixed strategy x, the entropic risk of the defender can be defined as follows:

$$E_\alpha(x) := \alpha \ln \sum_{i=1}^{n} y_i(x) \left(x_i e^{-\frac{R_i}{\alpha}} + (1 - x_i) e^{-\frac{P_i}{\alpha}} \right). \tag{4}$$

Notice that the defender wants to avoid high losses: consequently his objective is to minimize the risk of getting high negative payoffs. We consider that the situation in which no target is attacked can be represented as attacking a dummy target with moderate payoffs. In a risk measure this action would contribute little to the payoff of the defender. With this definition at hand the leader wants to solve the following optimization problem:

$$\min_{x \in [0,1]^n} \left\{ E_\alpha(x) : \sum_{i=1}^n x_i \leqslant m \right\}. \tag{5}$$

It can be proven that the expected value maximization model of Sect. 2 is a limiting case of the last problem as

$$\alpha \ln \mathbb{E}\left[e^{\frac{X}{\alpha}}\right] \xrightarrow[\alpha \to +\infty]{} \mathbb{E}[X].$$

We will see later that for theoretical complexity and computational tractability purposes, it is useful to use the following alternative form instead of (4) by factorizing the term inside the logarithm by $e^{-\bar{P}}$:

$$E_\alpha(x) = \alpha \ln \sum_{i=1}^n y_i(x) \left(x_i e^{-\frac{\tilde{R}_i - \bar{P}}{\alpha}} + (1 - x_i) e^{-\frac{\tilde{P}_i - \bar{P}}{\alpha}} \right) - \bar{P}.$$

Defining $\widetilde{R}_i := e^{-\frac{R_i - \Gamma}{\alpha}}$, $\widetilde{P}_i := e^{-\frac{P_i - \bar{P}}{\alpha}}$ and $\theta_i := \widetilde{P}_i - \widetilde{R}_i \geqslant 0$, the entropic risk of some mixed defense strategy x can be rewritten as:

$$E_\alpha(x) := \alpha \ln \left(\frac{\sum_{i=1}^n \beta_i e^{-\gamma_i x_i} \left(\widetilde{P}_i - \theta_i x_i \right)}{\sum_{i=1}^n \beta_i e^{-\gamma_i x_i}} \right) - \bar{P}$$

And given that $t \to \alpha \ln t$ is non decreasing and $\bar{P}$ is a constant, the general problem the defender solves is equivalent to the following problem:

$$\min_{x \in [0,1]^n} \left\{ \frac{\sum_{i=1}^n \beta_i e^{-\gamma_i x_i} \left(\widetilde{P}_i - \theta_i x_i \right)}{\sum_{i=1}^n \beta_i e^{-\gamma_i x_i}} : \sum_{i=1}^n x_i \leqslant m \right\}. \tag{6}$$

Which is very similar to the expected value maximization problem (2) described in the last section. We will refer to this problem as the basic model with only the resource constraint for Entropy minimization (ENT-B).

3.3 Solution Approach

Proposition 2. *We can solve problem (6) using a binary search in r that solves at each iteration the following problem:*

$$w(r) := \min_z \ \sum_{i=1}^n \beta_i z_i \left(\widetilde{P}_i - r + \frac{\theta_i}{\gamma_i} \ln z_i \right) \tag{7}$$

$$s.t.: \quad -\sum_{i=1}^n \frac{1}{\gamma_i} \ln z_i \leqslant m \tag{8}$$

$$z_i \in \left[e^{-\gamma_i}, 1 \right], \forall i \in \{1, \ldots, n\} \tag{9}$$

Which is a convex minimization problem.

Proof. Given that problem (6) is a fractional programming problem, we can solve it with binary search as in the risk neutral case of the last section. At each iteration of the binary search, we have to solve the following problem:

$$\min_{x \in [0,1]^n} \left\{ \sum_{i=1}^n \beta_i e^{-\gamma_i x_i} \left(\widetilde{P}_i - r - \theta_i x_i \right) : \sum_{i=1}^n x_i \leqslant m \right\}.$$

Using the invertible change of variables $z_i := e^{-\gamma_i x_i}$, the problem we have to solve is (7–9). As in Subsect. 2.2, it is easy to see that the feasible set is convex and given that $\theta_i, \gamma_i \geqslant 0$ and $t \longmapsto t \ln t$ is convex, the objective function is convex as well.

Proposition 3. *We can approximately solve Problem* (6) *to a precision ϵ in* $O\left(\ln \frac{\bar{R} - \bar{P}}{\epsilon} \right)$ *binary search iterations.*

Proof. Similar to the proof of Proposition 1,

$$L_r := e^{-\frac{\bar{R} - \bar{P}}{\alpha}} \text{ and } U_r := \frac{\sum_{i=1}^n \beta_i e^{-\gamma_i \frac{m}{n}} \left(\widetilde{P}_i - \theta_i \frac{m}{n} \right)}{\sum_{i=1}^n \beta_i e^{-\gamma_i \frac{m}{n}}}$$

are respectively lower and upper bounds for the optimal value of problem (6). Recalling that we want to minimize the complete entropic risk measure (see Problem (5)) and not solving problem (6) considered for convenience, the binary search stops after at most $O\left(\ln \frac{\alpha \ln U_r - \alpha \ln L_r}{\epsilon} \right)$ iterations to reach an ϵ-optimal solution for the original problem. Using again the fact that for any defense strategy x the $y_i(x)$ are probabilities, $e^{-\frac{\bar{P} - \bar{P}}{\alpha}} = 1$ is an upper bound of U_r, hence $\alpha \ln U_r \leqslant 0$. Moreover, $\alpha \log L_r = \bar{P} - \bar{R}$, thus obtaining the complexity bound.

In this section, we proved that we could find a ϵ-optimal strategy for risk averse and risk neutral defenders solving for both problems a succession of at most $O\left(\ln \frac{\bar{R} - \bar{P}}{\epsilon} \right)$ convex optimization problems, which can be done in polynomial time.

4   An Extended Model

In practice, we can face additional operational constraints. For example, some targets i and j cannot be defended at the same time, leading to the following additional constraint $x_i + x_j \leqslant 1$, or precedence constraints, allowing target i to be defended only if target j is defended: $x_i \leqslant x_j$.

Without loss of generality, in the remainder of this paper we will only consider minimization problems of the form

$$w(r) := \min_{x \in [0,1]^n} \left\{ w^r(x) := \sum_{i=1}^n w_i^r(x_i) : Ax \leqslant b \right\} \tag{10}$$

where in the case of Expected value maximization, we have:

$$w_i^r(x_i) := \beta_i e^{-\gamma_i x_i}\left(r - \bar{P}_i - \delta_i x_i\right)$$

and in the case of Entropy minimization, we have:

$$w_i^r(x_i) := \beta_i e^{-\gamma_i x_i}\left(\widetilde{P}_i - r - \theta_i x_i\right)$$

4.1   Linear Inequality Constraints with Positive Coefficients

First, lets consider a set of additional linear inequalities $Ax \leqslant b$ with positive coefficients $a_{ij} \geqslant 0$ and right-hand sides $b_j \geqslant 0$. The problem to solve is still a fractional programming problem, and as such we can use the same binary search based approach in r to guess its optimal value. The resulting problem to solve during each iteration of the binary search is then (10).

Proposition 4. *Computing $w(r)$ can be achieved solving a convex optimization problem. As a direct consequence, we can solve the original problem by solving $O\left(\ln\frac{\bar{R}-\bar{P}}{\epsilon}\right)$ convex optimization problems.*

Proof. We already proved that using the change of variables $x_i := -\frac{1}{\gamma_i}\ln z_i$, the objective function was convex. Let $\sum_{i=1}^{n} a_{ij}x_i \leqslant b_j$ be the j-th constraint of $Ax \leqslant b$. When applying the change of variables we obtain: $-\sum_{i=1}^{n}\frac{a_{ij}}{\gamma_i}\ln z_i \leqslant b_j$. Given that $a_{ij} \geqslant 0$ for every $i \in \{1,\ldots,n\}$, the constraint remains convex. As in the unconstrained case, it takes at most $O\left(\ln\frac{\bar{R}-\bar{P}}{\epsilon}\right)$ binary search iterations in r to reach an ϵ-optimal solution.

We will refer to this problem as the model with extra linear inequality constraints with positive coefficients for the Entropy minimization (ENT-PLC) and the Expected value maximization (EXP-PLC).

4.2   General Linear Constraints

We now consider a problem with general linear constraints $Ax \leqslant b$ with arbitrary signs. This model will be referred to as the model with extra linear constraints with arbitrary coefficients for the Entropy minimization (ENT-ALC) and the Expected value maximization (EXP-ALC). First, notice that we cannot use the change of variable $z_i := e^{-\gamma_i x_i}$ in problem (10) as some constraints would turn nonconvex. However, the only issue of Problem (10) is the nonconvexity of its objective function:

$$w^r(x) := \sum_{i=1}^{n} w_i^r(x_i)$$

where w^r is separable in x. There are several generic methods in the literature [15] that allow to piecewise linearly approximate it by approximating each w_i^r.

Proposition 5. *Given a partition $l = t_0 < t_1 < \ldots < t_K = u$ of $[l, u]$ we can approximate $f : [l, u] \to \mathbb{R}$ as follows:*

$$f(x) \approx \bar{f}(x) := \min_{x_k, z_k} \quad f(t_0) + \sum_{k=1}^{K} \left(f(t_k) - f(t_{k-1}) x_k \right)$$

$$s.t.: \qquad x = t_0 + \sum_{k=1}^{K} x_k (t_k - t_{k-1})$$

$$x_1 \leqslant 1$$

$$x_K \geqslant 0$$

$$x_{k+1} \leqslant z_k \leqslant x_k, \quad \forall k \in \{1, \ldots, K-1\}$$

$$z_k \in \{0, 1\}, \quad \forall k \in \{1, \ldots, K-1\}$$

This way of approximating by a piecewise linear function is known as an incremental model [3]. We will refer to this way of approximating the objective function as the Incremental Piecewise Linear approximation (IPL).

In the next proposition, we present a way to model the piecewise linear approximation with less binary variables on an arbitrary partition of $[0, 1]$, as described in [15].

Proposition 6. *Given a partition $l = t_0 < t_1 < \ldots < t_K = u$ of $[l, u]$ we can approximate $f : [l, u] \to \mathbb{R}$ as follows:*

$$f(x) \approx \bar{f}(x) := \min_{\lambda_k, z_l} \quad \sum_{k=0}^{K} \lambda_k f(t_k)$$

$$s.t.: \qquad x = \sum_{k=0}^{K} \lambda_k t_k$$

$$\sum_{k=0}^{K} \lambda_k = 1$$

$$\sum_{p \in S_K^+(l)} \lambda_p \leqslant z_l, \quad \forall l \in \{1, \ldots, L(K)\}$$

$$\sum_{p \in S_K^-(l)} \lambda_p \leqslant 1 - z_l, \quad \forall l \in \{1, \ldots, L(K)\}$$

$$z_l \in \{0, 1\}, \quad \forall l \in \{1, \ldots, L(K)\}$$

$$\lambda_k \geqslant 0, \quad \forall k \in \{1, \ldots, K\}$$

where $L(K) = \lceil \log_2 K \rceil$ and for any $l \in \{1, \ldots, L(K)\}$:

$$S_K^+(l) := \{p \in \{0, \ldots, K\} : \forall q \in Q_K(p), (B_K(q))_l = 1\}$$

$$S_K^-(l) := \{p \in \{0, \ldots, K\} : \forall q \in Q_K(p), (B_K(q))_l = 0\}$$

where $Q_K(p) := \{q \in \{1, \ldots, K\} : p \in \{q-1, q\}\}$, $B_K : \{1, \ldots, K\} \longmapsto \{0, 1\}^{L(K)}$ is a bijective mapping such that for all $q \in \{1, \ldots, K-1\}$, $B_K(q)$ and $B_K(q+1)$ differ in at most one component (See reflected binary or Gray code in [4]). Such a Gray code can be found quickly by the recursive algorithm of [8].

The main advantage of this formulation resides in the fact that it uses only $\lceil \log_2 K \rceil$ extra binary variables instead of K to model each piecewise linear approximation at the same precision. We will refer to this way of approximating the objective function as the Logarithmic Piecewise Linear approximation (LPL).

For each target $i \in \{1, \ldots, n\}$, we use a partition $0 = t_{i0} < t_{i1} < \ldots < t_{iK} = 1$ of $[0, 1]$.

Proposition 7. *The IPL approximation applied to problem (10) leads to the following MIP:*

$$\min_{x_i, x_{ik}, z_{ik}} \quad \sum_{i=1}^{n} \left(w_i^r(t_{i0}) + \sum_{k=1}^{K} \left(w_i^r(t_{ik}) - w_i^r(t_{i,k-1}) \right) x_{ik} \right)$$

$$s.t.: \qquad\qquad Ax \leqslant b$$

$$x_i \in [0, 1], \quad \forall i \in \{1, \ldots, n\}$$

$$x_i = t_{i0} + \sum_{k=1}^{K} x_{ik} (t_{ik} - t_{i,k-1}), \quad \forall i \in \{1, \ldots, n\}$$

$$x_{i1} \leqslant 1, \quad \forall i \in \{1, \ldots, n\}$$

$$x_{iK} \geqslant 0, \quad \forall i \in \{1, \ldots, n\}$$

$$x_{i,k+1} \leqslant z_{ik} \leqslant x_{ik}, \quad \forall k \in \{1, \ldots, K-1\}, \forall i \in \{1, \ldots, n\}$$

$$z_{ik} \in \{0, 1\}, \quad \forall k \subset \{1, \ldots, K-1\}, \forall i \in \{1, \ldots, n\}$$

Proposition 8. *The LPL approximation applied to problem (10) leads to the following MIP:*

$$\min_{x_i, \lambda_{ik}, z_{il}} \quad \sum_{i=1}^{n} \sum_{k=0}^{K} \lambda_{ik} w_i^r(t_{ik})$$

$$s.l.: \qquad\qquad Ax \leqslant b$$

$$x_i \in [0, 1], \quad \forall i \in \{1, \ldots, n\}$$

$$x_i = \sum_{k=0}^{K} \lambda_{ik} t_{ik}, \quad \forall i \in \{1, \ldots, n\}$$

$$\sum_{k=0}^{K} \lambda_{ik} = 1, \quad \forall i \in \{1, \ldots, n\}$$

$$\sum_{p \in S_K^+(l)} \lambda_{ip} \leqslant z_{il}, \quad \forall l \in \{1, \ldots, L(K)\}, \forall i \in \{1, \ldots, n\}$$

$$\sum_{p \in S_K^-(l)} \lambda_{ip} \leqslant 1 - z_{il}, \quad \forall l \in \{1, \ldots, L(K)\}, \forall i \in \{1, \ldots, n\}$$

$$z_{il} \in \{0, 1\}, \quad \forall l \in \{1, \ldots, L(K)\}, \forall i \in \{1, \ldots, n\}$$

$$\lambda_{ik} \geqslant 0, \quad \forall k \in \{1, \ldots, K\}, \forall i \in \{1, \ldots, n\}$$

5   Computational Results

The algorithms presented in this paper were coded in C programming language and run over a cluster node of 2.4 GHz with 6 Gb RAM. All the convex optimization problems were solved using the callable library of IPOPT [10,17], using as a subroutine the linear solver HSL [6]. All the Mixed Integer Linear Programming problems were solved using the callable library of CPLEX [2].

5.1   Instance Generation and Parameters

We solved all our problems at relative precision 10^{-8} for the subproblems in r, and at 10^{-6} relative precision for the binary search in r. and tested the algorithms with a rationality coefficient for the attacker $\lambda = 0.76$ as reported in [19].

The rewards (respectively penalties) of both defender and attacker were drawn uniformly in $]0,10]$ (respectively $[-10,0[$); notice that we did not assume zero sum games. We considered instances with a number of targets with $n \in \{1,2,5,7,10\} \cdot 10^3$ and a number of resources $m = \frac{n}{10}$. The parameter α captures an absolute risk aversion and penalizes greatly defense strategies whose bad realizations exceed α, so noticing that α has units - the same as the payoffs - we selected the parameter α of the Entropic risk measure $\alpha \in \{1,2,5,7\}$. Notice that α is a very subjective measure of the risk aversion of the decision maker and as such, it can be difficult to adjust in practice.

In the case of the more general problems with linear constraints with positive coefficients, we generated disjunctive constraints of the type $\sum_{i \in D} x_i \leqslant 1$. In the case of general linear constraints, we generated precedence constraints of the type $x_i \leqslant x_j$ for some pairs $(i,j) \in P \subset \{1,\ldots,n\}^2$. We generated $n/20$ such constraints with $|D| = n/20$ and the set of indices D in each constraint was drawn uniformly in $\{1,\ldots,n\}$. For the precedence constraints, we randomly generated $n/10$ pairs (i,j) uniformly in $\{1,\ldots,n\}^2$. Finally, we partitioned $[0,1]$ using uniform grids having $K \in \{16,32\}$ segments to piecewise linearly approximate the objective function in the most general model. It was proved in [20] that using uniform grids, the solution obtained is an $O\left(\frac{1}{K}\right)$-optimal solution for (10).

To analyze the influence of each parameter, we took as a base case $n = 10^3$, $m = 100$ and $\alpha = 5$. We then vary n and α independently and repeat the experiment 50 times.

5.2   Algorithmic Performances

We solved the following problems: (1) the basic model with only the resource constraint for the entropy (ENT-B) and the expected value (EXP-B), (2) the model with extra linear inequality constraints with positive coefficients for the entropy (ENT-PLC) and the expected value (EXP-PLC) and (3) the model with extra linear constraints with arbitrary coefficients for the entropy (ENT-ALC) and the expected value (EXP-ALC). We solved ENT-B and EXP-B with IPOPT, the incremental piecewise linear approximation (IPL) and the logarithmic piecewise linear approximation (LPL), ENT-PLC and EXP-PLC with IPOPT, IPL and LPL and ENT-ALC and EXP-ALC with IPL and LPL.

Basic model (B). In the most simple setting, we can see in Fig. 1 that IPOPT is always superior in terms of execution time and solution quality. Further, we can see that LPL is always faster than IPL by a wide margin. Finally, we can see that taking $K = 32$ pieces provides a significant precision advantage. The parameter α has no real effect on the execution time nor the precision achieved. For this basic model, we will use IPOPT as the reference method to solve it.

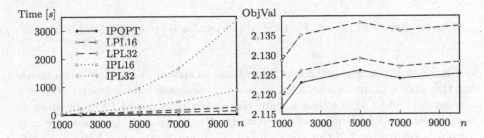

Fig. 1. Model B: execution time [s] (left) and objective value (right) vs. n

PLC model. Adding linear constraint with positive coefficients, we can see in Fig. 2 that IPOPT is also always superior in terms of execution time and solution quality. Surprisingly, IPL16 is faster than LPL32, even though the number of binary variables of the latter is three times lower than the former. Nevertheless, the numerical precision of LPL32 is higher than IPL16 and thus return better solutions. Again, the parameter α has no real effect on the execution time nor the precision achieved. For this model, we will use IPOPT as the reference method as well.

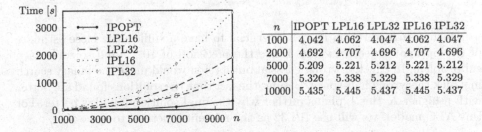

n	IPOPT	LPL16	LPL32	IPL16	IPL32
1000	4.042	4.062	4.047	4.062	4.047
2000	4.692	4.707	4.696	4.707	4.696
5000	5.209	5.221	5.212	5.221	5.212
7000	5.326	5.338	5.329	5.338	5.329
10000	5.435	5.445	5.437	5.445	5.437

Fig. 2. Model PLC: execution time [s] (left) and objective value (right) vs. n

ALC model. Adding linear constraint with arbitrary coefficients, we can obtain from the results depicted in Fig. 3 the same conclusions as in the PLC model: for the same level of precision, LPL is faster than IPL. The entropy parameter α

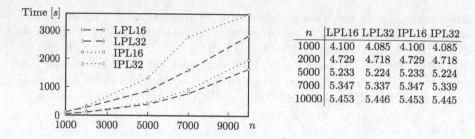

n	LPL16	LPL32	IPL16	IPL32
1000	4.100	4.085	4.100	4.085
2000	4.729	4.718	4.729	4.718
5000	5.233	5.224	5.233	5.224
7000	5.347	5.337	5.347	5.339
10000	5.453	5.446	5.453	5.445

Fig. 3. Model ALC: execution time [s] (left) and objective value (right) vs. n

does have an influence on the execution time, as shown in Fig. 4. We can notice that the more risk averse the defender is - i.e.: the lower the parameter α is - the incremental model IPL is faster than the logarithmic model LPL. In order to check which model is the most adequate to solve large scale instances, we may check in a future work if this tendency of IPL being faster than LPL still holds for larger instances than our current base case ($n = 1000$).

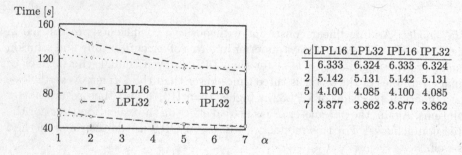

α	LPL16	LPL32	IPL16	IPL32
1	6.333	6.324	6.333	6.324
2	5.142	5.131	5.142	5.131
5	4.100	4.085	4.100	4.085
7	3.877	3.862	3.877	3.862

Fig. 4. Model ALC: execution time [s] (left) and objective value (right) vs. α

During each iteration in r, it is crucial to have a sufficiently close estimate of the optimal objective value, hence the precision of 10^{-8} used to solve each subproblem. In effect, a wrong guess about its sign would make the binary search in r pick the wrong half-space. Besides the fact that the solutions found are better with a higher K this explains further why the final solution found is better. For this ALC model, we will use IPL32 as the reference method to solve it.

5.3   Qualitative Results

To compare a risk neutral defense policy with a risk averse one, we want to see if there is some kind of stochastic dominance of a risk averse strategy versus a risk neutral one. To do so, we want to compare the payoffs distributions of the defender depending on its risk aversion. In a real situation, the defender can cover m targets out of n and the attacker targets a single place. The only possible outcomes for the defender are: (1) being attacked on a defended target i with payoff $V = \bar{R}_i > 0$ or (2) being attacked on an undefended target i with payoff $V = \bar{P}_i < 0$. Consequently, if we assume that all the payoffs $\bar{R}_i$ and $\bar{P}_i$ are different the only values possible are in

$$V \in \{V_1 < V_2 < \ldots < V_{2n-1} < V_{2n}\} = \bigcup_{i=1}^{n} \{\bar{R}_i, \bar{P}_i\}$$

Moreover, given a mixed defense strategy $x \in [0,1]^n$ and the associated QR $y(x) \in [0,1]^n$, the probability to block an attack at target i is:

$$\mathbb{P}\left[V = \bar{R}_i\right] = x_i y_i(x)$$

and the probability to undergo an attack at a defenseless target i is:

$$\mathbb{P}\left[V = \bar{P}_i\right] = (1 - x_i) y_i(x)$$

This way we can compute the probability distribution of the payoff of any defender with QR adversary without sampling a large number of simulations. For each instance solved, we report the expected value of the optimal solution, its variance, worst case payoff probability and Value at Risk (VaR) at level 10 %.

In Figs. 5 and 6, we compare some indicators of the minimizers of the entropic risk measure of parameter $\alpha = 5$ and the expected value maximizers. More specifically, in Fig. 5 we can see that the loss in expected value and the improvement in the payoff variance implied by the use of the entropy minimizers stay constant ($>-10\%$ and $<-35\%$) in the basic model (B) but decrease with the number of targets for the extended models PLC and ALC. This can be explained by the fact that the additional constraints are increasingly restraining and do not let a lot of slack with respect to the possible solutions attainable.

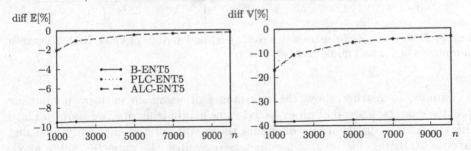

Fig. 5. Difference [%] in expected payoff (left) and payoff variance (right) between minimizers of $E_{\alpha=5}$ and minimizers of $\mathbb{E}$ vs. n

The same can be said of the results depicted in Fig. 6 where the Worst case probability and the $\mathrm{VaR}_{10\%}$ remain relatively constant ($>-50\%$ and $<-20\%$) in the basic model but get smaller with the number of targets for PLC and ALC.

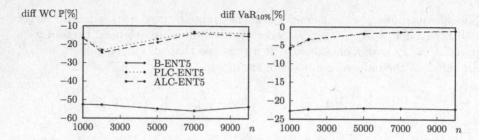

Fig. 6. Difference [%] in worst case probability (left) and $VaR_{10\%}$ (right) between minimizers of $E_{\alpha=5}$ and minimizers of $\mathbb{E}$ vs. n

In Figs. 7 and 8, we compare the same indicators of the minimizers of the entropic risk measure of different parameters α with the expected value maximizers' ones. In both figures we can see that for all three models the difference between the expected value maximizers and entropy minimizers gets smaller when α gets bigger. In effect, as mentioned in Sect. 3 the behavior of the entropic risk measure tends to that of an expected value. Still, we can see that it is possible to tune the desired degree of aversion quite easily by adjusting the parameter α depending on the amount of expected value we are ready to lose.

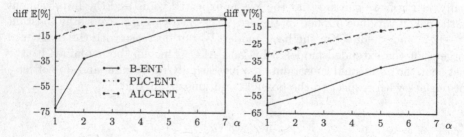

Fig. 7. Difference [%] in expected payoff (left) and payoff variance (right) between minimizers of E_α and minimizers of $\mathbb{E}$ vs. α

Finally, to further show the advantages of using an entropy minimizing defense strategy instead of an expected value maximizing one, we show in Fig. 9 the cumulative distributions of minus the payoffs of the entropy minimizing defense strategies for several parameters α against the expected value maximizing defense strategy for one instance of the base case. We can see that the expected value maximizing strategy is stochastically dominated on the worst cases by the entropy minimizing defense strategies. Furthermore, the higher the risk aversion factor α is, the lighter the 'tail' of bad realizations of the probability distribution will be. Once more, we can notice that in the ALC model, the effect of the risk aversion is attenuated in comparison with the basic model B.

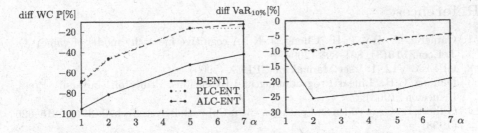

Fig. 8. Difference [%] in worst case probability (left) and $VaR_{10\%}$ (right) between minimizers of E_α and minimizers of $\mathbb{E}$ vs. α

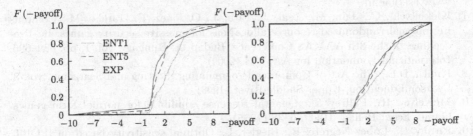

Fig. 9. Payoffs' cumulative distributions of the minimizers of E_α and maximizers of $\mathbb{E}$ in the basic model (left) and in ALC model (right)

6   Conclusions

In this paper, we extended the classic model of Stackelberg security games with quantal response (SSGQR) to a risk averse setting for the defender. We extended the model when linear inequalities with positive coefficients are added, and proved we could solve it solving a succession of convex minimization problems. We further extended the problem when arbitrary linear inequalities are added, and presented two ways of finding an approximate defense strategy solving a succession of MIPs. Computational results showed that minimizing an Entropic risk measure instead of maximizing the expected value can be advantageous from a qualitative point of view, allowing to significantly reduce the overall payoff variance and the probability of bad scenarios to occur.

In a future work, we would like to extend the model presented in this paper to a multiple attackers context and further improve the piecewise linear approximation models taking advantage of the convex parts of the objective function, and use smarter partitions than an uniform grid.

Acknowledgments. Funded by Conicyt through grant FONDEF No. D10I1002.

References

1. Camerer, C., Ho, T.-H., Chong, J.-K.: A cognitive hierarchy model of games. Q. J. Econ. **119**(3), 861–898 (2004)
2. CPLEX. V12. 1: User Manual for CPLEX (2009)
3. Dantzig, G.B.: Linear Programming and Extensions. Princeton University Press, Princeton (1963)
4. Gilbert, E.: Gray codes and paths on the n-Cube. Bell Syst. Tech. J. **37**(3), 815–826 (1958)
5. Haile, P., Hortaçsu, A., Kosenok, G.: On the empirical content of quantal response equilibrium. Am. Econ. Rev. **98**(1), 180–200 (2008)
6. HSL. A collection of fortran codes for large-scale scientific computation. http://www.hsl.rl.ac.uk/
7. Kiekintveld, C., Jain, M., Tsai, J., Pita, J., Ordóñez, F., Tambe, M.: Computing optimal randomized resource allocations for massive security games. In: Proceedings of the 8th AAMAS Conference, Budapest, Hungary, vol. 1, pp. 689–696. International Foundation for AAMAS (2009)
8. Knuth, D.E.: The Art of Computer Programming: Sorting and Searching, vol. 3. Pearson Education, Upper Saddle River (1998)
9. McKelvey, R., Palfrey, T.: Quantal response equilibria for normal form games. Games Econ. Behav. **10**(1), 6–38 (1995)
10. Pirnay, H., Lopez-Negrete, R., Biegler, L.: Optimal sensitivity based on IPOPT. Math. Program. Comput. **4**(4), 307–331 (2012)
11. Pita, J., Jain, M., Marecki, J., Ordóñez, F., Portway, C., Tambe, M., Western, C., Paruchuri, P., Kraus, S.: Deployed armor protection, the application of a game theoretic model for security at the los angeles international airport. In: Proceedings of the 7th AAMAS: Industrial Track, Cascais Miragem, Portugal, pp. 125–132. International Foundation for AAMAS (2008)
12. Pita, J., Jain, M., Ordóñez, F., Tambe, M., Kraus, S.: Solving stackelberg games in the real-world: addressing bounded rationality and limited observations in human preference models. Artif. Intell. J. **174**(15), 1142–1171 (2010)
13. Pratt, J.W.: Risk aversion in the small and in the large. Econometrica J. Econ. Soc. **32**(1/2), 122–136 (1964)
14. Stahl, D., Wilson, P.: Experimental evidence on players' models of other players. J. Econ. Behav. Organ. **25**(3), 309–327 (1994)
15. Vielma, J.: Mixed integer linear programming formulation techniques. SIAM Rev. **57**, 3–57 (2015)
16. Von Stackelberg, H.: The Theory of the Market Economy. William Hodge, London (1952)
17. Wachter, A., Biegler, L.: On the implementation of a primal-dual interior point filter line search algorithm for large-scale nonlinear programming. Math. Program. **106**(1), 25–57 (2006)
18. Wright, J., Leyton-Brown, K., Beyond equilibrium: predicting human behavior in normal-form games. In: Proceedings of the 24th AAAI Conference on Artificial Intelligence, Atlanta, GA (2010)
19. Yang, R., Kiekintveld, C., Ordóñez, F., Tambe, M., John, R.: Improving resource allocation strategy against human adversaries in security games. In: 22th IJCAI Proceedings, Barcelona, Spain, vol. 22, pp. 458–464. AAAI Press (2011)
20. Yang, R., Ordóñez, F., Tambe, M.: Computing optimal strategy against quantal response in security games. In: Proceedings of the 11th AAMAS, Valencia, Spain, vol. 2, pp. 847–854. International Foundation for AAMAS (2012)

Optimal and Game-Theoretic Deployment of Security Investments in Interdependent Assets

Ashish R. Hota[✉], Abraham A. Clements,
Shreyas Sundaram, and Saurabh Bagchi

School of Electrical and Computer Engineering,
Purdue University, West Lafayette, USA
{ahota,clemen19,sundara2,sbagchi}@purdue.edu

Abstract. We introduce a game-theoretic framework to compute optimal and strategic security investments by multiple defenders. Each defender is responsible for the security of multiple assets, with the interdependencies between the assets captured by an *interdependency graph*. We formulate the problem of computing the optimal defense allocation by a single defender as a convex optimization problem, and establish the existence of a pure Nash equilibrium of the game between multiple defenders. We apply our proposed framework in two case studies on interdependent SCADA networks and distributed energy resources, respectively. In particular, we investigate the efficiency loss due to decentralized defense allocations.

1 Introduction

Modern critical infrastructures have a large number of interdependent assets, operated by multiple stakeholders each working independently to maximize their own economic benefits. In these cyber-physical systems, interdependencies between the assets owned by different stakeholders have significant implications on the reliability and security of the overall system. For instance, in the electric grid, industrial control systems at the power generator are managed by a different entity (the generator) than the smart meters deployed by the distribution utility companies. If certain components of these assets are from a common vendor, then a sophisticated attacker can exploit potential shared vulnerabilities and compromise the assets managed by these different entities [21].

Security interdependencies are often modeled in varying degrees of abstractions. While the *attack graph* formalism [7] captures detailed models of how an attacker might exploit vulnerabilities within an enterprise network, representations of interdependencies in large-scale cyber-physical networks, such as the

Ashish R. Hota is supported by a grant from the Purdue Research Foundation (PRF). Abraham A. Clements is supported by Sandia National Laboratories. Sandia National Laboratories is a multi-program laboratory managed and operated by Sandia Corporation, a wholly owned subsidiary of Lockheed Martin Corporation, for the U.S. Department of Energy's National Nuclear Security Administration under contract DE-AC04-94AL85000. SAND2016-8085C.

© Springer International Publishing AG 2016
Q. Zhu et al. (Eds.): GameSec 2016, LNCS 9996, pp. 101–113, 2016.
DOI: 10.1007/978-3-319-47413-7_6

electric grid, are often captured in terms of coupled dynamical systems [12]. In addition to the interdependencies, individual stakeholders are often myopic and resource constrained, which makes identification and mitigation of vulnerabilities in a large number of cyber and physical assets prohibitively expensive. Furthermore, decentralized deployment of defense strategies by these self-interested defenders often leads to increased security risks for the entire system.

In this paper we present a systematic framework that can be used to efficiently compute optimal defense allocations under interdependencies. We model the network security problem as a game between multiple defenders, each of whom manages a set of assets. The interdependencies between these assets are captured by an *interdependency graph*. Each defender minimizes her own expected loss, where the attack probabilities of her assets are a function of her own defense strategies, strategies of other defenders, and the interdependency graph. In particular, attacker(s) are assumed to exploit the interdependencies to target valuable assets in the network. We first establish the existence of a pure Nash equilibrium in the game between self-interested defenders. For a general class of defense strategies, we show that the problem of computing an optimal defense allocation for a defender (i.e., her *best response*) is equivalent to solving a convex optimization problem.

We evaluate the inefficiency of decentralized decision-making in two case studies; the first is a SCADA system with multiple control networks managed by independent entities, and the second is a distributed energy resource failure scenario identified by the US National Electric Sector Cybersecurity Organization Resource (NESCOR). In both settings, we find that when entities have similar risks but disparate budgets, the total expected loss at a Nash equilibrium can be much larger than the total expected loss under the socially optimal solution. Furthermore, we show that it can be in the interest of a selfish actor to defend assets that belong to another entity due to mutual interdependencies.

1.1 Related Work

Security games on networks with multiple defenders have recently been considered within the broad framework of *Stackelberg security games* [9,22]. A Stackelberg security game is defined as an extensive form leader-follower game where a defender randomizes her defense allocations across multiple targets and an attacker observes the randomized strategies and chooses the target with highest successful attack probability. Several papers have considered multiple defenders and network interdependencies within this framework [14–16]. A recurring assumption in these papers is that the strategy space of a defender is discrete, e.g., a node is either fully protected or is vulnerable. In contrast, we consider defense strategies that are continuous variables. In addition, our work is related to recent explorations of attack graph games [3], though the defense strategies considered in that paper are very different from the ones explored here.

Our work is also related to the substantial body of literature on *interdependent security games*; [13] contains a comprehensive review. A common feature in this line of work is that each node is an independent decision maker, i.e., a player is responsible for the defense of a single node in the graph. We relax this assumption in this paper. In our formulation, a player is responsible for the defense of multiple assets (nodes) in the (interdependency) graph.

Our game-theoretic formulation and analysis borrows ideas and techniques from the literature on *network interdiction games* [8]. In the classical shortest path interdiction game [8], there is an underlying network; an attacker aims to find a path of shortest length from a given source to a target, while the defender can *interdict* the attack by increasing the lengths of the paths. Extensions to cases where multiple attackers and/or defenders operate on a given network are few, with the exception of our recent work [19]. The model we propose in this paper generalizes the formulation in [19] as we consider defenders who defend multiple nodes and with possibly nonlinear cost functions. Finally, our paper has a similar perspective as [5] as we develop a systems-theoretic framework that is readily applicable in a broad class of interdependent network security settings.

2   Security Game Framework

Interdependency Graph: We represent the assets in a networked (cyber-physical) system as nodes of a directed graph $\mathcal{G} = \{\mathcal{V}, \mathcal{E}\}$, i.e., each node $v_i \in \mathcal{V}$ represents an asset. The presence of a directed edge $(v_j, v_i) \in \mathcal{E}$ indicates that when the asset v_j is compromised, it can be used to launch an attack on asset v_i. This attack succeeds with a probability $p_{j,i} \in (0, 1]$, independent of analogous attack probabilities defined on the other edges. Without loss of generality, let s be the source node from which an attacker launches the attack from outside the network.[1] We refer to such a graph as an *interdependency graph*.[2]

For an asset $v_i \in \mathcal{V}$, let $\mathcal{P}_i$ be the set of directed paths from the source s to v_i on the graph; a path $P \in \mathcal{P}_i$ is a collection of edges $\{(s, u_1), (u_1, u_2), \ldots, (u_k, v_i)\}$. The probability that v_i is compromised due to an attacker exploiting a given path $P \in \mathcal{P}_i$ is $\prod_{(u_m, u_n) \in P} p_{m,n}$ which is the product of probabilities (due to our independence assumption) on the edges that belong to the path P.

Strategic Defenders: Let $\mathcal{D}$ be the set of defenders. A defender $D_k \in \mathcal{D}$ is responsible for the security of a set $\mathcal{V}_k \subseteq \mathcal{V} \setminus \{s\}$ of assets. For each asset $v_m \in \mathcal{V}_k$, there is a financial loss $L_m \in \mathbb{R}_{\geq 0}$ that defender D_k incurs if v_m gets compromised. The defender can allocate its resources to reduce the attack probabilities on the edges interconnecting different assets on the interdependency graph, subject to certain constraints. We denote the feasible (defense) strategy set of defender D_k as $\mathcal{X}_k \subset \mathbb{R}_{\geq 0}^{n_k}$, where $n_k < \infty$. We require that $\mathcal{X}_k$ is non-empty, compact and convex. The defense resources reduce the attack success probabilities on the edges. We will discuss the exact transformation of defense allocation into the reduction of attack probabilities in the next subsection.

---

[1] If there are multiple entry points to the network, we can add a source node s and add edges from s to all entry points with attack probabilities equal to 1.

[2] Interdependency graphs also capture essential features of *attack graphs* [3,7] where the nodes represent intermediate steps in multi-stage attacks.

Now, let $\mathbf{x} = [\mathbf{x}_1, \mathbf{x}_2, \ldots, \mathbf{x}_{|\mathcal{D}|}]$ be a joint defense strategy of the defenders, with $\mathbf{x}_k \in \mathcal{X}_k$ for every defender D_k. The attack success probability of an edge (v_j, v_i) under this joint defense strategy is denoted as $\hat{p}_{j,i}(\mathbf{x})$. The goal of each defender D_k is to minimize the cost function given by

$$C_k(\mathbf{x}) \triangleq \sum_{v_m \in \mathcal{V}_k} L_m \left(\max_{P \in \mathcal{P}_m} \prod_{(v_j, v_i) \in P} \hat{p}_{j,i}(\mathbf{x}) \right), \tag{1}$$

subject to $\mathbf{x}_k \in \mathcal{X}_k$. In other words, a defender minimizes her expected loss, where the probability of loss of an asset is given by the highest probability of attack on any path from the source to that asset on the interdependency graph.

Strategic Attacker(s): Cyber-physical systems in the field face multiple strategic adversaries with different objectives, capabilities and knowledge about the system. As a result, detailed modeling of strategic attackers is challenging. Nonetheless, a defender must be able to assess her security risks and allocate defense resources under inadequate information about the attackers. This motivates our choice of minimizing the worst case attack probabilities on an asset in (1), which implicitly captures strategic attackers who aim to compromise valuable assets and choose a plan of attack that has the highest probability of success for each asset. The defender can assess her risk profile against attackers of different capabilities by appropriately varying the probabilities on each edge.

As an example of a setting that can be modeled within our framework, consider the SCADA based control system shown in Fig. 1. There are two control subsystems, with interdependencies due to a shared corporate network and a common vendor for the remote terminal units (RTUs). Figure 2 shows the resulting interdependency graph. We further discuss this setting in Sect. 3.

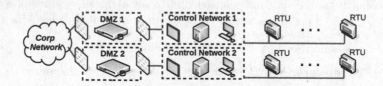

Fig. 1. A SCADA system diagram of two interacting control systems.

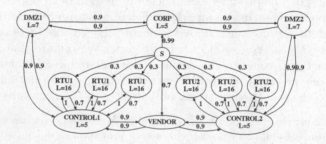

Fig. 2. Interdependency graph for the SCADA system in Fig. 1.

2.1   Defense Strategies

As noted above, the defense resources reduce the attack probabilities on the edges of the interdependency graph. Accordingly, we introduce a transformation matrix $T_k : \mathbb{R}^{n_k} \to \mathbb{R}^{|\mathcal{E}|}$ which maps a feasible defense strategy $\mathbf{x}_k$ to a defense allocation on edges. By appropriately defining the matrix T_k, we can capture very general class of defense strategies. We discuss two such examples.

Edge-based defense strategy: In this case, a defender D_k allocates defense resources on a subset of edges $\mathcal{E}_k \subseteq \mathcal{E}$ of the graph $\mathcal{G}$, and accordingly $n_k = |\mathcal{E}_k|$. For example, $\mathcal{E}_k$ can represent all the edges that are incoming to a node in $\mathcal{V}_k$, i.e., defender D_k can reduce the attack probabilities of all the edges that are incoming to the nodes under its ownership. Furthermore, an edge can potentially be managed by multiple defenders. Under edge-based defense scenarios, we will define the feasible strategy space of a defender D_k as $\mathcal{X}_k := \{x_{j,i}^k \in \mathbb{R}_{\geq 0}, (v_j, v_i) \in \mathcal{E}_k | \sum_{(v_j,v_i) \in \mathcal{E}_k} x_{j,i}^k \leq B_k\}$, where B_k is the total defense budget for defender D_k. In this case, T_k has a sub-matrix which is an identity matrix of dimension $|\mathcal{E}_k|$ and the other entities are equal to 0. An example of edge-based defense is when a device inspects the incoming traffic depending on the traffic source.

Node-based defense strategy: In this case, a defender D_k allocates defense resources to the set of nodes in $\mathcal{V}_k$, and accordingly, $n_k = |\mathcal{V}_k|$. Specifically, the defense resource x_i^k being allocated to node v_i implies that all the incoming edges to v_i in the graph $\mathcal{G}$ have a defense allocation x_i^k, i.e., $x_{j,i}^k = x_i^k$ for every $(v_j, v_i) \in \mathcal{E}$. Node-based defense strategy is motivated by *moving target defense* techniques [10]. Here x_i potentially represents the rate at which system configurations (such as the IP address) of a node v_i are being updated. Here T_k maps the allocation on a node into the edges that are incoming to it.

We now define the *length* or distance of an edge (v_j, v_i) in terms of the attack probability as,

$$l_{j,i} \triangleq -\log(p_{j,i}) \geq 0. \tag{2}$$

A higher probability of attack on an edge leads to smaller length for the edge.

In this paper, we assume that the defense allocations on an edge linearly increase the length of the edge. Mathematically, let $x_{j,i} = \sum_{(v_j,v_i) \in \mathcal{E}_k} x_{j,i}^k = \sum_{D_k \in \mathcal{D}} [T_k \mathbf{x}_k]_{(j,i)}$ denote the total defense allocation by all the defenders on the edge (v_j, v_i). Then, the modified length of the edge under a joint strategy profile $\mathbf{x}$ is given by

$$\hat{l}_{j,i}(\mathbf{x}) \triangleq l_{j,i} + x_{j,i} \implies -\log(\hat{p}_{j,i}(\mathbf{x})) = l_{j,i} + x_{j,i} \tag{3}$$

$$\implies \hat{p}_{j,i}(\mathbf{x}) \triangleq p_{j,i} e^{-x_{j,i}}, \tag{4}$$

i.e., the total defense allocation on an edge $x_{j,i}$ leads to a relative reduction of the corresponding attack success probability given by $e^{-x_{j,i}}$. This captures diminishing effectiveness of defense allocations and leads to a tractable formulation of the cost minimization problem (1). We denote the vector of modified lengths of the graph under joint defense strategy $\mathbf{x}$ as $\hat{\mathbf{L}}(\mathbf{x}) = \mathbf{L} + \sum_{D_k \in \mathcal{D}} T_k \mathbf{x}_k$, where $\mathbf{L}$ is the vector of lengths in the absence of any defense allocation, given by (2).

2.2 Existence of a Pure Nash Equilibrium (PNE)

We first show the existence of a PNE in the game between multiple defenders, each with a defender-specific transformation matrix T_k.

Proposition 1. *The strategic game with multiple defenders where a defender minimizes her cost defined in (1) possesses a pure Nash equilibrium.*

Proof. From our transformation of attack probabilities into lengths on edges given in (3) and (4), the probability of successful attack on a node $v_m \in \mathcal{V}_k$ due to a path $P \in \mathcal{P}_m$ and joint defense strategy $\mathbf{x}$ is equal to

$$\prod_{(u_j, u_i) \in P} \hat{p}_{j,i}(\mathbf{x}) = \exp\left(-\sum_{(v_j, v_i) \in P}\left[l_{j,i} + \sum_{D_r \in \mathcal{D}}[T_r \mathbf{x}_r]_{(j,i)}\right]\right),$$

where $\exp(\cdot)$ is the exponential function, i.e., $\exp(z) = e^z$. Accordingly, we can express the cost function of a defender D_k, defined in (1), as a function of her strategy $\mathbf{x}_k$ and the joint strategy of other defenders $\mathbf{x}_{-k}$ as

$$C_k(\mathbf{x}_k, \mathbf{x}_{-k}) = \sum_{v_m \in \mathcal{V}_k} L_m \exp\left(-\min_{P \in \mathcal{P}_m}\sum_{(v_j, v_i) \in P}\left[\hat{l}_{j,i}(\mathbf{x}_{-k}) + [T_k \mathbf{x}_k]_{(j,i)}\right]\right), \quad (5)$$

where $\hat{l}_{j,i}(\mathbf{x}_{-k}) = l_{j,i} + \sum_{D_r \in \mathcal{D}, D_r \neq D_k}[T_r \mathbf{x}_r]_{(j,i)}$ for an edge (v_j, v_i).

Note that $\sum_{(v_j, v_i) \in P}\left[\hat{l}_{j,i}(\mathbf{x}_{-k}) + [T_k \mathbf{x}_k]_{(j,i)}\right]$ is an affine and, therefore, a concave function of $\mathbf{x}_k$. The minimum of a finite number of concave functions is concave [1]. Finally, $\exp(-z)$ is a convex and decreasing function of z. Since the composition of a convex decreasing function and a concave function is convex, $C_k(\mathbf{x}_k, \mathbf{x}_{-k})$ is convex in $\mathbf{x}_k$ for any given $\mathbf{x}_{-k}$. Furthermore, the feasible strategy set $\mathcal{X}_k$ is non-empty, compact and convex for every defender D_k. As a result, the game is an instance of a *concave game* and has a PNE [18]. □

2.3 Computing the Best Response of a Defender

Let $\mathbf{x}_{-k}$ be the joint defense strategy of all defenders other than D_k. Then the *best response* of D_k is a strategy $\mathbf{x}_k^* \in \mathcal{X}_k$ which minimizes her cost $C_k(\mathbf{x}_k, \mathbf{x}_{-k})$ defined in (1). Let $\hat{\mathbf{L}}(\mathbf{x}_{-k}) = \mathbf{L} + \sum_{D_r \in \mathcal{D}, r \neq k} T_r \mathbf{x}_r$ be the vector of edge lengths under defense allocation $\mathbf{x}_{-k}$. We show that $\mathbf{x}_k^*$ can be computed by solving the following convex optimization problem:

$$\underset{y \in \mathbb{R}^{|\mathcal{V}|}, x_k \in \mathbb{R}^{n_k}}{\text{minimize}} \qquad \sum_{v_m \in \mathcal{V}_k} L_m e^{-y_m} \qquad (6)$$

$$\text{subject to} \qquad \mathcal{I}y - T_k \mathbf{x}_k \leq \hat{\mathbf{L}}(\mathbf{x}_{-k}), \qquad (7)$$

$$y_s = 0, \qquad (8)$$

$$\mathbf{x}_k \in \mathcal{X}_k, \qquad (9)$$

where $\mathcal{I}$ is the node-edge incidence matrix of the graph $\mathcal{G}$. Note that the constraint in (7) is affine. This formulation is motivated by similar ideas explored in the shortest path interdiction games literature [8, 19].

We refer to the vector $\{y_u\}_{u \in \mathcal{V}}$ as a *feasible potential* if it satisfies (7) for every edge in the graph. In graphs without a negative cycle, the well known Bellman-Ford algorithm for shortest paths corrects the inequality in (7) for an edge in every iteration and terminates with a feasible potential [2]. In our setting, the length of every edge is nonnegative. We now prove the following result.

Proposition 2. *A defense strategy* $\mathbf{x}_k^* \in \mathcal{X}_k$ *is the optimal solution of the problem defined in Eqs. (6)–(9) if and only if it is the minimizer of* $C_k(\mathbf{x}_k, \mathbf{x}_{-k})$ *defined in (1).*

Proof. Consider a feasible defense allocation vector $\mathbf{x}_k \in \mathcal{X}_k$. The joint strategy profile $(\mathbf{x}_k, \mathbf{x}_{-k})$ defines a modified length vector $\hat{\mathbf{L}}(\mathbf{x}_k, \mathbf{x}_{-k}) = \hat{\mathbf{L}}(\mathbf{x}_{-k}) + T_k \mathbf{x}_k$ on the edges of $\mathcal{G}$. Now consider a feasible potential $\{y_u\}_{u \in \mathcal{V}}$ which satisfies (7). A feasible potential exists, since the vector $y_u = 0$ for every $u \in \mathcal{V}$ satisfies (7).

Now consider a P from s to a node $v_m \in \mathcal{V}_k$. Then, the feasible potential at node v_m satisfies $y_{v_m} - y_s = y_{v_m} \leq \sum_{(u_j, u_i) \in P} \hat{l}_{j,i}(\mathbf{x}_k, \mathbf{x}_{-k})$. In other words, y_{v_m} is a lower bound on the length of every path (and consequently the shortest path) from s to v_m. Furthermore, in the absence of negative cycles, there always exists a feasible potential where y_{v_m} is *equal* to the length of the shortest path from s to v_m [2, Theorem 2.14] (the solution of the Bellman-Ford algorithm).

Now let $\{\mathbf{x}_k^*, \{y_u^*\}_{u \in \mathcal{V}}\}$ be the optimal solution of the problem defined in Eqs. (6)–(9) for a given $\mathbf{x}_{-k}^*$. The length of every edge (u_j, u_i) at the optimal defense allocation $\mathbf{x}_k^*$ is given by $\hat{l}_{j,i}(\mathbf{x}_k^*, \mathbf{x}_{-k})$. We claim that $y_{v_m}^*$ is equal to the length of the shortest path from s to v_m for every v_m with $L_m > 0$. Assume on the contrary that $y_{v_m}^*$ is strictly less than the length of the shortest path from s to v_m, under the defense allocation $\mathbf{x}_k^*$. From [2, Theorem 2.14] we know that there exists a feasible potential $\{\hat{y}_u\}_{u \in \mathcal{V}}$ such that $\hat{y}_{v_m}$ is equal to the length of the shortest path from s to v_m for every node $v_m \in \mathcal{V}_k$ with length of every edge (u_j, u_i) given by $\hat{l}_{j,i}(\mathbf{x}_k^*, \mathbf{x}_{-k})$. As a result, we have $y_{v_m}^* < \hat{y}_{v_m}$, and the objective is strictly smaller at $\hat{y}_{v_m}$, contradicting the optimality of $\{\mathbf{x}^*, \{y_u^*\}_{u \in \mathcal{V}}\}$.

Let P be a path from s to v_m in the optimal solution, and let P^* be a path of shortest length. The length of this path is given by

$$y_{v_m}^* \leq \sum_{(u_j, u_i) \in P} \hat{l}_{j,i}(\mathbf{x}_k^*, \mathbf{x}_{-k}) = -\sum_{(u_j, u_i) \in P} \log(\hat{p}_{j,i}(\mathbf{x}^*))$$

$$\implies e^{-y_{v_m}^*} \geq \prod_{(u_j, u_i) \in P} \hat{p}_{j,i}(\mathbf{x}^*),$$

with equality for the path P^*. Therefore the optimal cost of the problem defined in Eqs. (6)–(9) is equal to the cost in (1). $\square$

As a result, a defender can efficiently (up to any desired accuracy) compute her optimal defense allocation given the strategies of other defenders. Furthermore, the problem of social cost minimization, where a central planner minimizes the sum of expected losses of all defenders, can be represented in a form that is analogous to Eqs. (6)–(9) and can be solved efficiently.

However, proving theoretical guarantees on the convergence of best response-based update schemes is challenging for the following reasons. First, the expected loss of a defender represented in (5) is non-differentiable and we cannot apply gradient-based update schemes. Second, in the equivalent formulation Eqs. (6)–(9), the players' cost minimization problems are coupled through their constraints. As a result, the problem belongs to the class of *generalized Nash equilibrium problems* [4], which has very few general convergence results. We leave further theoretical investigations of convergence of different dynamics to PNE for future work.

3   Numerical Case Studies

We apply our proposed framework in two case studies. Our goal is to understand the loss of welfare due to decentralized decision making by the defenders with asymmetric defense budgets compared to the socially optimal defense allocation. The social optimum corresponds to the solution computed by a central authority as it minimizes the total expected loss of all the players. The ratio of the highest total expected loss at any PNE and the total expected loss at the social optimum is often used as a metric (*Price of Anarchy*) to quantify the inefficiency of Nash equilibrium. We consider PNE strategy profiles obtained by iteratively computing best responses of the players; the sequence of best response strategies converged to a PNE in all of the experiments in this section. We use the MAT-LAB tool CVX [6] for computing the best response of a defender and the social optimum. In both the experiments, we randomly initialize the attack success probabilities on the edges of the respective interdependency graphs.

3.1   An Interdependent SCADA Network with Two Utilities

We first consider the SCADA network shown in Fig. 1, based on NIST's guidelines for industrial control systems [20]. As discussed earlier, there are two control subsystems with interdependencies due to a shared corporate network and vendors for RTUs. Each subsystem is owned by a different defender. The resulting interdependency graph is shown in Fig. 2. The number in the name of a node indicates the defender who owns it and the amount of loss to its owner, if it is compromised. The corporate network is owned by both defenders. The compromise of the control network causes loss of the RTUs, and as a result, the corresponding edges have an attack success probability 1 and are indefensible.

In our experiments, we keep the total defense budget fixed for the overall system, and compare the resulting total expected loss (under this total budget) at the social optimum with the expected loss that arises at a PNE when each subsystem is defended independently. We consider an edge-based defense strategy for all our results. In the decentralized defense case, we consider two scenarios. First, the defenders can only defend edges that are incoming to a node under their ownership. We refer to this scenario as *individual defense*. Second, the defenders can *jointly defend* all the edges in the interdependency network, i.e., a defender can defend an edge within the subsystem of the other defender.

We plot our results in Fig. 3a for a SCADA network where each utility has 3 RTUs. The total budget is 20, and we vary the budget of defender 1 as shown in the x-axis of the plot. Defender 2 receives the remaining amount (20 minus the budget of defender 1). We observe that the joint defense case leads to a smaller total expected loss compared to the individual defense case at the respective PNEs. The difference between the two cases is most significant when the budgets of the two defenders are largely asymmetric. Our results show that it is beneficial for a selfish decision maker with a large budget to defend parts of the network owned by another defender with a smaller budget in presence of interdependencies. As the asymmetry in budgets decreases, the expected losses under joint defense and the individual defense approach the social optimum. This is because the network considered is symmetric for the defenders. In Fig. 3b, we plot analogous results when each utility has 30 RTUs with a total budget 40, and observe similar trends in the respective expected losses.

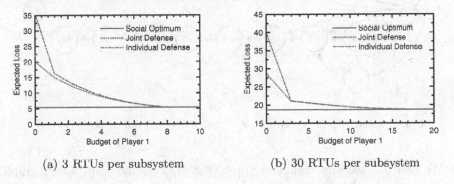

(a) 3 RTUs per subsystem   (b) 30 RTUs per subsystem

Fig. 3. Comparison of total expected loss under the social optimal defense allocation with the PNE strategies under joint and individual defense scenarios. The total budgets across both defenders are 20 and 40, respectively.

3.2 Evaluation of a Distributed Energy Resource Failure Scenario

In our second case study, we consider a distributed energy resource failure scenario, DER.1, identified by the US National Electric Sector Cybersecurity Organization Resource (NESCOR) [17]. We build upon the recent work by [11], where the authors developed a tool *CyberSAGE*, which represents NESCOR failure scenarios as a security argument graph to identify the interdependencies between different attack steps. We reproduce the security argument graph for the DER.1 scenario in Fig. 4. The authors of [11] note that applying all mitigations for the DER.1 failure scenario can be expensive. Our framework enables computing the optimal (and PNE) defense strategy under budget constraints.

Note that the nature of interdependency in Fig. 4 is qualitatively different from the setting in the previous subsection. In Fig. 4, the nodes in the interdependency graph capture individual attack steps (similar to the representation in attack graphs). In contrast, the nodes in Fig. 2 correspond to disparate devices in the SCADA network. Furthermore, multiple attack steps can occur within a single device; all the intermediate nodes that belong to a common device are shown to be within a box in Fig. 4. For example, nodes $w3, w4, w5, w6, w7$ belong to the Human-Machine Interface (HMI) of the photovoltaic (PV) system. The node S represents the entry point of an attack, the nodes G_0 and G_1 represent the final target nodes that compromise the PV and electric vehicle (EV) components of the DER. A more detailed description is available in [11].

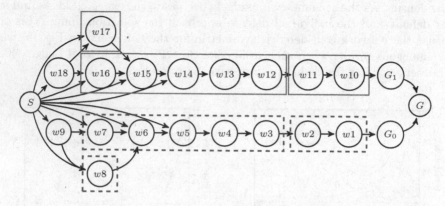

Fig. 4. Interdependency Graph of NESCOR DER.1 failure scenario [11]

We treat the security argument graph (Fig. 4) as the interdependency graph, and compute the globally optimal and Nash equilibrium strategies for two classes of defense strategies, (i) edge-based defense, where a player defends every edge independently, and (ii) device-based defense (such as IP-address randomization), where each device receives a defense allocation. In the second case, all the incoming edges to the nodes that belong to a specific device receive identical defense allocations. In the decentralized case, there are two players, who aim to protect nodes G_0 and G_1, respectively. In addition, each player experiences an additional loss if the other player is attacked successfully. This is captured by adding the extra node G which has edges from both G_0 and G_1 with attack probabilities equal to 1. Both players experience a loss if node G is compromised.

We plot the ratio of total expected losses under the socially optimal and PNE strategy profiles, for both edge-based and device-based defense strategies, in Figs. 5a and b, respectively. As the figures show, at a given total budget, the ratio of the expected losses at the social optimum and at a PNE is smaller when there is a larger asymmetry in the budgets of the individual players. In other words, when the individual players have similar defense budgets, the total

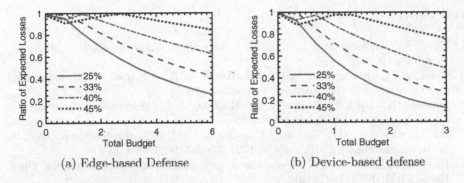

Fig. 5. The ratio of total expected losses of both defenders at the social optimum and a PNE in the DER.1 failure scenario under two different defense strategies. The total budget is divided among the two defenders, and defender 1 receives a percentage of the total budget as indicated in the legend.

expected loss at Nash equilibrium is not much larger than the expected loss under a globally optimal defense strategy.

4   Discussion and Conclusion

We presented a game-theoretic framework that enables systematic analysis of security trade-offs in interdependent networked systems. For a general class of defense strategies, the computation of optimal defense allocation for a defender is equivalent to solving a convex minimization problem. We also proved the existence of a pure Nash equilibrium for the game between multiple defenders. The SCADA network and DER.1 case studies illustrate how our framework can be used to study the security of interdependent systems at different levels of abstraction, from individual attack steps in the DER.1 scenario to an entire organization (vendor in the SCADA example) being abstracted to a single node. Our framework can be readily applied in practice by individual stakeholders to evaluate the effectiveness of different defense strategies and share information with other defenders to decide when and to what degree cooperative defense should be applied. The different levels of abstractions enable the creation of models with the available information a defender has. For example, the SCADA use case could be used to identify the degree to which the compromise of the vendor will affect the security of a system. This could translate into adding security requirements in procurement contracts with the vendors. In future, we will apply our framework in large-scale cyber-physical systems. Establishing convergence guarantees for best response dynamics and theoretical characterizations of inefficiencies at Nash equilibria remain as challenging open questions.

References

1. Boyd, S., Vandenberghe, L.: Convex Optimization. Cambridge University Press, New York (2004)
2. Cook, W.J., Cunningham, W.H., Pulleyblank, W.R., Schrijver, A.: Combinatorial Optimization, vol. 605. Springer, Heidelberg (1998)
3. Durkota, K., Lisý, V., Bošanský, B., Kiekintveld, C.: Approximate solutions for attack graph games with imperfect information. In: Khouzani, M.H.R., Panaousis, E., Theodorakopoulos, G. (eds.) GameSec 2015. LNCS, vol. 9406, pp. 228–249. Springer, Heidelberg (2015). doi:10.1007/978-3-319-25594-1_13
4. Facchinei, F., Kanzow, C.: Generalized Nash equilibrium problems. Ann. Oper. Res. **175**(1), 177–211 (2010)
5. Fielder, A., Panaousis, E., Malacaria, P., Hankin, C., Smeraldi, F.: Game theory meets information security management. In: Cuppens-Boulahia, N., Cuppens, F., Jajodia, S., Abou El Kalam, A., Sans, T. (eds.) SEC 2014. IAICT, vol. 428, pp. 15–29. Springer, Heidelberg (2014). doi:10.1007/978-3-642-55415-5_2
6. Grant, M., Boyd, S., Ye, Y.: CVX: Matlab software for disciplined convex programming (2008)
7. Homer, J., Zhang, S., Ou, X., Schmidt, D., Du, Y., Rajagopalan, S.R., Singhal, A.: Aggregating vulnerability metrics in enterprise networks using attack graphs. J. Comput. Secur. **21**(4), 561–597 (2013)
8. Israeli, E., Wood, R.K.: Shortest-path network interdiction. Networks **40**(2), 97–111 (2002)
9. Jain, M., Conitzer, V., Tambe, M.: Security scheduling for real-world networks. In: AAMAS, pp. 215–222 (2013)
10. Jajodia, S., Ghosh, A.K., Subrahmanian, V., Swarup, V., Wang, C., Wang, X.S.: Moving Target Defense II. Application of Game Theory and Adversarial Modeling. Advances in Information Security, vol. 100, p. 203. Springer, New York (2013)
11. Jauhar, S., Chen, B., Temple, W.G., Dong, X., Kalbarczyk, Z., Sanders, W.H., Nicol, D.M.: Model-based cybersecurity assessment with nescor smart grid failure scenarios. In: 21st Pacific Rim International Symposium on Dependable Computing, pp. 319–324. IEEE (2015)
12. Kundur, D., Feng, X., Liu, S., Zourntos, T., Butler-Purry, K.L.: Towards a framework for cyber attack impact analysis of the electric smart grid. In: IEEE SmartGridComm, pp. 244–249 (2010)
13. Laszka, A., Felegyhazi, M., Buttyan, L.: A survey of interdependent information security games. ACM Comput. Surv. (CSUR) **47**(2), 23:1–23:38 (2014)
14. Letchford, J., Vorobeychik, Y.: Computing randomized security strategies in networked domains. In: Applied Adversarial Reasoning and Risk Modeling 2011, vol. 06 (2011)
15. Letchford, J., Vorobeychik, Y.: Optimal interdiction of attack plans. In: AAMAS, pp. 199–206 (2013)
16. Lou, J., Smith, A.M., Vorobeychik, Y.: Multidefender security games. arXiv preprint arXiv:1505.07548 (2015)
17. Electric sector failure scenarios and impact analyses, National Electric Sector Cybersecurity Organization Resource, EPRI (2014)
18. Rosen, J.B.: Existence and uniqueness of equilibrium points for concave n-person games. Econometrica: J. Econometric Soc. **33**(3), 520–534 (1965)
19. Sreekumaran, H., Hota, A.R., Liu, A.L., Uhan, N.A., Sundaram, S.: Multi-agent decentralized network interdiction games. arXiv preprint arXiv:1503.01100 (2015)

20. Stouffer, K.: Guide to industrial control systems (ICS) security. NIST special publication 800-82, 16-16 (2011)
21. Emerging threat: Dragonfly/Energetic Bear - APT Group (2014). http://www.symantec.com/connect/blogs/emerging-threat-dragonfly-energetic-bear-apt-group, Symantec Official Blog. Accessed 15 Aug 2016
22. Tambe, M.: Security and Game Theory: Algorithms, Deployed Systems, Lessons Learned. Cambridge University Press, New York (2011)

Dynamics on Linear Influence Network Games Under Stochastic Environments

Zhengyuan Zhou[1]([✉]), Nicholas Bambos[1,2], and Peter Glynn[1,2]

[1] Department of Electrical Engineering, Stanford University,
Stanford, CA 94305, USA
zyzhou@stanford.edu

[2] Department of Management Science and Engineering, Stanford University,
Stanford, CA 94305, USA

Abstract. A linear influence network is a broadly applicable conceptual framework in risk management. It has important applications in computer and network security. Prior work on linear influence networks targeting those risk management applications have been focused on equilibrium analysis in a static, one-shot setting. Furthermore, the underlying network environment is also assumed to be deterministic.

In this paper, we lift those two assumptions and consider a formulation where the network environment is stochastic and time-varying. In particular, we study the stochastic behavior of the well-known best response dynamics. Specifically, we give interpretable and easily verifiable sufficient conditions under which we establish the existence and uniqueness of as well as convergence (with exponential convergence rate) to a stationary distribution of the corresponding Markov chains.

Keywords: Game theory · Networks · Security · Stochastic stability

1 Introduction

The application of game theory to networks has received much attention in the literature [1,2] in the past decade. The underlying model typically consists of agents, connected by physical or virtual links, who must strategically decide on some action given the actions of the other users and the network structure. The well-founded motivations for this study and the specific applications examined have spanned many fields such social or economic networks [3], financial networks [4,5], and a diverse range (packets, robots, virtual machines, sensors, etc.) of engineering networks [6–10]. These different contexts all have in common the presence of inter-agent influences: the actions of individual agents can affect others in either a positive or negative way, which are typically called externalities. As a simple example, consider two web-enabled firms [11] that have customers in common that use the same passwords on both sites. In this case, an investment in computer system security from one firm naturally strengthens the security of the other, resulting in larger effective investment of the other firm compared to

© Springer International Publishing AG 2016
Q. Zhu et al. (Eds.): GameSec 2016, LNCS 9996, pp. 114–126, 2016.
DOI: 10.1007/978-3-319-47413-7_7

its own, independent investment. On the other hand, this investment may shrink the effective investment of a third firm (a competitor) in the same business, as this enhanced security on the two firms makes the attack on the third firm more attractive.

As another example, technology innovation is another instance where a network of agents' actions can produce inter-dependent effects on one another. Here the concept of *innovation risk* glues together the investments made by each agent: if a social media company (e.g. Facebook) is seeking to expand its enterprise by innovating on new products, then a partnering video games (e.g. Zynga) company whose games are played on that social media platform will be benefited and will in turn benefit the social media company with its own investment. On the other hand, a similar effort made by another competing social media company (e.g. Myspace) will cause negative impact on both of the prior firms and will be negatively impacted by them as well.

In all these examples, this feature of inter-agent influences is captured by a linear influence network, which was first employed in [11, 12] to study and manage the risk in computer security settings. In a nutshell and in the words of [11], "[i]n this model, a matrix represents how one organization's investments are augmented by some linear function of its neighbors investments. Each element of the matrix, representing the strength of influence of one organization on another, can be positive or negative and need not be symmetric with respect to two organizations." [13] very recently generalized this interdependence model to an influence network, where every agent's action is augmented by some (arbitrary) function of its neighbors' joint action to yield a final, effective action, thus allowing for a general influence effect in terms of both directions and values.

We mention that in addition to the examples mentioned above, linear influence network model is a broadly applicable conceptual framework in risk management. The seminal work [14] provides more applications (one in security assets which generalizes [11] and another in vulnerabilities) and numerical examples to illustrate the versatility and power of this framework, to which the readers are referred to for an articulate exposition. On this note, we emphasize that the application of game theory to security has many different dimensions, to which the linear influence network model is but one. See [15] for an excellent and comprehensive survey on this topic.

On the other hand, all the prior work on linear influence networks and the applications [11–14, 16] have been focused on equilibrium analysis in a static, one-shot setting. Further, the influence matrix, which represents the underlying network environment, is assumed to be deterministic. Although prior analyses provide an important first step in gaining the insights, both of these assumptions are somewhat stringent in real applications: agents tend to interact over a course of periods and the influences are random and can fluctuate from period to period. Consequently, we aim to incorporate these two novel elements into our study.

In this paper, we consider a stochastic formulation of the best response dynamics by allowing the underlying network environment to be stochastic (and time-varying) and study its stochastic behavior. In the deterministic network environment case [11,12,16], it is known that the best response dynamics has the desired property of converging to the unique Nash equilibrium when the influence matrix is strictly diagonally dominant. The linear influence network represented by a strict diagonally dominant influence matrix has direct and intuitive interpretations [11,12,16] in the applications and constitutes an important class for study. Building on this observation, we aim to characterize the stochastic behavior of the best response dynamics when the influence matrix is sometimes strictly diagonally dominant and sometimes not. Our stochastic formulation is a rather broad framework in that we do not impose any exogenous bounds on each agent's action, nor on the randomness of the network environment. Of course, the same stochastic stability results hold should one wish to impose such constraints for a particular application.

We then give two sufficient conditions on the stochastic network environment that ensure the stochastic stability of the best response dynamics. These conditions have the merits of being both easily interpretable and easily verifiable. These two sufficient conditions (Theorem 3) serve as the main criteria under which we establish the existence and uniqueness of as well as convergence to a stationary distribution of the corresponding Markov chains. Furthermore, convergence to the unique stationary distribution is exponentially fast. These results are the most desired convergence guarantees that one can obtain for a random dynamical system. These sufficient conditions include as a special case the interesting and simultaneously practical scenario, in which we demonstrate that the best response dynamics may converge in a strong stochastic sense, even if the network itself is not strictly diagonally dominant on average.

2   Model Formulation

We start with a quick overview of the linear influence network model and the games induced therein. Our presentation mainly follows [14,16]. After disusing some of the pertinent results, we conclude this section with a motivating discussion in Sect. 2.4 on the main question and the modeling assumptions we study in this paper.

2.1   Linear Influence Network: Interdependencies and Utility

A linear influence network consists of a set of players each taking an action $x_i \in [0, \infty)$, which can be interpreted as the amount of investment made by player i. The key feature of an influence network is the natural existence of interdependencies which couple different players' investments. Specifically, player i's effective investment depends not only on how much he invests, but on how much each of his neighbors (those whose investments produce direct external effects on player i) invests. A linear influence network is an influence network where such interdependencies are linear.

We model the interdependencies among the different players via a *directed graph*, $\mathcal{G} = \{\mathcal{N}, \mathcal{E}\}$ of nodes $\mathcal{N}$ and edges $\mathcal{E}$. The nodes set $\mathcal{N}$ has N elements, one for each player i, $i = 1 \ldots N$. The edges set $\mathcal{E}$ contains all edges (i, j) for which a decision by i directly affects j. For each edge, there is an associated weight, $\psi_{ij} \in \mathbb{R}$, either positive or negative, representing the strength of player i's influence on player j (i.e. how much player j's effective investment is influenced per player i's unit investment). Consequently, the effective investment x_i^{eff} of player i is then $x_i^{eff} = x_i + \sum_{j \neq i} \psi_{ji} x_j$.

We can then represent the above linear influence network in a compact way via a single network matrix, $\mathbf{W} \in \mathbb{R}^{N \times N}$, as follows:

$$\mathbf{W}_{ij} = \begin{cases} 1 & \text{if } i = j \\ \psi_{ji} & \text{if } (j, i) \in \mathcal{E} \\ 0 & \text{otherwise.} \end{cases} \tag{1}$$

In particular, we call into attention that $\mathbf{W}_{ij}$ represents the influence of player j on player i. An example network and the associated $\mathbf{W}$ matrix are shown in Fig. 1.

We can therefore rewrite the effective investment of player i as $x_i^{eff} = x_i + \sum_{j \neq i} \psi_{ji} x_j = \sum_{j=1}^{N} \mathbf{W}_{ij} x_j$. Written compactly in matrix forms, if $\mathbf{x}$ denotes the vector of individual investments made by all the players, then the effective investment is given by $\mathbf{x}^{eff} = \mathbf{W} \mathbf{x}$.

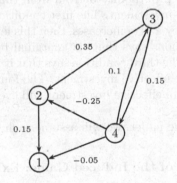

$$\mathbf{W} = \begin{bmatrix} 1 & 0.15 & 0 & -0.05 \\ 0 & 1 & -0.25 & 0.35 \\ 0 & 0 & 1 & 0.1 \\ 0 & 0 & 0.15 & 1 \end{bmatrix}$$

Fig. 1. An instance of a linear influence network.

Each player i has an utility function that characterizes his welfare:

$$U_i(\mathbf{x}) = V_i(x_i^{eff}) - c_i x_i. \tag{2}$$

The utility function has two components that admit a simple interpretation. V_i gives the value that player i places on the total effective investment made (resulting from player i's investment and the externalities coming from its neighbors.) The second component is the cost on player i's side for investing x_i amount, where c_i is the relative trade-off parameter that converts the value and the cost on the same scale.

Drawing from the literature on utility theory, we impose the following standard assumptions on the value function. Remark 1 gives an intuitive explanation on those assumptions.

Definition 1. *The continuously differentiable function* $V_i(\cdot) : [0, \infty) \to \mathbf{R}^1$ *is called admissible if the following conditions hold:*

1. *strictly increasing,*
2. *strictly concave,*
3. $V_i'(0) > c_i$ *and*
4. $\lim_{x \to \infty} V_i'(x) < c_i$.

Remark 1. Here we use the network security example to provide intuition on why the assumptions constituting an admissible value function are natural. Here the value function for each player (firm) can be viewed to represent the level of its network security or the profitability derived from that level of security, as a result of the total effective investment. The first condition says that if the total effective investment made by a firm increases, then this level of security increases as well. The second conditions says that the marginal benefit of more effective investment is decreasing. The third condition says that it is always in the interest of a firm to have a positive effective investment. The fourth condition says that the marginal benefit of more effective investment will eventually drop below the unit cost of investment.

For the remainder of the paper, we will assume each V_i is admissible.

2.2 Nash Equilibrium of the Induced Game: Existence, Uniqueness and Convergence

With the above utility (payoff) function U_i, a multi-player single-stage complete-information game is naturally induced. We proceed with the standard solution concept Nash equilibrium (NE), defined here below:

Definition 2. *Given an instance of the game* $(\mathcal{N}, \mathcal{E}, \mathbf{W}, \{V_i\}_{i \in \mathcal{N}}, \{c_i\}_{i \in \mathcal{N}})$, *the investment vector* $\mathbf{x}^*$ *is a (pure-strategy) Nash equilibrium if, for every player i,* $U_i(x_i^*, \mathbf{x}_{-i}^*) \geq U_i(x_i, \mathbf{x}_{-i}^*), \forall x_i \in [0, \infty)$, *where* $\mathbf{x}_{-i}^*$ *is the vector of all investments but the one made by player i.*

---

[1] If $x < 0$, we set $V_i(x) = -\infty$, representing the fact that negative effective investment is unacceptable.

The existence and uniqueness of a NE in the induced game on a linear influence network has been studied and characterized in depth by [11,12,14,16], where the connection is made between a NE and a solution to an appropriate Linear Complementarity Problem (LCP), the latter being an extensively studied problem [17]. As such, different classes of matrices (i.e. assumptions made on the network matrix $\mathbf{W}$) have been shown to lead to either existence or uniqueness (or both) of a NE.

As has been emphasized by the previous work [11,12,14], a particular class of network matrices $\mathbf{W}$ that deserve special and well-motivated attention is the class of strictly diagonally dominant matrices, defined next.

Definition 3. *Let* $\mathbf{W} \in \mathbf{R}^{N \times N}$ *be a square matrix.*

1. $\mathbf{W}$ *is a strictly diagonally row dominant matrix if for every row i:*
   $\sum_{j \neq i} |\mathbf{W}_{ij}| < |\mathbf{W}_{ii}|$.
2. $\mathbf{W}$ *is a strictly diagonally column dominant matrix if its transpose is strictly diagonally row dominant.*

A strictly diagonally dominant matrix is either a strictly diagonally row dominant matrix or a strictly diagonally column dominant matrix.

The class of strictly diagonally dominant matrices play a central role in linear influence networks because they are both easily interpretable and present. For instance, a strictly diagonally row dominant matrix represents a network where each player's influence on his own is larger than all his neighbors combined influence on him. A strictly diagonally column dominant matrix represents a network where each player's influence on his own is larger than his own influence on all of his neighbors combined. It turns out, as stated in the following theorem, a strictly diagonally dominant influence matrix ensures the existence of a unique Nash equilibrium. We mention in passing that although there are other classes of matrices that ensure the existence of a unique Nash equilibrium (such as the class of positive definite matrices), they do not have direct interpretations and do not easily correspond to practical scenarios of an linear influence network.

Theorem 1. *Let* $(\mathcal{N}, \mathcal{E}, \mathbf{W}, \{V_i\}_{i \in \mathcal{N}}, \{c_i\}_{i \in \mathcal{N}})$ *be a given instance of the game. If the network matrix $\mathbf{W}$ is strictly diagonally dominant, then the game admits a unique Nash equilibrium.*

Proof. See [12,16].

2.3 Convergence to NE: Best Response Dynamics

The existence and uniqueness results of NE play an important first step in identifying the equilibrium outcomes of the single-stage game on linear influence networks. The next step is to study dynamics for reaching that equilibrium (should a unique NE exist), a more important one from the engineering perspective. An important class of dynamics is that of best response dynamics, which in addition to being simple and natural, enjoys the attractive feature of being model-agnostic, to be described below. We first define the best response function.

Definition 4. *The best response function $g_i(\mathbf{x})$ for player i is defined as:* $g_i(\mathbf{x}) = \arg \max_{x_i \geq 0} U_i(x_i, \mathbf{x}_{-i})$. *The best response function for the network is denoted by* $g(\mathbf{x}) = (g_1(\mathbf{x}), g_2(\mathbf{x}), \ldots, g_N(\mathbf{x}))$

In the current setting, we can obtain an explicit form for the best response function. Let b_i represent the (single) positive value at which $V_i'(\cdot) = c_i$, which is always guaranteed to uniquely exist due to the assumption that V_i is admissible. Then, it can be easily verified that:

$$g(\mathbf{x}) = \max(\mathbf{0}, (\mathbf{I} - \mathbf{W})\mathbf{x} + \mathbf{b}). \tag{3}$$

With the above notation, we are now ready to state best response dynamics (formally given in Algorithm 1): it is simply a distributed update scheme where in each iteration, every player chooses its investment in response to the investments his neighbors have chosen in the previous iteration. Note that each player i, in order to compute its best response investment for the current iteration, need not know what his neighboring players' investments were in the previous iteration. Instead, it only needs to know the combined net investments $(\sum_{j \neq i} \psi_{ji} x_j)$ his neighbors has induced to him. This combined net investments can be inferred since player i knows how much he invested himself in the previous iteration and observes the current payoff he receives. This constitutes perhaps the single most attractive feature of best response dynamics in the context of linear influence network games: the model-agnostic property.

Algorithm 1. Best Response Dynamics

Given $\mathbf{x}(0) \geq \mathbf{0}$
$t \leftarrow 0$
for t = 1, 2, ... **do**
  Each player i: $x_i(t+1) = g_i(\mathbf{x}(t))$
end for

Writing the distributed update in Algorithm 1 more compactly, we have: $\mathbf{x}(t+1) = g(\mathbf{x}(t)) = \max(\mathbf{0}, (\mathbf{I} - \mathbf{W})\mathbf{x}(t) + \mathbf{b})$. It turns out that convergence for $\mathbf{x}(t)$ is guaranteed when $\mathbf{W}$ is strictly diagonally dominant.

Theorem 2. *If the network matrix $\mathbf{W}$ is strictly diagonally dominant, then the best response dynamics in Algorithm 1 converges to the unique NE.*

Proof. See [12,16].

2.4   Motivation and Main Question of the Paper

The best response dynamics, for reasons mentioned before, have enjoyed popularity in prior work. However, the convergence property of the best response dynamics rests on the crucial yet somewhat unrealistic assumption that the underlying network environment is fixed over time, i.e. $\mathbf{W}$ stays constant in every time period. In practice, $\mathbf{W}^t$ encode the influence of players over its neighbors and should be inherently dynamic and time-varying.

Here we lift this assumption by allowing $\mathbf{W}^t$ to be random and time-varying. Consequently, using $X(t)$ to denote the (random) investment vector at time t, the best response dynamics should be re-written as:

$$X(t+1) = \max(\mathbf{0}, (\mathbf{I} - \mathbf{W}^t)X(t) + \mathbf{b}). \tag{4}$$

Our principal goal in this paper is then to study the stochastic behavior of the resulting power iterate $\{X(t)\}_{t=0}^{\infty}$ (which is now a stochastic process) and to identify sufficient conditions under which stochastic stability is guaranteed. For simplicity, we assume that random network matrix $\mathbf{W}^t$ is **iid**. The **iid** case, although simple, provides illuminating structural insights and can be easily extended to the stationary and ergodic network environment environments case.

Observe that under **iid** assumption, the iterates $\{X(t)\}_{t=0}^{\infty}$ in the best response dynamics form a Markov chain. Our principal focus in the next section is to characterize conditions under which the Markov chain admits a unique stationary distribution with guaranteed convergence properties and/or convergence rates. These results are of importance because they establish the stochastic stability (in a strong sense to be formalized later) of the best response dynamics in the presence of random and time-varying network environments.

In addition, this stochastic process $\{X(t)\}_{t=0}^{\infty}$ can be generated in a variety of ways, each corresponding to a different practical scenario. Section 3.2 makes such investigations and presents two generative models and draw some interesting conclusions.

3   Main Criteria for Stochastic Stability Under Random Network Environment

In this section, we characterize the behavior of the best response dynamics under stochastic (and time-varying) network environment and give the main criteria for ensuring stochastic stability. Our focus here is to identify sufficient conditions that are broad enough while at the same time interpretable and efficiently verifiable.

Our assumption on the random network is rather mild: $\{\mathbf{W}^t\}_{t=1}^{\infty}$ is drawn **iid** according to some fixed distribution with bounded first moments from some support set $\mathcal{W} \subset \mathbf{R}^{N \times N}$, where $\mathcal{W}$ can be either discrete or continuous. Alternatively, this means that each individual influence term $\mathbf{W}_{ij}$ is absolutely integrable: $\mathbf{E}[\|\mathbf{W}_{ij}\|] < \infty$. It is understood that $\mathbf{W}_{ii} = 1$ for each $\mathbf{W} \in \mathcal{W}$ since one's own influence on oneself should not fluctuate.

3.1   Two Main Sufficient Conditions for Stochastic Stability

The state space for the Markov chain $\{X(t)\}_{t=1}^{\infty}$ will be denoted[2] by $\mathcal{X} = \mathbf{R}_+^N$. $(\mathcal{X}, \mathcal{B}(\mathcal{X}))$ is then the measurable space, where $\mathcal{B}(\mathcal{X})$ is the Borel sigma algebra on $\mathcal{X}$, induced by some vector norm. Since all finite-dimensional norms are equivalent (up to a constant factor), the specific choice of the norm shall not concern as here since they all yield the same Borel sigma algebra[3]. The transition kernel $K(x, A)$ denotes the probability of transitioning in one iteration from $x \in \mathcal{X}$ into the measurable set $A \in \mathcal{B}(\mathcal{X})$. $K^t(x, A)$ then denotes the t-step transition probability. We use $K^t(x, \cdot)$ to denote the probability measure (distribution) of the random variable $X(t)$ with the initial point at x.

Definition 5. *Let $\| \cdot \|$ be any vector norm on $\mathbf{R}^n$, A be a square matrix on $\mathbf{R}^{n \times n}$, w be a strictly positive weight vector in $\mathbf{R}^n$ and v be a generic vector in $\mathbf{R}^n$.*

1. *The induced matrix norm $\| \cdot \|$ is defined by $\|A\| = \max_{\|x\|=1} \|Ax\|$.*
2. *The weighted l_∞ norm with weight w is defined by $\|v\|_\infty^w = \max_i |\frac{v_i}{w_i}|$.*
3. *The weighted l_1 norm with weight w is defined by $\|v\|_1^w = \sum_{i=1}^n |\frac{v_i}{w_i}|$.*

Since we are using the same notation to denote both the vector norm and the corresponding induced matrix norm, the context shall make it clear which norm is under discussion. In the common special case where w is the all-one vector, the corresponding induced matrix norms will be denoted conventionally by $\| \cdot \|_1, \| \cdot \|_\infty$. The following proposition gives known results about the induced matrix norms $\| \cdot \|_1^w, \| \cdot \|_\infty^w$. See [18].

We are now ready to state, in the following theorem, our main sufficient conditions for the bounded random network environment case. The proof[4] is omitted here due to space limitation.

Theorem 3. *Let $\mathbf{W}$ be the random network matrix with bounded support from which $\mathbf{W}^t$ is drawn **iid**. Assume that either one of the following two conditions is satisfied:*

1. *There exists a strictly positive weight vector w such that $\mathbb{E}[\log \|\mathbf{I} - \mathbf{W}\|_\infty^w] < 0$;*
2. *There exists a strictly positive weight vector w such that $\mathbb{E}[\log \|\mathbf{I} - \mathbf{W}\|_1^w] < 0$.*

Then:

1. *The Markov chain $\{X(t)\}_{t=0}^{\infty}$ admits a unique stationary distribution $\pi(\cdot)$.*

---

[2] Note that here we do not impose any bounds on the maximum possible investment by any player. If one makes such an assumption, then $\mathcal{X}$ will be some compact subset of $\mathbf{R}_+^N$. All the results discussed in this section will still go through. For space limitation, we will not discuss the bounded investment case. Further, note that the unbounded investment case (i.e. without placing any exogenous bound on investments made by any player) which we focus on here is the hardest case.

[3] As we shall soon see, the two norms we will be using are weighted l_1 norm and weighted l_∞ norm.

[4] The proof utilizes the powerful framework presented in [19,20].

2. *The Markov chain converges to the unique stationary distribution in Prokhorov metric irrespective of the starting point:* $\forall X(0) \in \mathcal{X}, d_\rho(K^t(X(0), \cdot), \pi(\cdot)) \to 0$ *as* $t \to \infty$, *where* $d_\rho(\cdot, \cdot)$ *is the Prokhorov metric[5] induced by the Euclidean metric* ρ.

3. *The convergence has a uniform exponential rate: There exists an* r *(independent of* $X(0)$*), with* $0 < r < 1$*, such that* $\forall x \in \mathcal{X}$*, there exists a constant* $C_{X(0)} > 0$ *such that* $d_\rho(K^t(X(0), \cdot), \pi(\cdot)) \le C_{X(0)} r^t, \forall t$.

Remark 2. First, note that $\mathbb{E}[\log \|\mathbf{I} - \mathbf{W}\|_\infty^w] < 0$ is a weaker condition than $\mathbb{E}[\|\mathbf{I} - \mathbf{W}\|_\infty^w] < 1$, since by Jensen's inequality, $\mathbb{E}[\|\mathbf{I} - \mathbf{W}\|_\infty^w] < 1$ implies $\mathbb{E}[\log \|\mathbf{I} - \mathbf{W}\|_\infty^w] < 0$. Similarly, $\mathbb{E}[\log \|\mathbf{I} - \mathbf{W}\|_1^w] < 0$ is a weaker condition than $\mathbb{E}[\|\mathbf{I} - \mathbf{W}\|_1^w] < 1$.

Second, it follows from basic matrix theory that any square matrix satisfies $\|A\|_1^w = \|A^{\mathrm{T}}\|_\infty^{\frac{1}{w}}$. Consequently, the second sufficient condition can also be cast in the first by taking the transpose and inverting the weight vector: There exists a strictly positive weight vector w such that $\mathbb{E}[\log \|\mathbf{I} - \mathbf{W}\|_1^w] < 0$ if and only if there exists a strictly positive weight vector $\tilde{w}$ such that $\mathbb{E}[\log \|(\mathbf{I} - \mathbf{W})^{\mathrm{T}}\|_\infty^{\tilde{w}}] < 0$.

Third, for a deterministic matrix A, one can efficiently compute its induced weighted l_1 and weighted l_∞ norms. Further, for a fixed positive weight vector w and a fixed distribution on $\mathbf{W}$, one can also efficiently verify whether $\mathbb{E}[\log \|\mathbf{I} - \mathbf{W}\|_\infty^w] < 0$ holds or not (similar for $\mathbb{E}[\log \|\mathbf{I} - \mathbf{W}\|_1^w] < 0$). The most common and natural weight vector is the all-ones vector in the context of strictly diagonally dominant matrices. This point is made clear by the discussion in Sect. 3.1, which also sheds light on the motivation for the particular choices of the induced norms in the sufficient conditions.

3.2   A Generative Model: Discrete-Support Random Network Environment

Here we proceed one step further and give an interesting and practical generative model for the underlying random network for which there is a direct interpretation. In this generative model, we assume that each influence term $\mathbf{W}_{ij}$ ($i \ne j$) comes from a discrete set of possibilities. This is then equivalent to the network matrix $\mathbf{W}^t$ being drawn from a discrete support set $\mathcal{W} = \{W^1, W^2, \dots\}$. This generative model of randomness has the intuitive interpretation that one player's influence on another can be one of the possibly many different values, where it can be positive at one time and negative at another time.

In the special case that each W^i in $\mathcal{W}$ is either strictly diagonally row dominant or strictly diagonally column dominant, then stochastic stability is guaranteed, as given by the following statement.

---

[5] Thi is the well-known metric that is commonly used to characterize the distance between two probability measures [21]. Further, if (Ω, ρ) is a separable metric space, as is the case in our application, then convergence in Prokhorov metric implies weak convergence [21].

Corollary 1. *Let the support $\mathcal{W}$ be a set of strictly diagonally row dominant matrices. Then for any probability distribution on $\mathcal{W}$ from which $\mathbf{W}^t$ is sampled, the Markov chain $\{X(t)\}_{t=0}^{\infty}$ given by the best response dynamics satisfies the following:*

1. *The Markov chain $\{X(t)\}_{t=0}^{\infty}$ admits a unique stationary distribution $\pi(\cdot)$.*
2. *The Markov chain converges to the unique stationary distribution in Prokhorov metric irrespective of the starting point: $\forall X(0) \in \mathcal{X}, d_\rho(K^t(X(0), \cdot), \pi(\cdot)) \to 0$ as $t \to \infty$, where $d_\rho(\cdot, \cdot)$ is the Prokhorov metric induced by the Euclidean metric ρ.*
3. *The convergence has a uniform exponential rate: There exists an r (independent of $X(0)$), with $0 < r < 1$, such that $\forall x \in \mathcal{X}$, there exists a constant $C_{X(0)} > 0$ such that $d_\rho(K^t(X(0), \cdot), \pi(\cdot)) \leq C_{X(0)}r^t, \forall t$.*

The same conclusions hold if $\mathcal{W}$ is a set of strictly diagonally column dominant matrices.

Proof. Take the weight vector $w = \mathbf{1}$. Then since each $W^l \in \mathcal{W}$ is strictly diagonally row dominant, it follows that for each i, $\sum_{j \neq i} |W_{ij}^l| < 1$. Let i_l^* be the row that maximizes the row sum for W^l: $i_l^* = \arg\max_i \sum_{j \neq i} |W_{ij}^l|$, then for every l, $\sum_{j \neq i_l^*} |W_{ij}^l| < 1$.

Let $P(l)$ be the probability that $\mathbf{W} = W^l$. Then, we have

$$\mathbb{E}[\|\mathbf{I} - \mathbf{W}\|_\infty^w] = \sum_l P(l) \max_i \sum_{j \neq i} |W_{ij}^l| = \sum_l P(l) \sum_{j \neq i_l^*} |W_{ij}^l| < 1.$$

Consequently, by Jensen's inequality, $\mathbb{E}[\log \|\mathbf{I} - \mathbf{W}\|_\infty^w] < 0$. Theorem 3 implies the results.

The strictly diagonally column dominant case can be similarly established using the weighted l_1 norm.

It is important to mention that even if $\mathcal{W}$ does not solely consist of strictly diagonally row dominant matrices, the stochastic stability results as given in Corollary 1 may still be satisfied. As a simple example, consider the case where $\mathcal{W}$ only contains two matrices W^1, W^2, where: $W^1 = \begin{bmatrix} 1 & 2 \\ 2 & 1 \end{bmatrix}, W^2 = \begin{bmatrix} 1 & 0.45 \\ 0.45 & 1 \end{bmatrix}$, with the former and latter probabilities $0.5, 0.5$ respectively. Then one can easily verify that $\mathbb{E}[\log \|\mathbf{I} - \mathbf{W}\|_\infty] = -0.053 < 0$. Consequently, Theorem 3 ensures that all results still hold. Note that in this case, $\mathbb{E}[\|\mathbf{I} - \mathbf{W}\|_\infty] = 1.225 > 1$. Therefore, it is *not* a contraction on average.

So far, we have picked the particular all-ones weight vector $w = \mathbf{1}$ primarily because it yields the maximum intuition and matches with the strictly diagonally dominant matrices context. It should be evident that by allowing for an arbitrary positive weight vector w, we have expanded the applicability of the sufficient conditions given in the previous section, since in certain cases, a different weight vector may need to be selected.

4 Conclusions and Future Work

In addition to the conjecture we mentioned at the end of the previous section, we mention in closing that although there are different classes of network matrices that achieve existence and/or uniqueness of NE, it is not well-studied whether the best response dynamics will converge to a NE in other classes of network matrices (even if that class of network matrices guarantee the existence and uniqueness of a NE). For instance, best response dynamics may not converge when $\mathbf{W}$ is positive definite, although a unique NE is guaranteed to exist in that case. Expanding on such convergence results can be interesting and worthwhile and also shed additional light to the stochastic stability type of results.

References

1. Jackson, M.O.: Social and Economic Networks. Princeton University Press, Princeton (2008)
2. Mcnache, I., Ozdaglar, A.: Network games: theory, models, and dynamics. Synth. Lect. Commun. Netw. **4**(1), 1–159 (2011)
3. Molavi, P., Eksin, C., Ribeiro, A., Jadbabaie, A.: Learning to coordinate in social networks. Oper. Res. **64**(3), 605–621 (2015)
4. Allen, F., Babus, A.: Networks in finance (2008)
5. Acemoglu, D., Ozdaglar, A., Tahbaz-Salehi, A.: Systemic risk and stability in financial networks. Am. Econ. Rev. **105**(2), 564–608 (2015)
6. Bertsimas, D., Gamarnik, D.: Asymptotically optimal algorithms for job shop scheduling and packet routing. J. Algorithms **33**(2), 296–318 (1999)
7. Zhou, Z., Bambos, N.: Target-rate driven resource sharing in queueing systems. In: 2015 54th IEEE Conference on Decision and Control (CDC), pp. 4940–4945. IEEE (2015)
8. Moshtagh, N., Michael, N., Jadbabaie, A., Daniilidis, K.: Vision-based, distributed control laws for motion coordination of nonholonomic robots. IEEE Trans. Robot. **25**(4), 851–860 (2009)
9. Shakkottai, S.: Asymptotics of search strategies over a sensor network. IEEE Trans. Autom. Control **50**(5), 594–606 (2005)
10. Zhou, Z., Bambos, N.: A general model for resource allocation in utility computing. In: American Control Conference (2015, submitted)
11. Miura-Ko, R., Yolken, B., Bambos, N., Mitchell, J.: Security investment games of interdependent organizations. In: 2008 46th Annual Allerton Conference on Communication, Control, and Computing, pp. 252–260. IEEE (2008)
12. Miura-Ko, R.A., Yolken, B., Mitchell, J., Bambos, N.: Security decision-making among interdependent organizations. In: 2008 21st IEEE Computer Security Foundations Symposium, pp. 66–80, June 2008
13. Zhou, Z., Yolken, B., Miura-Ko, R.A., Bambos, N.: A game-theoretical formulation of influence networks. In: American Control Conference, 2016 (2016)
14. Alpcan, T., Başar, T.: Network Security: A Decision and Game-Theoretic Approach. Cambridge University Press, New York (2010)
15. Manshaei, M.H., Zhu, Q., Alpcan, T., Bacşar, T., Hubaux, J.-P.: Game theory meets network security and privacy. ACM Comput. Surv. (CSUR) **45**(3), 25 (2013)
16. Zhou, Z., Yolken, B., Miura-Ko, R.A., Bambos, N.: Linear influence networks: equilibria, monotonicity and free riding (2016)

17. Cottle, R., Pang, J., Stone, R.: The Linear Complementarity Problem. Academic Press, Boston (1992)
18. Horn, R.A., Johnson, C.R.: Matrix Analysis. Cambridge University Press, New York (2012)
19. Diaconis, P., Freedman, D.: Iterated random functions. SIAM Rev. **41**(1), 45–76 (1999)
20. Jarner, S., Tweedie, R.: Locally contracting iterated functions and stability of Markov chains. J. Appl. Prob. **38**, 494–507 (2001)
21. Billingsley, P.: Convergence of Probability Measures. Wiley Series in Probability and Mathematical Statistics: Tracts on Probability and Statistics. Wiley, New York (1968)

Special Track-Validating Models

Patrolling a Pipeline

Steve Alpern[1], Thomas Lidbetter[2(✉)], Alec Morton[3], and Katerina Papadaki[4]

[1] Warwick Business School, University of Warwick, Coventry CV4 7AL, UK
steve.alpern@wbs.ac.uk
[2] Department of Management Science and Information Systems,
Rutgers Business School, 1 Washington Park, Newark, NJ 07102, USA
t.r.lidbetter@lse.ac.uk
[3] Department of Management Science, University of Strathclyde,
199 Cathedral Street, Glasgow G4 0QU, UK
alec.morton@strath.ac.uk
[4] Deparment of Mathematics, London School of Economics, Houghton Street,
London WC2A 2AE, UK
k.p.papadaki@lse.ac.uk

Abstract. A pipeline network can potentially be attacked at any point and at any time, but such an attack takes a known length of time. To counter this, a Patroller moves around the network at unit speed, hoping to intercept the attack while it is being carried out. This is a zero-sum game between the mobile Patroller and the Attacker, which we analyze and solve in certain cases.

Keywords: Patrolling · Zero-sum game · Networks

1 Introduction

A game theoretic model of patrolling a graph was recently introduced in [1], in which an Attacker chooses a node of a graph to attack at a particular time and a Patroller chooses a walk on the nodes of the graph. The game takes place in discrete time and the attack lasts a fixed number of time units. For given mixed strategies of the players, the payoff of the game is the probability that the attack is intercepted by the Patroller: that is, the probability that the Patroller visits the node the Attacker has chosen during the period in which the attack takes place. The Patroller seeks to maximize the payoff and the Attacker to minimize it, so the game is zero-sum.

In [1], several general results of the game are presented along with solutions of the game for some particular graphs. This work is extended in [5], in which line graphs are considered. The game is surprisingly difficult to solve on the line graph, and the optimal policy for the Patroller is not always, as one might expect, the strategy that oscillates to and fro between the terminal nodes. Rather, depending on the length of time required for the attack to take place, it may be optimal for the Patroller to stay around the two ends of the line with some positive probability.

© Springer International Publishing AG 2016
Q. Zhu et al. (Eds.): GameSec 2016, LNCS 9996, pp. 129–138, 2016.
DOI: 10.1007/978-3-319-47413-7_8

In this paper we present a new continuous game theoretic model of patrolling, in a similar spirit to [1], but on a continuous network, so that the attack may take place at any point of the network (not just at nodes). We also model time as being continuous, rather than discrete. This is a better model for a situation in which a pipeline may be disrupted at any point.

At first glance, this might appear to be a more complicated game to analyze. However, it turns out that continuity simplifies matters, and we are able to solve the game for Eulerian networks (Sect. 3) and for line networks (Sect. 4). The solution of the game on the line network is considerably easier to derive than for the discrete analogue, and we also show that the value of the latter game converges to that of the former as the number of nodes of the graph approaches infinity.

A game theoretical approach to patrolling problems has been successful in real life settings, for example in [6,7]. Other work on game theoretic models of patrolling a network include [2,4].

2   Definition of the Game

We start by defining a continuous time patrolling game, where the Patroller moves at unit speed along a network Q with given arc lengths, and the Attacker can attack at any point of the network (not just at nodes). In this section we define the game formally and describe each of the players' strategy spaces.

The network Q can be viewed as a metric space, with $d(x, y)$ denoting the arc length distance, so we can talk about 'the midpoint of an arc' and other metric notions. We assume that the game has an infinite time horizon and that a Patroller pure strategy is a unit speed (Lipshitz continuous) path $w : [0, \infty) \rightarrow Q$, in particular, one satisfying

$$d(w(t), w(t')) \leq |t - t'|, \text{ for all } t, t' \geq 0.$$

For the Attacker, a pure strategy is a pair $[x, I]$, where $x \in Q$ and $I \subset [0, \infty)$ is an interval of length r. It is sometimes useful to identify I with its midpoint y, where $I = I_y = [y - r/2, y + r/2]$. Thus $y \in [r/2, \infty)$.

The payoff function, taking the Patroller as the maximizer, is given by

$$P(w, \{x, y\}) = \begin{cases} 1 \text{ if } w(t) = x \text{ for some } t \in I_y, \\ 0 \qquad\qquad \text{otherwise.} \end{cases} \tag{1}$$

Hence the value, if it exists, is the probability that the attack is intercepted. Note that in this scenario the pure strategies available to both players are uncountably infinite, so the von Neuman minimax theorem no longer applies. Furthermore, the payoff function is not continuous (in either variable), so minimax theorems using that property also don't apply. For example, if w is the constant function x, then $P(w, [x, I]) = 1$, however an arbitrarily small perturbation of w or x can have $P(w', [x', I]) = 0$. However, in the examples we study in this paper we show that the value exists by explicitly giving optimal strategies for the players.

3   General Results

We start by giving upper and lower bounds for the value of the game for general networks. First, we define the *uniform* attack strategy.

Definition 1. *The **uniform attack strategy** chooses to attack in the time interval* $[0, r]$ *at a uniformly random point on* Q. *More precisely, the probability the attack takes place in a region* A *of the network is proportional to the total length of* A.

We use the uniform attack strategy to deduce a simple lower bound on the value of the game. We denote the total length of Q by μ.

Lemma 1. *The uniform attack strategy guarantees that the probability* P *of interception is no more than* r/μ.

We also define a natural strategy for the Patroller. Recall that a *Chinese Postman Tour* (*CPT*) of the network Q is a minimum length tour that contains every point of Q. We denote the length of a CPT by $\bar{\mu}$. It is well known [3] that there are polynomial time algorithms (polynomial in the number of nodes of the network) that calculate $\bar{\mu}$. It is easy to see that $\bar{\mu} \leq 2\mu$, since doubling each arc of the network results in a new network whose nodes all have even degree and therefore contains an Eulerian tour.

Definition 2. *Fix a CPT,* $w : [0, \infty) \to Q$ *that repeats with period* $\bar{\mu}$. *The **uniform CPT strategy*** $\bar{w} : [0, \infty) \to Q$ *for the Patroller is defined by*

$$\bar{w}(t) = w(t + T),$$

where T *is chosen uniformly at random from the interval* $[0, \bar{\mu}]$. *In other words, the Patroller chooses to start the CPT at a random point along it.*

This strategy gives an upper bound on the value of the game.

Lemma 2. *The uniform CPT strategy guarantees that the probability* P *of interception is at least* $r/\bar{\mu}$.

Lemmas 1 and 2 give upper and lower bounds on the value of the game. If the network is Eulerian (that is, the network contains a tour that does not repeat any arcs) then $\mu = \bar{\mu}$ and Lemmas 1 and 2 imply that the value of the game is $r/\mu = r/\bar{\mu}$. We sum this up in the theorem below.

Theorem 1. *The value* V *of the game satisfies*

$$\frac{r}{\bar{\mu}} \leq V \leq \frac{r}{\mu}.$$

If the network is Eulerian then both bounds are tight, $V = r/\mu = r/\bar{\mu}$, *the uniform attack strategy is optimal for the Attacker and the uniform CPT strategy is optimal for the Patroller.*

Writing P^* for the probability the uniform CPT strategy intercepts the attack, we note that since it is true for any network that $\bar{\mu} \leq 2\mu$, we have

$$V \leq \frac{r}{\mu} \leq 2\left(\frac{r}{\bar{\mu}}\right) = 2P^*.$$

This shows that the value of the game is no more than twice the interception probability guaranteed by the uniform CPT strategy.

4   Solution on the Line Network

We now give a complete solution to the game on a line of unit length, that is the closed unit interval $[0, 1]$. The Attacker picks a point $x \in [0, 1]$ and an interval $I \subset [0, \infty)$ of length r. The Patroller picks a unit speed walk w on the unit interval, $w : \mathbb{R}^+ \to [0, 1]$. The attack is intercepted if $w(t) = x$, for some $t \in I$. We assume $0 \leq r \leq 2$, otherwise the Patroller can always intercept the attacks by oscillating between the endpoints of the unit interval.

4.1   The Case $r > 1$

We begin by assuming the attack interval r is relatively large compared to the size of the line, in particular when $r > 1$. We shall see that the following strategies are optimal.

Definition 3. *Let the **diametrical** Attacker strategy be defined as follows: choose y uniformly in $[0, 1]$ and attack equiprobably at one of the endpoints $x = 0$ or 1 during the time interval $I = [y, y + r]$.*

*For the Patroller, the **oscillation strategy** is defined as the strategy where the Patroller randomly picks a point x on the unit interval and a random direction and oscillates from one endpoint to the other.*

We note that the oscillation strategy is simply the uniform CPT strategy as defined in Definition 2, and thus ensures a probability $P \geq r/\bar{\mu} = r/2$ of interception, by Lemma 2.

We can show that the diametrical strategy ensures the attack will not be intercepted with probability any greater than $r/2$.

Lemma 3. *If $r \geq 1$ and the Attacker adopts the diametrical strategy then for any path w the attack is intercepted with probability $P \leq r/2$.*

We have the following corollary:

Theorem 2. *The diametric Attacker strategy and the oscillation strategy are optimal strategies and give value $V = r/2$.*

Proof. This follows directly from Lemmas 2 and 3.

4.2   The Case $r \leq 1$

Now we consider the case of $r \leq 1$. In this case r is small compared to 1 (the size of the unit interval), thus the Patroller stays at the end with some probability and oscillates between the endpoints of the unit interval with the remaining probability.

Let q be the quotient and ρ the remainder when r divides 1. Thus $1 = rq + \rho$, where q is an integer and $0 \leq \rho < r$. Let $k = r + \rho$. We first define the Attacker strategies.

Definition 4. *Consider the following Attacker strategy, which we call* r-**attack strategy***, that is performed at a random point in time, here we start it at time 0:*

1. *Attack at points* $\mathcal{E} = \{0, r, 2r, \ldots, (q-1)r, 1\}$, *starting attacks equiprobably between times* $[0, r]$, *each with total probability* $\frac{r}{1+r}$. *We call these the* **external attacks***.*
2. *Attack at the midpoint of* $(q-1)r$ *and* 1, *which is the point* $1 - \frac{r+\rho}{2} = 1 - \frac{r}{2}$, *starting the attack equiprobably between times* $\left[\frac{r-\rho}{2}; \frac{r+\rho}{2}\right]$ *with total probability* $\frac{\rho}{1+r}$. *We call this the* **internal attack***.*

The attacks are shown in Fig. 1. The horizontal axis is time and the vertical axis is the unit interval.

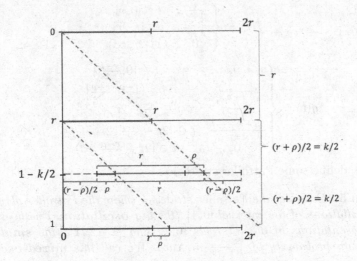

Fig. 1. The r-attack strategy is shown. The starting points of the attacks are shown in red. (Color figure online)

Let $f(t)$ be the probability of interception at an external attack point if the Patroller is present there at time t. Let $g(t)$ be this probability for the internal

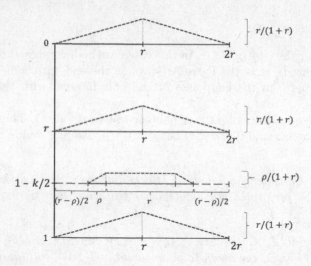

Fig. 2. The probability of interception at each point in time t is shown both for external attacks, $f(t)$, and for internal attacks, $g(t)$, for the r-attack strategy.

attack point. These probability functions for the r-attack strategy are shown in Fig. 2.

The functions f and g are as follows:

$$f(t) = \begin{cases} \frac{t}{1+r}, & t \in [0, r] \\ \frac{2r-t}{1+r}, & t \in [r, 2r] \\ 0, & t \in [2r, \infty) \end{cases} \tag{2}$$

$$g(t) = \begin{cases} 0, & t \in \left[0, \frac{r-\rho}{2}\right] \\ \frac{t-\frac{r-\rho}{2}}{1+r}, & t \in \left[\frac{r-\rho}{2}, \frac{r+\rho}{2}\right] \\ \frac{\rho}{1+r}, & t \in \left[\frac{r+\rho}{2}, 2r - \frac{r+\rho}{2}\right] \\ \frac{2r-\frac{r-\rho}{2}-t}{1+r}, & t \in \left[2r - \frac{r+\rho}{2}, 2r - \frac{r-\rho}{2}\right] \\ 0, & t \in \left[2r - \frac{r-\rho}{2}, \infty\right) \end{cases} \tag{3}$$

We now define some Patroller strategies.

Definition 5. *Consider the Patroller strategies where the Patroller plays a mixture of oscillations of the interval $[0, 1]$ (the **big oscillations**) with probability $\frac{1}{1+r}$, and oscillations of the intervals $\left[0, \frac{r}{2}\right]$ and $\left[1 - \frac{r}{2}, 1\right]$ (the **small oscillations**) with probability of $\frac{r}{2(1+r)}$ on each. We call this **mixed-oscillation strategy**.*

The mixed oscillation strategy is shown in Fig. 3. Note that the small oscillations have period r and thus intercept all attacks in the respective intervals. By attacking at 0 or 1 the Attacker secures $\frac{r}{2(1+r)} + \frac{r}{2} \times \frac{1}{1+r} = \frac{r}{1+r}$, since the big oscillation intercepts attacks at the endpoints with probability $\frac{r}{2}$. Any attacks in the open intervals $\left(0, \frac{r}{2}\right)$ and $\left(1 - \frac{r}{2}, 1\right)$, are dominated by attacks at endpoints.

Attacking in $\left[\frac{r}{2}, 1 - \frac{r}{2}\right]$ secures an interception probability of $\frac{2r}{2} \times \frac{1}{1+r} = \frac{r}{1+r}$, since at points in $\left[\frac{r}{2}, 1 - \frac{r}{2}\right]$, the big oscillation in each of its period time intervals of length 2, it intercepts attacks that start at two time intervals each of length r. Hence, $V \geq \frac{r}{1+r}$.

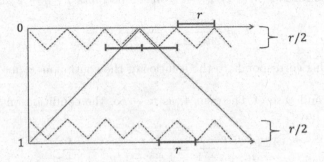

Fig. 3. The mixed oscillation strategy, where the horizontal axis is time and the vertical axis is the unit interval.

Theorem 3. *If $r \leq 1$, then the r-attack strategy and the mixed-oscillation strategy are optimal and the value of the game is $V = \frac{r}{1+r}$.*

4.3   Relation to Discrete Patrolling Game

The discrete analogue of our game, introduced in [1] was solved for line graphs in [5]. It is interesting (and reassuring) to find that the value of the discrete game converges to the value of the continuous game as the number of nodes tends to infinity.

We briefly describe the set-up of the discrete game. The game is played on a line graph with n nodes in a discrete time horizon $\mathcal{T} = \{1, 2, \ldots, T\}$. The Attacker chooses an *attack node* at which to attack and a set of m successive time periods in $\mathcal{T}$, which is when the attack takes place. The Patroller chooses a walk on the graph. As in the continuous case, the payoff of the game, which the Attacker seeks to minimize and the Patroller to maximize, is the probability that the Patroller visits the attack node while the attack is taking place.

The value of the game depends on the relationship between n and m, and the solution divides into 5 cases (see Theorem 6 of [5]). We are interested in fixing the ratio $r = m/n$ and letting n tend to infinity, therefore the solution of two of the cases of the game from [1] are irrelevant: in particular the case when $m = 2$, and the case when $n = m + 1$ or $n = m + 2$. The case $n < (m + 2)/2$ (corresponding to the case $r \geq 2$ in the continuous case) is also uninteresting, since then the value is 1. Therefore we are left with two cases, whose solutions we summarize below.

Theorem 4 (From Theorem 6 of [5]). *The value V of the discrete patrolling game on the line is*

1. $V = m/(2n - 2)$ *if* $(m + 1)/2 \leq n \leq m + 1$, *and*
2. $V = m/(n + m - 1)$ *if* $n \geq m + 3$, *or* $n = m + 2$ *and* $m \geq 3$ *is odd.*

We now consider the behaviour of the value of the discrete game as $n \to \infty$, assuming that the ratio $r = m/n$ is fixed. In the first case of Theorem 4, as $n \to \infty$, the condition $(m + 1)/2 \leq n \leq m + 2$ becomes $1 \leq r \leq 2$ and we have

$$V = \frac{m}{2n - 2} = \frac{r}{2 - 2/n} \to \frac{r}{2},$$

as $n \to \infty$. This corresponds to the solution of the continuous game as given in Theorem 2.

In the second case of Theorem 4, as $n \to \infty$, the condition on m becomes $r \leq 1$ and we have

$$V = \frac{m}{n + m - 1} = \frac{r}{1 + r - 1/n} \to \frac{r}{1 + r},$$

as $n \to \infty$. Again, this corresponds to the solution of the continuous game as given in Theorem 3.

5   Conclusion

We have introduced a new game theoretic model of patrolling a continuous network in continuous time, analagous to the discrete patrolling game introduced in [1]. We have given general bounds on the value of the game and solved it in the case that the network is Eulerian or if it is a line.

We are optimistic that our results on the line network can be extended to a larger class of networks, such as stars or trees, and we conjecture that the value of the game is $r/\bar{\mu}$ for any tree network with diameter D such that $D \leq r \leq \bar{\mu}$, where $\bar{\mu}$ is the length of a CPT of the network.

Appendix: Omitted Proofs

Proof of Lemma 1. The attack must be taking place during the time interval $[0, r]$. Let A be the set of points that the Patroller intercepts in this time interval. Then clearly A must have length no greater than r and so the probability the attack takes place at a point in A is r/μ. It follows that $P \leq r/\mu$.   □

Proof of Lemma 2. Suppose the attack starts at time t_0 at some point $x \in Q$. Then the attack is certainly intercepted if $\bar{w}$ is at x at time t_0. Let $t_x \in [0, \bar{\mu}]$ be such that $w(t_0 + t_x) = x$, so that the attack is intercepted by $\bar{w}$ if $T = t_x$. Let A be the set of times $t \in [0, \bar{\mu}]$ such that $t_x - r \leq t \leq t_x$ or $t \geq t_x + \bar{\mu} - r$, so if $T \in A$, then the attack is intercepted by $\bar{w}$. But the measure of A is r, so the probability that T is in A is $r/\bar{\mu}$ and hence $P \geq r/\bar{\mu}$.   □

Proof of Lemma 3. Take a Patroller path w. We can assume that w starts at an endpoint, otherwise it is weakly dominated by a strategy that does. To see this, suppose the Patroller starts at an interior point before traveling directly to an endpoint, arriving there at time $t < 1$. Now consider the Patroller strategy that is the same but in the time interval $[0, t]$ the Patroller remains at the endpoint. Then clearly the second strategy intercepts the same set of attacks as the first one. Without loss of generalization we assume w starts at $x = 0$.

We only need to consider the path in the time interval $[0, 1 + r]$, after which time the attack has been completed with probability 1. Since $r < 2$ the walk cannot go between the two ends more than twice, so there are three possibilities.

The first is that w stays at $x = 0$ for the whole time, in which case the probability the attack is intercepted is $P = 1/2 \leq r/2$.

The second possibility is that w stays at $x = 0$ for time t_1, then goes to $x = 1$ and stays there for time t_2. We can assume it takes the Patroller time 1 to go between the endpoints since any path taking longer than that would be dominated, so $t_1 + t_2 = r$. The attack is intercepted at $x = 0$ if it starts sometime during $[0, t_1]$, which has probability $(1/2)t_1$. It is intercepted at $x = 1$ if it ends sometimes during $[1 + r - t_2, 1 - r]$, which has probability $(1/2)t_2$. Hence $P = (1/2)(t_1 + t_2) = r/2$.

The final possibility is that w stays at $x = 0$ for time t_1, then goes directly to $x = 1$ for time t_2, then goes directly back to $x = 0$ for time t_3, in which case we must have $t_1 + t_2 + t_3 = r - 1$. This time the attack is intercepted at $x = 0$ in the case of either of the two mutually exclusive events that it starts in $[0, t_1]$ or ends in $[1 + r - t_3, 1 - r]$, which have total probability $(1/2)(t_1 + t_3)$. If the attack takes place at $x = 1$, it must be taking place during the whole of the time interval $[1, r]$. But w must reach $x = 1$ sometime during this time interval, since it must have time to travel from $x = 0$ to $x = 1$ and back again, and hence intercepts the attack with probability 1. So the overall probability the attack is intercepted is $(1/2)(t_1 + t_3) + 1/2 \leq (1/2)(t_1 + t_2 + t_3) + 1/2 = r/2$. □

Proof of Theorem 3. We already showed that $r/(1 + r)$ is a lower bound for the value and now we show that it is also an upper bound. Now, suppose that the Attacker plays the r-attack strategy. The Patroller could:

1. Stay at any attack point but will not win with probability greater than $\frac{r}{1+r}$.
2. Travel between consecutive external attacks and if possible try to reach the internal attack: Suppose the Patroller is at point 0 up to time t: If $t \in [0, r]$ and then leaves for point r, she will reach point r at times in the range $[r, 2r]$. This gives total interception probability $f(t) + f(t + r) = \frac{t}{1+r} + \frac{2r - (t+r)}{1+r} = \frac{r}{1+r}$. Note that if the Patroller continues to the next attack along the unit interval, if it is the internal attack she will reach it at times greater than $r + \frac{r+\rho}{2} = 2r - \frac{r-\rho}{2}$, when the internal attack has been completed, and if it is an external attack she will reach it at time greater than $2r$, where all external attacks have been completed. If $t \in [r, 2r]$ then all attacks at point 0 have been intercepted but the Patroller arrives at point r after all attacks have been completed, which gives interception probability of $\frac{r}{1+r}$.

3. Travel between last two external attacks, crossing internal attack in the middle (this is the same as doing a roundtrip from one of the last external attacks to the internal attack and back): Suppose the Patroller leaves point $(q-1)r$ at time t, toward the internal attack point and the last external attack point 1: If $t \in [0, r - \rho]$, she will reach the internal attack point at times $\left[\frac{r+\rho}{2}, r - \rho + \frac{r+\rho}{2}\right] = \left[\frac{r+\rho}{2}, 2r - \frac{r+\rho}{s}\right]$, and she will reach the external attack at point 1 at times $[r + \rho, 2r]$. This sums to a probability of $f(t) + g\left(t + \frac{r+\rho}{2}\right) + f(t + r + \rho) = \frac{t}{1+r} + \frac{\rho}{1+r} + \frac{2r-(t+r+\rho)}{1+r} = \frac{r}{1+r}$. If $t \in [r-\rho, r]$, she will reach the internal attack point at times $\left[2r - \frac{r+\rho}{2}, 2r - \frac{r-\rho}{2}\right]$, and the external attack point 1 at times greater than $2r$. This sums to a probability of $f(t) + g\left(t + \frac{r+\rho}{2}\right) = \frac{t}{1+r} + \frac{2r - \frac{r-\rho}{2} - \left(t + \frac{r+\rho}{2}\right)}{1+r} = \frac{r}{1+r}$. Finally, if $t \in [r, 2r]$, the Patroller will intercept all attacks at point $(q-1)r$ and will not make it in time for the internal attack nor the attack at point 1, this gives the desired probability. $\qquad\square$

References

1. Alpern, S., Morton, A., Papadaki, K.: Patrolling games. Oper. Res. **59**(5), 1246–1257 (2011)
2. Basilico, N., Gatti, N., Amigoni, F.: Patrolling security games: definition and algorithms for solving large instances with single patroller and single intruder. Artif. Intell. **184**, 78–123 (2012)
3. Edmonds, J., Johnson, E.L.: Matching, Euler tours and the Chinese postman. Math. Program. **5**(1), 88–124 (1973)
4. Lin, K.Y., Atkinson, M.P., Chung, T.H., Glazebrook, K.D.: A graph patrol problem with random attack times. Oper. Res. **61**(3), 694–710 (2013)
5. Papadaki, K., Alpern, S., Lidbetter, T., Morton, A.: Patrolling a Border: Oper. Res. (2016, in press)
6. Pita, J., Jain, M., Marecki, J., Ordóñez, F., Portway, C., Tambe, M., Western, C., Paruchuri, P., Kraus, S.: Deployed ARMOR protection: the application of a game theoretic model for security at the Los Angeles International Airport. In: Proceedings of the 7th International Joint Conference on Autonomous Agents and Multiagent Systems: Industrial Track, pp. 125–132. International Foundation for Autonomous Agents and Multiagent Systems (2008)
7. Yang, R., Ford, B., Tambe, M., Lemieux, A.: Adaptive resource allocation for wildlife protection against illegal poachers. In: Proceedings of the 2014 International Conference on Autonomous Agents and Multi-agent Systems, pp. 453–460. International Foundation for Autonomous Agents and Multiagent Systems (2014)

Optimal Allocation of Police Patrol Resources Using a Continuous-Time Crime Model

Ayan Mukhopadhyay[1], Chao Zhang[2], Yevgeniy Vorobeychik[1(✉)],
Milind Tambe[2], Kenneth Pence[1], and Paul Speer[1]

[1] Vanderbilt University, 2201 West End Ave, Nashville, TN 37235, USA
{ayan.mukhopadhyay,yevgeniy.vorobeychik}@vanderbilt.edu
[2] University of Southern California, Los Angeles, CA 90089-0894, USA

Abstract. Police departments worldwide are eager to develop better patrolling methods to manage the complex and evolving crime landscape. Surprisingly, the problem of spatial police patrol allocation to optimize expected crime response time has not been systematically addressed in prior research. We develop a bi-level optimization framework to address this problem. Our framework includes novel linear programming patrol response formulations. Bender's decomposition is then utilized to solve the underlying optimization problem. A key challenge we encounter is that criminals may respond to police patrols, thereby shifting the distribution of crime in space and time. To address this, we develop a novel iterative Bender's decomposition approach. Our validation involves a novel spatio-temporal continuous-time model of crime based on survival analysis, which we learn using real crime and police patrol data for Nashville, TN. We demonstrate that our model is more accurate, and much faster, than state-of-the-art alternatives. Using this model in the bi-level optimization framework, we demonstrate that our decision theoretic approach outperforms alternatives, including actual police patrol policies.

Keywords: Decision theoretic policing · Crime modeling · Survival analysis · Bender's decomposition

1 Introduction

Prevention, response and investigation are the three major engagements of police. Ability to forecast and then effectively respond to crime is, therefore, the holy grail of policing. In order to ensure that crime incidents are effectively handled, it is imperative that police be placed in a manner that facilitates quick response. Effective police placement, however, needs crime prediction as a prerequisite. This is one of the reasons why predicting crime accurately is of utmost importance. While a number of techniques have been proposed for characterizing and forecasting crime, optimizing response times has not been addressed so far, to the best of our knowledge.

© Springer International Publishing AG 2016
Q. Zhu et al. (Eds.): GameSec 2016, LNCS 9996, pp. 139–158, 2016.
DOI: 10.1007/978-3-319-47413-7_9

Our goal is to develop a rigorous optimization-based approach for optimal police placement in space in order to minimize expected time to respond to crime incidents as they occur. For the time being, we assume that a generative model for crime is available; we describe such a model, calibrated on real crime and police patrol data, in Sect. 4. The key challenge we face is that crime locations and timing are uncertain. Moreover, for a given placement of police resources in space, optimizing crime incident response for a collection of known incidents is itself a non-trivial optimization problem. What makes this problem particularly challenging is that criminals are affected by police, as they avoid committing crimes if the chances of being caught are high; consequently, we expect that police placement will impact spatial and temporal distribution of crime incidents. Our model, therefore, has both decision and game theoretic features, even though we make use of a data-driven generative model of crime that accounts for the impact of police locations, rather than relying on rationality as underpinning criminal behavior.

Formally, we frame the problem of police patrol optimization as a regularized two-stage stochastic program. We show how the second-stage program (computing optimal response to a fixed set of crime incidents) can be formulated as a linear program, and develop a Bender's decomposition method with sample average approximation for the overall stochastic program. To address the fact that the top-level optimization decisions actually influence the probability distribution over scenarios for the second-level crime response optimization problem, we propose a novel iterative stochastic programming algorithm, *IBRO*, to compute approximate solutions to the resulting bi-level problem of finding optimal spatial locations for police patrols that minimize expected response time. We show that our model outperforms alternative policies, including the response policy in actual use by a US metropolitan police department, both in simulation and on actual crime data.

In order to validate our model of police response, we develop a novel crime forecasting model that is calibrated and evaluated using real crime and police patrol data in Nashville, TN. Crime prediction has been extensively studied, and several models for it have been proposed. These include visualization tools, primarily focused on *hotspots*, or areas of high crime incidence [2], spatial cluster analysis tools [15,17], risk-terrain models [10], leading indicator models [4], and dynamic spatial and temporal models [9,19,23]. A major shortcoming of the existing methods is that they do not allow principled data-driven continuous-time spatial-temporal forecasting *that includes arbitrary crime risk factors*. For example, while risk-terrain modeling focuses on spatial covariates of crime, it entirely ignores temporal factors, and does not offer methods to learn a generative model of crime from data. The work by Short et al. [19] on dynamic spatial-temporal crime modeling, on the other hand, does not readily allow inclusion of important covariates of crime, such as locations of pawn shops and liquor stores, weather, or seasonal variations. Including such factors in a spatial-temporal model, however, is critical to successful crime forecasting: for example, these may inform important policy decisions about zoning and hours of operation for liquor stores, and will make the tool more robust to environmental changes that affect such variables. To address these concerns, validate our model, and forecast crimes, we propose a stochastic generative model of crime which is continuous in time and discretized in space, and readily incorporates crime covariates, bridging an important gap in prior art. Our model leverages survival analysis to learn a probability density over time for predicting crime. After creating a model to predict crime, we evaluate its performance by comparing it with a natural adaptation of the Dynamic Spatial Disaggregation Approach (DSDA) algorithm [9] and an Dynamic Bayes Network method [23] using automated abstraction [22].

1.1 Related Work

There has been an extensive literature devoted to understanding and predicting crime incidence, involving both qualitative and quantitative approaches. For example, a number of studies investigate the relationship between liquor outlets and crime [20, 21]. Many of the earlier quantitative models of crime focus on capturing spatial crime correlation (hot spots), and make use of a number of statistical methods towards this end [15, 17]; these are still the most commonly used methods in practice. An alternative approach, risk-terraine modeling, focuses on quantifiable environmental factors as determinants of spatial crime incidence, rather than looking at crime correlation [10]. These two classes of models both have a key limitation: they ignore the temporal dynamics of crime. Moreover, environmental risk factors and spatial crime analysis are likely complementary. Our approach aims to merge these ideas in a principled way.

Recently, a number of sophisticated modeling approaches emerged aiming to tackle the full spatio-temporal complexity of crime dynamics. One of these is based on a spatio-temporal differential equation model that captures both spatial and temporal crime correlation [16, 18]. These models have two disadvantages compared to ours: first, they do not naturally capture crime co-variates, and second, they are non-trivial to learn from data [16], as well as to use in making predictions [18]. Another model in this general paradigm is Dynamic Spatial Disaggregation Approach (DSDA) [9], which combines an autoregressive model to capture temporal crime patterns with spatial clustering techniques to model spatial correlations. The model we propose is significantly more flexible, and combines spatial and temporal predictions in a principled way by using well-understood survival analysis methods. Recently, an approach has been proposed for modeling spatial and temporal crime dynamics using Dynamic Bayes Networks [22, 23]. This approach necessitates discretization of time, as well as space. Moreover, despite significant recent advances, scalability of this framework remains a challenge.

2 Optimizing Police Placement

Our goal is to address a fundamental decision theoretic question faced by police: how to allocate limited police patrols so as to minimize expected response time to occurring crime. In reality, this is a high-dimensional dynamic optimization problem under uncertainty. In order to make this tractable in support of practical decision making, we consider a simplified two-stage model: in the first stage, police determines spatial location of a set of patrol vehicles, P, and in the second stage, vehicles respond to crime incidents which occur. The decisions in the first stage are made under uncertainty about actual crime incidents, whereas for second-stage response decisions, we assume that this uncertainty is resolved. *A key strategic consideration in police placement is its impact on crime incidence.* In particular, it is well known that police presence has some deterrence effect on crime, which in spatio-temporal domains takes two forms: reduced overall crime frequency, and spatial crime shift [12,19]. We assume below that the effect of police presence on crime distribution is captured in a stochastic crime model. Later, we describe and develop the stochastic crime model where we use real crime and police patrol data.

We present the problem formulation of allocating police given a stochastic generative model of crime. We divide the available area under police patrol into discrete grids. Formally, we define q as the vector of police patrol decisions, where q_i is the number of police vehicles place in grid i. Let s be a random variable corresponding to a batch of crime incidents occurring prior to the second stage. The two-stage optimization problem for police placement then has the following form:

$$\min_q \mathbb{E}_{s \sim f}[D(q; s)], \tag{1}$$

where $D(q; s)$ is the minimal total response time of police located according to q to crime incidents in realization s, which is distributed according to our crime distribution model f described in Sect. 4, associated with each grid (and the corresponding spatial variables). The model implicitly assumes that crime occurrence is distributed i.i.d. for each grid cell, conditional on the feature vector, where the said feature vector captures the inter-dependence among grids. While the crime prediction model is continuous in time, we can fix a second-stage horizon to represent a single *time zone* (4-hour interval), and simply consider the distribution of the crime incidents in this interval.

The optimization problem in Eq. (1) involves three major challenges. First, even for a given s, one needs to solve a non-trivial optimization problem of choosing which subset of vehicles to send in response to a collection of spatially dispersed crime incidents. Second, partly as a consequence of the first, computing the expectation exactly is intractable. Third, the probability distribution of future crime incidents, f, depends on police patrol locations q through the features that capture deterrence effects as well as spatial crime shift to avoid police. We address these problems in the following subsections.

2.1   Minimizing Response Time for a Fixed Set of Crime Incidents

While our goal is to minimize total response time (where the total is over the crime incidents), the information we have is only about spatial locations of crime and police in discretized space. As a result, we propose using distance traveled as a proxy. Specifically, if a police vehicle located at grid i is chosen to respond to an incident at grid j, the distance traveled is d_{ij}, distance between grids i and j. Assume that these distances d_{ij} are given for all pairs of grids i, j. Next, we assume that a single police vehicle is sufficient to respond to all crime incidents in a particular grid j. This is a reasonable assumption, since the number of crime incidents in a given cell over a 4-hour interval tends to be relatively small, and this interval is typically sufficient time to respond to all of them.

Given this set up, we now show how to formulate this response distance minimization problem as a linear integer program by mapping it to two classical optimization problems: the transportation [1] and k-server problems [3].

In the transportation problem, there are m suppliers, each with supply s_i, n consumers, each with demand r_j, and transportation cost c_{ij} between supplier i and consumer j. The goal is to transport goods between suppliers and consumers to minimize total costs. To map crime response to transportation, let police vehicles be suppliers, crime incidents be consumers, and let transportation costs correspond to distances d_{ij} between police vehicle and crime incident grids, with each grid being treated as a node in the network. While the transportation problem offers an effective means to compute police response, it requires that the problem is balanced: supply must equal demand. If supply exceeds demand, a simple modification is to add a dummy sink node. However, if demand exceeds supply, the problem amounts to the multiple traveling salesman problem, and needs a different approach.

To address excess-demand settings, we convert the police response to a more general k-server problem. The k-server problem setting involves k servers in space and a sequence of m requests. In order to serve a request, a server must move from its location to the location of the request. The k-server problem can be reduced to the problem of finding minimum cost flow of maximum quantity in an acyclic network [3]. Let the servers be $s_1, ..., s_k$ and the requests be $r_1, ..., r_m$. A network containing $(2 + k + 2m)$ nodes is constructed. In the formulation described in [3], each arc in the network has capacity one. The arc capacities are modified in our setting, as described later in the problem formulation. The total vertex set is $\{a, s_1, ..., s_k, r_1, ..., r_m, r'_1, ..., r'_m, t\}$. a and t are source and sink respectively. There is an arc of cost 0 from a to each of s_i. From each s_i, there is an arc of cost d_{ij} to each r_j, where d_{ij} is the actual distance between locations i and j. Also, there is an arc of cost 0 from each s_i to t. From each r_i, there is an arc of cost $-K$ to each r'_i, where K is an extremely large real number. Furthermore, from each r'_i, there is an arc of cost d_{ij} to each r_j where $i < j$ in the given sequence. In our setting, servers and requests correspond to grids with police and crime respectively. In the problem setting we describe, G is the set of all the nodes in the network. We term the set $\{s_i \ \forall i \in G\}$ as G^1, the set $\{r_i \ \forall i \in G\}$ as G^2 and the set $\{r'_i \ \forall i \in G\}$ as G^3. The structure of the network is shown in Fig. 1, which shows how the problem can be framed for a setting with 6 discrete locations. Shaded nodes represent the presence of police and crime in their respective layers.

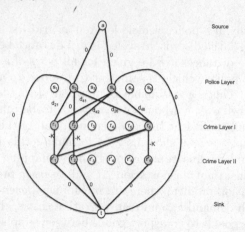

Fig. 1. Network structure

The problem of finding placement of k-servers in space to serve an unordered set of requests is the same as the multiple traveling salesperson problem (mTSP), a generalization of the TSP problem, which is NP-hard. The offline k-server problem gets around this by having a pre-defined sequence of requests. By sampling crimes from the spatio-temporal model, although we can create a sequence of crimes by ordering them according to their times of occurrence, this sequence need not necessarily provide the least time to respond to all the crimes. In order to deal with this problem, we leverage the fact that crimes are relatively rare events. In order to find the ordering of crimes that provides the least response time, we solve the problem for each possible ordering of crimes. Despite this, the k-server solution approach is significantly less scalable than the transportation formulation. Consequently, we make use of it only in the (rare) instances when crime incidents exceed the number of available police.

2.2 Stochastic Programming and Sample Average Approximation for Police Placement

Now that we have two ways of addressing the problem of minimizing response time given a *known* set of crime incidents, we consider the original problem of optimizing allocation of police patrols. As a first step, we point out that the resulting stochastic program is intractable in our setting because of the large space of possible crime incident realizations. We therefore make use of sample average approximation, whereby we estimate the expectation using a collection of i.i.d. crime incident realization samples (henceforth, scenarios) generated according to f. For each scenario, we represent the presence of crimes in the grids by a binary vector z and total available police by k. The decision variable, x_{ij}^s refers to the number of police vehicles traveling from grid i to grid j in scenario s. Under such a setting, the optimization program with transportation problem in the second level can be formulated as:

$$\min_q \sum_{s \in S} \left[\min_{x^s \geq 0} \sum_{ij} d_{ij} x_{ij}^s \right] \tag{2a}$$

$$\text{s.t.}: \quad q_i \in \mathbb{Z}_+ \; \forall i \in G$$

$$\sum_{i \in G} q_i = k \tag{2b}$$

$$\sum_{j \in G} x_{ij}^s = q_i, \; \forall i \in G, \forall s \in S \tag{2c}$$

$$\sum_{i \in G} x_{ij}^s = z_j^s, \; \forall j \in G, \forall s \in S, \tag{2d}$$

$$x_{ij}^s \geq 0 \; \forall i, j \in G \tag{2e}$$

The optimization program leveraging the k-server problem, on the other hand, can be formulated as:

$$\min_q \sum_{s \in S} \left[\min_{x^s \geq 0} \sum_{ij} d_{ij} x_{ij}^s \right] \tag{3a}$$

$$\text{s.t.}: \quad q_i \in \mathbb{Z}_+ \; \forall i \in G \tag{3b}$$

$$\sum_{j \in \{G^2, t\}} x_{ij}^s = q_i \quad \forall i \in G^1, \forall s \in S \tag{3c}$$

$$\sum_{i \in G} x_{ij}^s = z_j^s \quad \forall j \in G^2, \forall s \in S \tag{3d}$$

$$\sum_{j \in G} x_{ij}^s - \sum_{l \in G} x_{li}^s = s_i \quad \forall i \in G, \forall s \in S \text{ where } s_i = \begin{cases} k \text{ if } i = a \\ -k \text{ if } i = t \\ 0 \text{ otherwise} \end{cases} \tag{3e}$$

$$x_{ij}^s \leq 1 \quad \forall i, j \in \{i, j \in G\} \backslash \{\{i, j \in G \text{ and } i = a \text{ and } j \in G^1\}$$
$$\cup \{i, j \in G \text{ and } i \in G^1 \text{ and } j = t\}\}, \forall s \in S \tag{3f}$$

$$x_{ij}^s \geq 0 \; \forall i, j \in G, \forall s \in S \tag{3g}$$

The overall optimization problem then becomes

$$\min_{q \geq 0} \mathbb{E}_{s \sim f} \left[\mathbb{1}(k \geq m_s) \min_{x^s \in C_1^{s(q)}} \sum_{ij} d_{ij} x_{ij}^s + \mathbb{1}(k < m_s) \min_{x^s \in C_2^{s(q)}} \sum_{ij} d_{ij} x_{ij}^s \right] \tag{4}$$

where $C_1^{s(q)}$ includes the Constraints 2c, 2d, and $C_2^{s(q)}$ includes Constraints 3c, 3d and 3e, as well as the capacity constraints, for all realizations of crime incidents s, that are drawn from the distribution f.

We propose to solve this stochastic program using Bender's decomposition [1]. The first step is to represent the inner (lower-level) optimization problems using their duals, which for the transportation problem, is represented as:

$$\max_{\alpha,\beta} \sum_{i \in G} q_i \alpha_i^s + \sum_{j \in G} z_j^s \beta_j^s \tag{5a}$$

$$s.t.: \quad d_{ij} - \alpha_i^s - \beta_j^s \geq 0 \quad \forall i,j \in G, \tag{5b}$$

where $\{\alpha_1^s, ..., \alpha_g^s\}$ are the dual variables for Constraints 2c and $\beta_1^s, ..., \beta_g^1$ are dual variables for Constraints 2d. The dual for the k-server problem is represented as:

$$\max_{\lambda,\delta,f,c} - \sum_{i \in G^1} \lambda_i^s q_i - \sum_{j \in G^2} \delta_j^s z_j - \sum_{i,j \in C_c} c_{ij}^s - \sum_{i \in G} f_i^s s_i \tag{6a}$$

$$s.t. \tag{6b}$$

$$\mathbb{1}(i,j \in C_\lambda)\lambda_i^s + \mathbb{1}(i,j \in C_\delta)\delta_j^s + f_i^s - f_j^s + \mathbb{1}(i,j \in C_c)c_{ij}^s + d_{ij} \geq 0 \quad \forall i,j \in G \tag{6c}$$

where

$$i,j \in C_\lambda \quad \text{if} \quad i,j \in G \quad \text{and} \quad i \in G^1, j \in \{G^2, t\}$$

$$i,j \in C_\delta \quad \text{if} \quad i,j \in G \quad \text{and} \quad i \in G^2$$

$$i,j \in C_c \quad \text{if} \quad i,j \in \{i,j \in G\}\backslash\{\{i,j \in G \text{ and } i = a \text{ and } j \in G^1\}$$

$$\cup\{i,j \in G \text{ and } i \in G^1 \text{ and } j = t\}\}$$

We introduce dual variables $\lambda_i^s, ..., \lambda_k^s$ for constraints 3c, $\delta_i^s, ..., \delta_m^s$ for constraints 3d, $f_i^s, ..., f_n^s$ for constraints 3e and $c_{11}^s, c_{12}^s..., c_{nn}^s$ for constraints 3f.

By construction, the primal transportation problem always has a feasible solution as it is balanced, and the primal k-server problem always has a feasible solution provided $\sum_i q_i > 0$, which is ensured by always having a budget greater than 0. Consequently, there always exists an optimal dual solution which is one of the (finite number of) extreme points of the polyhedron comprised from Constraints 5b and 6c for the corresponding problems. Since these constraints do not depend on the police patrol allocation decisions q, the set of extreme points of the constraint polyhedra $E^s = \{(\lambda^s, \delta^s, f^s, c^s)\}$ and $E^s = \{\alpha^s, \beta^s\}$ for both the problems are independent of q. Thus, we can then rewrite the stochastic program as

$$\min_q \sum_{s \in S} \left[\mathbb{1}(k < m_s)\{ \max_{(\lambda^s, \delta^s, f^s, c^s) \in E^s} - \sum_{i \in G^1} \lambda_i^s q_i - \sum_{j \in G^2} \delta_j^s z_j - \sum_{i,j \in C_c} c_{ij}^s \right.$$

$$\left. - \sum_{i \in G} f_i^s s_i\} + \mathbb{1}(k \geq m_s)\{\max_{\alpha,\beta} \sum_{i \in G} q_i \alpha_i^s + \sum_{j \in G} z_j^s \beta_j^s\} \right] \tag{7}$$

Since E^s is finite, we can rewrite it as

$$\min_{q,u^s} \sum_s u^s \tag{8a}$$

s.t.: $\quad q_i \in \mathbb{Z}_+ \quad \forall\, i \in G$

$$u^s \geq -\sum_{i \in G^1} \lambda_i^s q_i - \sum_{j \in G^2} \delta_j^s z_j - \sum_{i,j \in C_c} c_{ij}^s - \sum_{i \in G} f_i^s s_i \quad \forall\, s, (\lambda^s, \delta^s, f^s, c^s) \in \tilde{E}^s$$

$$\tag{8b}$$

$$u^s \geq \sum_{i \in G} q_i \alpha_i^s + \sum_{j \in G} z_j^s \beta_j^s \quad \forall\, s, (\alpha^s, \beta^s) \in \tilde{E}^s \tag{8c}$$

where $\tilde{E}^s$ is a subset of the extreme points which includes the optimal dual solution and Constraints 8b and 8c are applicable based on whether the particular scenario is mapped to the transportation problem or the k-server problem. Since this subset is initially unknown, Bender's decomposition involves an iterative algorithm starting with empty $\tilde{E}^s$, and iterating solutions to the problem with this subset of constraints (called the *master* problem), while generating and adding constraints to the master using the dual program for each s, until convergence (which is guaranteed since E^s is finite).

A problem remains with the above formulation: if police vehicles significantly outnumber crime events, we only need a few of the available resources to attain a global minimum, and the remaining vehicles are allocated arbitrarily. In practice, this is unsatisfactory, as there are numerous secondary objectives, such as overall crime deterrence, which motivate allocations of police which are geographically diverse. We incorporate these considerations informally into the following heuristic objectives:

- There should be more police coverage in areas that observe more crime, on average, and
- Police should be diversely distributed over the entire coverage area.

We incorporate these secondary objectives by modifying the objective function in (3b) to be

$$\min_q -\gamma h_i q_i + \kappa q_i + \min_{x^s \geq 0} \sum_{s \in S} \sum_{ij} d_{ij} x_{ij}^s \tag{9}$$

where h_i is the observed frequency of crimes in grid i and γ and κ are parameters of our model. The first term $\gamma h_i q_i$ forces the model to place police in high crime grids. The second term κq_i penalizes the placement of too many police vehicles in a grid and thus forces the model to distribute police among grids.

2.3 Iterative Stochastic Programming

Bender's decomposition enables us to solve the stochastic program under the assumption that f is stationary. A key challenge identified above however, is that the distribution of future crime actually depends on the police placement policy q. Consequently, a solution to the stochastic program for a fixed set of samples s from a distribution f is only optimal if this distribution reflects the distribution of crime conditional on q, turning stochastic program into a fixed point problem. We propose to use an iterative algorithm, *IBRO (Iterative Bender's Response Optimization)* (Algorithm 1), to address this issue. Intuitively, the algorithm provides police repeated chances to react to crimes, while updating the distribution of crimes given current police positions. In the algorithm, MAX_ITER is an upper limit on the number of iterations, e is the set of all evidence (features) except police presence and $\tau|z$ is the response time to crime z. q and z, as before, refer to vectors of police placements and crime locations and $q_i|z_i$ refers to police placement given a particular set of crimes.

Algorithm 1. IBRO

1: **INPUT:** q_0: Initial Police Placement
2: **OUTPUT:** q^*: Optimal Police Placement
3: **for** $i = 1..MAX\_ITER$ **do**
4:     Sample Crime z_i from $f(t|e, q_{i-1})$
5:     Find Optimal Police Placement $q_i|z_i$ by Stochastic Programming.
6:     Calculate $\mathbf{E_i}(\tau|z_i)$
7:     **if** $\mathbf{E_i}(\tau|z_i) > \mathbf{E_{i-1}}(\tau|z_{i-1})$ **then**
8:         **Return** q_{i-1}
9:     **end if**
10:     **if** $|\mathbf{E_i}(\tau|z_i) - \mathbf{E_{i-1}}(\tau|z_{i-1})| \leq \epsilon$ **then**
11:         **Return** q_i
12:     **end if**
13: **end for**
14: **Return** q_i

---

3 Crime and Police Data

In order to validate the decision theoretic model above, we used the following data to learn the parametric model of crime described in Sect. 4. We use burglary data from 2009 for Davidson County, TN, a total of 4,627 incidents, which includes coordinates and reported occurrence times. Observations that lacked coordinates were geo-coded from their addresses. In addition, we used police vehicle patrol data for the same county, consisting of GPS dispatches sent by county police vehicles, for a total of 31,481,268 data points, where each point consists of a unique vehicle ID, time, and spatial coordinates. A total of 624 retail shops that sell liquor, 2494 liquor outlets, 41 homeless shelters, and 52 pawn shops were taken into account. We considered weather data collected at the county level. Additional risk-terrain features, included population density, housing density, and mean household income at a census tracts level.

4 Continuous-Time Crime Forecasting

4.1 Model

Crime models commonly fall into three categories: purely spatial models, which identify spatial features of previously observed crime, such as hot spots (or crime clusters), spatial-temporal models which attempt to capture dynamics of attractiveness of a discrete set of locations on a map, and risk-terrain models, which identify key environmental determinants (risk factors) of crime, and create an associated time-independent risk map. A key gap in this prior work is the lack of a spatial-temporal generative model that can capture both spatial and temporal correlates of crime incidents, such as time of day, season, locations of liquor outlets and pawn shops, and numerous others. We propose to learn a density $f(t|w)$ over time to arrival of crimes for a set of discrete spatial locations G, allowing for spatial interdependence, where w is a set of crime co-variates.

A natural choice for this problem is survival analysis [6] which allows us to represent distribution of time to events as a function of arbitrary features. Formally, the survival model is $f_t(t|\gamma(w))$, where f_t is a probability distribution for a continuous random variable T representing the inter-arrival time, which typically depends on covariates w as $\log(\gamma(w)) = \rho_0 + \sum_i \rho_i w_i$. A key component in a survival model is the survival function, which is defined as $S(t) = 1 - F_t(t)$, where $F_t(t)$ is the cumulative distribution function of T. Survival models can be parametric or non-parametric in nature, with parametric models assuming that *survival time* follows a known distribution. In order to model and learn $f(t)$ and consequently $S(t)$, we chose the exponential distribution, which has been widely used to model inter-arrival time to events and has the important property of being memoryless. We use Accelerated Failure Model (AFT) for the survival function over the semi-parametric Cox's proportional hazard model (PHM) and estimate the model coefficients using maximum likelihood estimation (MLE), such that in our setting, $S(t|\gamma(w)) = S(\gamma(w)\,t)$. While both the AFT and PHM models measure the effects of the given covariates, the former measures it with respect to survival time and the latter does so with respect to the hazard. The AFT model thus allows us to offer natural interpretations regarding how covariates affect crime rate.

A potential concern in using survival analysis in this setting is that grids can experience multiple events. We deal with this by learning and interpreting the model in a way that the multiple events in a particular grid are treated as single events from multiple grids and prior events are taken into consideration by updating the temporal and spatial covariates.

In learning the survival model above, there is a range of choices about its spatial granularity, from a single homogeneous model which captures spatial heterogeneity entirely through the model parameters w, to a collection of distinct models f_i for each spatial grid $i \in G$. For a homogeneous model it is crucial to capture most of the spatial variation as model features. Allowing for a collection of distinct models f_i, on the other hand, significantly increases the risk of overfitting, and reduces the ability to capture generalizable spatially-invariant

Table 1. Variables for crime prediction

Type of feature	Sub-type	Variable	Description
Temporal	Temporal cycles	Time of day	Each day was divided into 6 equal time zones with binary features for each.
		Weekend	Binary features to consider whether crime took place on a weekend or not.
		Season	Binary features for winter, spring, summer and fall seasons.
	Weather	Mean temperature	Mean temperature in a day
		Rainfall	Rainfall in a day
		Snowfall	Snowfall in a day
	Effect of police	Police presence	Number of police vehicles passing through a grid and neighboring grids over past 2 h
Spatial	Risk-Terrain	Population density	Population density (Census Tract Level)
		Household income	Mean household income (Census Tract Level)
		Housing density	Housing density (Census Tract Level)
Spatial-Temporal	Spatial correlation	Past crime	Separate variables considered for each discrete crime grid representing the number of crimes in the last two days, past week and past month. We also looked at same crime measures for neighbors of a grid.
	Effect of police	Crime spillover	Number of police vehicles passing in the past two hours through grids that are not immediately adjacent, but farther away

knowledge about crime co-variates. To balance these considerations, we split the discrete spatial areas into two coarse categories: high-crime and low-crime, and learned two distinct homogeneous models for these. We do this by treating the count of crimes for each grid as a data point and then splitting the data into two clusters using *k-means* clustering.

The next step in the modeling process is to identify a collection of features that impact crime incidence, which will comprise the co-variate vector w. In doing this, we divide the features into temporal (those that only change with time), spatial (those capturing spatial heterogeneity), and spatio-temporal (features changing with both time and space).

4.2 Temporal Features

Temporal Crime Cycles: Preliminary analysis and prior work [7,13] were used to identify the set of covariates, such as daily, weekly and seasonal cycles, that affect crime rate. Crime rates have also been shown to depend on seasons (with more crime generally occurring in the summer) [14]. Thus, we consider seasons as binary features. In order to incorporate crime variation throughout the day, each day was divided into six zones of four hours each, captured as binary features. Similarly, another binary feature was used to encode weekdays and weekends.

Temporal Crime Correlation: It has previously been observed that crime exhibits inter-temporal correlation (that is, more recent crime incidents increase thelikelihood of subsequent crime). To capture this aspect, we used recent crime counts in the week and month preceding time under consideration.

Weather: It is known that weather patterns can have a significant effect on crime incidence [5]. Consequently, we included a collection of weather-related features, such as rainfall, snowfall, and mean temperature.

Police Presence: The final class of features that are particularly pertinent to our optimization problem involves the effect of police presence on crimes. Specifically, it is often hypothesized that police presence at or near a location will affect future crime at that location [12]. We try to capture this relationship, by including a feature in the model corresponding to the number of police vehicles passing within the grid, as well as its immediate neighboring grid cells, over the previous two hours.

4.3   Spatial and Spatio-Temporal Features

Risk-Terrain Features: We leveraged the risk-terrain modeling framework [10], as well as domain experts, to develop a collection of spatial features such as population density, mean household income, and housing density at the census tract level. We used the location of pawn shops, homeless shelters, liquor stores, and retail outlets that sell liquor as the observed spatial-temporal variables (note that temporal variation is introduced, for example, as new shops open or close down).

Spatial Crime Correlation: One of the most widely cited features of crime is its spatial correlation (also referred to as *repeat victimization* [11]), a phenomenon commonly captured in hot-spotting or spatial crime clustering techniques. We capture spatial correlation as follows. For each discrete grid cell in the space we first consider the number of crime incidents over the past two days, past week, and past month, as model features, capturing repeat victimization within the same area. In addition, we capture the same features of past crime incidents for neighboring grid cells, capturing spatial correlation.

Spatial Effects of Police Presence: Aside from the temporal effect of police on crime (reducing its frequency at a particular grid cell), there is also a spatial effect. Specifically, in many cases criminals may simply commit crime elsewhere [8]. To capture this effect, we assume that the spillover of crime will occur between relatively nearby grid cells. Consequently, we add features which measure the number of police patrol units over the previous two hours in grid cells that are not immediately adjacent, but are several grid cells apart. In effect, for a grid cell, we hypothesize that cells that are very close push crime away or reduce it, whereas farther away grids spatially shift crime to the concerned grid, causing spillover effects. The list of all the variables is summarized in Table 1.

5 Results

5.1 Experiment Setup

We used python and R to learn the model parameters, with *rpy2* acting as the interface between the two. We make direct comparison of our model to the discrete-time non-parametric Dynamic Bayes Network model [22,23] and the DSDA continuous-time model [9]. We used CPLEX version 12.51 to solve the optimization problem described in Sect. 2. The experiments were run on a 2.4 GHz hyperthreaded 8-core Ubuntu Linux machine with 16 GB RAM.

5.2 Evaluation of Crime Prediction

Our first step is to evaluate the ability of our proposed continuous-time model based on survival analysis to forecast crime. Our parametric model is simpler (in most cases, significantly) than state-of-the-art alternatives, and can be learned using standard maximum likelihood methods for learning survival models. Moreover, it is nearly homogeneous: only two distinct such models are learned, one for low-crime regions, and another for high-crime regions. This offers a significant advantage both in interpretability of the model itself, as well as ease of use. Moreover, because our model incorporates environmental factors, such as locations of pawn shops and liquor stores, it can be naturally adapted to situations in which these change (for example, pawn shops closing down), enabling use in policy decisions besides police patrolling. On the other hand, one may expect that such a model would result in significant degradation in prediction efficacy compared to models which allow low-resolution spatial heterogeneity. As we show below, remarkably, our model actually outperforms alternatives both in terms of prediction efficacy, and, rather dramatically, in terms of running time.

For this evaluation, we divided our data into 3 overlapping datasets, each of 7 months. For each dataset, we used 6 months of data as our training set and 1 month's data as the test set. For spatial discretization, we use square grids of sides 1 mile throughout, creating a total of 900 grids for the entire area under consideration. While our model is continuous-time, we draw a comparison to both a continuous-time and a discrete-time models in prior art. However, since these are not directly comparable, we deal with each separately, starting with the continuous-time DSDA model. We refer to the DSDA model simply as *DSDA*, the model based on a Dynamic Bayes Network is termed *DBN*, and our model is referred to as *PSM* (parametric survival model).

Prediction Effectiveness Comparison with DSDA. Our first experiments involve a direct performance comparison to a state-of-the-art DSDA model due to Ihava et al. [9]. We chose this model for two reasons. First, DSDA provides a platform to make a direct comparison to a continuous time model. Second, it uses time series modeling and CrimeStat, both widely used tools in temporal and spatial crime analysis.

We introduce the underlying concept of the model before comparing our results. DSDA segregates temporal and spatial aspects of crime prediction and learns them separately. In the temporal model, days like Christmas, Halloween, and football match days that are expected to show deviation from the usual crime trend are modeled using hierarchical profiling (HPA) by using the complement of the gamma function:

$$y = a_p - b_p t^{c_p - 1} e^{-d_p t}$$

where y is observed count, t is time and a_p, b_p, c_p and d_p are the parameters to be estimated using ordinary least squares (OLS).

All other days are initially assumed to be part of a usual average weekly crime rate, which is modeled using the following harmonic function

$$y = a_a - b_a t + c_a t^2 + \sum_{i=1}^{26} \left[d_a \cos\left(\frac{i\pi t}{26}\right) + e_a \sin\left(\frac{i\pi t}{26}\right) \right]$$

where y_a is the weekly crime average, t is time and a_a, b_a, c_a, d_a and e_a are the parameters that are estimated using OLS. Then, the deviations are calculated from the observed data and these are again modeled using the harmonic function. This forms the deterministic part of the model $f(t)$. The error Z from the observed data is modeled using seasonal ARIMA, and the final model is $y = f(t) + Z$. The spatial component of DSDA was evaluated using STAC [9], which is now a part of CrimeStat [15].

In order to make a comparative analysis, we considered a natural adaptation of the HPA-STAC model, which enables us to compare likelihoods. We use the outputs (counts of crime) from the HPA model as a mean of a Poisson random variable, and sample the number of crimes from this distribution for each day. For the spatial model, HPA-STAC outputs weighted clusters in the form of standard deviation ellipses, a technique used commonly in crime prediction. Here, we consider that:

$$P(x_i) = P(c(x_i)) P(x_i^{c(x_i)})$$

where $P(x_i)$ is the likelihood of a crime happening at a spatial point x_i which belongs to cluster c_i, $P(c(x_i))$ is the probability of choosing the cluster to which point x_i belongs from the set of all clusters and $P(x_i^{c(x_i)})$ is the probability of choosing point x_i from its cluster c_i. We assume that $P(x_i^{c_i}) \propto \frac{1}{Area_{c(x_i)}}$. Finally, we assume that the total likelihood is proportional to the product of the spatial and temporal likelihoods.

Figure 2 shows the comparison of DSDA log-likelihood (on test data) for the three datasets described above. Indeed, our model outperforms DSDA in both the temporal and the spatial predictions by a large margin (overall, the improvement in log-likelihood is 25–30 %).

Prediction Effectiveness Comparison with the Dynamic Bayes-Network Model. Next, we compare our model to the framework proposed by Zhang et al. [23], which looks at crime prediction by learning a non-parametric

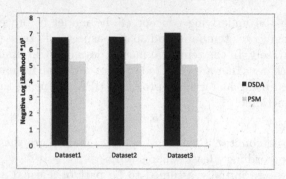

Fig. 2. Likelihood comparison of PSM vs. DSDA.

Dynamic Bayes Network (DBN) representation, and applying abstraction techniques to improve scalability [22]. The DBN includes three sets of state variables: numbers of police vehicles in each grid i at time t, denoted by D_{it}, the number of criminals in grid i at time t, X_{it}, and the number of crimes Y_{it} in each grid i at time t. The main assumptions of this DBN are that (a) police vehicle dynamics are known (so they are not random variables), (b) locations of criminals at time $t+1$ only depends on patrol and criminal (but not crime) locations at time t, and (c) crime incidents at time t only depend on locations of criminals and police at time t. Consequently, the problem involves learning two sets of transition models: $P(X_{i,t+1}|D_{1,t}, ..., D_{N,t}, X_{1,t}, ..., X_{N,t})$ and $P(Y_{i,t}|D_{1,t}, ..., D_{N,t}, X_{1,t}, ..., X_{N,t})$ for all grid cells i, which are assumed to be independent of time t. Since the model involves hidden variables X, Zhang et al. learn it using the Expectation-Maximization framework. While the model is quite general, Zhang et al. treat X, Y, and D as binary.

Since our proposed model is continuous-time, whereas Zhang et al. model is in discrete-time, we transform our model forecasts into a single probability of at least one crime event occurring in the corresponding interval. Specifically, we break time into 8-hour intervals (same temporal discretization as used by Zhang et al.), and derive the conditional likelihood of observed crime as follows. Given our distribution $f(t|w)$ over inter-arrival times of crimes, and a given time interval $[t_1, t_2]$, we calculate the probability of observing a crime in the interval as $F(t \leq t_2|w) - F(t \leq t_1|w)$, where F represents the corresponding cumulative distribution function (cdf).

To draw the most fair comparison to DBN, we use an evaluation metric proposed by Zhang et al. [22] which is referred to as *accuracy*. Accuracy is calculated as a measure of correct predictions made for each grid and each time-step. For example, if the model predicts a probability of crime as 60 % for a target, and the target experiences a crime, then the accuracy is incremented by 0.6. Formally, let p_i be the predicted likelihood of observing a crime count for data point i. Then accuracy is defined as $\frac{1}{m} \sum_i p_i$, where i ranges over the discrete-time sequence of crime counts across time and grids and m the total number of such time-grid items.

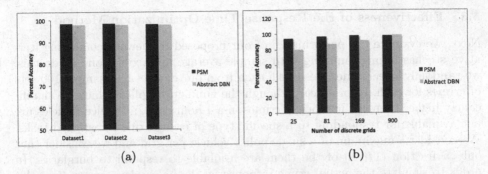

Fig. 3. Accuracy comparison between PSM and Abstract DBN. (a) Varying data subsets. (b) Varying the number of grids.

Figure 3(a) shows the results of accuracy comparison (with the accuracy measure defined above) between the DBN model and our model (PSM). We can observe that both models perform extremely well on the accuracy measure, with our model very slightly outperforming DBN. We also make comparisons by varying the number of grids, shown in Fig. 3(b), starting around downtown Nashville and gradually moving outwards. Our model outperforms DBN in all but one case, in which the accuracies are almost identical.

Runtime Comparison with DSDA and DBN. We already saw that our PSM model, despite its marked simplicity, outperforms two state-of-the-art forecasting models, representing continuous-time and discrete-time prediction methods, in terms of prediction efficacy. An arguably more important technical advantage of PSM over these is running time. Figure 4 shows running times (of training) for PSM, DSDA, and DBN (using the abstraction scheme proposed by Zhang et al. [22]). The DBN framework is significantly slower than both DSDA and PSM. Indeed, PSM running time is so small by comparison to both DSDA and DBN that it is nearly invisible on this plot.

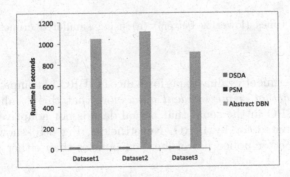

Fig. 4. Runtime comparison (seconds) between DSDA, Abstract DBN, and PSM.

5.3   Effectiveness of the Response Time Optimization Method

Next, we evaluate the performance of our proposed framework combining iterative stochastic programming with sample average approximation. To do this, we randomly select timezones of 4 h each from our dataset and sample 100 sets of crimes for each. In practice, although the number of police vehicles is significantly higher than the number of crimes in a 4-hour zone, all police vehicles are not available for responding to a specific type of crimes, due to assigned tasks. We consider a maximum of a single police vehicle per grid and we consider that only a fraction (1/6*th*) of the them are available to respond to burglaries. In order to simulate the actual crime response by the police department (in order to evaluate actual spatial allocation policy of police vehicles within our data), we greedily assign the closest police vehicle to a crime in consideration.

Our first evaluation uses our crime prediction model f to simulate crime incidents in simulation, which we use to both within the IBRO algorithm, as well as to evaluate (by using a distinct set of samples) the policy produced by our algorithm in comparison with three alternatives: a baseline stochastic programming method (using Bender's decomposition) which ignores the fact that distribution of crimes depends on the police allocation (*STOCH-PRO*), (b) actual police location in the data (*Actual*), and (c) randomly assigning police vehicles to grids (*Random*). Figure 5(a) demonstrates that IBRO systematically outperforms these alternatives, usually by a significant margin.

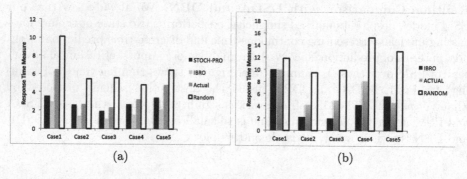

Fig. 5. Response times (lower is better): (a) using simulated crimes, (b) observed crimes.

Our next experiment evaluates performance of IBRO in comparison to others with respect to *actual crime incident data*. Note that this is inherently disadvantageous to IBRO in the sense that actual data is not adaptive to the police location as accounted for by IBRO. Nevertheless, Fig. 5(b) shows that IBRO typically yields better police patrol location policies than either actual (in the data) or random.

6   Conclusion

We develop a novel bi-level optimization method for allocating police patrols in order to minimize expected crime incident response time. Our approach makes use of stochastic programming, with a Bender's decomposition and constraint generation framework offering a scalable solution approach. Moreover, we introduce a novel iterative stochastic programming algorithm which allows us to account for the dependence of the spatio-temporal crime incidence distribution on police location. To evaluate this optimization framework, we presented a novel discrete-space continuous-time model for forecasting crime as a function of a collection of co-variates which include vehicular police deployment. Our model, which makes use of survival analysis, allows for spatial as well as temporal crime correlation, and effectively captures the effect of police presence both temporally and spatially. This model is learned from burglary incident data in a major US metropolitan area. Our experiments demonstrate that this model outperforms state of the art continuous- and discrete-time crime prediction models both in terms of prediction effectiveness and running time.

Acknowledgments. This research was partially supported by the NSF (IIS-1526860), ONR (N00014 15 1 2621), ARO (W011NF 16 1 0069), ARO MURI (W911NF-111-0332), and Vanderbilt University.

References

1. Bertsimas, D., Tsitsiklis, J.N.: Linear Optimization, 3rd edn. Athena Scientific, Belmont (1997)
2. Brantingham, P.J., Brantingham, P.L.: Patterns in Crime. Macmillan, New York (1984)
3. Chrobak, M., Karloof, H., Payne, T., Vishwnathan, S.: New results on server problems. SIAM J. Discrete Math. **4**(2), 172–181 (1991)
4. Cohen, J., Gorr, W.L., Olligschlaeger, A.M.: Leading indicators and spatial interactions: a crime-forecasting model for proactive police deployment. Geogr. Anal. **39**(1), 105–127 (2007)
5. Cohn, E.G.: Weather and crime. Br. J. Criminol. **30**(1), 51–64 (1990)
6. Cox, D.R., Oakes, D.: Analysis of Survival Data, vol. 21. CRC Press, Cleveland (1984)
7. Felson, M., Poulsen, E.: Simple indicators of crime by time of day. Int. J. Forecast. **19**(4), 595–601 (2003)
8. Hope, T.: Problem-oriented policing and drug market locations: three case studies. Crime Prev. Stud. **2**(1), 5–32 (1994)
9. Ivaha, C., Al-Madfai, H., Higgs, G., Ware, J.A.: The dynamic spatial disaggregation approach: a spatio-temporal modelling of crime. In: World Congress on Engineering, pp. 961–966 (2007)
10. Kennedy, L.W., Caplan, J.M., Piza, E.: Risk clusters, hotspots, and spatial intelligence: risk terrain modeling as an algorithm for police resource allocation strategies. J. Quant. Criminol. **27**(3), 339–362 (2011)
11. Kleemans, E.R.: Repeat burglary victimisation: results of empirical research in the Netherlands. Crime Prev. Stud. **12**, 53–68 (2001)

12. Koper, C.S.: Just enough police presence: reducing crime and disorderly behavior by optimizing patrol time in crime hot spots. Justice Q. **12**(4), 649–672 (1995)
13. Landau, S.F., Fridman, D.: The seasonality of violent crime: the case of robbery and homicide in Israel. J. Res. Crime Delinquency **30**(2), 163–191 (1993)
14. Lauritsen, J.L., White, N.: Seasonal Patterns in Criminal Victimization Trends, vol. 245959. US DOJ, Office of Justice Program, Bureau of Justice Statistics (2014)
15. Levine, N., et al.: Crimestat III: a spatial statistics program for the analysis of crime incident locations (version 3.0). Houston (TX): Ned Levine & Associates/Washington, DC: National Institute of Justice (2004)
16. Mohler, G.O., Short, M.B., Brantingham, P.J., Schoenberg, F.P., Tita, G.E.: Self-exciting point process modeling of crime. J. Am. Stat. Assoc. **106**(493), 100–108 (2011)
17. Murray, A.T., McGuffog, I., Western, J.S., Mullins, P.: Exploratory spatial data analysis techniques for examining urban crime implications for evaluating treatment. Br. J. Criminol. **41**(2), 309–329 (2001)
18. Short, M.B., D'Orsogna, M.R., Pasour, V.B., Tita, G., Brantingham, P.J., Bertozzi, A.L., Chayes, L.B.: A statistical model of criminal behavior. Math. Models Methods Appl. Sci. **18**, 1249–1267 (2008)
19. Short, M.B., D'Orsogna, M.R., Pasour, V.B., Tita, G.E., Brantingham, P.J., Bertozzi, A.L., Chayes, L.B.: A statistical model of criminal behavior. Math. Models Methods Appl. Sci. **18**(supp01), 1249–1267 (2008)
20. Speer, P.W., Gorman, D.M., Labouvie, E.W., Ontkush, M.J.: Violent crime and alcohol availability: relationships in an urban community. J. Public Health Policy **19**(3), 303–318 (1998)
21. Toomey, T.L., Erickson, D.J., Carlin, B.P., Quick, H.S., Harwood, E.M., Lenk, K.M., Ecklund, A.M.: Is the density of alcohol establishments related to nonviolent crime? J. Stud. Alcohol Drugs **73**(1), 21–25 (2012)
22. Zhang, C., Bucarey, V., Mukhopadhyay, A., Sinha, A., Qian, Y., Vorobeychik, Y., Tambe, M.: Using abstractions to solve opportunistic crime security games at scale. In: International Conference on Autonomous Agents and Multiagent Systems, 196–204. ACM (2016)
23. Zhang, C., Sinha, A., Tambe, M.: Keeping pace with criminals: designing patrol allocation against adaptive opportunistic criminals. In: International Conference on Autonomous Agents and Multiagent Systems. pp. 1351–1359 (2015)

A Methodology to Apply a Game Theoretic Model of Security Risks Interdependencies Between ICT and Electric Infrastructures

Ziad Ismail[1]([✉]), Jean Leneutre[1], David Bateman[2], and Lin Chen[3]

[1] Télécom ParisTech, Université Paris-Saclay, 46 rue Barrault, 75013 Paris, France
{ismail.ziad,jean.leneutre}@telecom-paristech.fr
[2] EDF, 1 Place Pleyel, 93282 Saint-Denis, France
david.bateman@edf.fr
[3] University of Paris-Sud 11, 15 Rue Georges Clemenceau, 91400 Orsay, France
lin.chen@lri.fr

Abstract. In the last decade, the power grid has increasingly relied on the communication infrastructure for the management and control of grid operations. In a previous work, we proposed an analytical model for identifying and hardening the most critical communication equipment used in the power system. Using non-cooperative game theory, we modeled the interactions between an attacker and a defender and derived the minimum defense resources required and the optimal strategy of the defender that minimizes the risk on the power system. In this paper, we aim at validating the model using data derived from real-world existing systems. In particular, we propose a methodology to assess the values of the parameters used in the analytical model to evaluate the impact of equipment failures in the power system and attacks in the communication infrastructure. Using this methodology, we then validate our model via a case study based on the polish electric power transmission system.

Keywords: Cyber-physical system · Non-cooperative game theory · SCADA security

1 Introduction

The power grid stands as one of the most important critical infrastructures on which depends an array of services. It uses a Supervisory Control and Data Acquisition (SCADA) system to monitor and control electric equipment. Traditionally, the reliability of the power grid and the security of the ICT infrastructure are assessed independently using different methodologies, for instance [1] and [2] respectively for electric and ICT infrastructures. More recently, a growing body of research has been dedicated to the modeling of interdependencies in critical infrastructures, focusing in particular on communication and electric systems. For example, Laprie et al. [3] proposed a qualitative model to address cascading, escalating, and common cause failures due to interdependencies between these infrastructures. In the case of quantitative models, we can

© Springer International Publishing AG 2016
Q. Zhu et al. (Eds.): GameSec 2016, LNCS 9996, pp. 159–171, 2016.
DOI: 10.1007/978-3-319-47413-7_10

distinguish two main categories: analytical-based and simulation-based models. In the first category of models, we find the work of Buldyrev et al. [4] in which a theoretical framework was developed to study the process of cascading failures in interdependent networks caused by random initial failures of nodes. In simulation-based models, the main techniques used include agent-based [5], petri nets [6] and co-simulation [7].

In complex interdependent systems, the interactions between the attacker and the defender play an important role in defining the optimal defense strategy. In this context, game theory offers a mathematical framework to study interactions between different players with the same or conflicting interests. For example, Law et al. [8] investigate false data injection attacks on the power grid and formulate the problem as a stochastic security game between an attacker and a defender. Amin et al. [9] present a framework to assess risks to cyber-physical systems when interdependencies between information and physical systems may result in correlated failures.

In [10], we proposed an analytical model based on game theory for optimizing the distribution of defense resources on communication equipment taking into account the interdependencies between electric and communication infrastructures. Due to the abstract nature of such analytical models, assessing their relevance in real-world scenarios is a challenging task. In this paper, we propose a methodology for assessing the values of the parameters in the analytical model related to the electric and communication infrastructures, and validate our approach on a case study based on the polish electric transmission system. Throughout the paper, the communication system refers to the telecommunication infrastructure responsible of controlling and monitoring the electrical system.

The paper is organized as follows. In Sect. 2, we present a slight adaptation of the analytical model presented in [10]. In Sect. 3, we propose an approach to evaluate the values of a number of parameters used in the analytical model. In Sect. 4, we validate our model via a case study based on the polish electric power transmission system. Finally, we conclude the paper in Sect. 5.

2   A Game Theoretical Model for Security Risk Management of Interdependent ICT and Electric Systems

In this section, we briefly recall our analytical model for identifying critical communication equipment used to control the power grid that must be hardened. The proofs are omitted, and we refer the reader to [10] for complete details.

2.1   Interdependency Model

We refer by initial risk, the risk on a node before the impact of an accident or an attack propagates between system nodes. We will denote by $r_i^e(0)$ and $r_j^c(0)$ the initial risk on electrical node i and communication equipment j respectively.

We assume that initial risk on a system node is a nonnegative real number and has been evaluated using risk assessment methods.

We use the framework proposed in [11] as a basis to represent the risk dependencies using a graph-theoretic approach. We model the interdependency between the electrical and the communication infrastructures as a weighted directed interdependency graph $\mathcal{D} = (V, E, f)$, where $V = \{v_1, v_2, ..., v_N\}$ is a finite set of vertices representing the set of electrical and communication nodes, E is a particular subset of V^2 and referred to as the edges of $\mathcal{D}$, and $f : E \rightarrow \mathbb{R}^+$ is a function where $f(e_{ij})$ refers to the weight associated with the edge e_{ij}.

Let $V = \{\mathcal{T}^e, \mathcal{T}^c\}$ where $\mathcal{T}^e - \{v_1, v_2, ..., v_{N_e}\}$ represents the set of electrical nodes in the grid and $\mathcal{T}^c = \{v_{N_e+1}, v_{N_e+2}, ..., v_{N_e+N_c}\}$ represents the set of communication nodes. Let $\mathcal{D}$ be represented by the weighted adjacency matrix $M = [m_{ij}]_{N \times N}$ defined as follows:

$$M = \begin{pmatrix} B & D \\ F & S \end{pmatrix}$$

where $B = [b_{ij}]_{N_e \times N_e}$, $D = [d_{ij}]_{N_e \times N_c}$, $F = [f_{ij}]_{N_c \times N_e}$, and $S = [s_{ij}]_{N_c \times N_c}$. Matrix M represents the effects of nodes on each other and is a block matrix composed of matrices B, D, F and S. Elements of these matrices are nonnegative real numbers. Without loss of generality, we assume that these matrices are left stochastic matrices. Therefore, for each node k, we evaluate the weight of other nodes to impact node k. For example, matrices B and S represent the dependency between electrical nodes and communication nodes respectively.

2.2   Risk Diffusion and Equilibrium

We consider that the first cascading effects of an attack on communication equipment take place in the communication infrastructure itself. We introduce a metric t_c in the communication system that refers to the average time for the impact of an attack on communication equipment to propagate in the communication infrastructure. In this model, as opposed to our model in [10], we do not consider the average time t_e in the electrical system that refers to the average time elapsed between the failure of a set of electric equipment and the response time of safety measures or operators manual intervention to contain the failures and prevent them from propagating to the entire grid.

Let $R^e(t) = [r_i^e(t)]_{N_e \times 1}$ and $R^c(t) = [r_i^c(t)]_{N_c \times 1}$ be the electrical and communication nodes risk vectors at time t respectively. We take discrete time steps to describe the evolution of the system. Let $S^l = [s_{ij}^l]_{N_c \times N_c}$ be the l-th power of the matrix S. At attack step r, the payoff is decreased by a factor of γ_c^r. In fact, we consider that each action of the attacker in the system increases the probability of him being detected. Let the matrix $S^{max} = [s_{ij}^{max}]_{N_c \times N_c}$ represents the maximum impact of an attack on communication equipment to reach communication nodes during time t_c, where $s_{ij}^{max} = \max_{l=1,...,\lfloor t_c \rfloor} \gamma_c^l s_{ij}^l$. Let S_n^{max} be the normalized matrices of S^{max} with respect to their rows s.t. $\forall j, \sum_i s_{n\,ij}^{max} = 1$.

We take a similar approach to [11] by balancing the immediate risk and the future induced one. Let β and τ refer to the weight of the initial risk on communication nodes and the weight of the diffused risk from electric nodes to communication nodes at time $t = 0$ respectively, and δ the weight of future cascading risk w.r.t. the value of the total risk on communication nodes. We can prove that the iterative system of the cascading risk converges and an equilibrium solution exists whenever $\delta < 1$ and is given by $R^{c*} = (I - \delta H)^{-1}(\beta R^c(0) + \tau D^T R^e(0))$, where $H = S_n^{max} FBD$, and β, τ, and δ are nonnegative real numbers and $\beta + \tau + \delta = 1$.

2.3 Security Game

We formulate the problem as a non-cooperative game and analyze the behavior of the attacker and the defender at the Nash equilibrium (NE), in which none of the players has an incentive to deviate unilaterally. The attacker's/defender's objective is to distribute attack/defense resources on the communication nodes in order to maximize/minimize the impact of attacks on the power system. We consider the worst-case scenario where both players have complete knowledge of the architecture of the system.

We associate for each communication equipment, a load l_i that represents the amount of computational work the equipment performs. Let $L = diag(l_i)_{N_c \times N_c}$ be the load matrix. Let $W = [w_{ij}]_{N_c \times N_c}$ be the redundancy matrix where $\forall i$, $w_{ii} = -1$ and $\sum_{j,j \neq i} w_{ij} \leq 1$. If $i \neq j$, w_{ij} represents the fraction of the load of node i, node j will be responsible of processing when node i is compromised.

The utility U_a and U_d of the attacker and the defender respectively are as follows:

$$U_a(p,q) = pR_D^{c*}(e^T - q^T) - pR_D^c(0)C^a p^T - \psi pL(Wq^T - I(e^T - 2q^T))$$

$$U_d(p,q) = -pR_D^{c*}(e^T - q^T) - qR_D^c(0)C^d q^T + \psi pL(Wq^T - I(e^T - 2q^T))$$

where $p = [p_i]_{1 \times N_c}$ refers to the attacker's strategy where $0 \leq p_i \leq 1$ is the attack resource allocated to target $i \in \mathcal{T}^c$, $q = [q_j]_{1 \times N_c}$ refers to the defender's strategy where $0 \leq q_j \leq 1$ is the defense resource allocated to target $j \in \mathcal{T}^c$, $R_D^c(0)$, R_D^{c*}, C^a and C^d are diagonal matrices and C^a and C^d refer to the cost of attacking and defending communication nodes respectively, I is the identity matrix, and $e = (1, ..., 1)_{1 \times N_c}$.

The players' utilities are composed of three parts: the payoff of an attack, the cost of attacking/defending, and the impact of redundant equipment in ensuring the control of the power system when a set of communication nodes is compromised. $\psi \in [0, 1]$ is a function of the probability that backup equipment are able to take charge of the load of compromised communication equipment.

We analyze the interactions between the attacker and the defender as a one-shot game [12] in which players take their decisions at the same time.

Theorem 1. *A unique NE of the one-shot game exists and is given by:*

$$q^* = \frac{1}{2}e(R_D^{c*} + \psi L)(R_D^c(0)C^a)^{-1}M[\frac{1}{2}M^T(R_D^c(0)C^a)^{-1}M + 2R_D^c(0)C^d]^{-1}$$

$$p^* = e(R_D^{c*} + \psi L)[\frac{1}{2}M(R_D^c(0)C^d)^{-1}M^T + 2R_D^c(0)C^a]^{-1}$$

where $M = R_D^{c*} + \psi L(W + 2I)$

We also analyze the interactions between players as a Stackelberg game [12]. In our case, the defender is the leader who tries to secure communication equipment in order to best protect the power system. We have the following theorem:

Theorem 2. *The game admits a unique Stackelberg equilibrium* (p^S, q^S) *given by:*

$$q^S = e(R_D^{c*} + \psi L)(R_D^c(0)C^a)^{-1}M(Q + 2R_D^c(0)C^d)^{-1}$$

$$p^S = \frac{1}{2}e(R_D^{c*} + \psi L)(R_D^c(0)C^a)^{-1}[I - M(Q + 2R_D^c(0)C^d)^{-1}M^T(R_D^c(0)C^a)^{-1}]$$

where $Q = M^T(R_D^c(0)C^a)^{-1}M$

3   Parameters Evaluation

In this section, we present our approach to assess the impact of attacks in the electric and communication infrastructures, and therefore evaluate matrices B and S respectively. While the problem of the assessment of the other parameters of the model remains, we discuss at the end of this section potential avenues for their evaluation.

3.1   Evaluation of Matrix B

We assess the impact of cascading failures in the power grid by solving power flow equations using the DC power flow approximation [13]. Following a similar approach as in [14], we simulate individual failures and assess their impact on the power grid such as identifying generators with insufficient capacities to meet the demand and overloaded lines.

In our model, we analyze the impact of tripping transmission lines or loosing generators on the power grid. The flowchart diagram in Fig. 1 shows the cascading algorithm used in our model to analyze the impact of tripping transmission lines. In general, this could have a significant impact on the power grid and could lead to the formation of islands in the electric system. In our algorithm, we shut down islands where the demand (denoted as d in Fig. 1) exceeds the maximum generation capacity in the island (denoted as $max(g)$ in Fig. 1). We then solve the DC power flow problem in the electric transmission system using MATPOWER [15] and check the existence of overloaded lines. These lines are tripped and the process is repeated until a balanced solution emerges. Similarly, we assess the impact of loosing generators on the power grid.

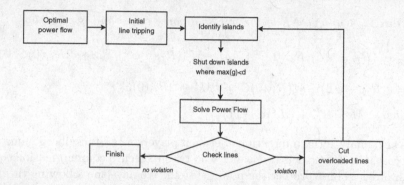

Fig. 1. Flowchart of the cascade algorithm in the case of tripped transmission lines

In our approach, we consider the worst-case scenario where load shedding is not an option when we conduct our analysis of the impact of cascading failures on the power grid. Further work taking into account more fine grained analysis of the behavior of the power grid will allow us to quantify more precisely the values of elements of matrix B.

3.2   Evaluation of Matrix S

To address the challenge of evaluating the impact of cyber attacks on the communication infrastructure, attack graphs [16] are a promising solution to generate all possible attack steps to compromise a target node. These graphs could be used in conjunction with risk assessment methods to evaluate the impact of each attacker action on the communication infrastructure.

Let $\mathcal{G} = (\mathcal{X}, \mathcal{E})$ be an attack graph where $\mathcal{X}$ refers to the set of nodes in the graph and $\mathcal{E}$ refers to the set of edges. In our case, a node $x \in \mathcal{X}$ in the graph refers to a state of the attacker in the system, and an edge $e = (x_i, x_j) \in \mathcal{E}$ refers to an action executed by the attacker after which the state of the attacker in the system transits from x_i to x_j. A state of the attacker refers to his knowledge at a particular time of the topology and the configuration of the system, the set of access levels acquired on equipment, and the set of credentials at his disposal. $\mathcal{G}$ represents all attack paths that can be used by the attacker to compromise a set of equipment or services in the system. In [17], we defined such graph and implemented a proof of concept for constructing it.

Let θ_{lm}^r be the number of paths of length r an attacker can use to compromise communication equipment m from communication equipment l. Let $\Theta_{lm} = \sum_r \gamma_c^r \theta_{lm}^r$ refer to the impact metric of a communication equipment l on a communication node m. Θ_{lm} is a measure of the cumulated impact on communication equipment m of an attack originating from equipment l. We consider that each action of the attacker in the system increases the probability of him being detected. Therefore, at attack step r, the payoff is decreased by a factor of γ_c^r representing the uncertainty for the attacker of getting the payoff of the rth future attack step. In this case, $s_{lm} = \dfrac{\Theta_{lm}}{\sum\limits_i \Theta_{im}}$, where $S = [s_{lm}]_{N_c \times N_c}$.

3.3   Other Parameters

In our case study, we rely on experts' knowledge to evaluate matrices D and F, which represent the dependency relation on communication nodes by electric nodes and vice versa respectively. However, at the end of the case study in the next section, we conduct a sensitivity analysis to evaluate errors in the outputs of our model to estimation errors on the values of the elements of matrix F.

   In our model, we introduced parameters β and τ, which represent the weight of the initial risk on communication nodes and the weight of the diffused risk from electric equipment to communication equipment at time $t = 0$ respectively, and δ which reflects the weight of future cascading risk w.r.t. the value of the total risk on communication equipment. These parameters can be evaluated as a result of the application of a risk assessment method coupled with quantitative metrics derived from the attack graph of the communication infrastructure. In fact, depending on the assessment of the efficiency of deployed defense mechanisms in thwarting threats, the value of β and τ w.r.t. δ can be adjusted. In particular, by analyzing the attack graph, we can evaluate the probability of compromising critical communication equipment given existing defense measures in the system.

4   Case Study

In this section, we validate our model on a case study based on the dataset of the polish electric transmission system at a peak load in the summer of 2004 provided in the MATPOWER computational package [15]. The dataset consists of 420 generators and 3504 transmission lines. The analysis of an electric system at a peak load is important, as it allows us to assess the maximum impact on the power grid as a result of a cyber attack.

4.1   System Architecture

We made a number of assumptions on the architecture of the communication infrastructure that we use in our case study to assess the impact of attacks on the power grid. In addition, to simplify our analysis, we combined a set of communication equipment in a single communication node depending on their functions, thus reducing the number of nodes to be represented in each electric transmission system control center. Let $\mathcal{Y}$ represent the polish electric transmission system. We assume that $\mathcal{Y}$ is controlled by 10 TSO (Transmission System Operator) control centers. Each center controls 42 generators and about 350 transmission lines in a specific area of the power grid. We assume that communication equipment in control centers are vulnerable to attacks, and the attacker has enough resources and both players know the architecture of the system. As we study the impact of attacks on the power grid in the worst-case scenario, this assumption holds. A unique TSO ICT control center is introduced to manage all communication equipment in TSO control centers.

TSO ICT Control Center. In the TSO ICT control center, four types of communication equipment are represented. A Time Server synchronizes the clocks in all communication equipment. A Domain and Directory Service manages access controls on communication equipment. The Remote Access Application is used by ICT administrators to access equipment remotely via secured connections. Finally, the Configuration Management System is responsible of pushing OS and software updates to equipment. Updates can be installed automatically or require specific authorizations on equipment performing critical operations.

TSO Area Control Centers. We represent four types of communication equipment in each TSO area control center: a SCADA HMI, a SCADA server, a SCADA frontend and a SCADA historian. The SCADA HMI is a human-machine interface that provides a graphics-based visualization of the controlled area of the power system. The SCADA server is responsible of processing data collected from sensors in the power grid and sending appropriate control commands back to electric nodes. The SCADA frontend is an interface between the SCADA server and electric nodes control equipment. It formats data in order to be sent through communication channels and to be interpreted when received by control equipment and vice versa. Finally, the SCADA historian is a database that records power state events.

Impact Matrix. We use the algorithm presented in the previous section to assess the impact of stopping generators or tripping transmission lines on the electric transmission system and compute matrix B. We rely on experts' knowledge to evaluate matrices F and D. In the communication infrastructure, we consider that each equipment in a TSO control center is also the backup of an equipment in another TSO control center.

In this case study, we assume that the values of the initial risk on communication equipment have been computed, and for each communication equipment, the cost to defend is always greater than the cost to attack. We fix $\beta = 0.4$, $\tau = 0$, $\delta = 0.6$, and $\psi = 0.5$. Therefore, the future cascading risk has more weight than initial risk w.r.t. the value of the total risk on communication equipment.

4.2   Results

Figure 2 shows the value of risk on communication equipment in each TSO area control center after the impact of attacks propagates in the interdependent communication and electric infrastructures. We can notice that the highest risk values in TSO control centers are on SCADA servers. In particular, risk values on SCADA servers in TSO 1 and TSO 2 control centers are significantly higher than risk values on SCADA servers in the other TSO control centers.

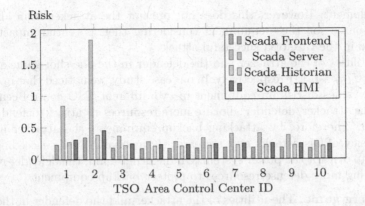

Fig. 2. Risk on communication equipment in TSO area control centers

Table 1 presents the results of the one-shot and Stackelberg games between the attacker and the defender for the TSO ICT and TSO area 1 and area 2 control centers.

Table 1. Nash Equilibrium

		r_i^{c*}	One-Shot game		Stackelberg game	
			p^*	q^*	p^S	q^S
TSO ICT	Time Server	2.547	0.287	0.972	0.146	0.986
	Domain Server	2.885	0.183	0.972	0.093	0.986
	Remote App.	2.089	0.202	0.966	0.103	0.9823
	Config. Manag.	3.073	0.21	0.985	0.106	0.992
TSO 1	SCADA Fontend	0.226	0.275	0.537	0.15	0.591
	SCADA Server	0.844	0.295	0.688	0.156	0.744
	SCADA Historian	0.266	0.315	0.515	0.177	0.584
	SCADA HMI	0.305	0.329	0.51	0.187	0.586
TSO 2	SCADA Fontend	0.339	0.302	0.648	0.162	0.697
	SCADA Server	1.888	0.213	0.895	0.108	0.909
	SCADA Historian	0.379	0.344	0.618	0.189	0.684
	SCADA HMI	0.451	0.358	0.631	0.197	0.7

One-Shot game. From Fig. 2 and Table 1, we notice that the Time, Configuration and Domain Servers have the highest risk values. These equipment are often connected to the internet which significantly increases their attack surface. In addition, given their functions, compromising these equipment could lead to important disruptions in the communication infrastructure. As a result, at equilibrium, the defender allocates a large amount of defense resources to protect these equipment. However, this does not prevent the attacker from allocating attack resources on these equipment considering their potential impact on the power grid in the case of a successful attack.

The utilities of the attacker and the defender in the one-shot game are $U_a = 0.941$ and $U_d = -6.151$ respectively. In our case study, we noticed that in the case where the values of risk on equipment in two different TSO control centers are similar, the attacker/defender allocate more resources to attack/defend backup equipment. Therefore, by attacking backup equipment, the attacker improves the efficiency of his attacks and increases the probability of succeeding in his attempts to disrupt the power system. On the other hand, the defender responds by allocating more defense resources to protect backup equipment.

Stackelberg game. The utilities of the attacker and the defender in the Stackelberg game are $U_a^S = 0.307$ and $U_d^S = -5.746$ respectively. Compared to the one-shot game, the defender allocates more defense resources on each communication equipment, which forces the attacker to reduce his attack resources on these equipment. In fact, an additional security investment by the defender by 2.908 reduced the attacker's allocated resources by 6.082. As a result, from the point of view of the defender, the benefits of operating at the Stackelberg equilibrium outweigh the additional cost of increasing security investments on communication equipment.

Impact of redundancies. Figure 3 shows the variation of total attack and defense resources w.r.t. the weight of the existence of redundancies in players' utility functions ψ. We notice that ψ has a negative effect on the total amount of resources allocated by the attacker. This is consistent with the fact that increasing the weight of redundancies in player's utilities leaves the attacker with fewer choices to achieve a better payoff since the defender will increase the protection of backup equipment. In addition, we notice that when the value of ψ increases, the difference between the one-shot and Stackelberg games total defense resources allocation decreases.

Figure 4 shows the variation of the attacker and the defender strategies on two communication equipment in TSO area 2 control center w.r.t. variation of elements of the redundancy matrix W. We analyze the behavior of the attacker and the defender when varying elements w_{ij}, the fraction of the load of node i, node j will be responsible of processing when node i is compromised. We notice that the behavior of the attacker and the defender depends on the type of the communication equipment. For example, the behavior of both players does not change significantly with respect to W for critical equipment such as the SCADA server. However, this behavior is different for the other equipment in TSO area 2 control center. Finally, increasing w_{ij} leads both the attacker and the defender to decrease their attack and defense resources on communication equipment.

Sensitivity Analysis. We conducted a sensitivity analysis of the diffused risk R^{c*}, the NE in the one-shot game, and the Stackelberg equilibrium w.r.t. the values of the initial risk $R^c(0)$ and the elements of matrices S and F. We averaged the results of 10000 iterations. At each iteration, we assume that a random

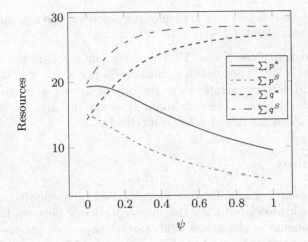

Fig. 3. Variation of attack and defense resources w.r.t. ψ

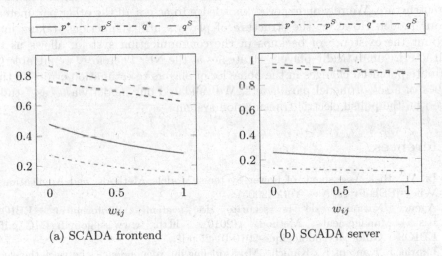

(a) SCADA frontend   (b) SCADA server

Fig. 4. Variation of attack and defense resources on TSO 2 w.r.t. redundancy matrix W

number of elements of $R^c(0)$ deviate from their correct values by $\pm 10\%$ (sign of the deviation is chosen randomly). We repeat the experiment taking into account errors in a random number of elements in matrices S and F.

Sensitivity to $\mathbf{R^c(0)}$. The maximum error on the values of R^{c*} was around 4%. The attacker strategy seems more sensitive than the defender strategy with respect to errors in $R^c(0)$ at equilibrium. In the one-shot game, the maximum error on the attacker strategy was about 4.1% whereas the error on the defender strategy was about 2.1%. However, in the Stackelberg game, we noticed that the maximum error on the attacker strategy has increased compared to the one-shot game and was about 5.1 %. In the case of the defender, the maximum error has decreased and was about 1.2 %.

Sensitivity to matrices S and F. The maximum error on the values of R^{c*} was around 3.4 %. We do not note a significant change in the maximum errors on the attacker and defender strategies in the case of the one-shot game compared to the Stackelberg game. The maximum error on the attacker and defender strategies was about 2.1 % and 1.3 % respectively.

5   Conclusion

In [10], we presented a quantitative model, based on game-theoretic analysis, to assess the risk associated with the interdependency between the cyber and physical components in the power grid. In this paper, we proposed a method to evaluate the values of parameters used in our model to assess the impact of equipment failures in the power system and attacks in the communication infrastructure. We rely on experts' knowledge to assess all the other parameters of our model. However, the structure of player's utility functions, taking into account the existence of backups in the communication system, allows us to characterize analytically players' strategies at the NE. Therefore, we are able to evaluate potential changes in the behavior of players to estimation errors on the values of a set of model parameters. We validated our model via a case study based on the polish electric transmission system.

References

1. Li, W.: Risk Assessment of Power Systems: Models, Methods, and Applications. Wiley-IEEE Press, New York (2005)
2. Agence Nationale de la sécurité des systèmes d'information. EBIOS Risk Management Method (2010). http://www.ssi.gouv.fr/IMG/pdf/EBIOS-1-GuideMethodologique-2010-01-25.pdf
3. Laprie, J., Kanoun, K., Kaniche, M.: Modeling interdependencies between the electricity and information infrastructures. In: SAFECOMP, pp. 54–67 (2007)
4. Buldyrev, S., Parshani, R., Paul, G., Stanley, H.E., Havlin, S.: Catastrophic cascade of failures in interdependent networks. Nature **464**, 1025–1028 (2010)
5. Casalicchio, E., Galli, E., Tucci, S.: Federated agent-based modeling and simulation approach to study interdependencies in it critical infrastructures. In: IEEE 11th International Symposium on Distributed Simulation and Real-Time Applications, pp. 182–189 (2007)
6. Chen, T., Sanchez-Aarnoutse, J., Buford, J.: Petri net modeling of cyber-physical attacks on smart grid. IEEE Trans. Smart Grid **2**(4), 741–749 (2011)
7. Lin, H., Veda, S.S., Shukla, S.K., Mili, L., Thorp, J.S.: GECO: global event-driven co-simulation framework for interconnected power system and communication network. IEEE Trans. Smart Grid **3**(3), 1444–1456 (2012)
8. Law, Y.W., Alpcan, T., Palaniswami, M.: Security games for voltage control in smart grid. In: 50th Annual Allerton Conference on Communication, Control, and Computing (Allerton), pp. 212–219 (2012)

9. Amin, S., Schwartz, G., Hussain, A.: In quest of benchmarking security risks to cyber-physical systems. IEEE Netw. **27**(1), 19–24 (2013)
10. Ismail, Z., Leneutre, J., Bateman, D., Chen, L.: A game-theoretical model for security risk management of interdependent ict and electrical infrastructures. In: IEEE 16th International Symposium on High Assurance Systems Engineering (HASE), pp. 101–109 (2015)
11. Alpcan, T., Bambos, N.: Modeling dependencies in security risk management. In: Proceedings of the 4th International Conference on Risks and Security of Internet and Systems (Crisis) (2009)
12. Osborne, M.J., Rubinstein, A.: A Course in Game Theory. MIT Press, Cambridge (1994)
13. Zhu, J.: Optimization of Power System Operation. Wiley-IEEE Press, Hoboken (2009)
14. Pfitzner, R., Turitsyn, K., Chertkov, M.: Statistical classification of cascading failures in power grids. In: 2011 IEEE Power and Energy Society General Meeting, pp. 1–8 (2011)
15. Zimmerman, R., Murillo-Snchez, C., Thomas, R.: Matpower: steady-state operations, planning, and analysis tools for power systems research and education. IEEE Trans. Power Syst. **26**(1), 12–19 (2011)
16. Ingols, K., Lippmann, R., Piwowarski, K.: Practical attack graph generation for network defense. In: 22nd Annual Computer Security Applications Conference (ACSAC), pp. 121–130 (2006)
17. Ismail, Z., Leneutre, J., Fourati, A.: An attack execution model for industrial control systems security assessment. In: Proceedings of the First Conference on Cybersecurity of Industrial Control Systems (CyberICS) (2015)

Decision Making for Privacy

On the Adoption of Privacy-enhancing Technologies

Tristan Caulfield[1]([✉]), Christos Ioannidis[2], and David Pym[1]

[1] University College London, London, UK
{t.caulfield,d.pym}@ucl.ac.uk
[2] Aston Business School, Aston University, Birmingham, UK
c.ioannidis@aston.ac.uk

Abstract. We propose a model, based on the work of Brock and Durlauf, which looks at how agents make choices between competing technologies, as a framework for exploring aspects of the economics of the adoption of privacy-enhancing technologies. In order to formulate a model of decision-making among choices of technologies by these agents, we consider the following: *context*, the setting in which and the purpose for which a given technology is used; *requirement*, the level of privacy that the technology must provide for an agent to be willing to use the technology in a given context; *belief*, an agent's perception of the level of privacy provided by a given technology in a given context; and the *relative value* of privacy, how much an agent cares about privacy in this context and how willing an agent is to trade off privacy for other attributes. We introduce these concepts into the model, admitting heterogeneity among agents in order to capture variations in requirement, belief, and relative value in the population. We illustrate the model with two examples: the possible effects on the adoption of iOS devices being caused by the recent Apple–FBI case; and the recent revelations about the non-deletion of images on the adoption of Snapchat.

1 Introduction

Recent high-profile events—such Snowden's revelations about surveillance and the dispute between Apple and the FBI—have demonstrated the increasing significance of privacy concerns for individuals, organizations, and governments. As privacy-enhancing technologies become more widely available, and are increasingly incorporated into consumer products such as messaging apps, it is interesting and important to understand the factors affecting the adoption by consumers of different technologies. In this paper, we propose a model of how agents make choices between competing technologies, as a framework for exploring aspects of the economics of the adoption of privacy-enhancing technologies.

Acquisti et al. [3] deliver an excellent up-to-date survey of the economics of privacy. They provide for historical evolution of the economic theory of privacy from its early beginnings—starting with Posner [15,16] and Stigler [19], arguing

© Springer International Publishing AG 2016
Q. Zhu et al. (Eds.): GameSec 2016, LNCS 9996, pp. 175–194, 2016.
DOI: 10.1007/978-3-319-47413-7_11

in favour of limiting privacy in the name of market efficiency—to the counterexamples where improved privacy (i.e., restrictions on the access to private information) may be welfare improving. According to Acquisti et al. [3]:

> 'Privacy is, after all, a process of negotiation between public and private, a modulation of what a person wants to protect and what she wants to share at any given moment and in any given context.'

Other work has considered the role of privacy in technology adoption (for example, [17]) or considered economic factors affecting privacy [20] or privacy-enhancing technology adoption [1,2].

We introduce a characterization of privacy based on four key factors: *context*, the setting in which, and the purpose for which, a given technology is used; *requirement*, the level of privacy that the technology must provide for an agent to be willing to use the technology in a given context; *belief*, an agent's perception of the level of privacy provided by a given technology in a given context; and the *relative value* of privacy, how much an agent cares about privacy in this context and how willing an agent is to trade off privacy for other attributes.

We introduce these concepts into the proposed model, admitting heterogeneity among agents in order to capture variations in requirement, belief, and relative value in the population.

In categorizing the agents' different attitudes to privacy we adopt the useful classification of Harris and Westin [9,14], who divide the agents into three groups based upon their own perceptions of the value of their own privacy:

The Fundamentalist. Fundamentalists are generally distrustful of organizations that ask for their personal information, worried about the accuracy of computerized information and additional uses made of it, and are in favour of new laws and regulatory actions to spell out privacy rights and provide enforceable remedies. They generally choose privacy controls over consumer-service benefits when these compete with each other. About 25 % of the public are privacy Fundamentalists.

The Pragmatist. Pragmatists weigh the benefits to themes of various consumer opportunities and services, protections of public safety or enforcement of personal morality against the degree of intrusiveness of personal information sought and the increase in government power involved. They look to see what practical procedures for accuracy, challenge and correction of errors the business organization or government agency follows when consumer or citizen evaluations are involved.

They believe that business organizations or government should "earn" the public's trust rather than assume automatically that they have it. And, where consumer matters are involved, they want the opportunity to decide whether to opt out of even non-evaluative uses of their personal information as in compilations of mailing lists. About 57 % of the public fall into this category.

The Unconcerned. The Unconcerned are generally trustful of organizations collecting their personal information, comfortable with existing organizational procedures and uses are ready to forego privacy claims to secure consumer-service benefits or public-order values and not in favour of the enactment of new privacy laws or regulations. About 18 % of public fall into this category.

Sharing personal information may be perceived as risky or costly— facilitating identity theft, inviting unwanted attention by individuals or institutions, and possibly introducing limited participation in certain activities (e.g., exclusion from health insurance). Such negative impacts are known and the degree of aversion to the loss of privacy will differ between individuals depending upon their preferences and context.

The model—taking into account the privacy characteristics of competing technologies and the preferences of agents—indicates the expected levels of adoption of the competing technologies in different contexts. For example, sending different types of content with different levels of sensitivity over a service, such as Snapchat. By varying the parameters of the model—reflecting the characteristics of the technologies and the attitudes of the decision-making consumers—we explore how these factors influence the adoption of the different technologies.

In Sect. 2, we introduce the basic Brock-Durlauf model upon which our work is based. We also explain briefly our extension, from previous work [7], of this model to encompass multiple attributes. In Sect. 3, we present our main theoretical contribution. Using our analysis of the key characteristics of privacy, together with Westin's characterization of attitudes towards privacy, we adapt our extended Brock-Durlauf set-up to model the adoption of privacy-enhancing technologies. In Sect. 4, we discuss two examples. First, the recent dispute between between Apple and the FBI [4] and, second, Snapchat, exploring the effects of the population's changing beliefs about and requirements for privacy. Finally, in Sect. 5, we summarize our analysis.

2   Background: The Brock–Durlauf Model

Brock and Durlauf model a market where various technologies compete for adoption by a number of agents. The agents choose which technology to adopt based on the technologies' relative profitabilties as well as the strength of the technologies' social externalities; that is, how much the value of a technology increases as the number of other agents choosing it increases. This last feature makes the model particularly useful for looking at communication technologies—which form a large part of PETS—because the value of a technology increases with the number of people you can communicate with using it. The model can also look at exongenously-imposed policy, in the form of incentives or taxation, as well as increasing profitabilities through technological progress.

2.1   The Basic Brock–Durlauf Model

The basic model consists of M different technologies competing in a market for adoption by N agents. The utility for an agent of a technology γ in time period t is given by

$$u_{\gamma,t} = \lambda_\gamma + \rho_\gamma x_{\gamma,t} \tag{1}$$

where λ_γ is the profitability of the technology, $x_{\gamma,t}$ is the fraction of agents using technology γ at time t, and $\rho_\gamma > 0$ gives the strength of the social externalities. A low value of ρ_γ means the utility of the technology will not increase much as adoption rises; a high value means that the social component, $\rho_\gamma x_{\gamma,t}$, can influence the utility of the technology significantly.

In the model, each agent i experiences their own utility from their choice, $\tilde{u}_{\gamma,i,t} = u_{\gamma,t} + \epsilon_{\gamma,i,t}$, plus noise, where the noise term $\epsilon_{\gamma,i,t}$ represents a random private component of utility and is independent and identically distributed across agents and known to the agent when it makes its decision. If the noise follows a double exponential distribution, then, as the number of agents tends to infinity, the probability that an agent will adopt technology γ at time t—which is equivalent to that technology's share of the market—converges to

$$x_{\gamma,t} = \frac{e^{\beta u_{\gamma,t-1}}}{\sum_{j=1}^{M} e^{\beta u_{j,t-1}}}. \tag{2}$$

See [7] for more explanation of this equation.

The parameter β is inversely proportional to the variance of the noise, ϵ, and characterises the degree to which choices made by the agents are determined by the deterministic components of utility. As $\beta \to 0$, choices are totally random and each technology will tend towards an equal share of the market; as $\beta \to \infty$, choices have no random component and the agents will all choose to adopt the technology providing the highest utility.

In Brock and Durlauf [5,6], the agents make decisions based on their expectations of the decisions of others in the same time period. The model can then be used to find the adoption equilibria. In contrast, we wish to look at the dynamics of adoption over time. Instead of using agents' expectations about others' decisions in the same time period, agents use information about the levels of adoption in the previous time period, as shown by the use of $u_{c,t-1}$ in Eq. 2.

The original definition of utility for a technology, in Eq. 1, can be expanded to include a component determined by a policy-maker. This can represent, for example, some form of taxation or incentive designed to increase the adoption of a particular technology.

$$u_{\gamma,t} = \lambda_\gamma + \rho_\gamma x_{\gamma,t} - \tau_\gamma(x_{1,t}, \ldots, x_{M,t}) \tag{3}$$

This policy component takes the form of a function, $\tau_\gamma(x_{1,t}, \ldots, x_{M,t})$, for each different technology γ and gives the level of incentive or taxation based on the adoption shares of all the technologies in the market. This means that, for example, a policy-maker could apply an incentive to a technology that decreases as it becomes more widely adopted. Policies that tax one technology and use the benefits to promote another can be modelled by using opposite-signed functions on the two technologies.

Switching costs can also be added to the model by introducing asynchronous updating. That is, a portion α of the agents do not switch technologies in each time period, simulating the retarding effect switching costs have on the speed with which new technologies are adopted:

$$x_{\gamma,t} = \alpha x_{\gamma,t-1} + (1 - \alpha) \frac{e^{\beta u_{\gamma,t-1}}}{\sum_{j=1}^{M} e^{\beta u_{j,t-1}}}. \tag{4}$$

Equilibria. The model allows for equilibria; that is, where the share of adoption in one time period is the same as the previous time period. For low values of ρ, there will only be one equilibrium point. For higher values, it is possible to have multiple equilibria. In general, the model will, over time, approach one of the equilibrium points.

Except in the case where $\beta = \infty$, a technology will never have all of the share of the market or become extinct: some (possibly very small) portion of the population will continue to choose it.

2.2   Extension to Multiple Attributes

In [7], we looked at how the Brock–Durlauf model could be applied to the adoption of encryption technologies. A key point from this work is that representing technologies with a single attribute, profitability, is not suitable for creating useful models about encryption adoption. Instead, it is necessary to use multiple attributes which better represent the technologies and the way decisions to use them are made. Multi-attribute utility theory is explained in [13] and applied to security in [11].

This is achieved by adapting the model to use a set of attributes, A. Now, the utility for each technology (Eq. 1) becomes

$$u_{\gamma,t} = \sum_{a \in A} v_{\gamma,a} + \rho_\gamma x_{\gamma,t}, \tag{5}$$

where $v_{\gamma,a}$ is the value of attribute a for technology γ.

Similarly, including policy, Eq. 3 becomes

$$u_{\gamma,t} = \sum_{a \in A} v_{\gamma,a} + \rho_\gamma x_{\gamma,t} - \tau_\gamma(x_{1,t}, \ldots, x_{M,t}). \tag{6}$$

The attributes used depend on the technologies being modelled and the purpose for which the models are intended. In [7], we used three attributes: monetary cost, functionality, and usability.

3 Modelling Privacy

The basic approach of the model as described in Sect. 2 is not adequate for modelling the adoption of privacy-enhancing technologies. The model must be extended to capture the characteristics of privacy. This represents a significant enrichment of the model to capture a more complex collection of interacting factors, including heterogeneity of agents.

In this section, we first discuss these characteristics; then, we describe how the model is extended to include them. Finally, we discuss the effects of different choices of parameters.

3.1 The Characteristics of Privacy

We consider a society of decision-making entities who wish to protect the privacy of certain information that they own in a range of contexts in which they interact with the providers of goods and services. These interactions are typically enabled by technologies with differing privacy-protecting characteristics.

Some transactions are more sensitive than others for some individuals. For example, some individuals will choose to use online banking services, in which private information is potentially exposed to the public internet, and some will prefer to perform their financial transactions in person at a branch of their bank, where the immediate exposure is limited to the specific bank employees involved.

We can deconstruct this situation in different ways. It may be the user of online banking simply does not place a high value on their privacy or it may be that they do place a high value on their privacy, but also believe that the bank's systems provide adequate protection for their judgement of value of their privacy. Similarly, the in-branch user may believe that the online privacy protections do not provide adequate protection for their judgement of the value of their privacy.

This set-up illustrates two characteristics that we need to incorporate into our model: first, that agents have a judgement of the value of their privacy; and, second, that they have beliefs about the ability of a given technology to protect their privacy given their judgement of its value.

These two examples illustrate the use of particular technologies to access services in specific contexts. In general, services, such as banking, will accessed in different contexts. For example, the user of online banking may be willing use the service from a personal computer at home, but not from a shared public computer: their belief about the level of protection is dependent on the context.

So, in order to formulate a model of decision-making among choices of technologies by these agents, we must consider what are the relevant characteristics of privacy in this context.

- *Context*: the setting in which and the purpose for which a given technology is used.
- *Requirement*: the level of privacy that the technology must provide for an agent to be willing to use the technology in a given context.

- *Belief*: an agent's perception of the level of privacy provided by a given technology in a given context.
- *Relative value of privacy*: how much an agent cares about privacy in this context and how willing an agent is to trade off privacy for other attributes

Attitudes to privacy have been classified into three groups—fundamentalist, pragmatist, and unconcerned—by Westin [9,21]. The final characteristic above, the relative value of privacy, includes the idea of a trade-off between privacy and other attribrutes. The Westin groups provide a convenient way in which to organize agents into groups with simliar trade-off preferences. The examples in Sect. 4 illustrate this organization.

3.2   An Adoption Model Using the Privacy Characteristics

We can capture these characteristics of privacy in the model by making some changes to its structure.

First, we can capture context by increasing the granularity of the model — instead of looking at technologies' share of the market, we can look at how adoption is shared between technologies' use in different contexts. Each technology is divided into multiple technology–context components, and the model now looks at how agents choose between these.

We introduce a set of all of these components, C, with subsets C_γ containing all of the components for technology γ. Now, we define $u_{c,t}$ to be the utility of *component* c, rather than a *technology* Similarly, $x_{c,t}$ is now the share of a component, not a technology, at time t.

The total share of a technology γ is now given by the sum of its components:

$$x_{\gamma,t} = \sum_{c \in C_\gamma} x_{c,t}. \tag{7}$$

As an example, consider a cloud storage technology, where users can keep backups of their files. This could be divided into three different contexts based on its use for different purposes: storing photos, storing documents, and using it to do both of these. Each context offers different advantages (and so has different values for its attributes), and for each context agents may have different requirements for privacy. One agent might feel that photos require more privacy their than documents do, whereas another might feel the opposite.

In the model up to this point, agents have been homogenous in terms of the utility they receive from a technology, with the only difference coming from $\epsilon_{c,i,t}$, the private utility they receive from the noise. Modelling privacy requires heterogeneity: each agent has different preferences towards privacy, different requirements, and a different willingness to trade privacy for other attributes.

We add this to the model by giving each agent i a value $b_{c,i} \in [0,1]$ for their belief about how well a component preserves or provides privacy, a value $r_{c,i} \in [0,1]$ for the agent's required level of privacy for a component, and a value $w_{c,i} > 0$ as a weight indicating the relative importance of privacy (in a component) to other attributes in the model. The utility function used in the model then becomes

$$u_{c,i,t} = \pi_{g(i)}(b_{c,i}, r_{c,i}, w_{c,i}) + \sum_{a \in A} v_{c,a} + \rho_c x_{c,t} - \tau_c(x_{1,t}, \ldots, x_{M,t}). \qquad (8)$$

where $g(i)$ gives the group an agent belongs to and $\pi_g(i)(b_{c,i}, r_{c,i}, w_{c,i})$ is a trade-off function that specifies how the utility an agent recieves from privacy changes for varying levels or belief, requirement, and value of privacy—essentially, how willing they are to trade privacy for other attributes.

Introducing the idea of a group here provides a convenient way of representing different attitudes towards security, and allows us to capture ideas such as Westin's [9,21] groups. In theory, each agent could belong to its own group, each with a different trade-off function, but it would be immensely difficult to get the data required to fit a function to each participant in a study, for example. Agents in a group share the same trade-off function, meaning that they respond to different values of belief and requirements about privacy in the same way.

In this paper, we divide the population of agents into three groups, based on Westin's classifications of attitudes about privacy. Each group has a different trade-off function, which are shown in Fig. 1. For those unconcerned about privacy, there is little difference between components that meet requirements and those that do not. For pragmatists, any component that satisfies requirements receives the full utility value, with a linear trade-off for those that do not. For fundamentalists, there is very steep decline in utility value—quickly going negative—for components for which beliefs about privacy do not meet requirments. The trade-off functions are

$$\pi_{fund}(b_{c,i}, r_{c,i}, w_{c,i}) = w_{c,i} \frac{0.5 + \tanh(10(b_{c,i} - r_{c,i} + 0.1))}{1.5} \qquad (9)$$

$$\pi_{prag}(b_{c,i}, r_{c,i}, w_{c,i}) = \begin{cases} w_{c,i} & b_{c,i} - r_{c,i} > 0 \\ w_{c,i}(b_{c,i} - r_{c,i} + 1) & b_{c,i} - r_{c,i} \leq 0 \end{cases} \qquad (10)$$

$$\pi_{unco}(b_{c,i}, r_{c,i}, w_{c,i}) = 0.1 w_{c,i}(b_{c,i} - r_{c,i}) + 0.9. \qquad (11)$$

We can update Eq. 3 to account for the heterogeniety by summing over the population of agents. Now, the share of each technology–context component is given by

$$x_{c,t} = \frac{1}{N} \sum_{i=1}^{N} \frac{e^{\beta u_{c,i,t-1}}}{\sum_{j=1}^{C} e^{\beta u_{j,i,t-1}}}. \qquad (12)$$

Each agent here has an equal weight $(1/N)$ and represents an equal share of the population, but this could easily be changed so that agents have different weights, making them representative of different proportions of the population. This might be useful, for example, when polling a population, where some agents' characteristics have a greater likelihood.

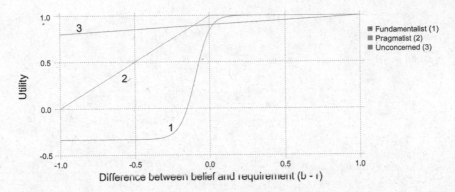

Fig. 1. Trade-off functions for each of the Westin groups. The figure shows the utility received from privacy given the difference in beliefs and requirements when privacy value is 1.

3.3   Parameters

Sample size. In the examples below, we approximate the distribution of preferences about privacy, including requirements, beliefs, and values by using beta distributions to represent the distribution of values in the population. We then sample from these distributions to create a collection of agents with heterogenous properties.

As the sampling is random, the points chosen can influence the behaviour of the model. We ran 100 trials for each of a number of different sample sizes in order to observe the magnitude of this influence. Figure 2 shows the mean and $\pm 2\sigma$ values for each of the sample sizes.

As expected, the variance of the low sample sizes is higher than for the larger sample sizes. For 100 samples, the 5 %–95 % range is 0.03491; for 5000 it is 0.0053, and for 10000 samples it is 0.0030. We use 10000 samples in all examples below.

The β parameter. The parameter β is inversely related to the variance of the noise $\epsilon_{\gamma,i,t}$, which is a private component of the utility an agent gets for a particular choice of technology γ. As the variance of the noise grows—so, as β grows smaller—the less the other, deterministic components of utility matter. Conversely, as β grows larger and the variance of the noise decreases, agents increasingly make their choice based on the deterministic parts of the utility function.

Figure 3 shows the adoption over time for a technology for different values of β. The technology shown is slightly more profitable than the competing technology, but all other values are the same. For low β, the more profitable technology shares the market with its competitor evenly. As β grows, more agents adopt the more profitable technology.

In the examples below, we use a value $\beta = 3.0$.

Social effects. The parameter ρ_c controls the strength of social effects for a component. Figure 4 shows the effect of different values of ρ on the adoption

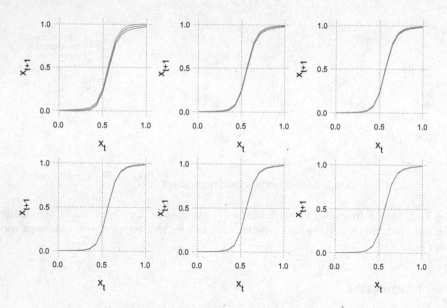

Fig. 2. Demonstration of the effect of sample size. The figure shows the mean and $\pm 2\sigma$ for samples sizes of 100, 500, 1000, 2500, 5000, and 10000.

curve (plotting x_t against x_{t+1}) of a technology. Both technologies use the same parameter values (including ρ), except for profitability, which is slightly higher for the technology shown.

As the value of ρ grows, the utility of a technology increases with increased adoption. High values of ρ amplify increases in utility from adoption.

Also shown in the figure is a diagonal line where $x_t = x_{t+1}$, meaning that the system is in equilibrium. For the lower values of ρ the adoption curves only have one equilibrium, meaning that adoption will approach this point over time. When $\rho = 1$, there are three equilibria: two stable points, low and high, and a third, unstable point near $x = 0.4$. If the initial state is below the unstable equilibrium, adoption will move towards the lower equilibrium; if it is higher than the unstable equilibrium, adoption will move towards to the higher stable equilibrium.

4   Examples

We discuss in detail two examples. First, we consider the recent dispute been Apple and the FBI [4], with the purpose of demonstrating how beliefs and requirements about privacy influence adoption. Second, we consider Snapchat (www.snapchat.com, accessed 03/03/2016), a picture-messaging app which promised that images were available only for brief periods time, but for which it transpired that images were in fact retained [18]. We use this example to demonstrate the role of context in privacy decision-making regarding the use of the technology.

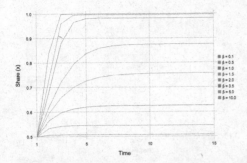
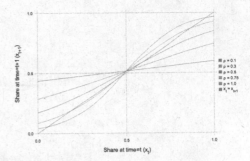

Fig. 3. Demonstration of the effect of parameter β. With a high β, the adoption of the more profitable technology is greater.

Fig. 4. Demonstration of the effect of parameter ρ.

Both of these examples are intended to be illustrative of the theoretical model. It would of course be valuable to condition the examples on empirical data. However, such data collection and analysis, requiring substantial dedicate effort, is beyond our current scope. The paper by Brock and Durlauf [6] shows how the basic model can be fitted by maximum likelihood estimation; in principle, our extensions can be given a similar analysis.

Building on our discussion in Sect. 3, we remark that the example discussed there—namely access to banking services—would also provide an examples of the issues discussed in this section.

The model is implemented using the julia language [12].

4.1   Apple v FBI

In California, there is an ongoing case between Apple and the FBI, where the FBI is investigating the San Bernardino killings and wishes to access one of the killer's locked and encrypted iPhones. The FBI is seeking a court order, under the All Writs Act, to compel Apple to assist in unlocking the device, possibly by creating and signing a custom firmware image that would allow the FBI to brute-force the password easily. Apple has argued against the FBI and the case has generated a great amount of media coverage.

For this example, we are interested in the effects this media coverage. Apple has publicly stated during the course of the case that it believes firmly in the privacy of its customers; this can be viewed as a strong signal about the level of privacy provided by Apple products and agents may update their beliefs in response, resulting in a change of technology choice. We will use the model to explore how adoption changes in response to shifting beliefs about the privacy a technology provides, shifting requirements, and shifts in both of these simultaneously.

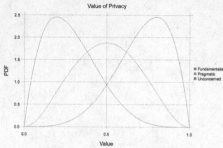 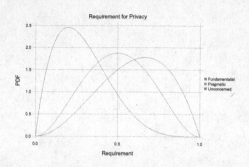

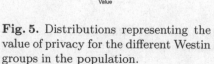

Fig. 5. Distributions representing the value of privacy for the different Westin groups in the population.

Fig. 6. Distributions representing the requirements for privacy of the different Westin groups in the population.

Set-up. In this example, we look at two technologies competing against each other, without considering any distinct contexts. The first technology is Apple's iPhone, and we look at its adoption when competing against Android phones.

For simplicity, we do not consider any attributes other than cost in this example; we assume that usability and functionality are essentially equivalent between the devices. Cost, on the other hand, differs: Apple devices tend to me more expensive than the bulk of android phones. Accordingly, we use a value of 1.1 for Apple and 1.5 for Android.

The value of ρ indicates how much the utility agents gain from adopting a technology increases as more agents begin to use it. In the case of mobile phones, they are largely interoperable with each other, and many of the applications written for them are present on both Apple and Android devices, suggesting that the value of ρ should be low. However, there are functions on the phone, such as Apple's iMessage, which increase in utility as more people use them, meaning that there is some social effect present. For this example, then, we use a value of $\rho = 0.5$ for both technologies.

We need to make some assumptions about the distributions of values, beliefs, and requirements of security in the population. First, Fig. 5 shows the distributions we are using for the value of privacy. There is a seperate distribution for each of the three Westin categories. We assume that fundamentalists are more likely to place a higher value on privacy than the pragmatic and the unconcerned. Similarly, we assume that it is more likely that the pragmatic have a higher value of privacy than the unconcerned. For this example, we say that privacy fundamentalists form 25 % of the population, pragmatists 55 %, and the unconcerned the remaining 20 %.

Next, Fig. 6 shows the distributions from which requirements about privacy are drawn. Again, we assume that fundamentalists are likely to have higher requirements than the pragmatic, and the pragmatic are likley to have higher requirements than the unconcerned.

Finally, we look at the distributions from which we sample values for belief about the privacy provided by the different technologies. These distributions of

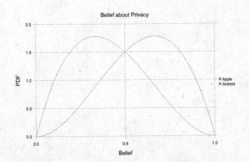

Fig. 7. Distributions representing the initial beliefs about the privacy provided by Apple and Android mobile phones.

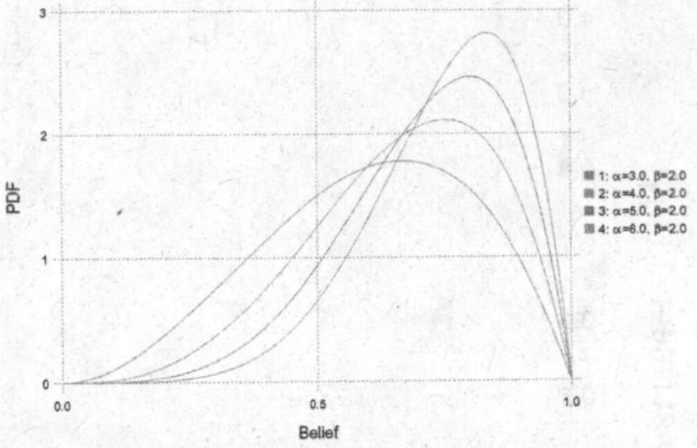

Fig. 8. Increasing beliefs about the privacy provided by Apple phones.

belief are shared by the entire population and are not segmented into Westin groups. Figure 7 shows the distributions; agents are more likely to believe that an Apple phone provides a greater level of privacy compared to Android than vice versa.

Changing Beliefs. Now, we will examine what are the likely effects on adoption of a shift in beliefs about the privacy provided by Apple phones. As stated above, the shift is a hypothetical one caused by the media attention around the Apple v FBI case and Apple's public stance on privacy. As such, we will look at how adoption changes for different magnitudes in shifts in belief to understand the range of possible effects.

We model the shifts in beliefs by changing the distribution of beliefs in the population and randomly sampling again. We look at four different distributions of beliefs about the privacy of Apple phones; we do not alter the distribution for Android phones. The different distributions are show in Fig. 8, labeled 1–4, each with increasing probability of a higher belief about privacy. The first is the same distribution shown in Fig. 7.

Table 1. Equilibrium Apple share values for shifts in belief.

Shift	Equilibrium
1 (orig.)	0.1402
2	0.1585
3	0.1667
4	0.1806

The resulting adoption curves are shown in Fig. 9. The shifts in belief about the privacy provided by Apple phones result in increased adoption. Table 1 shows

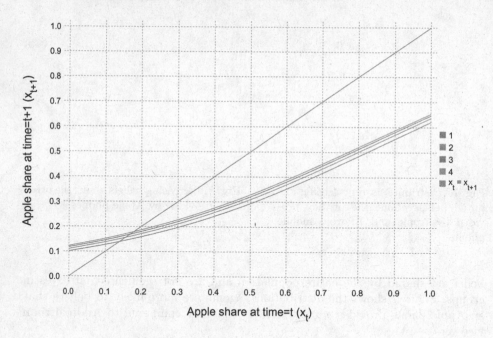

Fig. 9. Adoption curves for the different levels of belief.

Table 2. Equilibrium Apple share values for shifts in requirement.

Shift	Equilibrium
1 (orig.)	0.1417
2	0.1429
3	0.1451
4	0.1429

the equilibrium values for the four shifts. The base case, 1, shows Apple with a 14 % share of the market—intentionally close to the actual share in 2015 [10].

With each shift, the share of the market grows, showing that agents receive greater utility from technology that better meets their requirements and thus switch.

Changing Requirements. Next, we consider what happens if the media coverage increases agents' awareness of the need for privacy, resulting in a shift in requirements. As we are using different distributions of requirements for each of the Westin categories, we need to shift all three distributions to model the change in requirements. These are shown in Fig. 10. In each case, the distributions shift to the right, making higher values for the requirement for privacy more likely.

The adoption curves for the shifts in requirement are shown in Fig. 11. Unlike the shifts in beliefs, the shifts in requirements do not result in increased adoption.

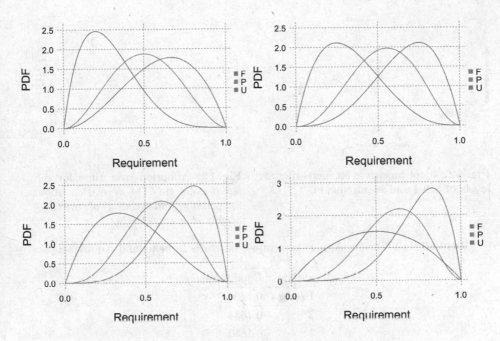

Fig. 10. Shifting requirements for each of the Westin groups.

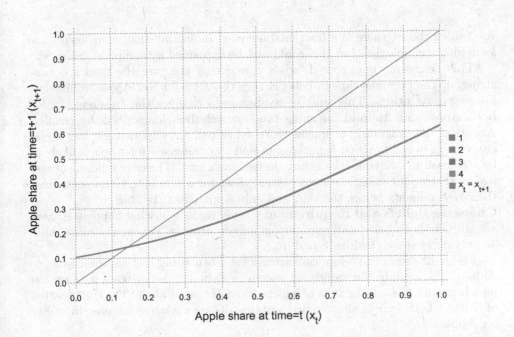

Fig. 11. Adoption curves for the different levels of requirements.

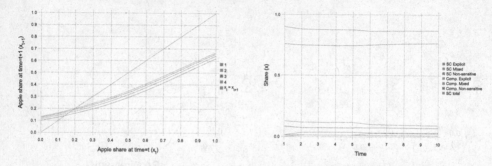

Fig. 12. Adoption curves for increasing levels of belief and requirements.

Fig. 13. Adoption over time for different contexts. The revelation about Snapchat's non-deletion of messages occurs after time $t = 5$.

Table 3. Equilibrium Apple share values for shifts in belief and requirement.

Shift	Equilibrium
1 (orig.)	0.1400
2	0.1634
3	0.1840
4	0.2011

As Table 2 shows, there is a fractional increase, indicating that some agents are switching technologies, but this could also be explained by sampling variance.

This behaviour is expected, when considering the way the model is constructed. The previous shift in belief change the value for just Apple technology, increasing its utility. This shift in requirements changes the requirements for both Apple and Android, meaning that any relative changes will be smaller. Agents of the pragmatic or unconcerned types will not experience a large relative change in utility when requirements shift—any increase for Apple is likely to be too small to overcome the utility derived from cost. The only fundamentalist agents that would change technologies are those for whom both technologies met their requirements before the shift and only Apple after the shift.

Changing Beliefs and Requirements. Here, we look at what happens if there are shifts in both belief and requirement simultaneously. We use the same shifts as previously shown in Figs. 8 and 10.

Figure 12 shows the adoption curves when both belief and requirements are shifted. The equilibrium values are shown in Table 3. The increase in adoption here is greater than in the shift of beliefs or requirements alone. The combination of shifting both beliefs and requirements results in a relative increase in utility for Apple.

4.2   Snapchat

In this example, we explore the use of contexts and how privacy affects in which contexts agents choose to use technology by looking at the ephemeral picture messaging application Snapchat. This is a widely used application that gives its users the ability to control how long the messages they send can be seen by the recipients, after which the messages are deleted. However, the messages are not deleted securely, and can still be recovered after they have disappeared from the application.

Set-up. Rocsner et al. [18] survey users of Snapchat, asking which types of content users send and how they feel about privacy. They give the breakdown of the study participants into Westin groups (39.4 % fundamentalist, 45.7 % pragmatist, 12.6 % unconcerned) and report how many users *primarily* send sexual content (1.6 %) and how many *have sent* sexual content (14.2 %).

We use these values directly in this example. We model three different contexts: sending only explicit content, sending only non-sensitive content, and using the technology for both. We say that Snapchat is competing against another similar application (which has the same contexts), but Snapchat initially has the majority share of adoption, around 90 %.

We assume that the values of usability and cost are the same for both technologies, but there is a difference in the utility received from functionality. For Snapchat, we assign using it for only explicit content the value 0.9; for mixed explicit and non-sensitive use 1.5, and for non-sensitive use only 1.54. For the competing technology, in the same order, we use the values 0.8, 1.2, and 1.44. These values were chosen so that the model roughly matches the values reported in Roesner et al. [18]. The values for the explicit-only and mixed-content use contexts are less than the non-sensitive context. This is because—even though an agent using the technology for both types of content technically has greater functionality—the proportion of agents who actually generate explicit content is very small and the attribute values reflect the utility received by the *population* of agents.

Since we are looking at messaging applications, the value of social effects is very high: the utility of such an application increases with the number of people you can contact using it. As such, we use a value of $\rho = 1$.

The distributions used for beliefs, requirements, and values are the same initially for the two technologies. Fundamentalists are likely to have very high requirements and to place a high value on privacy for the explicit and mixed-content messaging contexts, and higher-than-average requirements for non-sensitive messaging. The unconcerned have the lowest requirements and values, and the pragmatists are in between the two other groups.

Change in beliefs. We model the revelation that Snapchat messages are not securely deleted and can be recovered as a shock which causes a downward shift in belief about the privacy provided by Snapchat. Beliefs about the competing product do not change.

Figure 13 shows the share of adoption of the various components of Snapchat and its competitor over time, as well as the total market share for Snapchat. The shock occurs after time $t = 5$.

Table 4. Adoption of different components before and after Snapchat's non-deletion of messages is revealed.

	Before	After
Snapchat explicit	0.014	0.011
Snapchat mixed	0.119	0.094
Snapchat non-sensitive	0.742	0.764
Comp. explicit	0.014	0.028
Comp. mixed	0.038	0.032
Comp. non-sensitive	0.074	0.071
Total snapchat	0.874	0.869

The initial values, before the shock, are close to the values reported in Roesner [18]. Out of Snapchat's share—not the total share—1.5 % use it for explicit messages only, compared to 1.6 % in Roesner, and 13.2 % use it for mixed content, compared to 14.2 %.

Table 4 shows the values of adoption for the different components before and after the shock. The use of Snapchat for explicit messaging decreases from 1.4 % to 1.1 %. Similarly, the use of Snapchat for mixed explicit and non-sensitive messaging declines from 11.9 % to 9.4 %. The use of Snapchat in a non-sensitive context actually increases, from 74.2 % to 76.4 %, showing that agents who have a high level of privacy requirement in the explicit or mixed-conent messaging contexts no longer use the technology in those contexts when they believe that their privacy requirements are no longer being met.

Snapchat's total share declines post-shock from 87.4 % to 86.9 %. The agents that switched technologies to the competitor did so for explicit messaging, which grew from 1.4 % to 2.8 %. The beliefs about the security of the competing product did not change, so agents wishing to use the technology for explicit content were willing to swtich to a product with less functionality that met their privacy requirements.

5   Conclusions

We have discussed the characteristics of privacy from the point of view the economic agent:

– *Context*: the setting in which and the purpose for which a given technology is used;

- *Requirement*: the level of privacy that the technology must provide for an agent to be willing to use the technology in a given context;
- *Belief*: an agent's perception of the level of privacy provided by a given technology in a given context;
- *Relative value of privacy*: how much an agent cares about privacy in this context and how willing an agent is to trade off privacy for other attributes.

We have incorporated these characteristics into a model of technology adoption by a society of heterogenous decision-making agents.

Our analysis is based on Harris and Westin's classification of agents as Fundamentalist, Pragmatist, and Unconcerned. For each of these groups, we have assigned a function that determines the utility an agent derives from a technology, depending upon the agent's beliefs about how effectively the technology meets their requirements for protecting their privacy.

We have presented two main examples. First, to demonstrate the effects of changing beliefs and requirements, we have considered the signal of concern for privacy suggested by ongoing Apple v FBI dispute. Second, we have demonstrated the model's use to capture context by considering the change in types of messages that are prevalent on Snapchat before and after a change in beliefs about the level of privacy provided.

The literature on economic modelling of privacy and its role in technology adoption is quite limited, with [8] and the references therein providing a good guide. We believe the present paper represents a useful contribution in that we identify key characteristics, create a model that is capable of capturing them, and explore, with examples, their significance.

The model we have presented here allows preferred attributes for particular agents to be specified. Future work might employ empirical studies of the preferences, beliefs, and requirements of actual agents and incorporate this data into the model. Similarly, the trade-off functions used for the Westin groups might be derived from empirical studies.

The model as presented includes a policy component that is not exploited in this paper. Further work might explore the role of policy in the adoption of privacy-enhancing technologies.

References

1. Acquisti, A.: Protecting privacy with economics: economic incentives for preventive technologies in ubiquitous computing environments. In: Proceeding of Workshop on Socially-informed Design of Privacy-enhancing Solutions, 4th UBICOMP (2002)
2. Acquisti, A., Grossklags, J.: Privacy and rationality in individual decision making. IEEE Secur. Priv. **3**(1), 26–33 (2005)
3. Acquisti, A., Taylor, C., Wagman, L.: The economics of privacy. http://ssrn.com/abstract=2580411. Accessed 02 Mar 2016
4. FBI–Apple encryption dispute. https://en.wikipedia.org/wiki/FBI-Apple_encryption_dispute. Accessed 13 Apr 2016
5. Brock, W.A., Durlauf, S.N.: Discrete choice with social interactions. Rev. Econ. Stud. **68**(2), 235–260 (2001)

6. Brock, W.A., Durlauf, S.N.: A multinomial-choice model of neighborhood effects. Am. Econ. Rev. **92**(2), 298–303 (2002)
7. Caulfield, T., Ioannidis, C., Pym, D.: Discrete choice, social interaction, and policy in encryption technology adoption [Short Paper]. In: Grossklags, J., Preneel, B. (eds.) Proceedings of Financial Cryptography and Data Security 2016, Lecture Notes in Computer Science (2016, forthcoming). http://fc16.ifca.ai/program.html. http://fc16.ifca.ai/preproceedings/16_Caulfield.pdf
8. ENISA. Study on monetising privacy: an economic model for pricing personal information (2012). https://www.enisa.europa.eu/activities/identity-and-trust/library/deliverables/monetising-privacy. Accessed 04 Mar 2016
9. Harris, L., Westin, A.: The equifax Canada report on consumers and privacy in the information age. Technical Report (1992)
10. IDC: Smartphone OS Market Share 2015, 2014, 2013, and 2012. http://www.idc.com/prodserv/smartphone-os-market-share.jsp. Accessed 04 Mar 2016
11. Ioannidis, C., Pym, D., Williams, J.: Investments and trade-offs in the economics of information security. In: Dingledine, R., Golle, P. (eds.) FC 2009. LNCS, vol. 5628, pp. 148–166. Springer, Heidelberg (2009). doi:10.1007/978-3-642-03549-4_9
12. The julia language. http://julialang.org/. Accessed 30 Sept 2015
13. Keeney, R.L., Raiffa, H.: Decisions with Multiple Objectives: Preferences and Value Trade-offs. Wiley, New York (1976)
14. Kumaraguru, P., Cranor, L.F.: Privacy indexes: a survey of westins studiestechnical report, institute for software research, Carnegie Mellon University (2005). http://www.cs.cmu.edu/~ponguru/CMU-ISRI-05-138.pdf
15. Posner, R.A.: The right of privacy. Georgia Law Rev. **2**(3), 393–422 (1978)
16. Posner, R.A.: The economics of privacy. Am. Econ. Rev. **71**(2), 405–409 (1981)
17. Pu, Yu., Grossklags, J.: An economic model and simulation results of app adoption decisions on networks with interdependent privacy consequences. In: Poovendran, R., Saad, W. (eds.) GameSec 2014. LNCS, vol. 8840, pp. 246–265. Springer, Heidelberg (2014). doi:10.1007/978-3-319-12601-2_14
18. Roesner, F., Gill, B.T., Kohno, T.: Sex, lies, or kittens? investigating the use of snapchat's self-destructing messages. In: Christin, N., Safavi-Naini, R. (eds.) FC 2014. LNCS, vol. 8437, pp. 64–76. Springer, Heidelberg (2014). doi:10.1007/978-3-662-45472-5_5
19. Stigler, G.J.: An introduction to privacy in economics and politics. J. Legal Stud. **9**(4), 623–644 (1980)
20. Varian, H.R.: Economic aspects of personal privacy. In: Privacy and Self-regulation in the Information Age, US Department of Commerce (1997)
21. Westin, A.: Privacy and Freedom. Atheneum Publishers, New York (1967)

FlipLeakage: A Game-Theoretic Approach to Protect Against Stealthy Attackers in the Presence of Information Leakage

Sadegh Farhang$^{(\boxtimes)}$ and Jens Grossklags

College of Information Sciences and Technology,
The Pennsylvania State University, University Park, PA, USA
{farhang,jensg}@ist.psu.edu

Abstract. One of the particularly daunting issues in the cybersecurity domain is *information leakage* of business or consumer data, which is often triggered by multi-stage attacks and advanced persistent threats. While the technical community is working on improved system designs to prevent and mitigate such attacks, a significant residual risk remains that attacks succeed and may not even be detected, i.e., they are *stealthy*.

Our objective is to inform security policy design for the mitigation of stealthy information leakage attacks. Such a policy mechanism advises system owners on the optimal timing to reset defense mechanisms, e.g., changing cryptographic keys or passwords, reinstalling systems, installing new patches, or reassigning security staff.

We follow a game-theoretic approach and propose a model titled *FlipLeakage*. In our proposed model, an attacker will incrementally and stealthily take ownership of a resource (e.g., similar to advanced persistent threats). While her final objective is a complete compromise of the system, she may derive some utility during the preliminary phases of the attack. The defender can take a costly recovery move and has to decide on its optimal timing.

Our focus is on the scenario when the defender can only partially eliminate the foothold of the attacker in the system. Further, the defender cannot undo any information leakage that has already taken place during an attack. We derive optimal strategies for the agents in FlipLeakage and present numerical analyses and graphical visualizations.

1 Introduction

Security compromises which cause *information leakage* of business or consumer data are a particularly challenging problem in the cybersecurity domain. Affected businesses frequently struggle to recover once the consequences of a breach become apparent such as a competitor outpacing them in a race for the next innovation, or data troves appearing on cybercriminal marketplaces and eventually impacting consumer confidence. For example, data about small and medium-sized businesses suggests that approximately 60 % fail within six months after a data breach [3].

© Springer International Publishing AG 2016
Q. Zhu et al. (Eds.): GameSec 2016, LNCS 9996, pp. 195–214, 2016.
DOI: 10.1007/978-3-319-47413-7_12

Businesses struggle for multiple reasons to prevent information leakage. In particular, the increasing prevalence of well-motivated, technically capable, and well-funded attackers who are able to execute sophisticated multi-stage attacks and advanced persistent threats (APT) poses significant challenges to prevent information leakage. Such attacks may take time to execute, but they will eventually succeed with high likelihood. In a recent talk, the Chief of Tailored Access Operations, National Security Agency, characterized the mindset of these attackers in the following way: "We are going to be persistent. We are going to keep coming, and coming, and coming [12]."

Further, carefully orchestrated attacks as employed during corporate, cyber-criminal or nation-state sponsored cyber-espionage and sabotage (see Stuxnet [4]) change our understanding of the likelihood to reliably detect stealthy attacks before it is too late. Estimates for how long attacks remain undetected are dire. For example, a recent presentation by the CEO of Microsoft suggested that the time until detection of a successful attack is on average over 200 days [21].

All of these observations emphasize the need to reason about the suitable response to stealthy attacks which cause continued information leakage. We know that perfect security is too costly; and even air-gaped systems are vulnerable to insider risks or creative technical approaches. Another mitigation approach is to limit the impact of attacks by resetting system resources to a presumed safe state to lower the chances of a perpetual undetected leak. However, in most scenarios such actions will be costly. For example, they may impact productivity due to system downtime or the need to reissue cryptographic keys, passwords or other security credentials. As such, determining the best schedule to reset defense mechanisms is an economic question which needs to account for monetary and productivity costs, strategic and stealthy attacker behavior, and other important facets of information leakage scenarios such as the effectiveness of the system reset. To address this combination of factors, we propose a new game-theoretic model called *FlipLeakage*.

In our proposed model, an attacker has to engage in a sustained attack effort to compromise the security of a system. Our approach is consistent with two scenarios. On the one hand, the attacker may conduct surveillance of the system to collect information that will enable a security compromise, e.g., by pilfering traffic for valuable information, or by gathering information about the system setup. On the other hand, the attacker may incrementally take over parts of a system, such as user accounts, parts of a cryptographic key, or collect business secrets to enable further attack steps. In both scenarios, persistent activity and the accumulated information will then enable the attacker to reach her objective to compromise the system and to acquire the primary business secret; if the defender does not interfere by returning the system to a presumed safe state.

In Fig. 1, we provide an initial abstract representation of the studied strategic interaction between an attacker and a defender. The attacker initiates sustained attack efforts at t_1, t_2, and t_3 right after the defender's moves, where each time she also starts gaining information about the system. After accumulating sufficient information about the system, the attacker will be able to compromise it. The attacker's benefit until the security compromise is completed is represented as a triangle, which represents the value of the leaked information during the attack execution. After the compromise, the attacker continues to receive benefits from the compromised system which is represented as a rectangle.

The defender can take a recovery action (to reset the resource to a presumed safe state) and can thereby stop the attack. In our model, we consider the scenario when the defender only partially eliminates the foothold of the attacker in the system. In Fig. 1 those defensive moves occur at t_1, t_2, and t_3. Further, the defender cannot undo any information leakage that has already taken place during an attack.

In our model, we focus on periodic defensive moves for the defender. That means the time between any two consecutive moves is assumed the same motivated by practical observations for security policy updates of major software vendors such as Microsoft and Oracle which we will discuss in detail in Sect. 3. Within this context, we aim to determine the defender's best periodic defensive strategies when the moves of the attacker are unobservable to the defender, i.e., the attacks succeed to be stealthy. At the same time, we assume that the attacker can observe the defender's moves. The latter assumption rests on two observations. On the one hand, the attacker will be cut off from access to a partially compromised system when a recovery move takes place. On the other hand, many defensive moves may actually be practically observable for attackers, e.g., when a patch for a software system becomes available which makes a particular attack strategy impractical. The scenario under investigation is a security game of timing, e.g., we are studying *when* players should move to act optimally.

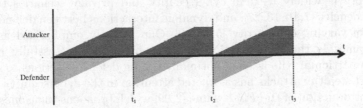

Fig. 1. FlipLeakage is a two-player game between an attacker and a defender competing with each other to control a resource. t_1, t_2, and t_3 represent the defender's move times. During the time when the attacker launches her attack, she incrementally benefits from information leakage which is shown as red triangles. (Color figure online)

In the following, we provide a brief summary overview over our contributions.

- We develop a game-theoretic model titled *FlipLeakage*. In our model, an attacker will incrementally take ownership of a resource (e.g., similar to advanced persistent threats). While her final objective is a complete compromise of the system, she may derive some utility during the preliminary phases of the attack. The defender can take a costly periodic mitigation move and has to decide on its optimal periodic timing.

- We consider the scenario when the defender only partially eliminates the foothold of the attacker in the system. Further, the defender cannot undo any information leakage that has already taken place during an attack.
- We derive optimal strategies for the agents in our model and present numerical analyses and graphical visualizations. One of our findings corroborates an intuition: the higher the defensive cost, the slower the defender's periodic move rhythm. Moreover, our numerical observations imply that the defender moves faster when the attacker's average time to totally compromise the defender's system is lower.

In the presence of stealthy attacks and information leakage, defenders have to set a schedule for updating and resetting their defense mechanisms without any feedback about the occurrence of attacks. This poses significant challenges for the design of new methods to mitigate such attacks. The objective of our theoretical model is to provide a systematic approach for the defender's best schedule to reset his system to a presumed safe state to lower the chances of a perpetually undetected leak. As such, our work provides important steps towards building a rigorous model for an optimal defender's response to these unknowns.

Roadmap: The rest of our paper is organized as follows. We discuss related work in Sect. 2. In Sect. 3, we develop the FlipLeakage model followed by payoff calculations in Sect. 4. We analyze our proposed model in Sect. 5. In Sect. 6, we present numerical examples. Finally, we conclude our paper in Sect. 7.

2   Related Work

Game theory is widely used in cybersecurity and privacy scenarios to study interdependencies [7,10,13,27], and dynamic interactions between defenders and attackers of varying complexity [5,17,19]. One recently emphasized aspect of security games is the consideration of *when* to act to successfully mitigate attacks. In particular, the issue of optimally timing defensive actions to successfully thwart stealthy attacks has attracted attention in the cybersecurity domain with the introduction of the *FlipIt* game [2,29] which broadens the *games of timing* literature initiated in the cold-war era [1,28]. In what follows, we provide a brief description of the FlipIt game as well as theoretical follow-up research.

FlipIt is a two-player game between a defender and an attacker competing with each other to control a resource which generates a payoff to the owner of the resource. Moves to take over the resource, i.e., *flips*, are costly [2,29]. In [29], the authors studied the FlipIt game with different choices of strategy profiles and aimed to calculate dominant strategies and Nash equilibria of the game in different situations. Pham and Cid [26] extended the FlipIt game by considering that players have the ability to check the state of the resource before their moves.

Feng et al. [6] and Hu et al. [9] modified the FlipIt game by considering insiders in addition to external adversaries. Zhang et al. [31] studied the FlipIt game with resource constraints on both players. Pawlick et al. extended the FlipIt game with characteristics of signaling games [25]. Wellman and Prakash developed a discrete-time model with multiple, ordered states in which attackers may compromise a server through cumulative acquisition of knowledge rather than in a one-shot takeover [30].

The original FlipIt paper assumed that the players compete with each other for *one* resource. Laszka et al. [14] addressed this limitation by modeling multiple contested resources in a game called *FlipThem*. Other authors extended this game by considering a threshold for the number of contested resources which need to be compromised to achieve the attacker's objective [18]. In a similar way, a variation of the game has been proposed with multiple defenders [24]. Laszka et al. [15,16] studied timing issues when the attacker's moves are non-instantaneous. Moreover, they considered that the defender's moves are non-covert and the attacker's type can be targeting and non-targeting. Johnson et al. [11] investigate the role of time in dynamic environments where an adversary discovers vulnerabilities based on an exogenous vulnerability discovery process and each vulnerability has its corresponding survival time.

Complementing these theoretical analyses, Nochenson and Grossklags [22] as well as Grossklags and Reitter [8] study human defensive players when they interact with a computerized attacker in the FlipIt framework.

Our work differs from the previous FlipIt literature regarding two key considerations. First, we take into account the problem of information leakage and propose a more realistic game-theoretic framework for defender's best time to update his defense mechanism. We propose a model in which an attacker will incrementally take ownership of a resource. Note that the attacker's goal is to compromise the defender's system completely, but she may acquire already some benefit during the initial steps of her attack. Second, we consider the possibility of the defender's defense strategy not being able to completely eliminate the attacker's foothold in the system. As a result, our work overcomes several significant simplifications in the previous literature which limited their applicability to realistic defense scenarios.

3   Model Definition

In this section, we provide a description of the FlipLeakage model which is a two-player game between a defender ($\mathcal{D}$) and an attacker ($\mathcal{A}$). We use the term **resource** for the defended system, but also for the target of the attack which will leak information during the attack and after the successful compromise. The attack progresses in a **stealthy** fashion. However, the defender can regain partial control over a compromised resource by taking a defensive recovery move (e.g., a variety of system updates).

In the FlipLeakage model, we emphasize the following aspects which we will discuss below: (1) **uncertainty** about the time of compromising the defender's resource entirely, (2) process of **information leakage**, (3) **quality** of defensive moves, (4) **strategies** of both players, and (5) other parameters which are necessary for our model.

Uncertainty About Attack Launch and Success Timings: In FlipLeakage, the defender is the owner of the resource at the beginning of the game. The resource is in a secure state, when it is completely controlled by the defender. However, due to the stealthy nature of many practically deployed attacks, e.g., related to cyber-espionage and advanced persistent threats, it is reasonable to assume that the defender cannot acquire any information about the time when an attack is launched as well as its success [21].

In contrast, we assume that the attacker can observe the time of a defender's move. One motivating practical example for this consideration is that many software companies publicly announce the arrival of new patches for previously discovered vulnerabilities. Hence, an attacker could infer when a certain system weakness is not available anymore. It follows that we model asymmetry with respect to knowledge between the two players.

Furthermore, we differentiate between the time of launching an attack and the time of an attack's full effectiveness (i.e., the resource is completely compromised). It is worth mentioning that the value of this time difference is not known to both the defender and the attacker. Hence, this time difference is represented by a random variable t_A with probability density function $f_A(t_A)$. The value of t_A depends on many factors such as the defender's defense strategy and the attacker's ability to compromise the defender's system.

The gap between these two factors can be interpreted as the attacker requiring a nontrivial amount of time and effort to control the resource completely, e.g., to gather leaked information from the resource and to conduct subsequent attack steps. Further, the time of launching an attack can be understood as the time that the attacker starts to gather information from the defender to execute the attack successfully (e.g., by conducting surveillance of the system setup or pilfering traffic to collect information that will enable a security compromise). For simplicity, we assume that the value of t_A is chosen according to a random variable, but it is constant during each round of the attack. For future work, we are going to consider the case where the values of t_A are different for each round of the attack. Note that we assume that other important parameters of the game are common knowledge between the players. The extension of the framework to uncertainty about game-relevant parameters is subject of future work

Process of Information Leakage: After initiation of the attack move, the attacker's reward until a complete compromise is accomplished is based on the percentage of the whole resource which is currently controlled by the attacker. For this purpose, we consider a function $g_A(t)$ (which is increasing on the range $[0, 1]$). $g_A(t)$ can also be interpreted as the normalized amount of leaked information accessible to the attacker over time which can be used by her to improve her attack effectiveness. Recall that the time of completing an attack successfully is represented by a random variable t_A. It follows that the function $g_A(t)$ should be dependent on t_A. In doing so, we define a general function $g_A(t)$ reaching to 1 (i.e., the amount at which the attacker would control the whole resource completely) at one unit of time. We represent, as an example, a simple version of this function in the left-hand side of Fig. 2. To represent the described dependency, we use then the function $g_A(t/t_A)$ for the reward calculation for

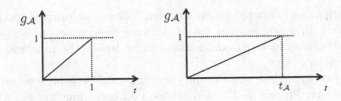

Fig. 2. The attacker's reward function during the time of completing an attack successfully depends on t_A. To show this dependency in our model, we define a function as shown on the left-hand side of this figure with one unit of time to reach 1. The figure on the right-hand side is $g_A(t/t_A)$ representing this dependence.

the attacker during the time of completing the attack successfully, i.e., as shown on the right-hand side of Fig. 2.

Defense Quality: In FlipLeakge, we consider the quality of the defender's recovery action (or alternatively the ability of the attacker to transfer information from a previous attack to the next attempt). That is, the defender's recovery action does not guarantee regaining complete control over the resource, so that the attacker has an initial advantage (during the next attack attempt) and retains a foothold in the system. In other words, the defender's defense strategy cannot entirely eliminate previous attacks' effects. Motivating examples for this imperfect recovery model are *advanced persistent threats*. These attacks are typically driven by human staff who intelligently make use of any available and gathered information during the next multi-stage attack step which may include an initial compromise, foothold establishment, reconnaissance, etc. In this scenario, any recovery move by the defender will frequently only partially remove the attacker from the system, or at the very least cannot eliminate any information advantage by the attacker. In the FlipLeakage game, we introduce a new random variable, i.e., α with range $[0,1]$, to represent the fraction of retained control over the previously compromised resource by the attacker after the defender's recovery move.

In the worst case, the defender's recovery move does not impact the level of the resource being controlled by the attacker (i.e., $\alpha = 1$). In contrast, $\alpha = 0$ represents the situation when the defender's recovery is perfect. Then, the attacker has to start with a zero level of knowledge during her next attack. We model α as a continuous random variable with PDF $f_\alpha(.)$ in which α chooses values between zero and one, i.e., $\alpha \in [0,1]$. Note that in the FlipLeakage model, the attacker never starts with a higher level than the level attained in the most recent compromise attempt, i.e., we assume that defense moves are not counterproductive. For simplicity, we assume that the random variable α takes its value after the first attack and it remains constant during the game. For future work, we will consider the case where the values of α are completely independent from each other in each step of the attack.

Players' Strategies: In FlipLeakage, we assume that the defender moves according to periodic strategies, i.e., the time interval between two consecutive moves is identical and denoted by $\delta_{\mathcal{D}}$. In what follows, we provide two examples to show that in practice, several major software vendor organizations update their security policies in a periodic manner to underline the practical relevance of this assumption.

The first example that we take into account are Microsoft's security policy updates which are known as *Patch Tuesday*, i.e., according to [20], "Microsoft security bulletins are released on the second Tuesday of each month." We visualize the time differences among security updates from March 14th, 2015, until March 12th, 2016, which is shown in Fig. 3(a). In this figure, the vertical axis represents the number of security updates for each update instance. On the horizontal axis, 0 represents the first security update we take into account which took place on March 14th, 2015. Based on this figure, Microsoft security policy updates are almost perfectly periodic. We only observe two dates with out-of-schedule security updates. These two security updates are corresponding to an update for Internet Explorer and a vulnerability in a Microsoft font driver which allowed remote code execution.

Another example are Oracle's critical patch updates. These updates occur in January, April, July, and October of each year. To visualize the time differences between updates, which are shown in Fig. 3(b), we consider Oracle's critical patch updates from 13 July, 2013, to January 19, 2016, based on available information at [23]. We calculate the time differences between two consecutive patch updates in terms of days and divided this number by 30 in order to calculate an approximate difference in months. In this figure, 1 along the vertical axis represents the occurrence of a patch update. We observe that Oracle's policy for critical patch updates is almost periodic.[1]

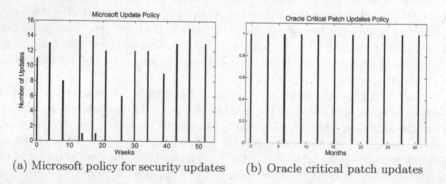

(a) Microsoft policy for security updates   (b) Oracle critical patch updates

Fig. 3. In practice, many organizations update their system according to periodic strategies. As examples, we provide two organizations: (1) Microsoft and (2) Oracle.

---

[1] Note that in our model, we do not consider the case where a software vendor has the ability to conduct out-of-schedule security updates. We are going to consider this issue in future work.

In the FlipLeakage model, we assume that the attacker moves right after the defender. We follow with this assumption the results of [16] who showed that in the scenario of a defender with a periodic strategy, the best strategy for the attacker, who has the ability to observe the defender's defense strategy, is to move right after the defender.

Other Parameters: The cost of the defender's recovery moves and the attacker's attack moves are represented by c_D and c_A, respectively, and we assume that they do not change over time. Examples of the defender's moves are changes of passwords, reinstallations of systems, and the application of new patches. Taking steps to incrementally infer cryptographic keys, brute-force passwords, or to inject malware are examples of the attacker's moves.

Once the attacker controls the resource completely, she receives an immediate reward which is represented by a constant value I_A. The rationale behind the introduction of this parameter is that once the attacker infers the defender's secret such as a cryptographic key, she can, for example, decrypt secret messages which she has collected.

For the time that the attacker (defender) controls the resource completely, we assume that the defender's (attacker's) reward is equal to zero and the attacker (defender) receives B_A (B_D) per unit of time controlling the resource. For example, these incremental earnings for the attacker represent newly arriving messages which can be decrypted with the compromised key. Note that the resource is controlled by the attacker completely after a successful attack and before the next recovery move by the defender.

4   Payoff Model

In this section, we develop the payoff functions for the FlipLeakage model based on what we presented in Sect. 3.

The time required to execute an attack successfully is defined by a continuous random variable with PDF f_A. We consider one of the realizations of this random variable as t_A. Moreover, the time between two consecutive defender's moves is represented by δ_D. Based on the t_A realization, we have two possible cases, i.e., $t_A \geq \delta_D$ and $t_A < \delta_D$. In what follows, we consider each of these two cases separately and then combine them according to the probability of each case to propose the payoff function.

Case 1: $t_A < \delta_D$

In this case, the attacker can complete her attack before the defender's recovery move. Hence, she receives the immediate reward for compromising the resource completely, i.e., I_A, as well as the reward for controlling the resource completely, i.e., B_A.

In our model, we assume that the attacker's control over the resource does not fall to zero right after the defender's recovery move. As discussed in Sect. 3, we have introduced a new parameter, α, and described the resulting changes to players' payoffs. For $t_A < \delta_D$, the attacker controls the resource completely before the next recovery move by the defender. Then, right after the defender's move, the attacker controls a fraction α of the resource. For the remainder of the resource to be taken over, i.e., $(1 - \alpha)$, the attacker can gain control based on $g_A(t/t_A)$. Hence, the attacker's benefit for this period is then based on $\alpha + (1 - \alpha)g_A(t/t_A)$. The attacker's payoff is as follows:

$$
u_A^1(t_A, \alpha, \delta_D) = \frac{\int_0^{t_A} \left(\alpha + (1 - \alpha) g_A \left(\frac{t}{t_A} \right) \right) dt + I_A + B_A(\delta_D - t_A) - c_A}{\delta_D}.
$$
(1)

In the above equation, based on our discussion in Sect. 3, the first term in the numerator represents the attacker's benefit due to information leakage. Note that the utility function is divided by δ_D, since this function is the average attacker's payoff over time.

Since the defender's move time is greater than the attacker's time of completing an attack successfully, the defender only receives a partial benefit during the period when the attacker is in the process of completing her attack. Therefore, the defender's payoff is as follows:

$$
u_D^1(t_A, \alpha, \delta_D) = \frac{\int_0^{t_A} \left(1 - \left(\alpha + (1 - \alpha) g_A \left(\frac{t}{t_A} \right) \right) \right) dt - c_D}{\delta_D}.
$$
(2)

Both payoff functions, i.e., Eqs. 1 and 2, are a function of t_A which is a random variable with PDF f_A as well as δ_D. Therefore, we need to calculate the expected value of both payoff functions. Note that these expected payoff functions are conditional, i.e., they are a function of a random variable t_A given that $t_A < \delta_D$. The conditional expected payoffs for these two functions are calculated as follows:

$$
u_A^1(\alpha, \delta_D) = \frac{\int_0^{\delta_D} u_A^1(t_A, \alpha, \delta_D) f_A(t_A) dt_A}{\int_0^{\delta_D} f_A(t_A) dt_A},
$$
(3)

$$
u_D^1(\alpha, \delta_D) = \frac{\int_0^{\delta_D} u_D^1(t_A, \alpha, \delta_D) f_A(t_A) dt_A}{\int_0^{\delta_D} f_A(t_A) dt_A}.
$$
(4)

Defender's and attacker's payoffs are both functions of α and δ_D. Finally, the probability of $t_A < \delta_D$ is calculated as follows:

$$
P[t_A < \delta_D] = \int_0^{\delta_D} f_A(t_A) dt_A.
$$
(5)

Case 2: $t_A \geq \delta_D$

In contrast to the previous case, the attacker cannot get the immediate reward as well as the benefit from controlling the resource completely. In this case, the attacker only reaches $g_A(\delta_D/t_A)$ level of control over the resource upon the defender's recovery move, and her reward is then equal to $\alpha g_A(\delta_D/t_A)$ right after the defender's move. The attacker gains her control for the rest of the resource, i.e., $(1 - \alpha)$, based on $g_A(t/t_A)$. Hence, during the time between two consecutive defender's moves, the attacker's benefit is equal to $\alpha g_A(\delta_D/t_A) + (1 - \alpha)g_A(t/t_A)$. Note that the upper integral bound changes into δ_D from t_A compared to the previous case.

$$u_A^2(t_A, \alpha, \delta_D) = \frac{\int_0^{\delta_D} \left(\alpha g_A\left(\frac{\delta_D}{t_A}\right) + (1 - \alpha)g_A\left(\frac{t}{t_A}\right)\right) dt - c_A}{\delta_D}. \tag{6}$$

The defender's payoff function is almost equivalent to Eq. 2 except the upper bound for the integral is changed into δ_D. Hence, the defender's payoff is as follows:

$$u_D^2(t_A, \alpha, \delta_D) = \frac{\int_0^{\delta_D} \left(1 - \left(\alpha g_A\left(\frac{\delta_D}{t_A}\right) + (1 - \alpha)g_A\left(\frac{t}{t_A}\right)\right)\right) dt - c_D}{\delta_D}. \tag{7}$$

Both players' payoffs are functions of t_A, α, and δ_D. We take the conditional expectation over parameter t_A in order to calculate the average payoffs with respect to t_A for this condition. The resulting equations are:

$$u_A^2(\alpha, \delta_D) = \frac{\int_{\delta_D}^{\infty} u_A^2(t_A, \alpha, \delta_D) f_A(t_A) dt_A}{\int_{\delta_D}^{\infty} f_A(t_A) dt_A}, \tag{8}$$

$$u_D^2(\alpha, \delta_D) = \frac{\int_{\delta_D}^{\infty} u_D^2(t_A, \alpha, \delta_D) f_A(t_A) dt_A}{\int_{\delta_D}^{\infty} f_A(t_A) dt_A}. \tag{9}$$

Furthermore, the probability that the required time by the attacker to compromise the resource entirely is greater than the time between two consecutive recovery moves is given by:

$$P[t_A \geq \delta_D] = \int_{\delta_D}^{\infty} f_A(t_A) dt_A. \tag{10}$$

By taking into account the probability of occurrence of each condition as well as their corresponding payoffs, we can calculate the defender's and the attacker's payoff functions which are represented by the following equations, respectively.

$$u_D(\alpha, \delta_D) = P[t_A \geq \delta_D]u_D^2(\alpha, \delta_D) + P[t_A < \delta_D]u_D^1(\alpha, \delta_D), \tag{11}$$

$$u_{\mathcal{A}}(\alpha, \delta_D) = P[t_{\mathcal{A}} \geq \delta_D] u_{\mathcal{A}}^2(\alpha, \delta_D) + P[t_{\mathcal{A}} < \delta_D] u_{\mathcal{A}}^1(\alpha, \delta_D). \tag{12}$$

In the above equation, each player's payoff is a function of α and $\delta_{\mathcal{A}}$. As mentioned before, α is a random variable whose range is in $[0, 1]$ with PDF $f_\alpha(.)$. Therefore, we can calculate the expected value of the defender's and the attacker's payoff functions with respect to α being represented in the following equations, respectively.

$$u_D(\delta_D) = \int_0^1 u_D(\alpha, \delta_D) f_\alpha(\alpha) d\alpha, \tag{13}$$

$$u_{\mathcal{A}}(\delta_D) = \int_0^1 u_{\mathcal{A}}(\alpha, \delta_D) f_\alpha(\alpha) d\alpha. \tag{14}$$

5   Analytical Results

In the previous section, we have developed the general payoff functions for the FlipLeakage model. Our payoff calculations are general and can be applied to many cybersecurity problems and we did not quantify any of the parameters being used in our model. For our analyses in this paper, we quantify $g_{\mathcal{A}}(.)$, $f_{\mathcal{A}}(.)$, and $f_\alpha(.)$, but we believe that the concrete functions we use still allow for meaningful insights about the stealthy information leakage scenarios. The instantiations of the other parameters in our proposed models would be specific to the concrete scenario under consideration, e.g., the corresponding cost for each player as well as the benefits.

To model the time of the attacker completing her attack successfully, we consider an *exponential* distribution with rate parameter $\lambda_{\mathcal{A}}$. The rationale behind choosing an exponential distribution for the random variable $t_{\mathcal{A}}$ is the memoryless feature of this distribution. Due to the memoryless condition, if the defender knows that his system is not compromised entirely at a specific time, it does not give any further information to the defender about the time of the next potential compromise. Moreover, the exponential distribution is a widely accepted candidate to model waiting times for event-driven models. The exponential distribution with rate parameter $\lambda_{\mathcal{A}}$ is as follows:

$$f_{\mathcal{A}}(t_{\mathcal{A}}) = \begin{cases} \lambda_{\mathcal{A}} e^{-\lambda_{\mathcal{A}} t_{\mathcal{A}}} & \text{if } t_{\mathcal{A}} \geq 0 \\ 0 & \text{if } t_{\mathcal{A}} < 0. \end{cases} \tag{15}$$

Moreover, for the random variable $\alpha \in [0, 1]$, we consider the uniform distribution, since the defender does not have any knowledge about the ability of the attacker to use previously leaked information and, accordingly, all values are possible with the same probability. The uniform distribution, $f_\alpha(.)$, is represented in Eq. 16.

$$f_\alpha(\alpha) = \begin{cases} 1 & \text{if } 0 \leq \alpha \leq 1 \\ 0 & \text{Otherwise.} \end{cases} \tag{16}$$

The attacker's reward function during the time to launch her attack successfully can be represented by a linear function:

$$g_{\mathcal{A}}\left(\frac{t}{t_{\mathcal{A}}}\right) = \begin{cases} \dfrac{t}{t_{\mathcal{A}}} & \text{if } 0 \leq t \leq t_{\mathcal{A}} \\ 0 & \text{Otherwise.} \end{cases} \tag{17}$$

In the following, we provide our lemmas and theorem based on our payoff calculation and the specification described above. First, the defender's and the attacker's best responses are stated in Lemma 1 and Lemma 2, respectively. Then, we propose the Nash equilibrium of the game being stated in Theorem 1.

Lemma 1. *The defender's best response is as follows:*

- *The defender plays a periodic strategy with period $\delta_{\mathcal{D}}^{\star}$ which is the solution of Eq. 18, if the corresponding payoff is non-negative, i.e., $u_{\mathcal{D}}(\delta_{\mathcal{D}}^{\star}) \geq 0$, and it yields a higher payoff compared to other solutions of Eq. 18.*

$$BR_{\mathcal{D}} = e^{-\lambda_{\mathcal{A}}\delta_{\mathcal{D}}}\left(\frac{1}{4} - \frac{3}{4}\lambda_{\mathcal{A}}\delta_{\mathcal{D}} + \frac{3}{4}\lambda_{\mathcal{A}} + \frac{1}{4\lambda_{\mathcal{A}}\delta_{\mathcal{D}}^2}\right)$$
$$+ \frac{1}{\delta_{\mathcal{D}}^2}\left(c_{\mathcal{D}} - \frac{1}{4\lambda_{\mathcal{A}}}\right) - \frac{3}{4}\lambda_{\mathcal{A}}\Gamma(0, \lambda_{\mathcal{A}}\delta_{\mathcal{D}}) = 0. \tag{18}$$

- *The defender drops out of the game (i.e., the player does not move anymore) if Eq. 18 has no solution for $\delta_{\mathcal{D}}$.*
- *The defender drops out of the game if the solutions of Eq. 18 yield a negative payoffs, i.e., $u_{\mathcal{D}}(\delta_{\mathcal{D}}) < 0$.*

Note that in Lemma 1, $\Gamma(0, \lambda_{\mathcal{A}}\delta_{\mathcal{D}})$ represents a Gamma function which is defined as follows:

$$\Gamma(s, x) = \int_x^{\infty} t^{s-1} e^{-t} dt. \tag{19}$$

Proof of Lemma 1 is provided in Appendix A.1.

Lemma 1 exhibits how we should calculate the defender's time between his two consecutive moves. As we see in Eq. 18, the defender's best response is a function of $c_{\mathcal{D}}$ and $\lambda_{\mathcal{A}}$.

Lemma 2 describes the attacker's best response in the FlipLeakage game.

Lemma 2. *In the FlipLeakage game model, the attacker's best response is:*

- *The attacker moves right after the defender if $c_{\mathcal{A}} < M(\delta)$ where*

$$M(\delta_{\mathcal{D}}) = \frac{3}{4}\delta_{\mathcal{D}}\lambda_{\mathcal{A}}\Gamma(0, \delta_{\mathcal{D}}\lambda_{\mathcal{A}}) + I_{\mathcal{A}} + B_{\mathcal{A}}\delta_{\mathcal{D}} + \frac{3}{4\lambda_{\mathcal{A}}} + B_{\mathcal{A}}\left(\delta_{\mathcal{D}} + \frac{1}{\lambda_{\mathcal{A}}}\right)e^{-\delta_{\mathcal{D}}\lambda_{\mathcal{A}}}$$
$$- \left(I_{\mathcal{A}} + B_{\mathcal{A}}\delta_{\mathcal{D}} + \frac{3}{4}\left(\delta_{\mathcal{D}} + \frac{1}{\lambda_{\mathcal{A}}}\right)\right)e^{-\delta_{\mathcal{D}}\lambda_{\mathcal{A}}} - \frac{B_{\mathcal{A}}}{\lambda_{\mathcal{A}}}. \tag{20}$$

- *The attacker drops out of the game if $c_{\mathcal{A}} > M(\delta)$.*

- *Otherwise, i.e., $c_A = M(\delta)$, dropping out of the game and moving right after the defender are both the attacker's best responses.*

The proof of Lemma 2 is provided in Appendix A.2. This lemma identifies conditions in which the attacker should move right after the defender, not move at all, and be indifferent between moving right after the defender and not moving at all. Note that the attacker's decision depends on c_A, δ_D, λ_A, I_A, and B_A.

The following theorem describes the Nash equilibria of the FlipLeakage game based on our described lemmas.

Theorem 1. *The FlipLeakage game's pure Nash equilibria can be described as follows.*

A. If Eq. 18 has a solution, i.e., $\delta_D^\star$, yielding the highest positive payoff for the defender compared to other solutions (if other solutions exist), then the following two outcomes apply:

1- If $c_A \leq M(\delta_D)$, then there is a unique pure Nash equilibrium in which the defender moves periodically with period $\delta_D^\star$ and the attacker moves right after the defender.

2- If $c_A > M(\delta_D)$, then there exists no pure Nash equilibrium.

B. If Eq. 18 does not have a solution or the solutions of this equation yield a negative payoff for the defender, i.e., $u_D(\delta_D) < 0$, then there exists a unique pure Nash equilibrium in which the defender does not move at all and the attacker moves once at the beginning of the FlipLeakage game.

The proof of Theorem 1 is provided in Appendix A.3.

In this theorem, in the first case, the defender's cost is lower than his benefit when he moves according to the solution of Eq. 18 and the attacker's cost is lower than Eq. 20. Hence, the attacker moves right after the defender's periodic move. In the second case, if the defender moves periodically, it is not beneficial for the attacker to move at all. Therefore, it is better for the defender to not move at all. But, if the defender does not move at all, the attacker can move once at the beginning of the game and control the resource for all time. However, as a result, the defender should move in order to hinder this situation. Because of this strategic uncertainty, in this scenario a Nash equilibrium does not exist. The third case represents the situation where the defender's benefit is lower than his cost for defending the resource. Then, it is beneficial for him to not move at all, and because of that the attacker has to move only once at the beginning of the game.

6 Numerical Illustrations

In this section, we provide selected numerical illustrations for our theoretical findings. First, we represent the defender's best response curves, i.e., Eq. 18, as well as the defender's payoff for different defender's cost values, i.e., c_D, which are depicted in Fig. 4. Then, we illustrate the defender's best responses for different values of c_D and λ_A in Fig. 5.

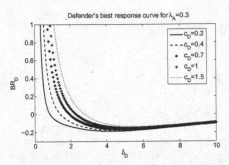
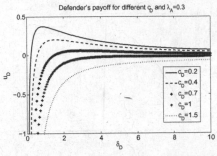

(a) Defender's best response curve for different values of $c_{\mathcal{D}}$.

(b) Defender's payoff as function of $\delta_{\mathcal{D}}$.

Fig. 4. The defender's best response curves and the corresponding payoff functions for different values of $c_{\mathcal{D}}$ are represented in (a) and (b), respectively. These two figures depict the situation that Eq. 18 has a solution, but the corresponding payoff may be negative.

We plot Eq. 18, i.e., the defender's best response curve, for different values of $c_{\mathcal{D}}$, i.e., $c_{\mathcal{D}} = \{0.2, 0.4, 0.7, 1, 1.5\}$, and $\lambda_{\mathcal{A}} = 0.3$ in Fig. 4(a). We illustrate the defender's payoff for these values in Fig. 4(b), as well. For all of these different $c_{\mathcal{D}}$s, Eq. 18 has a solution. But as we see in Fig. 4(b), the defender's payoffs are negative for $c_{\mathcal{D}} = \{0.7, 1, 1.5\}$ for all values of $\delta_{\mathcal{D}}$. Therefore, the defender will drop out of the game given these defense costs. For lower values of $c_{\mathcal{D}}$, i.e., $c_{\mathcal{D}} = \{0.2, 0.4\}$, the defender's best responses are to move periodically with period 0.8711 and 1.4681, respectively. This also provides us with the intuition that the higher the defender's costs are, the slower will be the defender's moves. To examine this intuition, we calculate the defender's best responses for different values of $c_{\mathcal{D}}$.

Figure 5(a) represents the defender's best response for different values of defense costs in which $\lambda_{\mathcal{A}} = 0.5$. This figure corroborates our intuition that the higher the defense costs are, the slower will be the defender's move period. When the cost of defense is high, the defender's best response is to drop out of the game which is represented as $\delta_{\mathcal{D}}^{\star} = 0$ in Fig. 5(a).

We are also interested to see the relation between $\lambda_{\mathcal{A}}$ and $\delta_{\mathcal{D}}$. We represent this relation in Fig. 5(b). It is worth mentioning that an exponential distribution with parameter $\lambda_{\mathcal{A}}$ has mean being equal to $1/\lambda_{\mathcal{A}}$. In the FlipLeakage game, a higher value of $\lambda_{\mathcal{A}}$ means that the attacker will successfully compromise the defender's system faster on average which is corresponding to $1/\lambda_{\mathcal{A}}$. Figure 5(b) represents the defender's best response for different values of $\lambda_{\mathcal{A}}$ for specific defender's cost, i.e., $c_{\mathcal{D}} = 0.3$. This figure shows that the faster the attacker can completely compromise the defender's system on average, the faster will be the defender's periodic move. In other words, the defender moves faster when the attacker's average time to successfully compromise the defender's system is faster. But if the attacker's average time to successfully compromise the

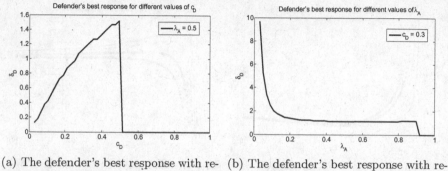

(a) The defender's best response with respect to his cost

(b) The defender's best response with respect to λ_A

Fig. 5. The impact of c_D and λ_A on the defender's best response

defender's system is too fast, the rational choice for the defender is to drop out of the game.

7   Conclusion

In this paper, we have proposed a novel theoretical model to provide guidance for the defender's optimal defense strategy when faced with a stealthy information leakage threat. In our model, an attacker will incrementally take ownership of a resource (e.g., as observed during advanced persistent threats). While her final objective is a complete compromise of the system, she may derive some utility during the preliminary phases of the attack. The defender can take a costly mitigation move and has to decide on its optimal timing.

In the FlipLeakage game model, we have considered the scenario when the defender only partially eliminates the foothold of the attacker in the system. In this scenario, the defender cannot undo any information leakage that has already taken place during an attack. We have derived optimal strategies for the agents in this model and present numerical analyses and graphical visualizations.

We highlight two observations from our numerical analyses which match well with intuition. First, the higher the defender's cost, the slower is the defender's periodic move. The second observation is that the faster the attacker's average time to compromise the defender's system completely (i.e., higher λ_A), the faster is the defender's periodic move. In addition, our model also allows for the determination of the impact of less-than-optimal strategies, and comparative statements regarding the expected outcomes of different periodic defensive approaches in practice, when information about the attacker and her capabilities is extremely scarce. As this problem area is understudied but of high practical significance, advancements that allow a rigorous reasoning about defense moves against stealthy attackers are of potentially high benefit.

In future work, we aim to conduct theoretical and numerical analyses using insights from data about practical information leakage scenarios. However, our current study is an important first step to reason about frequently criticized system reset policies to prevent information leakage in high-value systems. Reset policies have to provide an expected utility in the absence of concrete evidence due to the stealthiness of attacks which can be challenging to articulate. Our work also illustrates the positive deterrence function of system reset policies from a theoretical perspective. Further, we aim to consider a more general case in which the values of t_A and α are different in each step of the attack. In future work, we will also consider the case where a defender (e.g., a software vendor) has the ability to provide out-of-schedule security updates besides the periodic one.

Acknowledgments. We appreciate the comments from the anonymous reviewers. An earlier version of this paper benefited from the constructive feedback from Aron Laszka. All remaining errors are our own.

A   Proof

A.1   Proof of Lemma 1

Based on our payoff calculation, i.e., Eq. 13, as well as the quantified parameters, i.e., $g_A(.)$, $f_A(.)$, and $f_\alpha(.)$, the defender's payoff is:

$$u_D(\delta_D) = \frac{1}{\delta_D}\left(\frac{1}{4\lambda_A}\left(1 - e^{-\lambda_A\delta_D}\right) - c_D\right) + \frac{3}{4}e^{-\lambda_A\delta_D} - \frac{3}{4}\lambda_A\delta_D\Gamma(0, \lambda_A\delta_D) \tag{21}$$

To find the maximizing time between two consecutive defender's moves (if there exist any), we take the partial derivative of Eq. 21 with respect to δ_D and solve it for equality to 0 as follows:

$$\begin{aligned}
\frac{\partial u_D}{\partial \delta_D} = &-\frac{1}{\delta_D^2}\left(\frac{1}{4\lambda_A} - c_D - \frac{1}{4\lambda_A}e^{-\lambda_A\delta_D}\right) + \frac{1}{4}e^{-\lambda_A\delta_D} \\
&-\frac{3}{4}\lambda_A\delta_D e^{-\lambda_A\delta_D} - \frac{3}{4}\lambda_A\Gamma(0, \lambda_A\delta_D) + \frac{3}{4}\lambda_A e^{-\lambda_A\delta_D} = 0
\end{aligned} \tag{22}$$

Note that Eq. 18 is neither increasing nor decreasing on δ_D. Therefore, we have three possibilities for the above equation: (1) no solution, (2) one solution, and (3) more than one solution. When there is no solution, the defender's best response is to drop out of the game. In the case of one solution, the defender moves periodically with δ_D, i.e., the solution of Eq. 18 if the resulting payoff is non-negative. When there is more than one solution, the defender plays periodically with the solution with the highest non-negative payoff. Otherwise, the defender drops out of the game.   □

A.2   Proof of Lemma 2

In order to calculate the attacker's payoff, we first calculate the following based on Eq. 12.

$$u_{\mathcal{A}}(\alpha, \delta_{\mathcal{D}}) = \frac{1}{\delta_{\mathcal{D}}} \left(\left(\frac{1+\alpha}{2} - B_{\mathcal{A}} \right) \left(\frac{1}{\lambda_{\mathcal{A}}} - \frac{1}{\lambda_{\mathcal{A}}} e^{-\lambda_{\mathcal{A}} \delta_{\mathcal{D}}} - \delta_{\mathcal{D}} e^{-\lambda_{\mathcal{A}} \delta_{\mathcal{D}}} \right) \right.$$
$$\left. + (B_{\mathcal{A}} \delta_{\mathcal{D}} + I_{\mathcal{A}}) \left(1 - e^{-\lambda_{\mathcal{A}} \delta_{\mathcal{D}}} \right) + \frac{1+\alpha}{2} \delta_{\mathcal{D}} \lambda_{\mathcal{A}} \Gamma(0, \delta_{\mathcal{D}} \lambda_{\mathcal{A}}) - c_{\mathcal{A}} \right). \tag{23}$$

According to Eq. 14, the attacker's payoff is as follows.

$$u_{\mathcal{A}}(\delta_{\mathcal{D}}) = \frac{1}{\delta_{\mathcal{D}}} \left(\left(\frac{3}{4} - B_{\mathcal{A}} \right) \left(\frac{1}{\lambda_{\mathcal{A}}} - \frac{1}{\lambda_{\mathcal{A}}} e^{-\lambda_{\mathcal{A}} \delta_{\mathcal{D}}} - \delta_{\mathcal{D}} e^{-\lambda_{\mathcal{A}} \delta_{\mathcal{D}}} \right) \right.$$
$$\left. + (B_{\mathcal{A}} \delta_{\mathcal{D}} + I_{\mathcal{A}}) \left(1 - e^{-\lambda_{\mathcal{A}} \delta_{\mathcal{D}}} \right) + \frac{3}{4} \delta_{\mathcal{D}} \lambda_{\mathcal{A}} \Gamma(0, \delta_{\mathcal{D}} \lambda_{\mathcal{A}}) - c_{\mathcal{A}} \right). \tag{24}$$

The attacker moves right after the defender if her payoff is positive, i.e., $u_{\mathcal{A}}(\delta_{\mathcal{D}}) > 0$. If the attacker's payoff is negative, her reward is lower than her cost. Then, a rational player does not have any incentive to actively participate in the game. Hence, the attacker drops out of the game. If $u_{\mathcal{A}}(\delta_{\mathcal{D}}) = 0$, the attacker is indifferent between moving right after the defender or dropping out of the game. By considering Eq. 24 and $u_{\mathcal{A}}(\delta_{\mathcal{D}}) \geq 0$, we can derive Eq. 20.   □

A.3   Proof of Theorem 1

In Lemma 1, we have provided the best response for the defender. The defender has two choices: periodic move or dropping out of the game. Similarly, according to Lemma 2, the attacker has two choices for her best response: she moves right after the defender or drops out of the game. Note that Nash equilibrium is a mutual best response.

In doing so, we first consider the case where the defender's best response is to drop out of the game (this means that Eq. 18 does not have any solution(s) giving non-negative payoff(s)). Therefore, the attacker's best choice is to move only once at the beginning of the game.

The other choice for the defender, according to Lemma 1, is to move period-ically when Eq. 18 has a solution which yields a positive payoff. By calculating $\delta_{\mathcal{D}}^{\star}$ using this equation, we insert this value to Eq. 20 and compare it with $c_{\mathcal{A}}$. Based on Lemma 2, the attacker has two possible choices. First, if $c_{\mathcal{A}} \leq M(\delta_{\mathcal{D}})$, the attacker will initiate her attack right after the defender's move. Hence, the Nash equilibrium is to move periodically from the defender side and the attacker should initiate her attack right after the defender's move. Second, if $c_{\mathcal{A}} > M(\delta_{\mathcal{D}})$, the attacker will drop out of the game. In this case, the best response for the defender is to never move. Since he controls the resource all the time without spending any cost. But, if the defender never moves, then it is beneficial for the attacker to move at the beginning of the game. Hence, this situation is not a Nash equilibrium.   □

References

1. Blackwell, D.: The noisy duel, one bullet each, arbitrary accuracy. Technical report, The RAND Corporation, D-442 (1949)
2. Bowers, K.D., Dijk, M., Griffin, R., Juels, A., Oprea, A., Rivest, R.L., Triandopoulos, N.: Defending against the unknown enemy: applying FlipIt to system security. In: Grossklags, J., Walrand, J. (eds.) GameSec 2012. LNCS, vol. 7638, pp. 248–263. Springer, Heidelberg (2012). doi:10.1007/978-3-642-34266-0_15
3. Experian, Small business doesn't mean small data: Experian data breach resolution advises small businesses to be prepared for a data breach (2013). https://www.experianplc.com/media/news/
4. Falliere, N., Murchu, L., Chien, E.: W32.Stuxnet Dossier. Technical report, Symantec Corp., Security Response (2011)
5. Farhang, S., Manshaei, M.H., Esfahani, M.N., Zhu, Q.: A dynamic Bayesian security game framework for strategic defense mechanism design. In: Poovendran, R., Saad, W. (eds.) GameSec 2014. LNCS, vol. 8840, pp. 319–328. Springer, Heidelberg (2014). doi:10.1007/978-3-319-12601-2_18
6. Feng, X., Zheng, Z., Hu, P., Cansever, D., Mohapatra, P.: Stealthy attacks meets insider threats: a three-player game model. In: Proceedings of MILCOM (2015)
7. Grossklags, J., Christin, N., Chuang, J.: Secure or insure? A game-theoretic analysis of information security games. In: Proceedings of the 17th International World Wide Web Conference, pp. 209–218 (2008)
8. Grossklags, J., Reitter, D.: How task familiarity and cognitive predispositions impact behavior in a security game of timing. In: Proceedings of the 27th IEEE Computer Security Foundations Symposium (CSF), pp. 111–122 (2014)
9. Hu, P., Li, H., Fu, H., Cansever, D., Mohapatra, P.: Dynamic defense strategy against advanced persistent threat with insiders. In: Proceedings of the 34th IEEE International Conference on Computer Communications (INFOCOM) (2015)
10. Johnson, B., Grossklags, J., Christin, N., Chuang, J.: Uncertainty in interdependent security games. In: Alpcan, T., Buttyán, L., Baras, J.S. (eds.) GameSec 2010. LNCS, vol. 6442, pp. 234–244. Springer, Heidelberg (2010). doi:10.1007/978-3-642-17197-0_16
11. Johnson, B., Laszka, A., Grossklags, J.: Games of timing for security in dynamic environments. In: Khouzani, M.H.R., Panaousis, E., Theodorakopoulos, G. (eds.) GameSec 2015. LNCS, vol. 9406, pp. 57–73. Springer, Heidelberg (2015). doi:10.1007/978-3-319-25594-1_4
12. Joyce, R.: Disrupting nation state hackers (2016). https://www.youtube.com/watch?v=bDJb8WOJYdA
13. Laszka, A., Felegyhazi, M., Buttyan, L.: A survey of interdependent information security games. ACM Comput. Surv. 47(2), 23:1–23:38 (2014)
14. Laszka, A., Horvath, G., Felegyhazi, M., Buttyán, L.: FlipThem: modeling targeted attacks with FlipIt for multiple resources. In: Poovendran, R., Saad, W. (eds.) GameSec 2014. LNCS, vol. 8840, pp. 175–194. Springer, Heidelberg (2014). doi:10.1007/978-3-319-12601-2_10
15. Laszka, A., Johnson, B., Grossklags, J.: Mitigating covert compromises. In: Chen, Y., Immorlica, N. (eds.) WINE 2013. LNCS, vol. 8289, pp. 319–332. Springer, Heidelberg (2013). doi:10.1007/978-3-642-45046-4_26
16. Laszka, A., Johnson, B., Grossklags, J.: Mitigation of targeted and non-targeted covert attacks as a timing game. In: Das, S.K., Nita-Rotaru, C., Kantarcioglu, M. (eds.) GameSec 2013. LNCS, vol. 8252, pp. 175–191. Springer, Heidelberg (2013). doi:10.1007/978-3-319-02786-9_11

17. Laszka, A., Johnson, B., Schöttle, P., Grossklags, J., Böhme, R.: Secure team composition to Thwart insider threats, cyber-espionage. ACM Trans. Internet Technol. **14**(2–3), 19:1–19:22 (2014)

18. Leslie, D., Sherfield, C., Smart, N.P.: Threshold FlipThem: when the winner does not need to take all. In: Khouzani, M.H.R., Panaousis, E., Theodorakopoulos, G. (eds.) GameSec 2015. LNCS, vol. 9406, pp. 74–92. Springer, Heidelberg (2015). doi:10.1007/978-3-319-25594-1_5

19. Manshaei, M.H., Zhu, Q., Alpcan, T., Bacşar, T., Hubaux, J.-P.: Game theory meets network security and privacy. ACM Comput. Surv. **45**(3), 25:1–25:39 (2013)

20. Microsoft, Microsoft security bulletin. https://technet.microsoft.com/en-us/security/bulletin/dn602597.aspx

21. Nadella, S.: Enterprise security in a mobile-first, cloud-first world (2015). http://news.microsoft.com/security2015/

22. Nochenson, A., Grossklags, J.: A behavioral investigation of the FlipIt game. In: 12th Workshop on the Economics of Information Security (WEIS) (2013)

23. Oracle, Oracle critical patch updates. http://www.oracle.com/technetwork/topics/security/alerts-086861.html

24. Pal, R., Huang, X., Zhang, Y., Natarajan, S., Hui, P.: On security monitoring in SDNS: a strategic outlook. Technical report

25. Pawlick, J., Farhang, S., Zhu, Q.: Flip the cloud: cyber-physical signaling games in the presence of advanced persistent threats. In: Khouzani, M.H.R., Panaousis, E., Theodorakopoulos, G. (eds.) GameSec 2015. LNCS, vol. 9406, pp. 289–308. Springer, Heidelberg (2015). doi:10.1007/978-3-319-25594-1_16

26. Pham, V., Cid, C.: Are we compromised? Modelling security assessment games. In: Grossklags, J., Walrand, J. (eds.) GameSec 2012. LNCS, vol. 7638, pp. 234–247. Springer, Heidelberg (2012). doi:10.1007/978-3-642-34266-0_14

27. Pu, Y., Grossklags, J.: An economic model and simulation results of app adoption decisions on networks with interdependent privacy consequences. In: Poovendran, R., Saad, W. (eds.) GameSec 2014. LNCS, vol. 8840, pp. 246–265. Springer, Heidelberg (2014). doi:10.1007/978-3-319-12601-2_14

28. Radzik, T.: Results and problems in games of timing. In: Fergusons, T.S., Shapleys, L.S., MacQueen, J.B. (eds.) Statistics, Probability and Game Theory: Papers in Honor of David Blackwell. Lecture Notes-Monograph Series, vol. 30, pp. 269–292. Institute of Mathematical Statistics, Hayward (1996)

29. Van Dijk, M., Juels, A., Oprea, A., Rivest, R.: FlipIt: the game of "stealthy takeover". J. Cryptol. **26**(4), 655–713 (2013)

30. Wellman, M.P., Prakash, A.: Empirical game-theoretic analysis of an adaptive cyber-defense scenario (preliminary report). In: Poovendran, R., Saad, W. (eds.) GameSec 2014. LNCS, vol. 8840, pp. 43–58. Springer, Heidelberg (2014). doi:10.1007/978-3-319-12601-2_3

31. Zhang, M., Zheng, Z., Shroff, N.: Stealthy attacks and observable defenses: a game theoretic model under strict resource constraints. In: Proceedings of the IEEE Global Conference on Signal and Information Processing (GlobalSIP), pp. 813–817 (2014)

Scalar Quadratic-Gaussian Soft Watermarking Games

M. Kıvanç Mıhçak, Emrah Akyol, Tamer Başar$^{(\boxtimes)}$, and Cédric Langbort

Coordinated Science Laboratory, University of Illinois, Urbana-Champaign,
Urbana, IL 61801, USA
kivancmihcak@gmail.com, {akyol,basar1,langbort}@illinois.edu

Abstract. We introduce a zero-sum game problem of *soft watermarking:* The hidden information (watermark) comes from a continuum and has a perceptual value; the receiver generates an estimate of the embedded watermark to minimize the expected estimation error (unlike the conventional watermarking schemes where both the hidden information and the receiver output are from a discrete finite set). Applications include embedding a multimedia content into another. We study here the scalar Gaussian case and use expected mean-squared distortion. We formulate the problem as a zero-sum game between the encoder & receiver pair and the attacker. We show that for linear encoder, the optimal attacker is Gaussian-affine, derive the optimal system parameters in that case, and discuss the corresponding system behavior. We also provide numerical results to gain further insight and understanding of the system behavior at optimality.

1 Introduction

Watermarking (also termed as information or data hiding throughout the paper) refers to altering an input signal to transmit information in a hidden fashion while preserving the perceptual quality. The watermarked signal is then subject to an attack which "sabotages" the receiver. We focus here on "robust" watermarking (unlike steganography or fragile watermarking): The decoder aims to recover the embedded watermark as accurately as possible, even in the presence of (potentially malicious) attacks as long as they preserve the perceptual quality.

Robust watermarking has been an active area of research for nearly two decades, with applications ranging from security-related ones (such as copyright protection, fingerprinting and traitor tracing) to the ones aiming conventional tasks related to multimedia management and processing (such as database annotation, in-band captioning, and transaction tracking). Since the underlying task is to transmit the (hidden) information by means of a watermark, the resulting scheme falls within the category of information transmission problems and can be analyzed using techniques from communications and information theory - see [1] for an overview of data hiding from such a perspective.

Furthermore, the presence of an intelligent attacker enables a game theoretic perspective: Encoder, decoder, and attacker can be viewed as players where

© Springer International Publishing AG 2016
Q. Zhu et al. (Eds.): GameSec 2016, LNCS 9996, pp. 215–234, 2016.
DOI: 10.1007/978-3-319-47413-7_13

the encoder and decoder share a common utility and their gain is exactly the loss of the attacker, thereby resulting in a zero-sum game. In [2,3], the authors formulated the problem of (robust) information hiding as a game between the encoder & decoder and the attacker; using an information-theoretic approach, they derived expressions for the maximum rate of reliable information transmission (i.e., capacity) for the i.i.d. (independent identically distributed) setup. An analogous approach has been used to extend these results to colored Gaussian signals in [4].

To the best of our knowledge, so far all of the robust watermarking approaches have assumed that the information to be transmitted is an element of a discrete finite (usually binary) set. This is because of the fact that in most intended applications, the watermark is aimed to represent an identity for usage or ownership (or simply permission to use in case of verification problems). Consequently, the receiver is usually designed to decode the embedded watermark (or in case of verification problems detect the presence or absence of a watermark), resulting in a joint source-channel coding problem, where the channel coding counterpart refers to the reliable transmission of the watermark and the source coding counterpart refers to the lossy compression of the unmarked source. In [2–4], following the conventional joint source-channel coding paradigm of information theory, an error event is said to occur if the decoded watermark is not the same as the embedded watermark (hence a hard decision).

In contrast with prior art, in this paper we propose a setup where the information to be transmitted is from a continuum and there is an associated *perceptual value*. As such, the receiver acts as an estimator, whose goal is to produce an estimate of the hidden information from the same continuum (rather than a decoder or detector that reaches a hard decision). Applications include the cases where we hide one multimedia signal inside another (such as embedding one smaller low-resolution image inside another larger high-resolution image, or hiding an audio message inside a video, etc.). In such cases, the receiver output is from a continuum as well and there is no hard decision made by it (unlike the prior art in robust watermarking); hence the receiver provides a solution to a *soft decision* problem[1,2]. Accordingly, we use the term *"soft watermarking"* to refer to such data hiding problems. Therefore, unlike the prior art where the fundamental problem is joint source-channel coding, in this case we have a joint *source-source* coding problem where the encoder needs to perform lossy compression on both the unmarked source and the data to be hidden.

As a first step toward our long-term goal of studying the soft watermarking problem in its full generality, we consider here a simpler version of the broad

---

[1] This distinction is reminiscent of the differentiation between hard and soft decoding methods in classical communication theory.

[2] One alternative approach for this problem may involve using a separated setup, where we first apply lossy compression to the information to be hidden that possesses perceptual value, and subsequently embed the compression output into the unmarked host using a conventional capacity-achieving data hiding code. It is not immediately clear which approach is superior; a comparative assessment constitutes part of our future research.

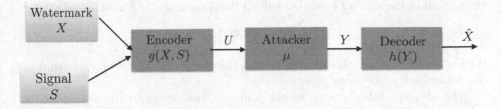

Fig. 1. The problem setting.

soft watermarking problem to gain insight: we confine ourselves to the scalar-Gaussian case where we use expected squared error as the distortion metric. In the future, our aim is to address the asymptotically high dimensional vector case (i.e., information-theoretic setting) with a general class of distributions associated with arbitrary distortion metrics.

In Sect. 2, we introduce the notation and provide the problem formulation. In Sect. 3, we present the main results: In Sect. 3.1, we show that Gaussian-affine attack mapping is optimal for the class of linear encoders; in Sect. 3.2, we derive optimal system parameters for such encoder and attacker classes; in Sect. 3.3, we discuss the system properties at optimality, provide bounds and analyze asymptotic behavior. We present numerical results in Sect. 4 and concluding remarks in Sect. 5.

2 Preliminaries

2.1 Notation

Let $\mathbb{R}$ and $\mathbb{R}^+$ denote the respective sets of real numbers and positive real numbers. Let $\mathbb{E}(\cdot)$ denote the expectation operator.

The Gaussian density with mean μ and variance σ^2 is denoted as $\mathcal{N}\left(\mu, \sigma^2\right)$. All logarithms in the paper are natural logarithms and may in general be complex valued, and the integrals are, in general, Lebesgue integrals.

2.2 Problem Definition

A generic model of the problem is presented in Fig. 1. We consider independent scalar Gaussian random variables $X \sim \mathcal{N}\left(0, \sigma_x^2\right)$ and $S \sim \mathcal{N}\left(0, \sigma_s^2\right)$ to denote the watermark (the data to be hidden) and the signal, respectively.

A deterministic mapping of X and S is transmitted over the channel generated by the encoder[3]. Let the transmitter strategy be denoted by $g(\cdot, \cdot)$, which is

---

[3] In classical (information-theoretic) watermarking literature, a pseudo-random key sequence is shared between the encoder and the decoder, mainly to render the attacker strategies memoryless. In this paper, we do not consider the key sequence in the problem formulation since our formulation is based on single-letter strategies.

an element of the space Γ_T, of real-valued Borel measurable functions satisfying:

$$\mathbb{E}\left\{(g(X,S) - S)^2\right\} \leq P_E. \tag{1}$$

We note that this is a classical constraint that limits the distortion incurred by the watermark embedding process.

The attacker (also termed as the jammer) has access to the output of the transmitter, $U = g(X, S)$, and outputs a random transformation of U, denoted by Y, i.e., assigns a probability measure, μ, on the output Y that satisfies

$$\int_{-\infty}^{\infty} \mathbb{E}\left\{(Y - U)^2 | Y\right\} \, d\mu(Y) \leq P_A \tag{2}$$

We denote the class of all associated probability measures μ for the jammer by $\mathcal{M}$. We note that the constraint (2) corresponds to the classical distortion constraint on the attacker used in the watermarking literature (see e.g., [2, Eq. 13], [3, Eq. 2.2]): It aims to guarantee that the attacker does not distort the watermarked signal beyond a perceptually acceptable level. Thus, in our framework, the attacker has two (possibly conflicting) objectives: (i) maximize the distortion between the watermark and its generated estimate by the decoder (cf. (3)), and (ii) maintain the usability of the attack output from a perceptual point of view (captured by (2)).

Note that, the constraint (2) differs from the traditional power constraint in classical communication (jamming) games, where the constraint on the attacker arises due to the physical limitations on the communication channel and can be formulated as a power constraint on the attack output Y (i.e., an upper bound on $\mathbb{E}(Y^2)$) [6]. Since such physical limitations do not exist in our current problem formulation, such a constraint is not immediately applicable to our setup[4]. Also note that, in assessing the perceptual quality of Y, as a first step and following [2,3], we take U as the benchmark for the attacker to compare. Alternatively, it is plausible to use S as the benchmark (see e.g., "type-S" constraint in [4, Eq. 2.3]), which implies imposing an upper bound on $\mathbb{E}(S - Y)^2$ and constitutes part of our future research as well.

We consider the power constraints (1, 2) in the expectation form, mainly for tractability purposes. Constraints for each realization (in almost sure sense) were also used in the literature [2], but are beyond the scope of our treatment here.

The receiver applies a Borel-measurable transformation $h(Y)$ on its input Y, so as to produce an estimate $\hat{X}$ of X, by minimizing the squared error distortion measure

$$J(g, h, \mu) = \int_{-\infty}^{\infty} \mathbb{E}\left\{(h(Y) - X)^2 | Y\right\} \, d\mu(Y) \tag{3}$$

---

[4] The investigation of a potential relationship between (2) and imposing an upper bound on $\mathbb{E}(Y^2)$ for the data hiding setup constitutes part of our future research.

We denote the class of all Borel-measurable mappings $h(\cdot)$ to be used as the receiver mapping by Γ_R. Joint statistics of X and S, and all objectives are common knowledge.

The common objective of the transmitter and the receiver is to minimize J by properly choosing $g \in \Gamma_T$ and $h \in \Gamma_R$, while the objective of the attacker is to maximize J over the choice of $\mu \in \mathcal{M}$. Since there is a complete conflict of interest, this problem constitutes a zero-sum game.

However, this game does not have a value, since the lower value of the game

$$\bar{J} = \sup_{\mu \in \mathcal{M}} \inf_{\substack{g \subset \Gamma_T \\ h \in \Gamma_R}} J(g,h,\mu)$$

is not well defined. This is because the attacker cannot guarantee that (2) is satisfied without knowing the encoder strategy g since the value of the left hand side of (2) depends on the joint distribution of U and Y which is impacted by g[5]. Hence our main focus is on the following minimax optimization problem which corresponds to the upper value of the game (which in fact safeguards the transmitter-receiver pair against worst attacks and is more relevant for the problem at hand)

$$J^* - \inf_{\substack{g \in \Gamma_T \\ h \in \Gamma_R}} \sup_{\mu \in \mathcal{M}} J(g,h,\mu). \tag{4}$$

Note that, the aforementioned formulation implies that the encoder mapping $g(\cdot)$ is known by the attacker. Note also that for each $g \in \Gamma_T$, we have a zero-sum game between the attacker and the receiver. This subgame has a well-defined value[6], and hence, inf and sup operators can be interchanged, and further they can be replaced by min and max, respectively, i.e., (4) is equivalent to

$$J^* = \inf_{g \in \Gamma_T} \max_{\mu \in \mathcal{M}} \min_{h \in \Gamma_R} J(g,h,\mu), \tag{5}$$

which we consider throughout the rest of the paper. In other words, there is no loss for the encoder-decoder team to determine and announce the decoder mapping before the attacker picks its own mapping, or there is no gain for the decoder to know the attacker mapping a priori.

---

[5] One way to get around this problem is to introduce soft constraints into the objective of the attacker. Then, the problem is no longer a zero-sum game. Another way is to define the attacker constraint for each realization, in almost sure sense, in which case the attacker can satisfy its constraint for any encoding strategy. These are beyond the scope of this paper.

[6] This is a zero-sum game where the objective is linear (hence, concave) in the attacker mapping for a fixed decoder map, and the optimal decoder mapping is unique (conditional mean) for a given attacker mapping. $\mathcal{M}$ is weak*-compact, and the minimizing h can be restricted to a compact subset of Γ_R (with (3) bounded away from zero); hence a saddle point exists due to the standard min-max theorem of game theory in infinite-dimensional spaces [7].

3   Main Results

Given Y, the encoder mapping $g(\cdot)$, and the attacker mapping $\mu(\cdot)$, the decoder's goal is to calculate the estimate $\hat{X}(Y)$ of X so as to minimize the expected squared error distortion measure (cf. (3)) $J = \mathbb{E}\left[X - \hat{X}(Y)\right]^2$. The minimizer here is the well-known MMSE (minimum mean squared error) estimate of X given Y: $\hat{X}_{MMSE}(Y) = \mathbb{E}(X|Y)$. Then, the resulting problem is

$$\inf_{g} \max_{\mu} J = \mathbb{E}\left[X - \hat{X}_{MMSE}(Y)\right]^2 \tag{6}$$

subject to the constraints (1, 2). In Sect. 3.1, we show that, in the sense of (6), the optimal functional form of μ is a "jointly-Gaussian-affine mapping" provided that the functional form of the encoder is linear. Using this result, in Sect. 3.2 we solve the problem (6) within the class of linear encoder mappings subject to $(1, 2)^7$ and characterize the parameters of the corresponding system. In Sect. 3.3, we provide a detailed discussion on the optimality results presented in Sect. 3.2.

3.1   On Optimal Functional Forms of the Encoder and Attacker

We first focus on a special case as an auxiliary step, where the encoder mapping is the identity operator, $g(X, S) = X$, together with a "generalized" version of the constraint (2), where an upper bound is imposed on $\mathbb{E}(Y - aU)^2$ for an arbitrary $a \in \mathbb{R}$. We present the corresponding optimality result of this special case in Lemma 1. We then use Lemma 1 as an auxiliary step to reach the main result of this section (Lemma 2), which states that the Gaussian-affine attack mapping is optimal for the class of linear encoders under the constraint (2).

Lemma 1. *Given the encoder strategy of $U = g(X, S) = X$, the solution to*

$$\max_{\mu} \mathbb{E}\left[X - \hat{X}_{MMSE}(Y)\right]^2 \tag{7}$$

subject to an upper bound on $\mathbb{E}(Y - aU)^2$ for some $a \in \mathbb{R}$ is of the form $Y = \kappa U + Z$ where $Z \sim \mathcal{N}(0, \sigma_z^2)$ is independent of U.

Proof. Define $\mathcal{C}(m_{XY}, m_{YY}) \triangleq \{\mu \,|\, \mathbb{E}_{\mu}(XY) = m_{XY}, \mathbb{E}_{\mu}(Y^2) = m_{YY}\}$, where $\mathbb{E}_{\mu}(\cdot)$ denotes expectation with respect to the joint distribution of X and Y, induced by the attack mapping $\mu(\cdot)$. Next, define $\hat{X}'(Y) \triangleq \frac{\mathbb{E}(XY)}{\mathbb{E}(Y^2)}Y$. Thus, by definition, $\hat{X}'(Y) := \hat{X}'_{\mathcal{C}}(Y)$ is the same for all $\mu \in \mathcal{C}(m_{XY}, m_{YY})$ given a pair (m_{XY}, m_{YY}). This further implies that, for any given $\mu \in \mathcal{C}(m_{XY}, m_{YY})$,

$$\mathbb{E}\left[X - \hat{X}_{MMSE,\mu}(Y)\right]^2 \le \mathbb{E}\left[X - \hat{X}'_{\mathcal{C}}(Y)\right]^2 \tag{8}$$

---

[7] As such, the result provided in Sect. 3.2 forms an upper bound on the solution of (6); see Remark 1 in Sect. 3.3 for a further discussion.

by the definition of the MMSE estimate. The upper bound of (8) is achieved when X and Y are jointly Gaussian, in which case $\hat{X}_{MMSE} = \hat{X}_{LMMSE} = \hat{X}'_C$, which is a well-known result. Thus, we conclude that among all the attack strategies which yield the same $\mathbb{E}(XY)$ and $\mathbb{E}(Y^2)$, the one (if exists) that renders X and Y jointly Gaussian achieves the maximum cost, thereby being the optimal choice for the attacker. A similar reasoning was used in Lemma 1 of [8] for a zero-delay jamming problem.

Next, note that, for any given $\mu \in \mathcal{C}(m_{XY}, m_{YY})$ and $\sigma_x^2 = \mathbb{E}(X^2)$, we have

$$\mathbb{E}_\mu [Y - aU]^2 = \mathbb{E}_\mu [Y - aX]^2 = m_{YY} - 2am_{XY} + \sigma_x^2,$$

implying that all elements of $\mathcal{C}(m_{XY}, m_{YY})$ yield the same $\mathbb{E}(Y - aU)^2$.

Let μ^* be a solution to (7) subject to an upper bound on $\mathbb{E}(Y - aU)^2$. Assuming existence, let μ' be an element of $\mathcal{C}(\mathbb{E}_{\mu^*}(XY), \mathbb{E}_{\mu^*}(Y^2))$ that renders X and Y jointly Gaussian. Due to the aforementioned arguments, we have $\mathbb{E}\left[X - \hat{X}_{MMSE,\mu^*}(Y)\right]^2 = \mathbb{E}\left[X - \hat{X}_{MMSE,\mu'}(Y)\right]^2 = \mathbb{E}\left[X - \hat{X}'_C(Y)\right]^2$ and that $\mathbb{E}_{\mu^*}[Y - aU]^2 = \mathbb{E}_{\mu'}[Y - aU]^2$. Hence, if μ' exists, it is optimal.

Existence: Consider the mapping $Y = \kappa U + Z = \kappa X + Z$ for some κ and $Z \sim \mathcal{N}(0, \sigma_z^2)$ independent of X. Then, straightforward algebra reveals that $\kappa = \mathbb{E}(XY)/\sigma_x^2$ and $\sigma_z^2 = \mathbb{E}(Y^2) - [\mathbb{E}(XY)]^2/\sigma_x^2$. Hence, given σ_x^2, there is a one-to-one mapping between the pairs of $(\mathbb{E}(XY), \mathbb{E}(Y^2))$ and (κ, σ_z^2). Consequently, for any given $(\mathbb{E}(XY), \mathbb{E}(Y^2))$, we can find (κ, σ_z^2) that guarantees the existence of μ' that renders X and Y jointly Gaussian. This completes the proof.   □

Lemma 2. *Given the linear encoder strategy of $U = g(X, S) = \alpha X + \beta S$ for some $\alpha, \beta \in \mathbb{R}$, the solution to*

$$\max_\mu \mathbb{E}\left[X - \hat{X}_{MMSE}(Y)\right]^2, \tag{9}$$

subject to $\mathbb{E}(Y - U)^2 \le P_A$, is of the form $Y = \kappa U + Z$, where $Z \sim \mathcal{N}(0, \sigma_z^2)$ is independent of U.

Proof. Let T denote the MMSE estimate of X given U: $T \overset{\triangle}{=} \hat{X}_{MMSE}(U)$. First, note that for any attack mapping μ and for any function $p(\cdot)$, $(X - T)$ is orthogonal to $p(Y)$:

$$\mathbb{E}[(X - T(U))p(Y)] = \mathbb{E}_U\{\mathbb{E}[Xp(Y)|U] - \mathbb{E}[T(U)p(Y)|U]\}$$
$$= \mathbb{E}_U\{\mathbb{E}[X|U]\mathbb{E}[p(Y)|U] - T(U)\mathbb{E}[p(Y)|U]\} \tag{10}$$
$$= 0, \tag{11}$$

where (10) follows from the fact that $X \leftrightarrow U \leftrightarrow Y$ forms a Markov chain in the specified order, and (11) follows from recalling that $\mathbb{E}[X|U] = T(U)$ by definition. This implies

$$J = \mathbb{E}\left[X - \hat{X}_{MMSE}(Y)\right]^2 = \mathbb{E}\left[X - T(U) + T(U) - \hat{X}_{MMSE}(Y)\right]^2$$

$$= \mathbb{E}\left[X - T(U)\right]^2 + \mathbb{E}\left[T(U) - \hat{X}_{MMSE}(Y)\right]^2, \tag{12}$$

where (12) follows from the fact that the estimation error $(X - T)$ is orthogonal to any function of U and Y (cf. (11)). Since $\mathbb{E}\left[X - T(U)\right]^2$ is invariant in μ, maximizing J over μ is equivalent to maximizing $\mathbb{E}\left[T(U) - \hat{X}_{MMSE}(Y)\right]^2$ over μ. Furthermore, since U is linear in X and they are jointly Gaussian, MMSE coincides with LMMSE, implying that $\theta T(U) = U$ for some $\theta \in \mathbb{R}$. Therefore, (9) is equivalent to

$$\max_{\mu} \mathbb{E}\left[T - \hat{X}_{MMSE}(Y)\right]^2, \tag{13}$$

subject to $\mathbb{E}(Y - \theta T)^2 \leq P_A$. By Lemma 1, we know the solution to (13): At optimality, $Y = \kappa' T + Z$ where $Z \sim \mathcal{N}(0, \sigma_z^2)$ is independent of T. But since T is linear in U, this is equivalent to the statement of the lemma. $\square$

3.2   Characterization of Optimal System Parameters

Motivated by Lemma 2, throughout the rest of the paper we confine ourselves to the class of linear mappings for the encoder (14) and jointly-Gaussian-affine mappings for the attacker (15):

$$U = g(X, S) = \alpha X + \beta S, \tag{14}$$

$$Y = \kappa U + Z, \tag{15}$$

where $X \sim \mathcal{N}(0, \sigma_x^2)$, $S \sim \mathcal{N}(0, \sigma_s^2)$, $Z \sim \mathcal{N}(0, \sigma_z^2)$ are all independent of each other. The decoder generates the (L)MMSE estimate of X given Y:

$$\hat{X}_{LMMSE}(Y) = \hat{X}_{MMSE}(Y) = \mathbb{E}(X \mid Y) = \frac{\mathbb{E}(XY)}{\mathbb{E}(Y^2)}Y, \tag{16}$$

with the corresponding mean-squared error cost function

$$J := J(g, h, \mu) = \mathbb{E}\left[X - \hat{X}_{LMMSE}(Y)\right]^2 = \mathbb{E}(X^2) - \left[\mathbb{E}(XY)\right]^2 / \mathbb{E}(Y^2). \tag{17}$$

Using (14, 15, 17) in (1, 2, 5), the resulting equivalent problem to (5) is given by

$$\min_{\alpha, \beta \in \mathbb{R}} \quad \max_{\kappa \in \mathbb{R}, \sigma_z^2 \in \mathbb{R}^+} \quad J \tag{18}$$

$$\text{s.t. } \mathbb{E}(U - S)^2 \leq P_E \quad \text{and} \quad \mathbb{E}(Y - U)^2 \leq P_A, \tag{19}$$

where we have replaced "inf" with "min", since g is restricted to linear maps and the cost function is bounded from below by zero (thus restricting g to a compact set without any loss of generality). Note that, the parameters α, β (resp. the parameters κ, σ_z^2) constitute the degrees of freedom for the encoder (resp, the attacker) given its linear (resp. affine) nature. Also, the first (resp. the second) constraint in (19) represents the power constraint for the encoder (resp. the attacker), equivalent to (1) (resp. (2)) to ensure the perceptual fidelity of the marked signal U (resp. the attacked signal Y) under the aforementioned parametrization. In Theorem 1, we provide the solution to the minimax problem (18) under the constraints (19). Our results are given using the parametrization via $\sigma_u^2 \triangleq \mathbb{E}\left(U^2\right)$. A summary of the results of Theorem 1 is given in Table 1.

Theorem 1. *The solution to the minimax soft watermarking problem (18) subject to the constraints (19) is as follows:*
(a) *For $P_A \leq \left(\sigma_s + \sqrt{P_E}\right)^2$, at optimality σ_u^2 is the unique positive root of the depressed cubic polynomial*

$$f\left(\sigma_u^2\right) \triangleq \sigma_u^6 - \sigma_u^2\left[\left(\sigma_s^2 - P_E\right)^2 + 2P_A\left(\sigma_s^2 + P_E\right)\right] + 2P_A\left(\sigma_s^2 - P_E\right)^2, \quad (20)$$

in the interval of $\left[\max\left(P_A, \left(\sigma_s - \sqrt{P_E}\right)^2\right), \left(\sigma_s + \sqrt{P_E}\right)^2\right]$. The corresponding optimal values of the system parameters are

$$\beta = \frac{1}{2}\frac{\sigma_u^2 + \sigma_s^2 - P_E}{\sigma_s^2}, \quad (21)$$

$$\alpha = \sqrt{\frac{\left[\left(\sigma_s + \sqrt{P_E}\right)^2 - \sigma_u^2\right]\left[\sigma_u^2 - \left(\sigma_s - \sqrt{P_E}\right)^2\right]}{4\sigma_s^2\sigma_x^2}}, \quad (22)$$

$$\kappa = 1 - \frac{P_A}{\sigma_u^2}, \quad \sigma_z^2 = P_A\kappa. \quad (23)$$

leading the corresponding optimal value of the cost function as

$$J = \sigma_x^2 - \frac{\sigma_x^4\alpha^2\left(\sigma_u^2 - P_A\right)}{\sigma_u^4} \quad (24)$$

(b) *If $P_A > \left(\sigma_s + \sqrt{P_E}\right)^2$, then at optimality we have $\kappa = 0$, $\sigma_z^2 \in \left[0, P_A - \sigma_u^2\right]$ is arbitrary, where*

$$\sigma_u^2 = \alpha^2\sigma_x^2 + \beta^2\sigma_s^2 < P_A$$

for any $\alpha, \beta \in \mathbb{R}$ such that $\alpha^2\sigma_x^2 + \left(\beta - 1\right)^2\sigma_s^2 \leq P_E$. In that case, the corresponding value of the cost function is given by $J = \sigma_x^2$.

Proof. We first characterize the cost function J (cf. (17)) and the power constraints (19) under the given parameterization. Given (14, 15) and the independence of X, S, Z, we have

$$\mathbb{E}\left(XY\right) = \kappa\mathbb{E}\left(XU\right) = \kappa\alpha\sigma_x^2, \quad (25)$$

$$\mathbb{E}\left(Y^2\right) = \mathbb{E}\left(\kappa U + Z\right)^2 = \kappa^2\sigma_u^2 + \sigma_z^2. \quad (26)$$

Using (25, 26) in (17), we get

$$J = \sigma_x^2 - \sigma_x^4 \frac{\kappa^2 \alpha^2}{\kappa^2 \sigma_u^2 + \sigma_z^2}. \tag{27}$$

Furthermore, using (14, 15), we can rewrite (19) as

$$\mathbb{E}\left(U - S\right)^2 = \mathbb{E}\left[\alpha X + (\beta - 1) S\right]^2 = \alpha^2 \sigma_x^2 + (\beta - 1)^2 \sigma_s^2 \le P_E, \tag{28}$$

$$\mathbb{E}\left(Y - U\right)^2 = \mathbb{E}\left[(\kappa - 1) U + Z\right]^2 = (\kappa - 1)^2 \sigma_u^2 + \sigma_z^2 \le P_A, \tag{29}$$

where

$$\sigma_u^2 = \mathbb{E}\left(U\right)^2 = \mathbb{E}\left[\alpha X + \beta S\right]^2 = \alpha^2 \sigma_x^2 + \beta^2 \sigma_s^2. \tag{30}$$

Step 1 (Inner Optimization): For any given fixed α, β, we focus on the innermost maximization problem of (18) subject to the constraint (29). Using (27), we observe that an equivalent problem is

$$\min_{\kappa \in \mathbb{R}, \sigma_z^2 \in \mathbb{R}^+} J_A, \tag{31}$$

subject to (29), where $J = \sigma_x^2 - \sigma_x^4 J_A$, and

$$J_A \triangleq \frac{\kappa^2 \alpha^2}{\kappa^2 \sigma_u^2 + \sigma_z^2}. \tag{32}$$

First, consider the special case of $\kappa = 0$: In that case, the attacker erases all the information about the original signal and the watermark and we would have $J_A = 0$. Since J_A is lower-bounded by zero by definition, this would be the optimal policy for the attacker as long as (29) can be satisfied for $\kappa = 0$. In that case, we have $\mathbb{E}\left(Y - U\right)^2 \big|_{\kappa=0} = \sigma_u^2 + \sigma_z^2$ per (29). Since the attacker can choose σ_z^2 arbitrarily small but cannot alter σ_u^2, we arrive at

if $\alpha, \beta \in \mathbb{R}$ are such that $P_A \ge \sigma_u^2$, then

$$\kappa_{opt} = 0, \sigma_{z,opt}^2 \in \left[0, P_A - \sigma_u^2\right] \text{ is arbitrary, and } J_{A,opt} = 0, J_{opt} = \sigma_x^2. \tag{33}$$

Thus, for the rest of the proof, we consider $P_A < \sigma_u^2$ and $\kappa \ne 0$. In that case, we can rewrite $J_A = \frac{\alpha^2}{\sigma_u^2 + \sigma_z^2/\kappa^2}$, and an equivalent problem to (31) is

$$\max_{\kappa \in \mathbb{R} \backslash \{0\}, \sigma_z^2 \in \mathbb{R}^+} J_A', \tag{34}$$

subject to (29) where $J_A = \frac{\alpha^2}{\sigma_u^2 + J_A'}$ and $J_A' \triangleq \sigma_z^2/\kappa^2$. Next, note that, for all $\kappa \in \mathbb{R} \backslash \{0\}$, J_A' and left hand side of (29) are both monotonic increasing in σ_z^2. Thus, the constraint (29) is active at optimality, which yields

$$\sigma_z^2 = P_A - (\kappa - 1)^2 \sigma_u^2. \tag{35}$$

Using (35) and the re-parametrization of $t \triangleq 1/\kappa$ in the definition of J_A', we get

$$J_A' = \frac{\sigma_z^2}{\kappa^2} = \frac{P_A - (\kappa - 1)^2 \sigma_u^2}{\kappa^2} = \left(P_A - \sigma_u^2\right) t^2 + 2\sigma_u^2 t - \sigma_u^2. \tag{36}$$

Since $P_A < \sigma_u^2$, J_A' is concave in t. Hence, maximization of (36) subject to (35) admits a unique solution, given by

$$t_{opt} = -\frac{\sigma_u^2}{P_A - \sigma_u^2} \quad \Leftrightarrow \quad \kappa_{opt} = \frac{\sigma_u^2 - P_A}{\sigma_u^2}, \tag{37}$$

which is equivalent to the first equation in (23). Using this result in (35) yields

$$\sigma_{z,opt}^2 = P_A - (\kappa_{opt} - 1)^2 \sigma_u^2 = P_A - \frac{P_A^2}{\sigma_u^2}, \tag{38}$$

which is equivalent to the second equation in (23). Note that, $\sigma_u^2 > P_A$ implies the positivity of κ_{opt} and $\sigma_{z,opt}^2$ per (37) and (38), respectively. Using (37, 38) in the cost function definitions, we get

$$J_{A,opt}' \triangleq J_A'\big|_{\kappa=\kappa_{opt},\sigma_z^2=\sigma_{z,opt}^2} = \frac{\sigma_{z,opt}^2}{\kappa_{opt}^2} = \frac{P_A}{\kappa_{opt}} = \frac{P_A \sigma_u^2}{\sigma_u^2 - P_A}, \tag{39}$$

$$J_E \triangleq J_{A,opt} = J_A\big|_{\kappa=\kappa_{opt},\sigma_z^2=\sigma_{z,opt}^2} = \frac{\alpha^2}{\sigma_u^2 + J_{A,opt}'} = \frac{\alpha^2 (\sigma_u^2 - P_A)}{\sigma_u^4}, \tag{40}$$

$$J = \sigma_x^2 - \sigma_x^4 J_E. \tag{41}$$

Note that (24) directly follows from using (40) in (41).

Step 2 (Outer Optimization): Next, given the solution to the inner optimization problem of (18) in (40), we proceed with solving the corresponding outer optimization problem, given by

$$\max_{\alpha,\beta \in \mathbb{R}} J_E, \tag{42}$$

subject to the constraint (28) and $\sigma_u^2 > P_A$.

First, we show that, without loss of generality (w.l.o.g.) we can assume $\alpha, \beta \geq 0$. Define the left hand side of (28) as a bivariate function of (α, β):

$$q(\alpha, \beta) \triangleq \alpha^2 \sigma_x^2 + (\beta - 1)^2 \sigma_s^2. \tag{43}$$

Note that both J_E and q are even functions of α. So, w.l.o.g. we can assume $\alpha \geq 0$. Furthermore, for any $\beta < 0$, we have $(\beta - 1)^2 = (|\beta| - 1)^2 + 2|\beta|$. Hence, for any $\alpha \in \mathbb{R}$ and $\beta < 0$, we have $q(\alpha, \beta) = q(\alpha, |\beta|) + 2|\beta|\sigma_s^2$. This implies that for any $\beta < 0$, $[q(\alpha, \beta) \leq P_E] \Rightarrow [q(\alpha, |\beta|) \leq P_E]$. Combining this observation with the fact that J_E is an even function of β, we reach the conclusion that w.l.o.g. we can assume $\beta \geq 0$.

Next, we show that for $\sigma_u^2 > P_A$, the constraint (28) is active at optimality. In order to do that, we examine the behavior of both J_E and q with respect to α^2 (noting that there is a one-to-one mapping between α and α^2 since we assume $\alpha \geq 0$ w.l.o.g.). We have

$$\frac{\partial q}{\partial \alpha^2} = \sigma_x^2 > 0, \tag{44}$$

$$\frac{\partial J_E}{\partial \alpha^2} = \frac{\sigma_u^2 - P_A}{\sigma_u^4} + \alpha^2 \frac{\partial \sigma_u^2}{\partial \alpha^2} \frac{\sigma_u^4 - 2\sigma_u^2 (\sigma_u^2 - P_A)}{\sigma_u^8},$$

$$= \frac{1}{\sigma_u^6} \left[(\sigma_u^2 - P_A) \beta^2 \sigma_s^2 + \alpha^2 \sigma_x^2 P_A \right] > 0, \tag{45}$$

where (45) follows from using (30) and recalling that $\sigma_u^2 > P_A$ by assumption. Now, the monotonicity results (44, 45) jointly imply that, at optimality the encoder will choose α as large as possible for any fixed β provided that the power constraint (28) is satisfied. Therefore, the constraint (28) is active at optimality.

Using the fact that the constraint is active at optimality, we have

$$P_E = \mathbb{E}(U - S)^2 = q(\alpha, \beta) = \alpha^2 \sigma_x^2 + \beta^2 \sigma_s^2 + (1 - 2\beta) \sigma_s^2,$$

$$= \sigma_u^2 + (1 - 2\beta) \sigma_s^2, \tag{46}$$

where (46) follows from (30). Using (46), we get (21).

Our next goal is to rewrite J_E (cf. (40)) as a function of σ_u^2, and accordingly formulate the problem (42) as a univariate maximization problem in terms of σ_u^2. Using (21) in (30), we obtain

$$\sigma_u^2 = \alpha^2 \sigma_x^2 + \frac{1}{4} \left(\frac{\sigma_u^2 + \sigma_s^2 - P_E}{\sigma_s^2} \right)^2 \sigma_s^2 = \alpha^2 \sigma_x^2 + \frac{(\sigma_u^2 + \sigma_s^2 - P_E)^2}{4\sigma_s^2}. \tag{47}$$

Using (47), we get

$$\alpha^2 = \frac{4\sigma_s^2 \sigma_u^2 - (\sigma_u^2 + \sigma_s^2 - P_E)^2}{4\sigma_s^2 \sigma_x^2}$$

$$= -\frac{\left[(\sigma_u - \sigma_s)^2 - P_E \right] \left[(\sigma_u + \sigma_s)^2 - P_E \right]}{4\sigma_s^2 \sigma_x^2} \tag{48}$$

$$= -\frac{1}{4\sigma_s^2 \sigma_x^2} \left[\sigma_u^2 - \left(\sigma_s + \sqrt{P_E} \right)^2 \right] \left[\sigma_u^2 - \left(\sigma_s - \sqrt{P_E} \right)^2 \right] \tag{49}$$

$$= \frac{-\left[\sigma_u^2 - (\sigma_s^2 + P_E) \right]^2 + 4 P_E \sigma_s^2}{4\sigma_s^2 \sigma_x^2}. \tag{50}$$

Using (49), we get (22). Also, the analysis of the constraint $\alpha^2 \geq 0$ yields

$$\left[\alpha^2 \geq 0 \right] \Leftrightarrow \left[\left(\sigma_s - \sqrt{P_E} \right)^2 \leq \sigma_u^2 \leq \left(\sigma_s + \sqrt{P_E} \right)^2 \right], \tag{51}$$

which directly follows from (49). Moreover, (51) and the constraint $(\sigma_u^2 > P_A)$ jointly imply that, if $P_A > \left(\sigma_s + \sqrt{P_E} \right)^2$, the feasible set for the problem (42) is empty. In that case, the encoder cannot generate a powerful enough marked signal U such that $\sigma_u^2 > P_A$. Then, at optimality the attacker chooses $\kappa = 0$ and erases U (i.e., (33) is valid). As a result, if $P_A > \left(\sigma_s + \sqrt{P_E}\right)^2$ (which is equivalent to $\sqrt{P_A} - \sqrt{P_E} > \sigma_s$), Theorem 1(b) holds.

On the other hand, if

$$\left[P_A < \left(\sigma_s + \sqrt{P_E}\right)^2\right] \quad \Leftrightarrow \quad \left[\sqrt{P_A} - \sqrt{P_E} < \sigma_s\right], \tag{52}$$

the problem (42) reduces to

$$\max_{\sigma_u^2} J_E\left(\sigma_u^2\right) \tag{53}$$

$$\text{s.t. } \max\left[P_A, \left(\sigma_s - \sqrt{P_E}\right)^2\right] \le \sigma_u^2 \le \left(\sigma_s + \sqrt{P_E}\right)^2, \tag{54}$$

where

$$J_E\left(\sigma_u^2\right) = -\frac{\left\{\left[\sigma_u^2 - \left(\sigma_s^2 + P_E\right)\right]^2 - 4P_E\sigma_s^2\right\}\left(\sigma_u^2 - P_A\right)}{4\sigma_s^2\sigma_x^2\sigma_u^4}, \tag{55}$$

which follows from using (50) in (40).

Next we quantify asymptotic behavior of J_E which will be useful in characterizing properties of its extrema. We proceed by first defining the numerator of J_E as $N\left(\sigma_u^2\right) \triangleq -\left\{\left[\sigma_u^2 - \left(\sigma_s^2 + P_E\right)\right]^2 - 4P_E\sigma_s^2\right\}\left(\sigma_u^2 - P_A\right)$. Then, we reach

$$\lim_{\sigma_u^2 \to 0} J_E\left(\sigma_u^2\right) = \lim_{\sigma_u^2 \to 0} \frac{\left(\sigma_s^2 - P_E\right)^2 P_A}{4\sigma_s^2\sigma_x^2\sigma_u^4} \quad \to \quad \infty, \tag{56}$$

$$\lim_{\sigma_u^2 \to \infty} J_E\left(\sigma_u^2\right) = \lim_{\sigma_u^2 \to \infty} \frac{\partial^2 N/\partial\left(\sigma_u^2\right)^2}{8\sigma_s^2\sigma_x^2} \quad \to \quad -\infty, \tag{57}$$

$$\lim_{\sigma_u^2 \to -\infty} J_E\left(\sigma_u^2\right) = \lim_{\sigma_u^2 \to -\infty} \frac{\partial^2 N/\partial\left(\sigma_u^2\right)^2}{8\sigma_s^2\sigma_x^2} \quad \to \quad \infty, \tag{58}$$

where (56) follows from (55), and (57, 58) follow from noting $\frac{\partial^2 N}{\partial(\sigma_u^2)^2} = -6\sigma_u^2 + 4\left(\sigma_s^2 + P_E\right) + 2P_A$.

Next note that, $J_E\left(\sigma_u^2\right)$ has 3 roots: P_A, $\left(\sigma_s - \sqrt{P_E}\right)^2$, $\left(\sigma_s + \sqrt{P_E}\right)^2$. The first one is obvious with a direct inspection of (55); the second and third roots directly follow from noting the equality of (49) and (50), and using that in (55).

Assuming that the feasible set for the problem (42) is non-empty, i.e., (52) holds, (52, 56, 57, 58) jointly imply

$$\left\{0 < \sigma_u^2 < \min\left[P_A, \left(\sigma_s - \sqrt{P_E}\right)^2\right]\right\} \Rightarrow J_E\left(\sigma_u^2\right) > 0,$$

$$\left\{\min\left[P_A, \left(\sigma_s - \sqrt{P_E}\right)^2\right] < \sigma_u^2 < \max\left[P_A, \left(\sigma_s - \sqrt{P_E}\right)^2\right]\right\} \Rightarrow J_E\left(\sigma_u^2\right) < 0,$$

$$\left\{\max\left[P_A, \left(\sigma_s - \sqrt{P_E}\right)^2\right] < \sigma_u^2 < \left(\sigma_s + \sqrt{P_E}\right)^2\right\} \Rightarrow J_E\left(\sigma_u^2\right) > 0, \tag{59}$$

$$\left\{\left(\sigma_s + \sqrt{P_E}\right)^2 < \sigma_u^2\right\} \Rightarrow J_E\left(\sigma_u^2\right) < 0.$$

Table 1. Summary of Theorem 1: Characterization of the scalar-Gaussian soft water-marking system at optimality. The first (leftmost) column indicates the condition that leads to the operation regime specified in the second column (see Remark 2 of Sect. 3.3 for the description and a discussion on "trivial" and "non-trivial" policies); the third and fourth columns show the corresponding values of the encoder and attacker parameters at optimality, respectively. Note that, in case of Theorem 1(b), while there are infinitely many choices of encoder parameters, the choice of $\alpha = 0$, $\beta = 1 + \sqrt{P_E/\sigma_s^2}$ is a sensible one and maintains continuity between regions (see Remark 3 of Sect. 3.3 for details).

Condition	Operation regime and cost	Encoder parameters	Attacker parameters
$\left[P_A \le \left(\sigma_s + \sqrt{P_E}\right)^2\right] \Longleftrightarrow \left[\sqrt{P_A} - \sqrt{P_E} \le \sigma_s\right]$	Non-trivial Theorem 1(a) $J = \sigma_x^2 - \frac{\sigma_x^4 \alpha^2 \left(\sigma_u^2 - P_A\right)}{\sigma_u^4}$	σ_u^2 unique root of $f\left(\sigma_u^2\right)$ (20) s.t. $\max\left(P_A, \left(\sigma_s - \sqrt{P_E}\right)^2\right) \le \sigma_u^2 \le \left(\sigma_s + \sqrt{P_E}\right)^2$ α given by (22), $\beta = \frac{1}{2}\left(1 + \frac{\sigma_u^2 - P_E}{\sigma_s^2}\right)$	$\kappa = 1 - \frac{P_A}{\sigma_u^2}$, $\sigma_z^2 = P_A \kappa$.
$\left[P_A > \left(\sigma_s + \sqrt{P_E}\right)^2\right] \Longleftrightarrow \left[\sqrt{P_A} - \sqrt{P_E} > \sigma_s\right]$	Trivial Theorem 1(b) $J = \sigma_x^2$	Any $\alpha, \beta \in \mathbb{R}$ s.t. $\alpha^2 \sigma_x^2 + (\beta - 1)^2 \sigma_s^2 \le P_E$, $\alpha^2 \sigma_x^2 + \beta^2 \sigma_s^2 < P_A$	$\kappa = 0$, $\sigma_u^2 = \sigma_s^2 \le P_A - \frac{\sigma_z^2}{\sigma_u^2}$

Thus, there are a total of 3 extrema of $J_E(\cdot)$. The one that is of interest to us, i.e., the one which satisfies (54), is a maximizer by (59). Furthermore, there is a unique such stationary point within the feasible region of (54). In order to calculate this maximizer, we take the derivative of (55) with respect to σ_u^2. After some straightforward algebra we get $\left(-4\sigma_s^2 \sigma_x^2 \sigma_u^4\right) \frac{dJ_E}{d\sigma_u^2} = f\left(\sigma_u^2\right)$, where $f(\cdot)$ is a depressed cubic polynomial and is given by (20). The solution of (53) is then given by the unique positive root of (20) which falls in the region specified by (54). Recall that this is the solution if (52) are satisfied. This completes the proof of part (a) of Theorem 1.   □

3.3   Discussion on the System Behavior at Optimality

Remark 1 (On Optimality of Theorem 1). Per Lemma 2, the results reported in Theorem 1 describe the optimal system characterization when the encoder is confined to the class of *linear* mappings (cf. (14)). Hence, in the sense of (17), our results form an *upper bound* on the optimal system performance within an arbitrary class of encoder mappings. Investigation of the tightness of this bound constitutes part of our future research. Throughout the rest of the paper, when we refer to optimality, we mean optimality in the sense of Theorem 1.

Remark 2 (Trivial and Non-trivial Policies)
(a) We say that "the (optimal) system is trivial" if, at optimality, the attacker erases all the information on the mark-embedded signal U and only retains

information on the additive noise Z. This coincides with having $Y = Z$ (i.e., $\kappa = 0$). Conversely, we have a "non-trivial system" if the attacked signal Y contains information on the marked signal U (i.e., $\kappa > 0$) at optimality.

(b) The case of *non-trivial system* happens if we have $\sigma_u^2 > P_A$ at optimality (due to the arguments leading to (33)). This is possible if and only if $P_A \leq \left(\sigma_s + \sqrt{P_E}\right)^2$ (or equivalently $\sqrt{P_A} - \sqrt{P_E} \leq \sigma_s$) (cf. (52)). In this case, given σ_s^2 and P_E, the encoder is able to design α and β such that the power of the marked signal U is larger than the power constraint P_A and is able to transmit information about X through the channel. This case is covered in part (a) of Theorem 1. Conversely, if $P_A > \left(\sigma_s + \sqrt{P_E}\right)^2$ (or equivalently $\sqrt{P_A} - \sqrt{P_E} > \sigma_s$), we have the case of *trivial system*, and it is impossible for the encoder to design U to exceed P_A. Then, the optimal attacker can afford to erase U, thus essentially sending noise to the decoder. This case is covered in part (b) of Theorem 1.

Corollary 1 *(Power Constraints)*. *In the non-trivial regime (Theorem 1(a)), the encoder and decoder power constraints (19) are both active.*

Corollary 2 *(Cost Ordering)*. *At optimality, the cost of the non-trivial regime is upper-bounded by the cost of the trivial regime, σ_x^2.*

Corollary 3 *(Existence and Uniqueness)*. *In the non-trivial regime, the optimal system parameters specified in Theorem 1(a) are guaranteed to exist and they are essentially unique.*[8]

Corollaries 1, 2 and 3 directly follow from the proof Theorem 1. Specifically, Corollary 1 is a consequence of arguments following (34, 45), Corollary 2 follows from using $P_A \leq \sigma_u^2$ in (24), and Corollary 3 is because of (59) and the arguments following it.

Remark 3 (On Optimal Encoding Parameters in case of the Trivial System). In case of a trivial system, by Theorem 1(b), there are infinitely many combinations of system parameters that achieve optimality. Among those, one choice for the encoder that is intuitively meaningful is to choose $\alpha = 0$ and $\beta = 1 + \sqrt{P_E / \sigma_s^2}$. This choice corresponds to having $U = \beta S$ such that $\mathbb{E}\left(U - S\right)^2 = (\beta - 1)^2 \sigma_s^2 = P_E$, i.e., the encoder chooses not to send any information on the watermark X and chooses the scaling factor β such that the first power constraint in (19) is satisfied with equality. Note that, such a choice ensures continuity between the values of the system parameters at optimality across non-trivial and trivial regions (cf. Remark 4).

Remark 4 (On Optimal Operation Regimes)
(a) Variation with respect to system inputs: For $P_E \geq P_A$, the system always operates at the non-trivial mode at optimality since $\sqrt{P_A} - \sqrt{P_E} \leq 0 \leq \sigma_s$. For

---

[8] In the proof of Theorem 1, all expressions that involve α are, in fact, functions of α^2; therefore if α^* is optimal, so is $-\alpha^*$. To account for such multiple trivial solutions, we use the term "essential uniqueness".

fixed σ_s^2 and P_E, as P_A increases, the strength of the attack channel increases and the attacker is able to act with a better "budget". As P_A increases, when we reach $P_A = \left(\sigma_s + \sqrt{P_E}\right)^2 + \epsilon$ for an arbitrarily small $\epsilon > 0$, the feasible region for the encoder becomes the empty set (cf. (54)) and a transition from the non-trivial mode to the trivial mode occurs. Conversely, for fixed P_A and for fixed P_E (resp. σ_s^2), as σ_s^2 (resp. P_E) increases, we observe a transition from the non-trivial region to the trivial region.

(b) Continuity between the regions: Suppose we have a system that is initially in the non-trivial region with $P_A < \left(\sigma_s + \sqrt{P_E}\right)^2$. Then, we have

$$\lim_{P_A \uparrow \left(\sigma_s + \sqrt{P_E}\right)^2} \sigma_u^2 = P_A = \left(\sigma_s + \sqrt{P_E}\right)^2, \tag{60}$$

$$\lim_{P_A \uparrow \left(\sigma_s + \sqrt{P_E}\right)^2} \beta = \frac{1}{2} \frac{\sigma_s^2 - P_E + \left(\sigma_s + \sqrt{P_E}\right)^2}{\sigma_s^2} = 1 + \sqrt{\frac{P_E}{\sigma_s^2}}, \tag{61}$$

$$\lim_{P_A \uparrow \left(\sigma_s + \sqrt{P_E}\right)^2} \alpha = 0, \tag{62}$$

$$\lim_{P_A \uparrow \left(\sigma_s + \sqrt{P_E}\right)^2} \kappa = \lim_{P_A \uparrow \left(\sigma_s + \sqrt{P_E}\right)^2} \sigma_z^2 = 0, \tag{63}$$

$$\lim_{P_A \uparrow \left(\sigma_s + \sqrt{P_E}\right)^2} J = \sigma_x^2, \tag{64}$$

where (60) follows from noting that the feasible region (54) converges to the singleton $P_A = \left(\sigma_s + \sqrt{P_E}\right)^2$, (61, 62) follow from using $\sigma_u^2 = \left(\sigma_s + \sqrt{P_E}\right)^2$ in (21, 22), respectively, and (63, 64) follow from using $\sigma_u^2 = P_A$ in (23, 24), respectively. Note that, the attack parameters (63) and the optimal cost value (64) readily satisfy continuity with their unique counterparts of Theorem 1(b). Furthermore, it can be shown that the encoder parameters (61, 62) achieve optimality in the trivial regime (cf. Remark 3).

Remark 5 (Performance Bounds). We focus here on deriving bounds on the cost for the more interesting case of Theorem 1(a) when $\left(\sqrt{P_A} - \sqrt{P_E}\right) < \sigma_s$. Note that (24), and problem construction clearly imply the bounds of $0 \leq J \leq \sigma_x^2$. The upper bound is tight and can be attained for $P_A = \left(\sigma_s + \sqrt{P_E}\right)^2$. In order to obtain a potentially tighter lower bound, we initially proceed with deriving an upper bound on α. Consider the polynomial $g\left(\sigma_u^2\right) \triangleq \left[\left(\sigma_s + \sqrt{P_E}\right)^2 - \sigma_u^2\right] \cdot \left[\sigma_u^2 - \left(\sigma_s - \sqrt{P_E}\right)^2\right]$. Then, it is straightforward to show that $g\left(\cdot\right)$ is concave and that $\frac{\partial g}{\partial \sigma_u^2} = -2\sigma_u^2 + \left(\sigma_s - \sqrt{P_E}\right)^2 + \left(\sigma_s + \sqrt{P_E}\right)^2 = -2\sigma_u^2 + 2\left(\sigma_s^2 + P_E\right)$. Hence $g\left(\cdot\right)$ is maximized for $\sigma_u^2 = \sigma_s^2 + P_E$. Using this in (22) yields the following upper bound on α:

$$\alpha \leq \sqrt{\frac{g\left(\sigma_u^2\right)\Big|_{\sigma_u^2 = \sigma_s^2 + P_E}}{4\sigma_s^2 \sigma_x^2}} = \sqrt{\frac{P_E}{\sigma_x^2}}. \tag{65}$$

Furthermore, using (65) on (24), we get

$$J \geq \sigma_x^2 \left[1 - P_E \frac{\sigma_u^2 - P_A}{\sigma_u^4} \right]. \tag{66}$$

Define $h\left(\sigma_u^2\right) \triangleq \frac{\sigma_u^2 - P_A}{\sigma_u^4}$. It is straightforward to show that for $\sigma_u^2 \geq P_A$, $h\left(\cdot\right)$ is concave and the maximizer is $2P_A$, which yields $h\left(\sigma_u^2\right) \leq \frac{1}{4P_A}$. Using this result in (66) and combining it with the previously mentioned bounds, we get

$$\max \left(0, \sigma_x^2 \left(1 - \frac{P_E}{4P_A} \right) \right) \leq J \leq \sigma_x^2. \tag{67}$$

We thus have a non-trivial lower bound for J if $P_A > P_E/4$.

Remark 6 (Role of the Watermark Power σ_x^2). A quick inspection of Theorem 1 reveals that the optimal operational mode of the resulting system depends on σ_s^2, P_A, and P_E, and is independent of the watermark power σ_x^2. Intuitively, this is because of two reasons: First, because of the nature of the underlying problem, we do not impose any distortion constraint between X and any other variable in the system. Next, the scaling parameter α can be used to adjust the contribution of the watermark (cf. (14)) to make it arbitrarily large or arbitrarily small. Indeed, (22) implies the following: For fixed σ_s^2, P_A, P_E, if a pair of $(\alpha_1, \sigma_{x,1}^2)$ is optimal in the sense of Theorem 1, then so is another pair $(\alpha_2, \sigma_{x,2}^2)$ if and only if $\alpha_1 \sigma_{x,1} = \alpha_2 \sigma_{x,2}$. On the other hand, σ_x^2 directly affects the value of the resulting cost J.

Remark 7 (Asymptotics - Large Signal Case). It is possible to obtain some asymptotics when $\sigma_s^2 \gg P_E, P_A$. In that case, the system will operate at the non-trivial mode at optimality (governed by Theorem 1(a)). Then, (20) can be written as

$$f\left(\sigma_u^2\right) \sim \sigma_u^6 - \sigma_u^2 \left(\sigma_s^4 + 2P_A \sigma_s^2\right) + 2P_A \sigma_s^2 = \left(\sigma_u^2 - \sigma_s^2\right) \left(\sigma_u^4 + \sigma_u^2 \sigma_s^2 - 2P_A\right). \tag{68}$$

Thus, at optimality, using $\sigma_s^2 \gg P_E, P_A$, we have

$$\sigma_u^2 \sim \sigma_s^2, \tag{69}$$

$$\beta \sim \left[\frac{1}{2} \frac{\sigma_u^2 + \sigma_s^2}{\sigma_s^2} \right] \sim 1, \tag{70}$$

$$\alpha \sim \sqrt{ \frac{\left[\left(\sigma_s + \sqrt{P_E} \right)^2 - \sigma_s^2 \right] \left[\sigma_s^2 - \left(\sigma_s - \sqrt{P_E} \right)^2 \right]}{4 \sigma_s^2 \sigma_x^2} } = \sqrt{ \frac{P_E}{\sigma_x^2} }, \tag{71}$$

$$J \sim \left[\sigma_x^2 - \sigma_x^4 \frac{\alpha^2}{\sigma_s^2} \right] \sim \left[\sigma_x^2 \left(1 - \frac{P_E}{\sigma_s^2} \right) \right] \tag{72}$$

where (69) follows from the fact that σ_s^2 is the unique root of (68) in the region of interest, (70) and (71) follow from using (69) in (21) and (22), respectively, (72) follows from using (69, 71) in (24).

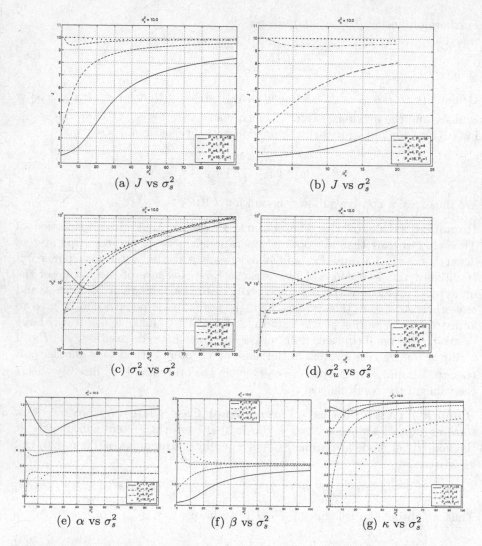

Fig. 2. System behavior at optimality as a function of σ_s^2 for the case of $\sigma_x^2 = 10$; solid, dashed, dash-dotted, and dotted lines represent the cases of $(P_A = 1, P_E = 16)$, $(P_A = 1, P_E = 4)$, $(P_A = 4, P_E = 1)$, $(P_A = 16, P_E = 1)$, respectively. By Remark 3 of Sect. 3.3, we use $\alpha = 0$ and $\beta = 1 + \sqrt{P_E/\sigma_s^2}$ in case of a trivial system.

4   Numerical Results

In this section, we numerically illustrate the behavior of the optimal scalar-Gaussian soft watermarking system as a function of the power of the unmarked signal, σ_s^2. The results are presented in Figs. 2 and 3 for a fixed watermark power $\sigma_x^2 = 10$ owing to the discussion in Remark 6. Because of the linear relationship between κ and σ_z^2, we do not show σ_z^2 vs σ_s^2 plots since κ vs σ_s^2 plots are already present.

In Fig. 2, solid, dashed, dash-dotted and dotted lines correspond to the cases of $(P_A = 1, P_E = 16)$, $(P_A = 1, P_E = 4)$, $(P_A = 4, P_E = 1)$, and $(P_A = 16, P_E = 1)$, respectively. Here,

- (a) and (b) show J vs σ_s^2 for the whole range and for σ_s^2 small, respectively;
- (c) and (d) show σ_u^2 vs σ_s^2 for the whole range and for σ_s^2 small, respectively;
- (e), (f), and (g) show α, β and κ as functions of σ_s^2, respectively.

By Theorem 1, the system always operates in the non-trivial mode for $P_A \leq P_E$ (solid and dashed lines), and it operates in the trivial mode for $\sigma_s^2 \leq \left(\sqrt{P_A} - \sqrt{P_E}\right)^2$ for $P_A > P_E$, i.e., trivial mode for $\sigma_s^2 \leq 1$ for $(P_A = 4, P_E = 1)$ (dash-dotted line) and $\sigma_s^2 \leq 9$ for $(P_A = 16, P_E = 1)$ (dotted line), which is clearly observable in panel (b). Note the continuity in the behavior of all system elements during the transition between trivial and non-trivial regions.

In Fig. 3, solid and dashed lines represent true values and large-signal approximations for $P_A = P_E = 1$. As expected, the large-signal approximation becomes more accurate as σ_s^2 increases.

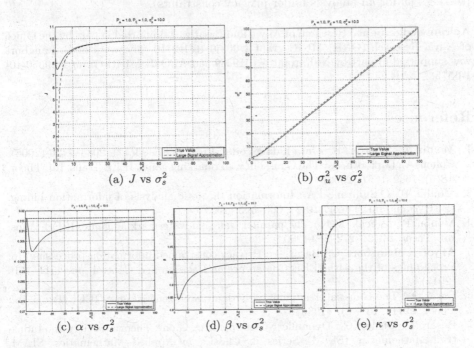

(a) J vs σ_s^2 (b) σ_u^2 vs σ_s^2

(c) α vs σ_s^2 (d) β vs σ_s^2 (e) κ vs σ_s^2

Fig. 3. System behavior at optimality as a function of σ_s^2 for the case of $\sigma_x^2 = 10$, $P_A = P_E = 1$; solid and dashed lines represent true values and large signal approximation (cf. Remark 7 of Sect. 3.3), respectively.

5   Conclusions

We have introduced the zero-sum game problem of *soft watermarking* where the hidden information has a perceptual value and comes from a continuum, unlike the prior literature on robust data hiding where the focus has been on cases where the hidden information is an element of a discrete finite set. Accordingly, the receiver produces a soft estimate of the embedded information in the proposed setup. As a first step toward this new class of problems, we focus in this paper on the scalar-Gaussian case with the expected squared estimation error as the cost function and analyze the resulting zero-sum game between the encoder & decoder pair and the attacker. Expected distortion constraints are imposed both on the encoder and the attacker to ensure the perceptual quality. Restricting the encoder mapping to be linear in the watermark and the unmarked host, we show that the optimal attack mapping is Gaussian-affine. We derive closed-form expressions for the system parameters in the sense of minimax optimality. We further discuss properties of the resulting system in various aspects, including bounds and asymptotic behavior, and provide numerical results.

Our future work includes an information-theoretic analysis of the problem considered here, and extensions to settings with privacy and security constraints (see e.g., [5] for an analysis under privacy constraints).

Acknowledgement. Research of Akyol and Başar was supported in part by the Office of Naval Research (ONR) MURI grant N00014-16-1-2710. The work of C. Langbort was supported in part by NSF grant #1619339 and AFOSR MURI Grant FA9550-10-1-0573.

References

1. Moulin, P., Koetter, R.: Data hiding codes. Proc. IEEE **93**(12), 2083–2126 (2005)
2. Cohen, A., Lapidoth, A.: The Gaussian watermarking game. IEEE Trans. Inf. Theor. **48**(6), 1639–1667 (2002)
3. Moulin, P., O'Sullivan, J.A.: Information theoretic analysis of information hiding. IEEE Trans. Inf. Theor. **49**(3), 563–593 (2003)
4. Moulin, P., Mihcak, M.K.: The parallel Gaussian watermarking game. IEEE Trans. Inf. Theor. **50**(2), 272–289 (2004)
5. Akyol, E., Langbort, C., Başar, T.: Privacy constrained information processing. In: Proceedings of the 54th IEEE Conference on Decision and Control, pp. 4511–4516 (2015)
6. Başar, T.: The Gaussian test channel with an intelligent jammer. IEEE Trans. Inf. Theor. **29**(1), 152–157 (1983)
7. Başar, T., Olsder, G.: Dynamic Noncooperative Game Theory. Society for Industrial Mathematics (SIAM) Series in Classics in Applied Mathematics. SIAM, Philadelphia (1999)
8. Akyol, E., Rose, K., Başar, T.: Optimal zero-delay jamming over an additive noise channel. IEEE Trans. Inf. Theor. **61**(8), 4331–4344 (2015)

Strategies for Voter-Initiated Election Audits

Chris Culnane(✉) and Vanessa Teague

Department of Computing and Information Systems,
Melbourne School of Engineering, University of Melbourne,
Parkville, VIC 3010, Australia
{christopher.culnane,vjteague}@unimelb.edu.au

Abstract. Many verifiable electronic voting systems are dependent on voter-initiated auditing. This auditing allows the voter to check the construction of their cryptographic ballot, and is essential in both gaining assurance in the honesty of the constructing device, and ensuring the integrity of the election as a whole. A popular audit approach is the Benaloh Challenge [5], which involves first constructing the complete ballot, before asking the voter whether they wish to cast or audit it.

In this paper we model the Benaloh Challenge as an inspection game, and evaluate various voter strategies for deciding whether to cast or audit their ballot. We shall show that the natural strategies for voter-initiated auditing do not form Nash equilibria, assuming a payoff matrix that describes remote voting. This prevents authorities from providing voters with a sensible auditing strategy. We will also show that when the constructing device has prior knowledge of how a voter might vote, it critically undermines the effectiveness of the auditing. This is particularly relevant to internet voting systems, some of which also rely on Benaloh Challenges for their auditing step.

A parallel version, in which the voter constructs multiple ballots and then chooses which one to vote with, can form Nash equilibria. It still relies on some uncertainty about which one the voter will choose.

1 Introduction

Verifiable electronic voting systems aim to provide strong integrity guarantees and protection from tampering. In order to deliver this, they provide a number of verifiability properties, namely, cast-as-intended and counted-as-cast. Cast-as-intended means that the cast ballot accurately reflects the intentions of the voter, it is verifiable if the voter has the opportunity to gain assurance that the vote was cast in keeping with their intentions. Counted-as-cast means that the cast ballots are correctly counted.

In this paper we are only interested in the first of these properties, cast-as-intended. A popular technique for providing cast-as-intended verifiability is to

C. Culnane—Partially funded by Australian Research Council Discovery project DP140101119 'More information for better utility; less information for better privacy'.

Q. Zhu et al. (Eds.): GameSec 2016, LNCS 9996, pp. 235–247, 2016.
DOI: 10.1007/978-3-319-47413-7_14

provide an auditing step of ballots. Such a step aims to assure the voter that ballot is correctly constructed, and will therefore accurately reflect their vote. The exact methodology of the audit is dependent on the system, but broadly falls into two categories, cut-and-choose [2] and the Benaloh Challenge [5]. The cut-and-choose approach is applicable to systems that pre-construct ballot papers. Such systems allow a voter to choose whether to audit or vote with the ballot they are given. If they choose to audit, the ballot is opened and the voter may check the construction of any cryptographic values. However, once a ballot has been audited it cannot be used for voting, since this would break the privacy of the voter and the secrecy of the ballot. Provided a sufficient number of audits are performed, and assuming none fail and that the constructing device did not know whether a particular ballot would be audited, there is a very high probability that the ballots were honestly constructed. Crucially, the audit takes place prior to the voter expressing any preferences. Such an approach is used in Prêt à Voter [13] and Scantegrity [8].

The Benaloh Challenge is similar, except it is used in systems where ballots are not pre-constructed. In such systems, a voter first enters their preferences and constructs their encrypted ballot on a voting device, they are then given the choice of whether they wish to vote or audit the constructed ballot. If they choose to audit it, the ballot is opened, allowing the voter to check the cryptographic construction of the ballot. Crucially, the audit takes place after the voter has expressed their preferences. Like the cut-and-choose approach, once a ballot has been opened it cannot be used for voting, and therefore the voter must construct a new ballot to vote with. Such an approach is used in Helios [1], Wombat [4] and Star-Vote [3]. Both approaches allow voters to repeat the audit step as many times as they like—the protocol ends when the voter decides to cast their ballot. As such, their final cast ballot will not be audited, and their assurance that it has been cast-as-intended is based on having run a number of successful rounds of auditing previously, or in the general case, that enough other people have run successful rounds of auditing.

In this paper, we will evaluate Benaloh Challenges from a game theoretic point of view using a game that describes the payoffs of a remote voting setting. We analyze the effectiveness of various voter strategies when choosing whether to cast or audit a constructed ballot, and the corresponding device strategies for constructing a dishonest ballot. We will show that none of the natural strategies for voter-initiated auditing, using Benaloh Challenges, form Nash equilibria. This presents a particular problem for voting systems relying on such auditing, since it precludes providing the voter with instructions on a sensible auditing strategy. The provision of such advice, when it does not form a Nash equilibria, can do more harm than good, creating a potential advantage for a cheating device. This calls into question the validity of the cast-as-intended auditing in verifiable remote electronic voting systems that utilise Benaloh Challenges. Modelling an attendance voting setting, in which there is a higher penalty for device misbehaviour, is important future work.

A simple parallel variant, in which voters are instructed to make multiple ciphertexts and then choose one to vote with, can form a Nash equilibrium. However, this too needs careful analysis of the cheating machine's ability to guess which vote will be cast. The estimate must be correct or what seems to be a Nash equilibrium might not be.

2   Voter-Initiated Auditing

We are primarily interested in voter-initiated auditing used in schemes that construct encrypted ballots. As such, we shall focus on Benaloh Challenges [5], which have been widely adopted as the auditing technique for such schemes.

2.1   Purpose of Auditing

Arguably the purpose of audits is not just to detect cheating, but to provide an evidence trail after the fact to support an announced election outcome. For example, Risk Limiting Audits [12] of a voter-verifiable paper trail provide a statistical bound on the likelihood that an undetected error might have changed the election outcome. We might hope to conduct a statistical assessment of the transcript of a voter-initiated electronic auditing procedure, in order to produce the same sort of guarantee. However, this work shows that such an assessment would be very difficult to perform. In particular, a naive computation of the probability of detection given the rate of auditing would give an incorrectly high degree of confidence.

2.2   Origins of Benaloh Challenges

Benaloh Challenges were first introduced in [5], and later refined in [6]. Benaloh Challenges are an auditing technique that can be used by voting systems that construct encrypted ballots. They are commonly referred to as "cast-or-audit", on account of the technique involving first constructing an encrypted ballot, followed by asking the voter whether they wish to cast or audit it. If the voter chooses to cast the ballot it will be signed, or otherwise marked for voting, and included in the tally. If the voter chooses to audit the ballot, the encryptions are opened to allow the voter to verify that the ballot was correctly constructed from their preferences. The Benaloh style of auditing has been widely adopted in the verifiable voting field, in both theory and practice, including in Helios [1], VoteBox [15], Wombat [4], and StarVote [3]. Of particular interest is Helios [1], which is a web-based open-audit voting system, which has been used in binding elections, notably, the International Association for Cryptologic Research (IACR) elections.

2.3 Making Audit Data Public

Benaloh, in [6], makes no mention of whether the audit information is made public. However, there is a discussion on providing assurance of integrity to a wider population from a smaller random sample of audits. This would seem to indicate the necessity that the auditing is made public, so as to enable that wider population to inspect it. The original version of Helios [1] did not mention the posting of audit data, however, in Helios V3 there is a provision for posting audit data to the public bulletin board [10]. In Wombat [4] the audited ballot must be shown to an election worker, and taken home to be checked, all of which threatens the secrecy of the vote.

2.4 Revealing Voter Intent via an Audit

The auditing step, by its very nature, reveals a set of preferences and the corresponding ciphertext construction. If those preferences are a true reflection of the voters intent, the audit will reveal the voters intent, and thus break ballot secrecy. This is equally problematic whether the information is posted publicly, or shown to an election official for checking.

If the voter is deciding after construction whether to vote or audit, as described in [6], the voter will be obliged to always construct a ballot with their true preferences, and as a result, any audit will break ballot secrecy. A simple counter strategy is for the voter to construct a ballot with fake preferences to audit. Crucially, this requires the voter to decide whether to cast or audit prior to ballot construction. It is critical that the machine cannot distinguish between a voter creating a genuine ballot and a voter constructing an audit ballot.

2.5 Indistinguishability of Real and Fake Ballots

The requirement for indistinguishability between a ballot that will be audited and one that will be voted with is implicitly covered by an assumption in [6], which states that it is crucial that the ballot encryption device does not receive any information that may indicate the likelihood of a ballot being audited. However, realising this assumption presents a significant challenge, even in a secure polling place. Whilst it seems possible that the voters identity could be hidden from the machine, it seems impossible to exclude global information from being used to indicate whether a ballot will be cast or audited. Such information could include voting preference patterns, geographical voting patterns and election wide voter guidelines, all of which could be used to infer information about whether a ballot is being constructed for voting or audit.

For example, it is easy to see how voters could easily fall into a pattern of auditing one ballot, and if that succeeds, voting with the next. Such a pattern has been seen in real-world elections using Helios, in [11] the authors analyse the first IACR election, showing a clear pattern for performing zero or one audit, but very rarely anymore.

2.6   Benaloh Challenges in Remote Voting

Benaloh Challenges were proposed for the supervised voting setting, and not for remote voting. Supervised voting refers to voting that takes place in a controlled environment, for example, at a polling location on election day. Remote voting is considered to be any voting that takes places outside of a controlled environment, for example, voting from home over the internet. Helios [1] is a remote voting system which constructs ballots on a voter's own device. Such a device is likely to be able to infer significant information about the behaviour, and therefore voting preferences, of the user. In particular, since Helios was first designed in 2008, there has been a great increase in the intrusiveness of privacy invasion via identifying individuals' online behaviour [9]. It is clear that in the remote setting it is feasible for the device to be able to predict the voting intention of the voter. In the supervised setting, identifying an individual voter is unlikely, however, identifying groups of voters, or general patterns, is still feasible.

The payoffs for cheating and the penalties for failure are also different in the remote vs attendance setting. In the remote setting, typically only one or two voters use the same device, and there is no independent check when cheating is detected; in the attendance setting, a successfully cheating device could take hundreds or thousands of votes, and the penalties for cheating could be severe. For the rest of the paper, we consider only the remote setting, leaving the attendance game for future work.

3   The Game Theory Model - Inspection Game

We model the interaction as an inspection game in which the voter is the inspector and the device wins only if it cheats and is not inspected. Voters incur a small cost for inspecting, a benefit from successfully casting the vote of their choice, and a large cost for having their vote inaccurately recorded. The device (which aims to cheat) benefits from getting away with casting a vote other than the voter's intention.

The voter begins with a true vote v_t chosen from a publicly-known probability distribution Π.

In the first step, the voter (V) chooses a vote from the range of Π and sends it to the device (D). The device then chooses whether to encode it truthfully (T) or falsely (F), but this choice cannot be observed by V. Next, V may cast the vote (C), in which case the game ends without revealing the device's choice, or she may audit the vote (A), so the device's choice is revealed. If she audits a truthfully-encoded vote, the process begins again. Otherwise, the game ends. Payoffs for one step of the game are shown in Fig. 1. G_V is a positive payoff reflecting the voter's successful casting of her intended ballot; $-B_V$ is the negative payoff when she is tricked into casting a vote other than v_t. For the device, G_D is the positive payoff associated with successfully casting a vote other than the voter's intention; $-B_D$ is the negative payoff associated with being caught cheating. The voter incurs a small cost $-c_{audit}$ for each audit.

		Voter(V)	
		$Cast(C)$	$Audit(A)$
Device (D)	$Truthful(T)$	$(0, G_V)$	$Add(0, -c_{audit})$; repeat.
	$False(F)$	$(G_D, -B_V)$	$(-B_D, -c_{audit})$

Fig. 1. Payoffs for one step of the game. If the device is truthful and the voter audits, the game repeats.

	Voter Payoff	Device Payoff	Description
$n_{cast} > n_{false}$	$-n_{false}c_{audit}$	$-B_D$	Voter catches cheating device.
$n_{cast} = n_{false}$	$-(n_{cast} - 1)c_{audit} - B_V$	G_D	Device successfully cheats.
$n_{cast} < n_{false}$	$-(n_{cast} - 1)c_{audit} + G_V$	0	Voter votes as intended.

Fig. 2. Payoffs for the extended game. The voter casts at step n_{cast}; the device encodes falsely (for the first and only time) at step n_{false}

In order to model the repeated nature of the game, the voter's strategy is a sequence of n votes, followed by n choices to audit, then a final $n + 1$-th vote that is cast. The device's strategy is a sequence of choices to encode truthfully or falsely, which may be random or may depend on what the voter has chosen.

Assumptions.

1. that $c_{audit} < B_V$,
2. that (it's common knowledge that) the voter never casts a vote other than v_t,

Whatever the voter's strategy, the first false encoding by the device ends the game. We can therefore describe D's strategy completely as a choice of n_{false}, the first step at which D encodes falsely, preceded by truthful rounds. Of course this can be random, or can depend on the votes that the voter has requested before then. The game's outcome depends on whether n_{false} is earlier, later, or exactly equal to the round n_{cast} in which the voter chooses to cast. This gives us the payoff table, shown in Fig. 2, for the extended game.

3.1 Negative Results: Simple Sequential Strategies Do Not Form Nash Equilibria

As expected in an inspection game, it is immediately clear that there is no pure strategy equilibrium. Indeed, there is no equilibrium with a fixed value of n.

Lemma 1. *If $c_{audit} < B_V$, there is no Nash equilibrium in which the voter's number of audits is fixed.*

Proof. Suppose V always audits $n_{cast} - 1$ times, and suppose this is a Nash equilibrium with some strategy S_D by the device D. Then S_D must specify encoding untruthfully in the n_{cast}-th round—otherwise there would be a strict unilateral improvement by doing so. But this gives V a payoff of $nc_{audit} - B_V$, which is bad. This could be improved to $(n-1) * c_{audit}$ by auditing at round n_{cast}, which is strictly better assuming that $c_{audit} < B_V$. $\square$

Also backward induction applies:

Lemma 2. *Suppose there is a common-knowledge upper bound $n_{\max}$ on n_{cast}. If $c_{audit} < B_V$, then there is no Nash equilibrium in which the voter votes as intended.*

Proof. Backward induction. The device's best response is to cheat at round $n_{\max}$, whenever the game gets that far, thus giving V the worst possible payoff. But then V improves her payoff by never waiting until $n_{\max}$, and instead casting at round $n_{\max} - 1$. The induction step is similar: if V is guaranteed not to audit at round n_i, then D should cheat at round n_i, and V would improve her payoff by casting at round $n_i - 1$. $\square$

Lemma 3. *There is no Nash equilibrium in which, for any n, the probability that D encodes falsely at round n is zero.*

Proof. V should always cast then, so D should always cheat then. $\square$

Now we can address our main question: whether information about the true vote can influence the device's success in getting away with cheating (and hence both parties' payoffs in the game).

Lemma 4. *If $-B_D < 0$, there is no Nash Equilibrium in which, with nonzero probability, D falsely encodes a vote outside the support of Π.*

Proof. Suppose S_D is a device strategy and let n be the first round at which, with nonzero probability, V chooses and D falsely encodes a vote outside the support of Π. Then V will certainly audit this vote (by Assumption 2), leading the device to a payoff of $-B_D$, the worst possible. If D was always truthful, it could guarantee a payoff of 0. $\square$

Lemma 5. *If $-B_D < 0$, then every device strategy in which, with nonzero probability, D falsely encodes a vote outside the support of Π, is weakly dominated.*

Proof. Similar. Weakly dominated by the always-truthful strategy. $\square$

So whether we take the solution concept to be Nash Equilibrium or survival of iterated deletion of weakly dominated strategies, we see that there is no solution in which the device falsely encodes a vote that the voter could not possibly intend to cast. This has important implications, particularly for voting from home, where the device may have very accurate information about the voter's intentions. In many practical scenarios, Π is a point function—the device knows exactly how the person will vote.

Note that the argument does not hold if $B_D = 0$, meaning that there is no downside to being caught cheating.

The strategy most commonly recommended to voters is to toss a coin at each stage and then, based on the outcome, to either cast their true vote v_t or choose some vote and audit it. We distinguish between a few such strategies:

TRUTHANDCOINTOSS. Always request v_t; toss a coin to decide whether to cast or audit.

PIANDCOINTOSS. Toss a coin to decide whether to cast or audit; in the case of audit, choose a vote at random according to Π.

These two are the same if Π has only one nonzero probability.

On the device side, recall that the strategy is determined entirely by the choice of which round to encode untruthfully in. We have already argued that there is no Nash equilibrium in which there is an upper bound on n_{false} (Lemma 3). We first examine the equilibria in which the voter plays TRUTHAND-COINTOSS. There is no new information communicated to the device as this strategy plays out: S_D consists entirely of a (static) probability distribution for choosing n_{false}.

We begin with the surprising result that there is no Nash equilibrium in which V plays TRUTHANDCOINTOSS—the probability of detecting cheating falls off too quickly.

Theorem 1. *There is no Nash equilibrium in which V plays* TRUTHANDCOIN-TOSS.

Proof. We're assuming that $Pr(n_{cast} = i) = 1/2^i$. Crucially, the game tree has only one branch, the one in which the voter always requests the same vote. The device's strategy is therefore simply a probability distribution P_D for n_{false}. Its expected payoff is

$$\mathbb{E}(D\text{'s payoff}) = \sum_{i=1}^{\infty} \left(G_D Pr(n_{cast} = i) Pr_D(n_{false} = i) - B_D Pr(n_{cast} > i) Pr_D(n_{false} = i) \right)$$

$$= (G_D - B_D) \sum_{i=1}^{\infty} Pr_D(n_{false} = i)/2^i$$

(Note that the case in which $n_{cast} > i$ and $n_{false} = i$ gives D a zero payoff.)
This is strictly maximised by setting $Pr_D(n_{false} = 1) = 1$, that is, falsely encoding always on the first round. But then V could improve her payoff by always auditing in round 1 (by Assumption 1, $c_{audit} < B_V$).   □

The following corollary shows that, if the device knows exactly how the voter is going to vote, the coin-tossing advice doesn't produce a Nash equilibrium.

Corollary 1. *If Π is a point function, then there is no Nash equilibrium in which V plays* PIANDCOINTOSS.

Proof. Immediate from Theorem 1

This easily generalises to any exponentially-decreasing auditing strategy with any coefficient. Suppose the voter, at round i, chooses to audit the vote with probability r, and otherwise to cast. The generalised strategies are

TRUTHANDRANDOMCHOICE(r). Always request v_t; audit with probability r, otherwise cast.

PIANDRANDOMCHOICE(r). Audit with probability r, otherwise cast. In the case of audit, choose a vote at random according to Π.

Again these are not part of any Nash equilibrium.

Lemma 6. *There is no Nash equilibrium in which V plays* TRUTHAND RANDOMCHOICE(r) *for any $r \in (0,1)$.*

Proof. First compute the probabilities of casting at or after round i:

$$Pr(n_{cast} = i) = r^{i-1}(1-r).$$

$$Pr(n_{cast} > i) = r^i.$$

So we can recompute D's expected payoff as

$$\mathbb{E}(D's\ payoff) = \sum_{i=1}^{\infty} \left(G_D r^{i-1}(1-r) Pr_D(n_{false} = i) - B_D r^i Pr_D(n_{false} = i) \right)$$

$$= [(1-r)G_D - rB_D] \sum_{i=1}^{\infty} Pr_D(n_{false} = i) r^{i-1}$$

$$= [(1-r)G_D - rB_D] \left(Pr_D(n_{false} = 1) + r \sum_{i=2}^{\infty} Pr_D(n_{false} = i) r^{i-2} \right)$$

$$\leq [(1-r)G_D - rB_D] \left(Pr_D(n_{false} = 1) + r(1 - Pr_D(n_{false} = 1)) \right)$$

$$\leq [(1-r)G_D - rB_D] Pr_D(n_{false} = 1) \text{ because } r < 1.$$

Again, this is strictly maximised, to $[(1-r)G_D - rB_D]$, when $Pr_D(n_{false} = 1) = 1$. In other words, the device always cheats in the first round. This is clearly not part of any Nash equilibrium in which the voter does not always audit. $\square$

This result generalises to Π being a more general distribution over possible votes. Suppose the Voter's strategy is PIANDRANDOMCHOICE(r). Suppose also that the voter's true vote v_t is chosen according to Π. One way to think of it is that Π represents the voter's guess about what the machine guesses V's distribution to be. In equilibrium, they should match.

Theorem 2. *There is no Nash equilibrium in which V plays* PIANDRANDOM CHOICE(r) *for any $r \in (0,1)$ or any probability distribution Π, assuming that the true vote v_t is also chosen according to Π.*

Proof. Think about the tree of all possible sequences of vote requests, shown in Fig. 3. The device's strategy is described by a probability P_D that takes a node N in the tree and outputs a probability of playing F for the first time at N.

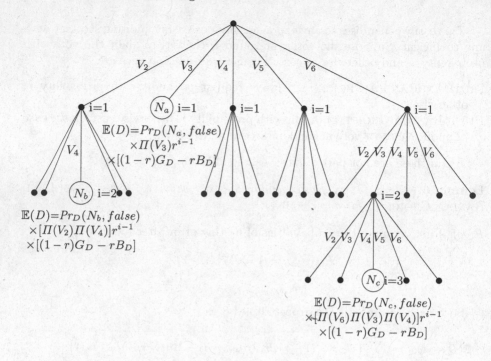

Fig. 3. Game tree

To be a meaningful probability distribution, we require that, along any (possibly infinite) path p down the tree, $\sum_{N \in p} P_D(N, false) \leq 1$.

The probability of reaching node N at all, assuming that D is truthful until then, is determined by V's strategy S_V. The probability that a particular node N is reached is simply the product of all the vote choices along that path, times r^{i-1}, where i is its depth in the tree (starting at 1).

Since a false encoding ends the game (one way or another), we can attach an expected payoff to each node, representing D's expected payoff from the game ending at that node. Remember that when D is truthful it derives no payoff. For example, in Fig. 3, the probability of reaching node N_b is $[\Pi(V_2)\Pi(V_4)]r$ and the probability the device plays false at that node is $Pr_D(N_b, false)$. In general:

$$\mathbb{E}(D\text{'s payoff from node } N) = [(1-r)G_D - rB_D]Pr_D(N, false)Pr_{S_V}(N \text{ is reached})$$

We claim that D's best response to PiAndRandomChoice(r) is to play *false* always at $i = 1$. In other words, to cheat at the first opportunity. To see why, suppose instead that there is some best response P_{D-best}, in which there is some (parent) node N_p at level $i \geq 1$ such that

$$\sum_{N_c \, a \, child \, of \, N_p} P_{D-best}(N_c, false) > 0.$$

But now D's payoff can be strictly improved by shifting to strategy P'_{D-best} in which all the probabilities in N_p's children are shifted up to N_p. Let $\alpha = \sum_{N_c \, a \, child \, of \, N_p} P_{D-best}(N_c, false)$. The improved strategy is:

$$P'_{D-best}(N, false) = \begin{cases} P_{D-best}(N, false) + \alpha, & \text{when } N = N_p; \\ 0 & \text{when } N \text{ is a child of } N_p; \\ P_{D-best}(N, false) & \text{otherwise.} \end{cases}$$

This is strictly better than P_{D-best} because the total probability of reaching any of N_p's children is at most r (conditioned on having reached N_p), which is less than 1. The expected payoff is increased by at least $(1-r)\alpha[(1-r)G_D - rB_D]$.

Hence there is no best response to PIANDRANDOMCHOICE(r) other than always playing *false* at the first opportunity. Hence PIANDRANDOMCHOICE(r) is not part of any Nash equilibrium.   □

3.2 Positive Results: Parallel Strategies Can Form Nash Equilibria, but only if the Device's Probability of Guessing the Voter's Choice Is Small Enough

Now consider an apparently-slight variation: instead of auditing sequentially, the voter makes some fixed number (k) of ciphertexts, chooses one at random to cast, then audits the other $k - 1$. Again, if they're all the same, the device has no information about which one will be cast, but privacy is compromised; if they're not all the same then the voter has to simulate some distribution for the $k - 1$ that are audited. In either case, if the device's probability of guessing correctly which vote will be cast is α, its expected payoff for cheating is

$$\mathbb{E}(D's \; payoff \; from \; cheating) = [\alpha G_D - (1 - \alpha)B_D]$$

If all the votes are identical, or if the device has no information about how V will vote, then $\alpha = (k - 1)/k$. Depending on whether its expected payoff for cheating is positive or negative, it will be a Nash equilibrium either to cheat on the most-likely-voted ciphertext, or not to cheat, and for the voter to audit as instructed.

4 Conclusion

We have shown that none of the natural sequential strategies for voter-initiated auditing form Nash equilibria in a game that captures remote (Internet) voting.

This is significant because voter-initiated auditing is probably the most promising of strategies for verifiable Internet voting. The only alternatives are codes [7,14], which require a secrecy assumption and hence a threshold trust assumption on authorities, and which anyway don't work for complex ballots. Preprinted auditable ballots [8,13] only work in polling places. We have shown that voter-initiated auditing must be conducted with multiple parallel ballots, rather than sequential challenges.

The next step is to repeat the analysis for a game that captures the payoffs for polling-place voter-initiated auditing. This setting has a significantly higher cost to the device for cheating, so probably has very different equilibria.

Acknowledgments. Thanks to Wojtek Jamroga, Ron Rivest, and Josh Benaloh for interesting conversations about this work.

References

1. Adida, B.: Helios: web-based open-audit voting. USENIX Secur. Symp. **17**, 335–348 (2008)
2. Adida, B., Rivest, R.L.: Scratch & vote: self-contained paper-based cryptographic voting. In: Proceedings of the 5th ACM Workshop on Privacy in Electronic Society, pp. 29–40. ACM (2006)
3. Bell, S., Benaloh, J., Byrne, M.D., Debeauvoir, D., Eakin, B., Kortum, P., McBurnett, N., Pereira, O., Stark, P.B., Wallach, D.S., Fisher, G., Montoya, J., Parker, M., Winn, M.: Star-vote: a secure, transparent, auditable, and reliable voting system. In: 2013 Electronic Voting Technology Workshop/Workshop on Trustworthy Elections (EVT/WOTE 2013). USENIX Association, Washington, D.C. https://www.usenix.org/conference/evtwote13/workshop-program/presentation/bell
4. Ben-Nun, J., Fahri, N., Llewellyn, M., Riva, B., Rosen, A., Ta-Shma, A., Wikström, D.: A new implementation of a dual (paper and cryptographic) voting system. In: Electronic Voting, pp. 315–329 (2012)
5. Benaloh, J.: Simple verifiable elections. EVT **6**, 5 (2006)
6. Benaloh, J.: Ballot casting assurance via voter-initiated poll station auditing. EVT **7**, 14 (2007)
7. Chaum, D.: Surevote: technical overview. In: Proceedings of the workshop on trustworthy elections (WOTE 2001) (2001)
8. Chaum, D., Carback, R., Clark, J., Essex, A., Popoveniuc, S., Rivest, R.L., Ryan, P.Y., Shen, E., Sherman, A.T.: Scantegrity II: end-to-end verifiability for optical scan election systems using invisible ink confirmation codes. EVT **8**, 1–13 (2008)
9. Eckersley, P.: How unique is your web browser? In: Atallah, M.J., Hopper, N.J. (eds.) PETS 2010. LNCS, vol. 6205, pp. 1–18. Springer, Heidelberg (2010). doi:10.1007/978-3-642-14527-8_1
10. Karayumak, F., Olembo, M.M., Kauer, M., Volkamer, M.: Usability analysis of helios-an open source verifiable remote electronic voting system. In: EVT/WOTE 2011 (2011)
11. Kiayias, A., Zacharias, T., Zhang, B.: Ceremonies for end-to-end verifiable elections (2015)
12. Lindeman, M., Stark, P.B.: A gentle introduction to risk-limiting audits. IEEE Secur. Priv. **5**, 42–49 (2012)
13. Ryan, P.Y., Bismark, D., Heather, J., Schneider, S., Xia, Z.: Prêt à voter: a voter-verifiable voting system. IEEE Trans. Inf. Forensics Secur. **4**(4), 662–673 (2009)

14. Ryan, P.Y.A., Teague, V.: Pretty good democracy. In: Christianson, B., Malcolm, J.A., Matyáš, V., Roe, M. (eds.) Security Protocols 2009. LNCS, vol. 7028, pp. 111–130. Springer, Heidelberg (2013). doi:10.1007/978-3-642-36213-2_15
15. Sandler, D., Derr, K., Wallach, D.S.: Votebox: a tamper-evident, verifiable electronic voting system. In: USENIX Security Symposium, vol. 4, p. 87 (2008)

Security Games

Combining Graph Contraction and Strategy Generation for Green Security Games

Anjon Basak[1]([envelope]), Fei Fang[2], Thanh Hong Nguyen[2],
and Christopher Kiekintveld[1]

[1] University of Texas at El Paso, 500 W University Ave., El Paso, TX 79902, USA
abasak@miners.utep.edu, cdkiekintveld@utep.edu
[2] University of Southern California,
941 Bloom Walk, SAL 300, Los Angeles, CA 90089, USA
{feifang,thanhhng}@usc.edu

Abstract. Many real-world security problems can be modeled using Stackelberg security games (SSG), which model the interactions between a defender and attacker. *Green security games* focus on environmental crime, such as preventing poaching, illegal logging, or detecting pollution. A common problem in green security games is to optimize patrolling strategies for a large physical area such as a national park or other protected area. Patrolling strategies can be modeled as paths in a graph that represents the physical terrain. However, having a detailed graph to represent possible movements in a very large area typically results in an intractable computational problem due to the extremely large number of potential paths. While a variety of algorithmic approaches have been explored in the literature to solve security games based on large graphs, the size of games that can be solved is still quite limited. Here, we introduce abstraction methods for solving large graph-based security games and integrate these methods with strategy generation techniques. We demonstrate empirically that the combination of these methods results in dramatic improvements in solution time with modest impact on solution quality.

Keywords: Security · Green security · Abstraction · Contraction · Game theory

1 Introduction

We face many complex security threats with the need to protect people, infrastructure, computer systems, and natural resources from criminal and terrorist activity. A common challenge in these security domains is making the best use of limited resources to improve security against intelligent, motivated attackers. The area of *green security* focuses on problems related to protecting wildlife and natural resources against illegal exploitation, such as poaching and illegal logging. Resource limitations are particularly acute in fighting many types of environmental crime, due to a combination of limited budgets and massive areas

© Springer International Publishing AG 2016
Q. Zhu et al. (Eds.): GameSec 2016, LNCS 9996, pp. 251–271, 2016.
DOI: 10.1007/978-3-319-47413-7_15

that need surveillance and protection. For example, it is common for small numbers of rangers, local police, and volunteers to patrol protected national parks that may cover thousands of square miles of rugged terrain [22].

Work on *green security games* [7,11] has proposed formulating the problem of finding optimal patrols to prevent environmental crime as a Stackelberg security game [25]. In these games, the defender (e.g., park ranger service) must decide on a randomized strategy for patrolling the protected area, limited by the geographic constraints and the number of available resources. The attacker (e.g., poacher) selects an area of the park to attack based on the intended target and knowledge of the typical patrolling strategy (e.g., from previous observations and experience). Green security games are used to find randomized patrolling strategies that maximize environmental protection given the resources available.

Green security games typically model the movement constraints for the defender patrols using a graph representing the physical terrain. Unfortunately, this leads to a major computational challenge because the number of possible paths for the defender grows exponentially with the size of the graph. Enumerating all possible combinations of paths for multiple resources makes the problem even more intractable [29,35]. Several algorithms have been proposed in the literature to solve these games more efficiently [24,28]. Most of these rely on incremental strategy generation (known as double oracle algorithms, or column/constraint generation) to solve an integer programming formulation of the problem without enumerating the full strategy space. The most recent application called PAWS [10] approaches the scalability issue by incorporating cutting plane and column generation techniques.

Here, we take a new approach that *combines* strategy generation methods with automated game abstraction methods based on graph contraction. The idea of using automated abstraction has been very successful in solving other types of very large games, such as computer poker [16,17,19,20,40]. The basic idea of our game abstraction is motivated by graph contraction techniques used to speed up pathfinding and other computations on graphs. When we apply graph contraction to a green security game, it dramatically reduces the strategy space for the defender, leading to lower solving time. To improve scalability even further we integrate graph contraction with strategy generation to create a new class of algorithms capable of solving very large green security games. We evaluate our new algorithms on graph-based security games motivated by the problems encountered in green security domains, including some based on real world data sets. The experiments show that we can dramatically improve solution times by using abstraction in combination with strategy generation, leading to high-quality approximations within seconds even for graphs with a thousand nodes.

2   Related Work

The first approach to compute security resource allocations was to find a randomized strategy after enumerating all possible resource allocations [29], which is used by the Los Angeles Airport Police in an application called ARMOR [30]. A more compact form of security game representation was used [25] to develop a faster algorithm (IRIS [35]), which is used for scheduling by the Federal Marshal Service (FAMS). ASPEN [24] was introduced to deal with the exponential size of games with complex scheduling constraints by using a branch-and-price app- roach. Most recently, to tackle more massive games an approach based on cutting planes was introduced [38] to make the solution space more manageable. Game theoretic algorithms are also used to secure ports [32] and trains [39]. Recently, successful deployment of game theoretic applications motivated researchers to use game theory in green security domains [7,21,37]. This led to new game model called GSG [11]. Assumptions about the attacker being able to fully observe the defender strategy can be unrealistic in some cases, so partial observability and bounded rationality have been introduced to make the attacker model better fit the practice. Defender payoff uncertainty has also been addressed with these issues in an algorithm called ARROW [28]. Despite the models and algorithms introduced, how to handle the large strategy space in GSGs remains a challenge. In this paper, we introduce abstraction techniques to address this problem. Many abstraction techniques have been developed for extensive form games with uncer- tainty including both lossy [31] and lossless [18] abstraction. There has been some work which gives bounds on the error introduced by abstraction [26]. There are also imperfect recall abstractions that consider hierarchical abstraction [8] and Earth mover's distance [13].

Graph contraction techniques [14] have been used to achieve fast routing in road networks, where contraction acts as a pre-processing step. This method has been improved using fast bidirectional Dijkstra searches [15,34]. A time- dependent contraction algorithm has also been introduced for time-dependent road networks [5]. Graph contraction has also been used in imperfect informa- tion security games with infinite horizon where the area is patrolled by a single robot [4]. In this paper, we leverage insights from graph contraction to handle the large strategy space in GSGs. Another recent closely related work [23] uses cut-based graph contraction and also column generation approach for restricting the strategy space, but for a different type of security model based on checkpoint placement for urban networks.

3 Domain Motivation

Illegal activities such as poaching pose a major threat to biodiversity across all types of habitats, and many species such as rhinos and tigers. A report [1] from the Wildlife Conservation Society (WCS) on May 2015 stated that the elephant population in Mozambique has shrunk from $20,000$ to $10,300$ over the last five years. Elephants were recently added to the IUCN Red List [2]. Marine species also face danger due to illegal fishing and overfishing, causing harm to the peo- ple of coastal areas who depend on fishing for both sustenance and livelihood. According to World Wide Fund for Nature (WWF), the global estimated finan- cial loss due to illegal fishing is $23.5 billion [3]. Organizations like WCS are

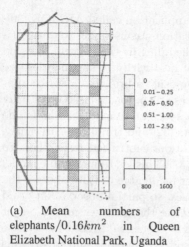

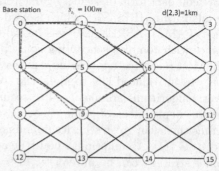

(a) Mean numbers of elephants/$0.16km^2$ in Queen Elizabeth National Park, Uganda

(b) A graph representation of a grid-based GSG (a patrolling path is shown in red).

Fig. 1. Domain example and game model.

studying strategies for combating environmental crime that include patrols of both land and sea habitats to detect and deter poaching. PAWS [10] is a new application based on green security games that helps to design patrolling strategies to protect wildlife in threatened areas. The area of interest is divided into grid cells that capture information about the terrain, animal density, etc. Each grid cell is a potential target for the poachers. The patroller plans a route to protect the targets along a path. However, if the grid cell is too large (e.g., 1 km by 1 km) or the terrain is complex, it is very difficult for the patroller to patrol even a single grid cell without any detailed path provided in the cell. Therefore, a fine-grained discretization is often required, leading to a large number of targets and a exponential number of patrol routes that existing solvers cannot handle. PAWS handles this problem by pre-defining a limited set of routes based on domain knowledge of features like ridgelines and streams, which can be found based on elevation changes. We also observe that in many green security domains, there is a high variance in the importance of the targets. For example, Fig. 1(a) shows the mean number of elephants in each area of a grid representing the Queen Elizabeth National Park in Uganda [12]. There are many cells that have no animal count at all, and if there is minimal activity it is very inefficient to consider these areas as targets to patrol (or poach). This motivates our abstraction-based approach to make it computationally feasible to directly analyze high-fidelity maps for green security without preprocessing.

4   Game Model and Basic Solution Technique

A typical green security game (GSG) model is specified by dividing a protected wildlife area into grid based cells, as shown in Fig. 1(a). Each cell is considered a potential target t_i where an attacker could attempt a poaching action. We transform this grid-based representation into a graph as shown in Fig. 1(b). Each node represents a target t_i.

Definition 1. *A GSG Graph is a graph $G = (V, E)$ where each node $t_i \in V$ is associated with a patrolling distance s_{t_i} and each edge $e_{ij} \in E$ is associated with a traveling distance $d(i, j)$. There exists a base node $B \in V$. A feasible patrolling path is a sequence of consecutive nodes that starts and ends with B, with a total distance that does not exceed the distance limit d_{max}.*

For example, in Fig. 1(b), $s_{t_1} = 100\,\text{m}$. This means that to protect target t_1, the patroller needs to patrol for a distance $100\,\text{m}$ within target t_1. $d(2, 3) - 1\,\text{km}$ indicates the distance from target t_2 to t_3. The defender patrols to protect every target on the patrolling path. Therefore, the total distance of a path is the sum of patrolling and travel distance. Typically the patrol starts in a base station and ends in the same base station. For example, a patrolling path is shown in Fig. 1(b) where the patrol starts at t_0 and traverses through targets $t_1 \rightarrow t_6 \rightarrow t_9 \rightarrow t_4$ and ends back in target t_0.

The defender has a limited number of resources R, each of which can be assigned to at most one patrolling path that covers a set of targets $t \in T$. So the defender's pure strategies are the set of joint patrolling paths $J_m \in J$. Each joint patrolling path J_m assigns each resource to a specific path. We denote a patrolling path by p_k and the base target by t_h. The length of p_k is constrained by d_{max}.

We use a matrix $P = P_{J_m t} = (0, 1)^n$ to represent the mapping between joint patrolling paths and the targets covered by these paths, where $P_{J_m t}$ represents whether target t is covered by the joint patrolling path J_m. We define the defender's mixed strategy x as a probability distribution over the joint patrolling paths J where x_m is the probability of patrolling a joint patrolling path J_m. The coverage probability for each target is $c_t - \sum_{J_m} P_{J_m t} x_m$.

If target t is protected then the defender receives reward $U_d^c(t)$ when the attacker attacks target t, otherwise a penalty $U_d^u(t)$ is given. The attacker receives reward $U_a^u(t)$ if the attack is on an area where the defender is not patrolling, or penalty $U_a^c(t)$ if the attack is executed in a patrolled area. These values can be based on the density of the animals in the area attacked, as a proxy for the expected losses due to poaching activities. We focus on the zero-sum game case where $U_d^c(t) = U_a^c(t) = 0$ and $U_d^u(t) = -U_a^u(t)$. In the rest of the paper, we also refer to $U_a^u(t)$ as the utility of target t.

We use the Stackelberg model for GSG. In this model, the patroller, who acts as defender, moves first and the adversary observes the defender's mixed strategy and chooses a strategy afterwards. The defender tries to protect targets $T = t_1, t_2, ..., t_n$ from the attackers by allocating R resources. The attacker attacks one of the T targets. We focus on the case where the attacker is perfectly rational and compute the Strong Stackelberg Equilibrium (SSE) [6,27,36], where the defender selects a mixed strategy (in this case a probability distribution x over joint patrolling paths J_m), assuming that the adversary will be able to observe the defender's strategy and will choose a best response, breaking ties in favor of the defender. Given a defender's mixed strategy x and the corresponding coverage vector c, the expected payoff for the attacker is

$$U_a(c, t) = \max_{t \in T}\{(1 - c_t)U_a^u(t)\} \tag{1}$$

It is possible to solve this problem by enumerating all feasible joint patrolling paths [24]. In the case of zero-sum games, the optimal patrolling strategy for the defender can be determined by solving the following linear program (LP).

$$\min_{x,k} \quad k \tag{2}$$

$$(1 - Px)U_a^u \leq k \tag{3}$$

$$\sum_i x_i \leq 1 \tag{4}$$

$$x \geq 0 \tag{5}$$

Equation 2 represents the objective function, which minimizes the expected payoff for the attacker, or equivalently, maximizes the expected payoff for the defender. Constraint 4 makes sure that the probability distribution over the joint patrolling paths does not exceed one. The solution of the LP is a probability distribution x over the joint patrolling paths J, and this is the strategy the defender commits to. The attacker will choose the target with highest expected utility, as shown in Constraints 3. This formulation does not scale well to large games due to the exponential number of possible joint paths as the graph grows larger.

5   Solving GSG with Abstraction

Our approach combines the key ideas in double oracle methods and graph contraction. There are often relatively few important targets in a GSG. For example, the key regions of high animal density are relatively few, and many areas have low density, as shown in Fig. 1(a). This suggests that many targets in the game can be removed to simplify the analysis while retaining the important features of the game.

We describe our approach in three stages. First, we describe our method for contracting a graph by removing nodes and calculating a new set of edges to connect these nodes that retains the shortest path information. This contracted graph can be solved using any existing algorithm for GSG; as a baseline, we use the LP on the full set of paths. Second, we describe a single-oracle approach for finding the set of targets that must be included in the contracted game. This method restricts the set of targets to a small set of the highest-valued targets, and iteratively adds in additional targets as needed. Finally, we describe the double-oracle algorithm. This uses the same structure as the single oracle, but instead of solving each restricted game optimally, we restrict the defender's strategy space and use heuristic oracles to iteratively generate paths to add to the restricted game.

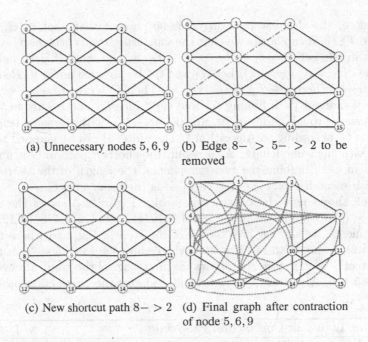

(a) Unnecessary nodes 5, 6, 9   (b) Edge 8− > 5− > 2 to be removed

(c) New shortcut path 8− > 2   (d) Final graph after contraction of node 5, 6, 9

Fig. 2. Instant Contraction procedure for different nodes

5.1 Graph Contraction

We first describe how we construct an abstracted (simplified) graph for a restricted set of target nodes. Essentially, we remove all of the nodes except the restricted set, and then add additional edges to make sure the shortest paths are preserved.

Many graph contraction procedures used in pathfinding remove nodes one by one, but we use a contraction procedure that removes the nodes in one step. Suppose we have decided to remove the set of nodes $T_u \in T$. We find all the neighbors of set T_u, denoted as V. Next we try to find the shortest paths between each pair of nodes $(v_i, v_j) \in V$ that traverse through nodes T_u where v_i and v_j are not adjacent. We use Floyd-Warshall algorithm [9] to find the shortest paths for all the nodes in V using only nodes T_u. If the length of the shortest path does not exceed d_{max}, we add an edge (v_i, v_j) in the contracted graph, with distance equals the length of the shortest path.

Theorem 1. *The contraction process described in Algorithm 1 preserves the shortest paths for any pair of nodes that are not removed in the original graph. Formally, given a graph $G = (T, E)$ and a subset of nodes T_u, Algorithm 1 provides a contracted graph $G' = (T \setminus T_u, E')$ and the length of the shortest path for any pair of nodes $(v_i, v_j) \in T \setminus T_u$ in G' is the same as in G.*

Proof sketch: First, $\forall (v_i, v_j) \in T \setminus T_u$, the shortest path in G' can be easily re-mapped to a path in G, and thus is a candidate for the shortest path in

G. Therefore, the shortest path in G' is no shorter than that in G. Second, $\forall (v_i, v_j) \in T \setminus T_u$, it can be shown that the shortest path in G can also be mapped to a path in G'. Let $P = t_1 \rightarrow t_2 \rightarrow \dots \rightarrow t_K$ be the shortest path between v_i and v_j in G ($t_1 = v_i, t_K = v_j$). Let t_{k1} and t_{k2} be any two nodes in P such that $t_{k1} \in V$, $t_{k2} \in V$ and $t_k \in T_u, \forall k1 < k < k2$. Then $\bar{P} = t_{k1} \rightarrow t_{k1+1} \rightarrow \dots \rightarrow t_{k2}$ has to be a shortest path linking t_{k1} and t_{k2}. Since t_{k1} and t_{k2} are in V and $\bar{P}$ only traverses through nodes in T_u, an edge (t_{k1}, t_{k2}) with the same length of $\bar{P}$ is added to G' according to Algorithm 1. Therefore, P can be mapped to a path P' in G' with the same length. As a result, the shortest path in G is no shorter than that in G'. Combine the two statements, the length of the shortest path for any pair of nodes $(v_i, v_j) \in T \setminus T_u$ in G' is the same as in G. □

Figure 2 shows how the contraction works. Figure 2(a) shows the removed nodes $T_u = (5, 6, 9)$. The neighbor set of T_u is $V = (0, 1, 2, 4, 7, 8, 10, 12, 13, 14)$. For convenience we show a breakdown of the step in Fig. 2(b) where the edge $(8 \rightarrow 5 \rightarrow 2)$ is shown and in Fig. 2(c) where the edge $(8 \rightarrow 5 \rightarrow 2)$ is replaced with shortcut $8 \rightarrow 2$. Figure 2(d) shows the final stage of the graph after contracting nodes $5, 6, 9$. Algorithm 1 shows pseudocode for the contraction procedure.

Algorithm 1. Instant Contraction Procedure

```
1:  procedure INSTANTCONTRACTGRAPH                                    ▷
2:      G ← Graph()                              ▷ Initiate the graph to contract
3:      n_d ← ContractedNodes()                  ▷ Get the nodes to contract
4:      n_nei ← ComputeNeighbors(n_d)
5:      apsp ← AllPairShortestPath(G, n_d, paths)
6:      for v ← neighbors.pop() do
7:          for v' ← neighbors.pop() do
8:              if v ≠ v' & notadjacent(v, v') then
9:                  d ← apsp[v][v']
10:                 path ← getPath(paths, v, v')
11:                 if d ≤ dmax then              ▷ if d is less than the distance limit
12:                     UpdateNeighbors(v, v', path, d)
13:                     v.AddNeighbor(v', path)
14:                     v'.AddNeighbor(v, path)
15:      RemoveAllContractedNodes(G, n_d)
```

Reverse Mapping. When we solve a GSG with a contracted graph (e.g., using the standard LP), the paths found in the solution must be mapped back to the paths in the original graph so they can be executed. This is because a single edge in the abstract path can correspond to a path of several nodes in the original graph. In Algorithm 1, when the contracted graph is constructed, the corresponding path in the original graph of each edge being added is already recorded, and is the basis the reverse mapping.

5.2   Single-Oracle Algorithm Using Abstraction

We begin by describing a basic "single oracle" algorithm that restricts only the attacker's strategy space (i.e., the number of targets). The basic observation that leads to this approach is based on the notion of an attack set. In the Stackelberg equilibrium solution to a security game, there is a set of targets that the attacker is willing to attack; this is the set that the defender must cover with positive probability. All other target have maximum payoffs lower than the expected payoff for the attacker in the equilibrium solution, so the attacker will never prefer to attack one of these targets, even though it is left unprotected. If we could determine ahead of time which set of targets must be covered in the solution, we could simply apply graph contraction to this set of targets, solve the resulting game, and be guaranteed to find the optimal solution.

Algorithm 2. Single Oracle With Abstraction (SO)

Input: original graph G, target utility U_i, $\forall i \in V$
Output: defender mixed strategy x and coverage vector c
1: $\bar{T}$=GreedyCoverR(G)   ▷ Find initial set of targets to be considered in the restricted graph
2: Set current graph $G_c = G$
3: **repeat**
4:    G_t =**Contract**$(G_c, \bar{T})$ ▷ Contract graph
5:    $(u, x_t, c_t) = Solve(G_t)$ ▷ Solve restricted graph, get attacker's expected utility u, defender strategy x_t, coverage vector c_t
6:    $v = AttEU(G_c, c_t)$ ▷ Calculate actual attacker's expected utility on current graph
7:    **if** $v == u$ **then**
8:       Break
9:    G_c =**ContractWithThreshold**(G_c, u) ▷ Remove targets with utility $< u$
10:   **if** G_c is small enough **then**
11:      $(u, x, c) = Solve(G_c)$ ▷ Solve G_c directly
12:      Break
13:   Add at least one additional target into $\bar{T}$
14: **until** $1 < 0$

Our approach is to start by considering only a small set of targets $\bar{T}$, perform contraction, and solve the abstracted game for this small set of targets. If the attacker expected value in the solution is lower than the value the attacker can get from attacking the best target that was not included in the restricted game, we add at least one (and possible more than one) additional target to the restricted game and repeat the process. Targets are added in decreasing order of the attacker's payoff for attacking the target if it is not protected at all. If we solve a restricted game and the attacker's expected value is greater than the unprotected values of all remaining targets, we can terminate having found the correct attack set and the optimal solution.

The initial set of targets to be considered is determined by *GreedyCoverR* (GCR). First consider the case where there is only one patroller. We use an algorithm named GC1 to find a greedy patrolling path. GC1 greedily inserts targets to the path and asks the patroller to take the shortest path to move from one target to the next target. The targets are added sequentially in a descending order of the target utility. GC1 terminates when the distance limit constraint is violated. GCR calls GC1 R times to find greedy paths for R patrolling resources. If the greedy paths can cover the top K targets, GCR returns the set of targets whose utility is no less than the utility of the $(K + 1)^{th}$ target. This is because a restricted graph with the top K targets can be perfectly protected given the greedy paths, and therefore the patroller can try to protect more targets.

Algorithm 2 shows psuedocode for this procedure. Clearly, u is non-decreasing and v is non-increasing with each iteration. For a value of u in any iteration, we can claim that any target whose utility is smaller than u can be safely removed as those targets will never be attacked (attacker will not deviate if those targets are added to the small graph). The function $\text{Contract}(G, \bar{T})$ completes two tasks. First, it removes targets that are not in $\bar{T}$, and second, refine the graph by removing dominated targets. In each iteration, u provides a lower bound of the attacker's expected utility in the optimal solution (optimal defender strategy) and v provides an upper bound. If $v == u$, it means current solution is the optimal. Line 13 adds at least one target to the set $\bar{T}$. Figure 3 illustrates the algorithm on an example graph. Figure 3 illustrates Algorithm 2 with an example.

5.3   Double Oracle Graph Contraction

The single oracle methods can prevent us from having to solve the full graph with the complete set of targets. However, it still assumes that we use an exact, exhaustive method to solve the smaller abstracted graphs. For very large problems, this may still be too slow and use too much memory. To address this we introduce the Double Oracle method that also restricts the defender's strategy space when solving the abstracted graphs. This basic idea (a version of column generation) has been widely used in security games literature [24,33]. Algorithm 3 outlines the procedure.

The outer loop is based on the single oracle method, and gradually adds targets to the restricted set. However, each time we solve the problem for a new contracted graph, we also start from a restricted set of possible paths for the defender. We then solve the "Master" problem (i.e., the original LP), but only with this restricted set of paths. If the solution to this restricted problem already implies that we need to add more targets (because the attacker's payoff is lower than the next best target), we do so and start over with a new, larger contracted graph. Otherwise, we solve a "Slave" problem to find at least one new path to add to the restricted problem, and then go back to solve the Master again. This process terminates when we cannot add any additional paths to the Master that would improve the payoff for the defender (and lower it for the attacker).

To guarantee that we have found the optimal solution, the slave should always return a new path to add that has the minimum reduced cost. The reduced cost

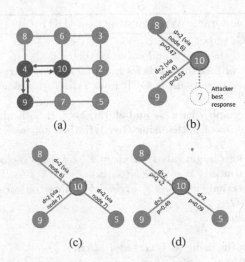

(a)   (b)

(c)   (d)

Fig. 3. Example of Single-oracle Algorithm. The numbers shown in the nodes represent the index and the utility of the target. Node 10 is the base node and the defender has only one patrol resource. (a): Original graph (distance limit = 4), which is also the initial current graph G_c. Red lines indicate the greedy route, which determines $\bar{T} = \{10, 9, 8\}$. (b): First restricted graph G_t and the corresponding optimal defender strategy (taking the route $10 \rightarrow 8 \rightarrow 10$ with probability 0.47), which leads to $u = 4.23$ and $v = 7$. (c): Updated current graph G_c, which is achieved by removing all nodes with utility $\leq u$ (i.e., nodes 2, 3, 4) and then removing dominated targets (node 7 is dominated by node 9 and node 6 is dominated by node 8). (d): Second restricted graph G_t with updated $\bar{T} = \{10, 9, 8, 5\}$, which leads to $u = v = 4.58$ and the termination of the algorithm. (Color figure online)

of a new joint path J_m is $r_{J_m} = -\sum_i y_i U_a^u(i) P_{J_m,i} - \rho$, where y_i refers to the dual variable of the ith constraint in the original LP (3), and ρ is the dual variable of constraint 4. The joint path with the most negative reduced cost improves the objective the most. If the reduced cost of the best new joint path is non-negative, then the current solution is optimal. In fact, finding the joint path with the lowest reduced cost is equivalent to solving the following combinatorial optimization problem:

Definition 2. *In the coin collection problem, a GSG graph $G = (V, E)$ is given, and each node t_i is associated with a number of coins, denoted as Y_i. When a node is covered by a patrolling path, the coins on the node will be collected and can be collected at most once. The goal is to find a feasible joint path that collects the most number of coins.*

When $Y_i = y_i U_a^u(i)$, the optimal solution of the coin collection problem is the joint path with the lowest reduced cost. The coin collection problem is NP-hard based on a reduction form the hamiltonian cycle problem (details omitted for space). Designing efficient algorithms for finding the optimal or a near-optimal solution of the coin collection problem can potentially improve the scalability

Algorithm 3. Double Oracle With Abstraction (DO)

Input: original graph G, target utility U_i, $\forall i \in V$
Output: defender mixed strategy x and coverage vector c
1: Sort the targets according to attacker's reward T_{srt}=sortTargets()
2: Get the list of initial targets using GCR from T_{srt}, T_{cur} = GreedyCoverR()
3: **repeat**
4:     Set temporary graph where G_t and all targets $t_i \in G_t$ is also in T_{cur}
5:     Generate initial set of paths using GreedyPathR, $s_{cur} = GPR(G_t)$
6:     **repeat**
7:         Solve SSG for G_t, get mixed strategy x_t, coverage vector c_t, and attacker's
            expected utility $u = AttEU(G_t, c_t)$
8:         Calculate actual attacker's expected utility on original graph $v =$
            $AttEU(G, c_t)$
9:         **if** $u < v$ **then**
10:            Break
11:        Generate paths using $s_t = GreedyPathR()$
12:        Append paths $s_{cur} = s_{cur} U s_t$
13:        **if** $s_t == 0$ **then**
14:            Break
15:    **until** $1 < 0$
16:    Find attack target in G $attackTarget(G, c_t)$
17:    Add next n e.g. $n = 5$ targets to T_{cur} from $T_{srt} - T_{cur}$
18: **until** $u \geq v$ and no more path can be added to s_{cur}

of using the double oracle method to find the exact optimal solution to GSG. However, here we are interested in maximizing scalability for the DO approach combined with abstraction, so we designed heuristic methods for the slave that are very fast, but will not necessarily guarantee the optimal solution. More specifically, we use Algorithm 4 as a heuristic approach for solving the coin collection problem.

6   Experimental Evaluation

We present a series of experiments to evaluate the computational benefits and solution quality of our solution methods. We begin by evaluating the impact of abstraction in isolation, and then provide a comparison of many different variations of our methods on synthetic game instances. Finally, we test our best algorithms on large game instances using real-world data, demonstrating the ability to scale up to real world problems.

6.1   Graph Abstraction

We begin by isolating the effects of abstraction from the use of strategy generation (using either the single or double-oracle framework). The baseline method solves a graph-based security game directly using the standard optimization formulation, enumerating all joint patrolling paths directly on the full graph.

Algorithm 4. GreedyPathR (GPR)

```
 1: procedure GREEDYCOVER-COINCOLLECTION
 2:     Initialize best joint path set J_best
 3:     for iter = 0 to 99 do
 4:         if iter == 0 then
 5:             Tlist ← sort(T \ B, Y)              ▷ Get a sorted list with decreasing Y_i
 6:         else
 7:             Tlist ← shuffle(T \ B)                   ▷ Get a random ordered list
 8:         Y_r ← Y                                ▷ Initialize the coins remained
 9:         for j = 1 to R do
10:             Initialize the current patrol route Q_j
11:             for each target t_i in Tlist with Y_r(i) > 0 do    ▷ Check all uncovered
                     targets
12:                 Insert t_i to Q_j while minimizing the total distance
13:                 if total distance of Q_j exceeds d_max then
14:                     remove t_i from Q_j
15:             for each target t_i in Q_j do
16:                 Y_r(i) = 0
17:             if {Q_1, ..., Q_R} collects more coins than J_best then
18:                 update J_best
19:     return J_best
```

We compare this to first applying our graph abstraction method to the game, and then using the same solver to find the solution to the abstracted graph. We compare the methods on both solution quality and runtime. To measure the amount of error introduced we introduce an error metric denoted by epsilon(ϵ) = $\frac{[U_d(c,a) - U'_d(c,a)]}{U_d(c,a)*100}$, where $U'_d(c,a)$ is the expected payoff for defender when using contraction and $U_d(c,a) \geq U'_d(c,a)$.

For our experiments we used 100 randomly generated, 2-player security games intended to capture the important features of green security games. Each game has 25 targets (nodes in the graph). Payoffs for the targets are chosen uniformly at random from the range -10 to 10. The rewards for the defender or attacker are positive and the penalties are negative. We set the distance constraint to 6. In the baseline solution the is no contraction. For different levels of abstraction the number of contracted nodes (#CN) varies between the values: $(0, 2, 5, 8, 10)$. Figure 4 shows us how contraction affects contraction time (CT), solution time (ST) and reverse mapping time (RMT). CT only consider the contraction procedure, ST considers the construction of the P matrix and the solution time for the optimization problem, and RMT considers time to generate the P matrix for the original graph from the solution to the abstracted game.

We first note that as the graph becomes more contracted ST takes much less time, as shown in Fig. 4. The next experimental result presented in Fig. 5 shows how much error is introduced as we increase the amount of contraction and the amount of time we can save by using contraction.

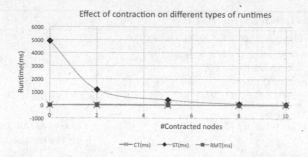

Fig. 4. Effect of contraction on times CT, ST and RMT

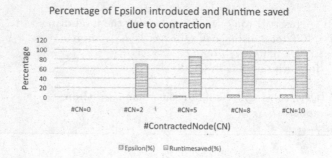

Fig. 5. Effect of contraction on Epsilon and runtime saved

6.2 Comparison of Solution Algorithms

We now present results comparing the solution quality and runtimes of different versions of our solution algorithm on graph-based security games of increasing size. We focus on grid-based graphs, which are typical of open-area patrolling problems like those in wildlife protection domains. For the experiments we generated 20 sample games for each size of game. For simplicity, the distance between every node and it's neighbors is set to 1. The patroller has two resources to conduct patrols in each case, and the distance constraint on the paths varies depending on the game size.

All of the games are zero-sum. We randomly assign payoffs to the targets. In wildlife protection, it is typical for there to be a relatively small number of areas with high densities of animal/poaching activity. To reflect this low density of high-valued targets, we partition the targets into high and low value types, with values uniformly distributed in the ranges of [0, 4] and [8, 10], respectively. We assign 90 % of the targets values from the low range, and 10 % values from the high range.

We break down the runtime into three different components: (1) The time to contract graphs, ContractionTime (CT), (2) The time to solve optimization problems, SolvingTime (ST), and (3) the total runtime, TotalTime (TT). All runtimes are given in milliseconds. EPd denotes the expected payoff for defender.

We compare out methods to two baselines that solve the original optimization problem with no contraction by enumerating joint patrolling paths. The first one enumerates all paths and directly solves the problem, while the second algorithm uses column generation to iteratively add joint paths (but does not use contraction). All algorithms that use the path sampling heuristic generate 1000 sample paths. We considered different combinations of the heuristics for both the Single Oracle (SO) and Double Oracle (DO) formulations. In Double Oracle, there are three modules where heuristic approaches can be used: (1) selecting the initial set of targets for the restricted graph; (2) selecting initial paths for solving the restricted graph; (3) in column generation, adding paths that can improve the solution for the restricted graph. The first two modules are also needed in Single Oracle. We discuss the heuristic approaches tested for these three modules. First, for selecting the initial set of targets, we test *GreedyCover1* (GC1) and *GreedyCoverR* (GCR). Second, for selecting initial paths for the restricted graph, we enumerate all the paths (denoted as All paths) for small scale problems. In addition, we test *GreedyPathR* (GPR) and *GreedyPath3* (GP3). When using GPR for selecting initial paths, we use target utility as the number of coins on the targets. *GreedyPath3* (GP3) initialize the set of paths by listing the shortest paths from the base to a target and back to base for each target. Third, to add new paths in column generation, we test GPR and random sampling of paths (denoted as sample path).

We present the runtime and solution quality data for our algorithms as we increase the size of the game, considering game sizes of 25, 50, 100 and 200 targets. Tables 1, 2, 3 and 4 show the results for each of these four cases, respectively. We had a memory limitation of 16 GB, and many of the algorithms were not able to solve the larger problems within this memory limit. We include only the data for algorithms that successfully solved all of the sample games for a given size within the memory limit.

We note that the baseline algorithms are only able to solve the smallest games within the memory limit. Even for these games, the single and double

Table 1. Performance comparison, #target = 25 and dmax = 8

Algorithm	#targets	dmax	#remaining targets	EPd	CT	ST	TT
DO + GC1 + GPR + LP + GPR	25	8	13	6.759	2	11	69
DO + GCR + GPR + LP + GPR	25	8	14	5.8845	2	12	65
DO + GC1 + GP3 + LP + GPR	25	8	12	7.2095	3	22	44
DO + GCR + GP3 + LP + GPR	25	8	10	7.1865	2	15	38
DO + GC1 + GPR + LP + Sample Paths	25	8	14	7.481	2	14	165
DO + GCR + GPR + LP + Sample Paths	25	8	14	7.3955	2	14	205
DO + GC1 + GP3 + LP + Sample Paths	25	8	14	7.605	3	97	267
DO + GCR + GP3 + LP + Sample Paths	25	8	14	7.587	2	99	283
SO + GC1 + IC + All paths + LP	25	8	12	7.702	1	105	632
SO + GCR + IC + All paths + LP	25	8	14	7.702	2	135	827
SO + GCR + IC + GP3 + LP	25	8	11	2.05	4	10	33
No contraction + No column generation	25	8	25	7.702	0	1417	14140
No contraction + Column generation	25	8	25	7.702	0	1480	14661

Table 2. Performance comparison, #target = 50 and dmax = 20

Algorithm	#targets	dmax	#remaining targets	EPd	CT	ST	TT
DO + GC1 + GPR + LP + GPR	50	20	30	5.018	9	1313	1981
DO + GCR + GPR + LP + GPR	50	20	29	5.8195	7	461	790
DO + GC1 + GP3 + LP + GPR	50	20	28	7.4945	14	187	292
DO + GCR + GP3 + LP + GPR	50	20	27	7.6415	8	162	261
DO + GC1 + GPR + LP + Sample Paths	50	20	30	5.794	9	280	4154
DO + GCR + GPR + LP + Sample Paths	50	20	29	6.4185	6	167	3925
DO + GC1 + GP3 + LP + Sample Paths	50	20	20	6.8935	6	2194	4499
DO + GCR + GP3 + LP + Sample Paths	50	20	23	6.777	6	1570	4330
BA + GCR + IC + GP3 + LP	50	20	22	0.75	5	26	1113

Table 3. Performance comparison, #target = 100 and dmax = 29

Algorithm	#targets	dmax	#remaining targets	EPd	CT	ST	TT
DO + GC1 + GPR + LP + GPR	100	29	51	6.5135	51	5753	8433
DO + GCR + GPR + LP + GPR	100	29	51	6.193	38	2170	3392
DO + GC1 + GP3 + LP + GPR	100	29	48	7.0545	52	766	1084
DO + GCR + GP3 + LP + GPR	100	29	47	7.2435	37	659	1017
DO + GC1 + GPR + LP + Sample Paths	100	29	50	6.098	46	1792	25017
DO + GC1 + GP3 + LP + Sample Paths	100	29	20	5.4735	13	2200	4420
DO + GCR + GP3 + LP + Sample Paths	100	29	30	5.0745	12	2596	5864

oracle methods using abstraction are all dramatically faster, and many of the variations come close to finding the optimal solutions. As we scale up the game size, the single oracle methods are not able to solve the game within the memory limit. For the largest games, the double oracle methods without sampled paths are still able to solve the problems to find good solutions, and do so very quickly. The third and fourth variation consistently show the best overall performance, with a good tradeoff between solution quality and speed.

We conduct a second experiment on large, 200-target graphs with the same distance and resource constraints but a different distribution of payoffs. For this experiment we have three payoff partitions, with the value ranges: $[0, 1], [2, 8], [9, 10]$. The ratio of target values in these ranges is 80 %, 10 % and 10 %, respectively. Table 5 shows the results. In comparison with Table 5, the DO algorithms (especially variations 3 and 4) are even faster, though in this case variation 1 and 2 do result in higher solution qualities. The distribution of payoffs has a significant effect on the algorithm performance, and as expected,

Table 4. Performance comparison, #target = 200 and dmax = 45

Algorithm	#targets	dmax	#remaining targets	EPd	CT	ST	TT
DO + GC1 + GPR + LP + GPR	200	45	85	6.5355	345	10360	17904
DO + GCR + GPR + LP + GPR	200	45	83	6.501	287	5307	9657
DO + GC1 + GP3 + LP + GPR	200	45	72	6.551	270	2658	4156
DO + GCR + GP3 + LP + GPR	200	45	70	6.656	189	2274	3603

Table 5. Performance comparison with 3 partition in payoff, #target = 200 and dmax = 45

Algorithm	#targets	dmax	#remaining targets	EPd	CT	ST	TT
DO + GC1 + GPR + LP + GPR	200	45	44	8.621	110	8177	12363
DO + GCR + GPR + LP + GPR	200	45	43	8.6085	73	3796	5275
DO + GC1 + GP3 + LP + GPR	200	45	40	7.7445	96	595	906
DO + GCR + GP3 + LP + GPR	200	45	40	7.7075	70	721	1058

the DO variations with abstraction are most effective when there is a relatively small fraction of important targets and a large number of unimportant ones.

Next we present figures to visualize the runtime differences among different solution algorithms. Again, only algorithms that were able to complete within the memory bound are shown. Figures 6(a)–(c) show the TotalTime, Contraction-Time and SolvingTime comparison respectively among Double Oracle methods and Basic Abstraction Methods with the baseline algorithms. The figures show the same patterns of scalability discussed previously.

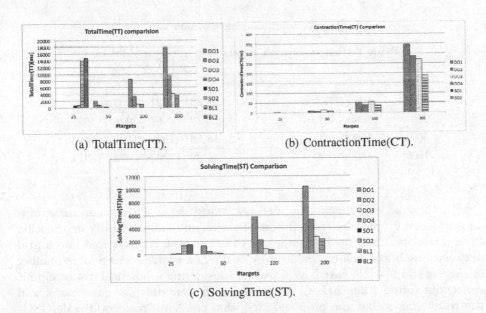

(a) TotalTime(TT).   (b) ContractionTime(CT).

(c) SolvingTime(ST).

Fig. 6. Runtime comparison among solvers.

Next we visualize the solution quality of our proposed algorithms in comparison with the baseline algorithms. The experiment setup is the same as the previous experiment. Figure 7 shows that we were able to compare the solution quality properly for #target = 25 since the baseline algorithms were able to finish. The Basic Abstraction methods except the one which uses GP3 were able to compute the exact solution. All of the Double Oracle methods are suboptimal, but typically provide good approximations.

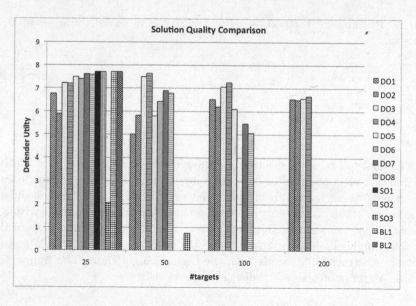

Fig. 7. Solution quality evaluation

Table 6. Results of using abstraction in real world data

#targets	dmax	#remaining targets	EPd	CT	ST	TT
100	5000	56	25.83	121	303	789
200	8000	88	25.79	67	926	1678
500	15000	92	18.56	1928	1107	4403
1000	18000	100	16.29	12302	2072	18374

For the final experiments we used real world data. We test our algorithms on grid-based graphs constructed from elevation and animal density information from a conservation area in Southeast Asia. The area is discretized into a grid map with each grid cell of size 50 m by 50 m. The problem has a large number of targets and feasible patrol routes when considering a practical distance limit constraint (often 5 km–20 km). We tested with four different game sizes, and the result shows that the proposed algorithm can solve real-world scale GSGs efficiently (see Table 6). Only DO4 was used for this experiment since it provides superior performance than others. The payoff range for the targets were [0, 90].

7 Conclusion

Green security games are being used to help combat environmental crime by improving patrolling strategies. However, the applications of GSG are still limited due to the computational barriers of solving large, complex games based on underlying graphical structures. Existing applications require manual pre-processing to come up with suitably abstract games that can be solved by existing solvers. We address this problem by designing the first algorithm for solving graph-based security games that integrates automated abstraction techniques with strategy generation methods. Our algorithm is the first to be able to provide high-quality solutions to very large green security games (thousands of nodes) in seconds, potentially opening up many new applications of GSG while avoiding the need for some of the arbitrary, manual abstraction stages when generating game models. With additional work to develop fast exact slave algorithms, we should also be able to provide exact solutions using this approach to large GSG. We also plan to investigate approximate slave formulations with performance bounds, using abstraction to compute solution concepts from behavioral game theory such as quantal response equilibrium, and applying our algorithms to real-world applications in green security games.

Acknowledgement. We would like to thank to our partners from Rimba and Panthera for providing the real world data set. This work was supported by the NSF under Grant No. IIS-1253950.

References

1. Govt. of Mozambique announces major decline in national elephant population, May 2015
2. The IUCN Red List of threatened species, April 2015
3. Estimate of global financial losses due to illegal fishing, February 2016
4. Basilico, N., Gatti, N., Amigoni, F.: Patrolling security games: definition and algorithms for solving large instances with single patroller and single intruder. Artif. Intell. **184**, 78–123 (2012)
5. Batz, G.V., Geisberger, R., Neubauer, S., Sanders, P.: Time-dependent contraction hierarchies and approximation. In: Festa, P. (ed.) SEA 2010. LNCS, vol. 6049, pp. 166–177. Springer, Heidelberg (2010). doi:10.1007/978-3-642-13193-6_15
6. Breton, M., Alj, A., Haurie, A.: Sequential Stackelberg equilibria in two-person games. J. Optim. Theory Appl. **59**(1), 71–97 (1988)
7. Brown, M., Haskell, W.B., Tambe, M.: Addressing scalability and robustness in security games with multiple boundedly rational adversaries. In: Poovendran, R., Saad, W. (eds.) GameSec 2014. LNCS, vol. 8840, pp. 23–42. Springer, Heidelberg (2014). doi:10.1007/978-3-319-12601-2_2
8. Brown, N., Ganzfried, S., Sandholm, T.: Hierarchical abstraction, distributed equilibrium computation, and post-processing, with application to a champion no-limit Texas Hold'em agent. Technical report (2014)
9. Cormen, T.H., Leiserson, C.E., Rivest, R.L.: The Floyd-Warshall algorithm. In: Introduction to Algorithms, pp. 558–565 (1990)
10. Fang, F., Nguyen, T.H., Pickles, R., Lam, W.Y., Clements, G.R., An, B., Singh, A., Tambe, M., Lemieux, A.: Deploying PAWS: field optimization of the protection assistant for wildlife security. In: Proceedings of the Innovative Applications of Artificial Intelligence (2016)
11. Fang, F., Stone, P., Tambe, M.: When security games go green: designing defender strategies to prevent poaching and illegal fishing. In: International Joint Conference on Artificial Intelligence (IJCAI) (2015)

12. Field, C., Laws, R.: The distribution of the larger herbivores in the Queen Elizabeth National Park, Uganda. J. Appl. Ecol. **7**, 273–294 (1970)
13. Ganzfried, S., Sandholm, T.: Potential-aware imperfect-recall abstraction with earth mover's distance in imperfect-information games. In: Conference on Artificial Intelligence (AAAI) (2014)
14. Geisberger, R., Sanders, P., Schultes, D., Delling, D.: Contraction hierarchies: faster and simpler hierarchical routing in road networks. In: McGeoch, C.C. (ed.) WEA 2008. LNCS, vol. 5038, pp. 319–333. Springer, Heidelberg (2008). doi:10.1007/978-3-540-68552-4_24
15. Geisberger, R., Sanders, P., Schultes, D., Vetter, C.: Exact routing in large road networks using contraction hierarchies. Transp. Sci. **46**(3), 388–404 (2012)
16. Gilpin, A., Sandholm, T.: A competitive Texas Hold'em poker player via automated abstraction and real-time equilibrium computation. In: Proceedings of the National Conference on Artificial Intelligence (AAAI), vol. 21, p. 1007 (2006)
17. Gilpin, A., Sandholm, T.: Better automated abstraction techniques for imperfect information games, with application to Texas Hold'em poker. In: AAMAS, p. 192 (2007)
18. Gilpin, A., Sandholm, T.: Lossless abstraction of imperfect information games. J. ACM (JACM) **54**(5), 25 (2007)
19. Gilpin, A., Sandholm, T., Sørensen, T.B.: Potential-aware automated abstraction of sequential games, and holistic equilibrium analysis of Texas Hold'em poker. In: Proceedings of the Conference on Artificial Intelligence (AAAI), vol. 22, p. 50 (2007)
20. Gilpin, A., Sandholm, T., Sørensen, T.B.: A heads-up no-limit Texas Hold'em poker player: discretized betting models and automatically generated equilibrium-finding programs. In: AAMAS, pp. 911–918 (2008)
21. Haskell, W.B., Kar, D., Fang, F., Tambe, M., Cheung, S., Denicola, E.: Robust protection of fisheries with COmPASS. In: AAAI, pp. 2978–2983 (2014)
22. Holmern, T., Muya, J., Røskaft, E.: Local law enforcement and illegal bushmeat hunting outside the Serengeti National Park, Tanzania. Environ. Conserv. **34**(01), 55–63 (2007)
23. Iwashita, H., Ohori, K., Anai, H., Iwasaki, A.: Simplifying urban network security games with cut-based graph contraction. In: Proceedings of the 2016 International Conference on Autonomous Agents and Multiagent Systems, pp. 205–213 (2016)
24. Jain, M., Kardes, E., Kiekintveld, C., Ordóñez, F., Tambe, M.: Security games with arbitrary schedules: a branch and price approach. In: AAAI (2010)
25. Kiekintveld, C., Jain, M., Tsai, J., Pita, J., Ordóñez, F., Tambe, M.: Computing optimal randomized resource allocations for massive security games. In: Proceedings of the 8th International Conference on Autonomous Agents and Multiagent Systems, vol. 1, pp. 689–696 (2009)
26. Kroer, C., Sandholm, T.: Extensive-form game abstraction with bounds. In: Proceedings of the Fifteenth ACM Conference on Economics and Computation, pp. 621–638 (2014)
27. Leitmann, G.: On generalized Stackelberg strategies. J. Optim. Theory Appl. **26**(4), 637–643 (1978)
28. Nguyen, T.H., et al.: Making the most of our regrets: regret-based solutions to handle payoff uncertainty and elicitation in green security games. In: Khouzani, M.H.R., Panaousis, E., Theodorakopoulos, G. (eds.) GameSec 2015. LNCS, vol. 9406, pp. 170–191. Springer, Heidelberg (2015). doi:10.1007/978-3-319-25594-1_10

29. Paruchuri, P., Pearce, J.P., Marecki, J., Tambe, M., Ordonez, F., Kraus, S.: Playing games for security: an efficient exact algorithm for solving Bayesian Stackelberg games. In: Proceedings of the 7th International Joint Conference on Autonomous Agents and Multiagent Systems, vol. 2, pp. 895–902 (2008)

30. Pita, J., Bellamane, H., Jain, M., Kiekintveld, C., Tsai, J., Ordóñez, F., Tambe, M.: Security applications: lessons of real-world deployment. In: ACM SIGecom Exchanges, vol. 8, no. 2, p. 5 (2009)

31. Sandholm, T., Singh, S.: Lossy stochastic game abstraction with bounds. In: Proceedings of the 13th ACM Conference on Electronic Commerce, pp. 880–897 (2012)

32. Shieh, E., An, B., Yang, R., Tambe, M., Baldwin, C., DiRenzo, J., Maule, B., Meyer, G.: Protect: a deployed game theoretic system to protect the ports of the united states. In: Proceedings of the 11th International Conference on Autonomous Agents and Multiagent Systems, vol. 1, pp. 13–20 (2012)

33. Shieh, E., Jain, M., Jiang, A.X., Tambe, M.: Efficiently solving joint activity based security games. In: AAAI, pp. 346–352. AAAI Press (2013)

34. Skiena, S.: Dijkstra's algorithm. In: Implementing Discrete Mathematics: Combinatorics and Graph Theory with Mathematica, pp. 225–227. Addison-Wesley, Reading (1990)

35. Tsai, J., Kiekintveld, C., Ordonez, F., Tambe, M., Rathi, S.: IRIS-a tool for strategic security allocation in transportation networks (2009)

36. Von Stengel, B., Zamir, S.: Leadership with commitment to mixed strategies (2004)

37. Yang, R., Ford, B., Tambe, M., Lemieux, A.: Adaptive resource allocation for wildlife protection against illegal poachers. In: Proceedings of the 2014 International Conference on Autonomous Agents and Multi-agent Systems, pp. 453–460 (2014)

38. Yang, R., Jiang, A.X., Tambe, M., Ordonez, F.: Scaling-up security games with boundedly rational adversaries: a cutting-plane approach. In: IJCAI (2013)

39. Yin, Z., Jiang, A.X., Tambe, M., Kiekintveld, C., Leyton-Brown, K., Sandholm, T., Sullivan, J.P.: TRUSTS: scheduling randomized patrols for fare inspection in transit systems using game theory. AI Mag. **33**(4), 59 (2012)

40. Zinkevich, M., Johanson, M., Bowling, M., Piccione, C.: Regret minimization in games with incomplete information. In Advances in Neural Information Processing Systems, pp. 1729–1736 (2007)

Divide to Defend: Collusive Security Games

Shahrzad Gholami$^{(\boxtimes)}$, Bryan Wilder, Matthew Brown, Dana Thomas,
Nicole Sintov, and Milind Tambe

Computer Science Department, University of Southern California,
Los Angeles, USA
{sgholami,bwilder,matthew.a.brown,danathom,sintov,tambe}@usc.edu

Abstract. Research on security games has focused on settings where
the defender must protect against either a single adversary or multiple, independent adversaries. However, there are a variety of real-world
security domains where adversaries may benefit from colluding in their
actions against the defender, e.g., wildlife poaching, urban crime and
drug trafficking. Given such adversary collusion may be more detrimental for the defender, she has an incentive to break up collusion by playing off the self-interest of individual adversaries. As we show in this
paper, breaking up such collusion is difficult given bounded rationality of
human adversaries; we therefore investigate algorithms for the defender
assuming both rational and boundedly rational adversaries. The contributions of this paper include (i) collusive security games (COSGs),
a model for security games involving potential collusion among adversaries, (ii) SPECTRE-R, an algorithm to solve COSGs and break collusion assuming rational adversaries, (iii) observations and analyses of
adversary behavior and the underlying factors including bounded rationality, imbalanced- resource-allocation effect, coverage perception, and
individualism/collectivism attitudes within COSGs with data from 700
human subjects, (iv) a learned human behavioral model that incorporates these factors to predict when collusion will occur, (v) SPECTRE-
BR, an enhanced algorithm which optimizes against the learned behavior model to provide demonstrably better performing defender strategies
against human subjects compared to SPECTRE-R.

Keywords: Stackelberg security game · Collusion · Human behavior
model · Amazon mechanical turk

1 Introduction

Models and algorithms based on Stackelberg security games have been deployed
by many security agencies including the US Coast Guard, the Federal Air Marshal Service, and Los Angeles International Airport [23] in order to protect
against attacks by strategic adversaries in counter-terrorism settings. Recently,
security games research has explored new domains such as wildlife protection,
where effective strategies are needed to tackle sustainability problems such as
illegal poaching and fishing [4].

© Springer International Publishing AG 2016
Q. Zhu et al. (Eds.): GameSec 2016, LNCS 9996, pp. 272–293, 2016.
DOI: 10.1007/978-3-319-47413-7_16

Crucially, though, most previous work on security games assumes that different adversaries can be modeled independently [10,11,18]. However, there are many real-world security domains in which adversaries may collude in order to more effectively evade the defender. One example domain is wildlife protection. Trade in illicit wildlife products is growing rapidly, and poachers often collude both with fellow poachers and with middlemen who help move the product to customers [26]. These groups may coordinate to gain better access to information, reduce transportation costs, or reach new markets. This coordination can result in higher levels of poaching and damage to the environment. Additionally, connections have been observed between illicit wildlife trade and organized crime as well as terrorist organizations, and thus activities such as poaching can serve to indirectly threaten national security [27].

Another example domain is the illegal drug trade where international crime syndicates have increased collusive actions in order to facilitate drug trafficking, expand to distant markets, and evade local law enforcement [1]. In some cases, drug traders must collude with terrorist organizations to send drugs through particular areas. More broadly, expansion of global transportation networks and free trade has motivated collusion between criminal organizations across different countries [20]. A third example of a domain with collusive actions is the "rent-a-tribe" model in the payday lending industry. Authorities in the US attempt to regulate payday lenders which offer extremely high interest rates to low-income borrowers who cannot obtain loans from traditional banks. Recently, payday lenders have begun to operate in partnership with Native American tribes, which are exempt from state regulations. Thus, regulators seek policies which prevent collusion between payday lenders and Native American tribes [8].

Despite mounting evidence of the destructive influence of collusive behavior, strategies for preventing collusion have not been explored in the security games literature (there are some recent exceptions, which we discuss in Sect. 2). Furthermore, analysis of collusive adversary behaviors is complicated by the bounded rationality of human adversaries; such analysis with data from human players is also missing in the security games literature. To address these limitations and improve defender performance by combating collusion between adversaries, this paper (i) introduces the COllusive Security Game (COSG) model with three players: one defender and two adversaries with the potential to collude against the defender, (ii) provides a baseline algorithm, SPECTRE-R, which optimizes against collusive adversaries assuming them to be perfectly rational, (iii) analyzes data from an experiment involving 700 human subjects, (iv) proposes a data driven human behavioral model based on these factors to predict the level of collusion between human adversaries, and (v) develops a novel algorithm, SPECTRE-BR, which optimizes against the learned behavior model to better prevent collusion between adversaries (and as a result, outperforms SPECTRE-R). Indeed, we find that human adversaries are far from perfectly rational when deciding whether or not to collude. Our experiments show that defenders can improve their utility by modeling the subjective perceptions and attitudes which shape this decision and crafting strategies tuned to prevent collusion.

2 Background and Related Work

The Stackelberg Security Game model, introduced almost a decade ago, has led to a large number of applications and has been discussed widely in the literature [12,17,23]. All of these works consider adversaries as independent entities and the goal is for a defender (leader) to protect a set of targets with a limited set of resources from a set of adversaries (followers)[1]. The defender commits to a strategy and the adversaries observe this strategy and each select a target to attack. The defender's pure strategy is an assignment of her limited resources to a subset of targets and her mixed strategy refers to a probability distribution over all possible pure strategies. This mixed strategy is equivalently expressed as a set of coverage probabilities, $0 \leq c_t \leq 1$, that defender will protect each target, t [12]. Defender's utility is denoted by $U_\Theta^u(t)$ when target t is uncovered and attacked by the adversary and by $U_\Theta^c(t)$ if t is covered and attacked by the adversary. The payoffs for the attacker are analogously written by $U_\Psi^u(t)$ and $U_\Psi^c(t)$. The expected utilities of the defender, $U_\Theta(t, C)$, and attacker, $U_\Theta(t, C)$ for the defender coverage vector C, are then computed as follows:

$$U_\Theta(t, C) = c_t \cdot U_\Theta^c(t) + (1 - c_t)U_\Theta^u(t) \tag{1}$$
$$U_\Psi(t, C) = c_t \cdot U_\Psi^c(t) + (1 - c_t)U_\Psi^u(t) \tag{2}$$

The solution concept for security games involves computing a strong Stackelberg equilibrium (SSE) which assumes that the adversaries maximize their own expected utility and break ties in favor of the defender [11,23].

Given this basic information about SSG, we next start a discussion of related work. Security game models where an adversary is capable of attacking multiple targets simultaneously have been explored in [13,29]. To address cooperation between adversaries, [7] introduced a communication network based approach for adversaries to share their skills and form coalitions in order to execute more attacks. However, no previous work on security games has conducted behavioral analysis or considered the bounded rationality of human adversaries in deciding whether to collude in the first place.

Another area of related work, as well as one that provides concepts that we will use in this paper for modeling and analyzing adversary behaviors in COSGs is that of behavioral models in game theory [3]. This area is particularly relevant given our focus on modeling human adversaries in this paper. In real-world settings, it is useful to model human adversaries as not strictly maximizing their expected utility, but rather, as their choosing strategies stochastically [14]. Quantal response equilibrium (QRE) is a solution concept based on the assumption of bounded rationality [15]. The intuition behind the QR model is that the higher the expected utility for an action, the higher the probability of the adversary selecting that action. SUQR [19] has been proposed as an extension of QR and seen to outperform QR in modeling human adversaries [28].

---

[1] We use the convention in the security game literature where the defender is referred as "she" and an adversary is referred to as "he".

This model is used in this paper to predict the probability of attack at each target. The logit function shown in Eq. 3 is the most common specification for QR and SUQR functional form where q_t is the probability of choosing strategy t among all possible strategies in set of T.

$$q_t = \frac{e^{\hat{U}_\Psi(t,C)}}{\sum_{t \in T} e^{\hat{U}_\Psi(t,C)}} \tag{3}$$

In SUQR model, $\hat{U}_\Psi(t, C)$ refers to subjective utility, and it replaces expected utility. Subjective utility in SUQR is defined as a linear combination of key domain features including the defender's coverage probability and the adversary's reward and penalty at each target which are respectively weighted by ω_1, ω_2 and ω_3. These are assumed to be the most salient features in the adversary's decision-making process.

$$\hat{U}_\Psi(t, C) = \omega_1 \cdot c_t + \omega_2 \cdot U_\Psi^u(t) + \omega_3 \cdot U_\Psi^c(t) \tag{4}$$

Another relevant aspect of bounded rationality is how humans weight probabilities. Prospect Theory (PT) proposes that individuals overweight low probabilities and underweight high probabilities; essentially, probabilities are transformed by an inverse S-shaped function [9,25]. Various functional forms have been proposed to capture this relationship [9,25]. Later work, specific to security games, has found the opposite of what Prospect Theory suggests: human players underweight low probabilities and overweight high probabilities [10]. This corresponds to an S-shaped weighting function. In either case, incorporating a model of probability perception allows the defender to exploit inaccuracies in the adversary's reasoning. Human subject experiments have been conducted for security games to test both bounded rationality and probability weighting [10], but have never included the collusive actions investigated in this paper.

Additionally, humans' decisions in strategic settings can be influenced by the relative advantage of participants. According to Inequity Aversion (IA) theory humans are sensitive to inequity of outcome regardless of whether they are in the advantaged or disadvantaged situation and they make decisions in a way that minimizes inequity [5]. Inequity aversion has been widely studied in economics and psychology and is consistent with observations of human behavior in standard economic experiments such as the dictator game and ultimatum game in which the most common choice is to split the reward 50-50 [2]. Along these lines and contrary to the theoretical predictions, IA theory also supports our analyses in the security game domain.

Finally, the personal attitudes and attributes of participants can also influence their interactions in strategic settings. A key characteristic is the well-established individualism-collectivism paradigm, which describes cultural differences in the likelihood of people to prioritize themselves versus their in-group [22]. This paper is the first to provide analysis of human adversary behavior in security games using individualism-collectivism paradigm. Specifically, those who identify as part of collectivistic cultures, compared to people in individualistic cultures, tend to identify as part of their in-groups, prioritize group-level goals, define most relationships with in-group members as communal, and are more self-effacing. Individualism-collectivism can be reliably measured using psychometrically-validated survey instruments [24].

3   Illustrative Motivating Domain: Wildlife Poaching Game

As an illustrative motivating domain for the work reported in this paper, we focus on the challenge of wildlife poaching. Wildlife poaching poses a serious threat to the environment as well as national security in numerous countries around the world and is now estimated to be worth $5 billion annually. The most common types of illicitly poached and traded wildlife products include elephant ivory, rhino horn, tiger parts, and caviar [16]. Biodiversity loss, invasive species introduction, and disease transmission resulting from illicit wildlife trade can all have disastrous impacts on the environment. Evidence [26] confirms that collusive actions (e.g., cost sharing for storage, handling, and transportation of goods) among adversaries can increase the rate of poaching and cause further damage to the environment. Modeling this as a security game, the defender is a ranger whose goal is to allocate patrolling resources optimally over the targets. The adversaries are poachers or illegal traders who execute attacks, possibly in collusion with one another. To better understand collusion in the wildlife poaching domain, we designed a game for human subjects to play on Amazon Mechanical Turk (AMT). Participants were asked to play our game in different settings and answer survey questions. Afterwards, their actions were analyzed using the theories explained above, allowing us to test assumptions about the rationality of human adversaries.

3.1   Game Overview

In our game, human subjects are asked to play the role of a poacher in a national park in Africa. The entire park area (see Fig. 1) is divided into two sections (right and left) and each human subject can only attack in one section (either right or left); however, they can explore the whole park to obtain information about the other player's situation. To ensure repeatability of the experiments, the other side is played by a computer, not a real player. Since our goal is to study human adversaries, we do not reveal the identity of the other player to the human subjects. This creates a more realistic environment since the subjects believe that they are playing against another human. Each section of the park is divided into a 3 × 3 grid, giving each player nine potential targets to attack.

There are different numbers of hippopotamus distributed over the area which indicate the animal density at each target. The adversary's reward at each target is equal to the animal density at that target; hereafter, reward and animal density are used interchangeably. Players are able to view the probability of success

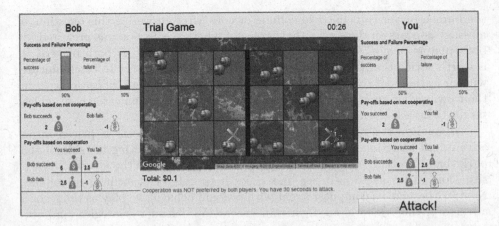

Fig. 1. Poachers vs. Rangers game: Right side of the park is assigned to the player and the left side is assigned to Bob who is the other fellow poacher. Payoffs for each marked target are shown.

and failure, as well as the reward and penalty, at any target on either section of the park as shown on the sides of the Fig. 1. To help the human subjects better visualize the success/failure percentages (i.e., defender coverage) for each sub-regions, we overlaid a heat-map of the success probability on Google Maps imagery of the park. Also, to help the players understand the collusion mechanism, we provided a table that summarizes all possible payoffs for both colluding and not colluding. The human subjects may decide to attack "individually and independently" or "in collusion" with the other player. In both situations, they will attack different sections separately but if both agree to attack in collusion, they will share all of their payoffs with each other equally.

3.2 Experimental Procedure

To enhance understanding of the game, participants were provided with a background story and detailed instructions about the game and then asked to play one trial game to become familiar with the game interface and procedures. After the trial game, participants played a validation game to ensure that had they read the instructions and were fully aware of the rules and options of the game. For our analysis, we included only players whose performance in the validation game passed a set of baseline criteria. Lastly, subjects played the main game for the analysis. After finishing all of the games, participants answered a set of survey questions.

In each individual game, the human player is given a set amount of time to explore the park and make decisions about: (i) whether to collude with the other player or not and (ii) which region of the park to place their snare. While the other player is a computer, it is suggested that they are actually another human. To make the first decision, a question appears on the screen which asks whether the human player is inclined to collude or not. After answering this question, a message appears on the screen that indicates whether collusion was preferred by both players or not. Collusion occurs only if it is preferred by both players. It is worth noting that the human participant has no opportunity to communicate with or learn about the other player. Next, players are asked to choose a target in their own region to attack. As before, players cannot communicate about which target to attack.

We analyze two situations: one where the human attacker is placed in an advantaged situation, with fewer defender resources protecting his side of the park than the other; and a disadvantaged situation, which is the reverse. In each situation, as we mentioned, we first check if the player is inclined to collude. Next, we designed a computer agent with rational behavior to play as the second adversary; thus there is an algorithm generating defender strategies, and two adversaries (one a human and one a computer agent). This computer agent seeks collusion when it is placed on the disadvantaged side and refuses collusion when it is in advantaged situation (Choosing a computer agent as a second player let us to avoid requiring coordination between two human players in the experiments). To simplify the analysis, we assume that the second stage of decision making (where each adversary chooses a target to attack) depends on his own inclination for collusion and does not depend on the attitude of the other adversary.

Consequently, there are four possible types of human adversaries in this game: (i) a disadvantaged attacker who decides to collude, DA-C, (ii) a disadvantaged attacker who decides not to collude, DA-NC, (iii) an advantaged attacker who decides to collude, A-C, and (iv) an advantaged attacker who decides not to collude, A-NC.

We tested different defender mixed strategies based on both the assumption of rationality and bounded rationality given by a behavioral model introduced in Sect. 6. For each strategy deployed on AMT, we recruited a new set of participants (50 people per setup) to remove any learning bias and to test against a wider population. Using the rational model for adversaries, four different defender strategies were deployed for each reward structure. The data sets collected from rational model deployments were used to learn the parameters of the bounded rationality model. This learning mimics the fact that in the real world, often data about past poaching incidents is available to build models of poacher behavior [18]. Players were given a base compensation of $0.50 for participating in the experiment. In order to incentivize the players to perform well, we paid each player a performance bonus based on the utility that they obtained in each game. This bonus had a maximum total value of $1.32 and a minimum of $0.04.

3.3   Game Payoff Design

This "Poachers vs. Rangers" game is a three-player security game with 9 targets available to each adversary. There is one defender with m resources to cover all the 18 targets (sub-regions in the park) and there are two adversaries that can attack a side of the park. An adversary's reward at each cell for an uncovered

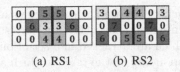

(a) RS1   (b) RS2

Fig. 2. Reward (animal density) structures deployed on AMT. Darker green shows higher reward. (Color figure online)

attack is equal to the animal density at that cell and the penalty at each cell for a covered attack is equal to -1. We deployed two different reward structures, $RS1$ and $RS2$, shown in Figs. 2(a) and (b). In both of these symmetric structures, both players have an identical 3×3 reward distribution. In $RS1$ animal density is concentrated along the central axis of the park and is covered by 3 defender resources and in $RS2$ animal density is concentrated toward the center of each half of the park and is covered by 4 defender resources. We assumed a bonus of 1 for collusion in both set-ups; this bonus is added to the payoff for each successful attack if both attackers decide to collude. Section 4 gives further mathematical description and motivates the introduction of this bonus. This game is zero-sum, i.e., at each target the uncovered payoffs for the attacker and defender sum to zero.

4 Collusive Security Game Model

In the collusive security game which we study in this paper, there is one defender, Θ, and multiple adversaries, $\Psi_1, ..., \Psi_N$, where N is the total number of attackers. Similarly to standard Stackelberg Security Games [23], the defender is the leader and the attackers are the followers. In this subsection, we focus on the games with one defender and two adversaries, such that adversaries can attack separate targets, but they have two options: (i) attack their own targets individually and earn payoffs independently or (ii) attack their own targets individually but collude with each other and share all of the payoffs equally. If the attackers decide to collude, the utility for a successful attack increases by ϵ. This reward models many of the example domains where adversaries operate in different geographic areas or portions of a supply chain, and so do not directly compete over the same targets. Instead, they choose to combine their operations or share information in some way which produces extra utility exogenous to the targets themselves.

To precisely define the model, let $T = \{t_1, ..., t_n\}$ be a set of targets. T is partitioned into disjoint sets T_1 and T_2 which give the targets accessible to the first (resp. second) attacker. The defender has m resources, each of which can be assigned to cover one target. Since we consider games with no scheduling constraints [29], the set of defender pure strategies is all mappings from each of the m resources to a target. A mixed strategy is a probability distribution over such schedules, and can be compactly represented by a coverage vector C which gives the probability that each target is covered. Each attacker pure strategy is the combination of a choice of target to attack and the decision of whether or not to collude. Since the attackers choose their strategies after the defender, there is always an equilibrium in which they play only pure strategies [11]. Hence, we encapsulate the targets which are attacked in a set of binary variables a_t, $t \in T$, where the variables corresponding to the targets which are attacked are set to 1.

We denote the utility that the defender receives when target t is attacked by $U_\Theta^u(t)$ if t is uncovered, and $U_\Theta^c(t)$ if t is covered. The payoffs for the ith attacker are analogously written $U_{\Psi_i}^u(t)$ and $U_{\Psi_i}^c(t)$. Suppose that the attackers select target $t_1 \in T_1$ and $t_2 \in T_2$. Since each may be covered or uncovered, four different outcomes are possible. Table 1 summarizes the players' payoffs in all possible cases when the attackers do not collude (the first two columns) and collude (the last two columns). In this table the first row indicates the payoffs when both targets are uncovered and both adversaries are successful. The second and third rows show the payoffs when only one attacker succeeds and the last row indicates the case of failure for both attackers.

Table 1. Payoffs table for individual and collusive attacks

Payoffs for individual attacks		Payoffs for collusive attacks	
Attackers: Ψ_1, Ψ_2	Defender: Θ	Each attacker: Ψ_1 or Ψ_2	Defender: Θ
$U_{\Psi_1}^u(t_1), U_{\Psi_2}^u(t_2)$	$U_\Theta^u(t_1) + U_\Theta^u(t_2)$	$(U_{\Psi_1}^u(t_1) + U_{\Psi_2}^u(t_2) + 2\epsilon)/2$	$U_\Theta^u(t_1) + U_\Theta^u(t_2) - 2\epsilon$
$U_{\Psi_1}^u(t_1), U_{\Psi_2}^c(t_2)$	$U_\Theta^u(t_1) + U_\Theta^c(t_2)$	$(U_{\Psi_1}^u(t_1) + U_{\Psi_2}^c(t_2) + \epsilon)/2$	$U_\Theta^u(t_1) + U_\Theta^c(t_2) - \epsilon$
$U_{\Psi_1}^c(t_1), U_{\Psi_2}^u(t_2)$	$U_\Theta^c(t_1) + U_\Theta^u(t_2)$	$(U_{\Psi_1}^c(t_1) + U_{\Psi_2}^u(t_2) + \epsilon)/2$	$U_\Theta^c(t_1) + U_\Theta^u(t_2) - \epsilon$
$U_{\Psi_1}^c(t_1), U_{\Psi_2}^c(t_2)$	$U_\Theta^c(t_1) + U_\Theta^c(t_2)$	$(U_{\Psi_1}^c(t_1) + U_{\Psi_2}^c(t_2))/2$	$U_\Theta^c(t_1) + U_\Theta^c(t_2)$

If the attackers collude with each other, they share all of their utility equally. Additionally, they receive a bonus reward, ϵ, for any successful attack. As we focus on zero-sum games for the experiments, this bonus value is deducted from the defender's payoff. Further, while we assume that adversaries who choose to collude split their combined payoff equally, it is important to note that the algorithms we present are easily generalized to accommodate arbitrary payoff splits. There are two principal reasons as to why we specify a 50-50 split in this work. First, this division is motivated by inequity aversion theory, as outlined earlier. Second, our focus here is on the factors which lead individuals to collude in the first place, not on the bargaining process which decides their allocation of the rewards (a topic which is itself the subject of a great deal of work in game theory and psychology). Since the reward structures we consider are symmetric between the players, an equal distribution of rewards is a natural assumption. Thus, we can isolate the factors which lead subjects to enter into collusion instead of confounding the decision to collude with an additional bargaining process.

For a given coverage vector C defender's utility at each target t_i attacked individually by attacker i is defined by Eq. 1. By replacing Θ with Ψ, the same notation applies for the expected utility of the attacker.

$$U_\Theta(t_i, C) = c_{t_i} \cdot U_\Theta^c(t_i) + (1 - c_{t_i})U_\Theta^u(t_i) \tag{5}$$

Now we introduce our solution concept for COSGs, the Collusive Security Equilibrium (CSE), which generalizes the SSE to the case of multiple attackers. Let the defender's strategy be a coverage vector C, and the attackers' strategies g_1 and g_2 be functions from coverage vectors to $T \times \{collude, not\ collude\}$. Recall that a strategy profile forms an SSE if (1) the attacker and defender play mutual best responses and (2) the attacker breaks ties in favor of the defender. In COSGs, each attacker's best response depends on the other, since the decision of whether or not to collude depends on the utility the other attacker will obtain. Essentially, any fixed C induces a game between the attackers; the defender sets the attackers' payoff at each target via their resource allocation. The following conditions define a CSE:

1. C is a best response to g_1 and g_2.
2. $g_1(C)$ and $g_2(C)$ form a Nash equilibrium in the game where each target's utility is $U_\Psi(t, C)$.
3. Both attackers play *collude* if they obtain strictly greater utility in a (*collude, collude*) equilibrium than (*not collude, not collude*) equilibrium.
4. The attackers break ties between equilibria which satisfy (1)–(3) in favor of the defender.

   The first two conditions are analogous to the best response conditions for SSE. In particular, when the followers play a Nash equilibrium (Condition 2), each is playing a best response to the fixed strategies of the other two players. Condition 3 removes the trivial equilibrium where neither attacker chooses to collude because they cannot gain unless the other attacker also decides to collude. Condition 4 enforces the normal SSE condition that remaining ties are broken in favor of the defender.

5   SPECTRE-R: Optimal Defender Strategy for Rational Adversaries

SPECTRE-R (Strategic Patrolling to Extinguish Collaborative ThREats from Rational adversaries) takes a COSG as input and solves for an optimal defender coverage vector corresponding to a CSE strategy through a mixed integer linear program (MILP). This MILP is based on the ERASER formulation introduced by Kiekintveld et al. [11]. The original formulation was developed for SSGs with one defender and one adversary. We extend these ideas to handle collusion between two adversaries via the MILP in Eqs. 6–20. It is important to note that while the rewards structures we consider in the experiments are zero sum, the MILP we give applies to general sum games. Additionally, our methods are not restricted to the case of two adversaries. In the online appendix[2], we provide a generalization of this MILP to COSGs with N adversaries. Since a naive extension would entail a number of constraints which is exponential in N, we conduct more detailed analysis of the structure of the game, which allows

---

[2] https://www.dropbox.com/s/kou5w6b8nbvm25o/nPlayerAppendix.pdf?dl=0.

us to formulate a MILP with only $O(N^3)$ constraints. However, this analysis is also deferred to the appendix as our experimental focus is on COSGs with two adversaries.

$$\max d \quad \text{s.t.} \tag{6}$$

$$a_t^{nc}, a_t^c, \alpha_1, \alpha_2, \beta \in \{0,1\} \tag{7}$$

$$c_t \in [0,1] \tag{8}$$

$$\sum_{t \in T} c_t \leq m \quad \sum_{t_i \in T_i} a_{t_i}^{nc} = 1 \quad \sum_{t_i \in T_i} a_{t_i}^c = 1 \tag{9}$$

$$U_\Theta^c(t_1, t_2, C) = U_\Theta(t_1, C) + U_\Theta(t_2, C) - \\ (1 - c_{t_1})\epsilon - (1 - c_{t_2})\epsilon \tag{10}$$

$$U_\Theta^{nc}(t_1, t_2, C) = U_\Theta(t_1, C) + U_\Theta(t_2, C) \tag{11}$$

$$d - U_\Theta^c(t_1, t_2, C) \leq (1 - a_{t_1}^c)Z + (1 - a_{t_2}^c)Z + (1 - \beta)Z \tag{12}$$

$$d - U_\Theta^{nc}(t_1, t_2, C) \leq (1 - a_{t_1}^{nc})Z + (1 - a_{t_2}^{nc})Z + \beta Z \tag{13}$$

$$U_{\Psi_i}^c(t_i, C) = U_{\Psi_i}(t_i, C) + (1 - c_{t_i})\epsilon \tag{14}$$

$$U_{\Psi_i}^{nc}(t_i, C) = U_{\Psi_i}(t_i, C) \tag{15}$$

$$0 \leq k_i^c - U_{\Psi_i}^c(t_i, C) \leq (1 - a_{t_i}^c)Z \tag{16}$$

$$0 \leq k_i^{nc} - U_{\Psi_i}^{nc}(t_i, C) \leq (1 - a_{t_i}^{nc})Z \tag{17}$$

$$-\alpha_i Z \leq k_i^{nc} - \frac{1}{2}(k_1^c + k_2^c) \leq (1 - \alpha_i)Z \tag{18}$$

$$\beta \leq \alpha_i \tag{19}$$

$$(\alpha_1 + \alpha_2) \leq \beta + 1 \tag{20}$$

We now proceed to an explanation of the above MILP which is named as SPECTRE-R algorithm in this paper and optimizes defender utility, d, against collusive adversaries. In all equations, nc stands for not colluding cases and c stands for colluding cases, and Z is a large constant. Additionally, constraints with free indices are repeated across all possible values, e.g. $i = 1, 2$ or $t \in T$. Equation 7 defines the binary decision variables. a_t^c and a_t^{nc} whether each target would be attacked if the corresponding adversary chooses to collude or not collude, respectively. α_1 and α_2 indicate each adversary's decision of whether to collude. β is indicates whether collusion actually occurs; it is one if and only if both α_1 and α_2 are one. c_t, introduced in Eq. 8 is the defender coverage probability at target t. Equation 9 enforces the defender resource constraint, and that the attackers each select exactly one target. Equations 10 and 11 calculate the defender expected utilities at each target in the case of collusion and no collusion. Equations 12 and 13 define the defender's final expected payoff based on which target is attacked in each case.

Equations 14 and 15 define the expected utility of the attackers in colluding and non-colluding situations. Equations 16 and 17 constrain the attackers to select a strategy in attack set of C in each situation. Equation 18 requires each attacker to collude whenever they obtain higher utility from doing so. Lastly, Eqs. 19 and 20 set $\beta = \alpha_1 \wedge \alpha_2$.

Proposition 1. *Any solution to the above MILP is a CSE.*

Proof. We start by showing that the followers play a Nash equilibrium as required by condition (2). Let $(a_{t_i}^*, \alpha_i^*)$ be the action of one of the followers produced by the MILP where t_i is the target to attack and α_i is the decision of whether to collude. Let (a_{t_i}, α_i) be an alternative action. We need to show that the follower cannot obtain strictly higher utility by switching from $(a_{t_i}^*, \alpha_i^*)$ to (a_{t_i}, α_i). If $\alpha_i^* = \alpha_i$, then Eqs. 16 and 17 imply that a_{t_i} already maximizes the follower's utility. If $\alpha_i^* \neq \alpha_i$ then Eq. 18 implies that $(a_{t_i}^*, \alpha_i^*)$ yields at least as much utility as $(a_{t_i}, 1 - \alpha_i^*)$, for the a_{t_i} which maximizes the follower's utility given that they make the opposite decision about collusion. So, $(a_{t_i}^*, \alpha_i^*)$ yields at least as much utility as (a_{t_i}, α_i), and condition (2) is satisfied. For condition (3), note that in Eq. 18, both followers compute the utility for collusion assuming that the other will also collude. So, if follower i would be best off with $\beta = 1$, the MILP requires that $\alpha_i = 1$. Thus, if both followers receive strictly highest utility in an equilibrium with $\beta = 1$, both will set $\alpha = 1$. In all other cases, the objective is simply maximizing d, which satisfies conditions (1) and (4) by construction.

The following observations and propositions hold for the games with symmetric reward distribution between the two adversaries.

Observation 1. *The defender optimizes against rational adversaries by enforcing an imbalance in resource allocation between the sides and preventing collusion.*

In SPECTRE-R, the key idea for preventing collusion between two adversaries is to impose a resource imbalance between their situations. This places one adversary in an advantaged condition and the other in a disadvantaged condition. Assuming perfectly rational adversaries, we expect that the disadvantaged adversary will always seek to collude, and the advantaged attacker will always refuse (provided the imbalance outweighs the bonus ϵ). In other words, the optimal solution provided by SPECTRE-R satisfies $\theta \neq 0$ where $\theta = |x_1 - x_2|$, $x_i = \sum_{t_i \in T_i} c_{t_i}$ is difference in total resource allocation to the two sides. This approach incentivizes one attacker to refuse to collude by putting them in a better position than the other.

To analyze the effect of the imbalance in resource allocation on defender expected payoff, we added another constraint to the MILP formulation shown in Eq. 21 forces a resource imbalance at an arbitrary level, δ. For the case of symmetric reward distribution, WLOG, we can fix the first attacker to be the one who receives higher payoff and simply linearize the following equation; however generally, we can divide the equation into two separate linear constraints.

$$|k_1^{nc} - k_2^{nc}| = \delta \tag{21}$$

Observation 2. *By varying δ, the following cases can occur:*

1. *For $\delta < \delta^*$, $k_i^{nc} - \dfrac{1}{2}(k_1^c + k_2^c) < 0$ for both attackers and consequently $\alpha_i = 1$ for $i = 1, 2$. In other words, the defender is not able to prevent collusion between the attackers and $\beta = 1$.*

2. *For $\delta = \delta^*$, $k_1^{nc} - \dfrac{1}{2}(k_1^c + k_2^c) = 0$ for one of the attackers and $k_2^{nc} - \dfrac{1}{2}(k_1^c + k_2^c) < 0$ for the other one, so consequently α_1 can be either 0 or 1 and $\alpha_2 = 1$. In this case, the followers break ties in favor of the leader, so $\alpha_1 = 0$ and $\beta = 0$.*

3. *For $\delta > \delta^*$, $k_1^{nc} - \dfrac{1}{2}(k_1^c + k_2^c) > 0$ for one of the attackers and consequently $\alpha_1 = 0$. For the other attacker $k_2^{nc} - \dfrac{1}{2}(k_1^c + k_2^c) < 0$ and $\alpha_2 = 1$. In other words, the defender is able to prevent collusion between the attackers and $\beta = 0$.*

Proposition 2. *The switch-over point, δ^*, introduced in the Observation 2 is lower bounded by 0 and upper bounded by 2ϵ.*

Proof. Using Eq. 17, we know that at any target t_i, $k_i^{nc} \geq U_{\Psi_i}^{nc}(t_i, C)$. If we assume that the attacker attacks target t_i^c with coverage $c_{t_i}^c$ by adding and subtracting a term as $\epsilon(1 - c_{t_i}^c)$, we can conclude that $k_i^{nc} \geq k_i^c - \epsilon(1 - c_{t_i}^c)$. Consequently, $k_1^c + k_2^c \leq k_1^{nc} + k_2^{nc} + \epsilon(1 - c_{t_1}^c) + \epsilon(1 - c_{t_2}^c)$. On the other hand, according to Observation 2.2, at $\delta = \delta^*$, we have $k_1^{nc} - \dfrac{1}{2}(k_1^c + k_2^c) = 0$. Combining these last two equations, we will get $(k_1^{nc} - k_2^{nc}) \leq \epsilon(1 - c_{t_1}^c) + \epsilon(1 - c_{t_2}^c)$. The LHS is equal to δ^* and the RHS can be rearranged as $2\epsilon - \epsilon(c_{t_1}^c + c_{t_2}^c)$, so we will have $\delta^* \leq 2\epsilon - \epsilon(c_{t_1}^c + c_{t_2}^c)$. Given the fact that coverage at each target is in range $[0, 1]$, the upper bound for $-(c_{t_1}^c + c_{t_2}^c)$ will be zero. Finally, by aggregating these results, we can conclude that $\delta^* \leq 2\epsilon$. Following the same analysis, the lower bound for δ^* can be found starting from $k_1^c + k_2^c \geq k_1^{nc} + k_2^{nc} + \epsilon(1 - c_{t_1}^{nc}) + \epsilon(1 - c_{t_2}^{nc})$ and as a result, $0 \leq \delta^*$.

Given the facts presented in Proposition 2, by enforcing an imbalance of maximum 2ϵ, the defender will be able to prevent collusion. These bounds can be tighter, if we have more information about the distribution of reward at targets. For instance, if reward distribution over targets is close enough to uniform distribution, then the average coverage on each side will be $\bar{c}_{t_1} = \dfrac{2x_1}{n}$ and $\bar{c}_{t_2} = \dfrac{2x_2}{n}$, where x_1 and x_2 are fraction of resources assigned to each side and there are $\dfrac{n}{2}$ targets on each side. As a result, $-(c_{t_1}^c + c_{t_2}^c) \approx -(\bar{c}_{t_1} + \bar{c}_{t_2})$. So we will be able to find an approximate upper bound of $2\epsilon(1 - \dfrac{m}{n})$, where $m = x_1 + x_2$. This implies that when the ratio of $\dfrac{m}{n}$ is large, less imbalance in resource allocation is needed to prevent collusion. In the human subject experiments that will be discussed in the next section, we also observed that with a wider range of rewards ($RS2$ compared to $RS1$ in Fig. 5(a) in OBSERVATION A) over targets, it becomes harder to prevent collusion between attackers.

SIMULATION 1. Simulation results of SPECTRE-R algorithm for the two games introduced in Sect. 3 are shown in Figs. 3(a) and (b) for different values of the bonus ϵ. We vary δ along the x axis, and show the defender loss on the

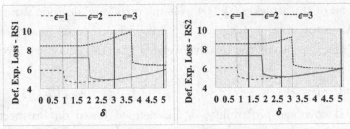

(a) RS1: Def. Exp. Loss vs. δ vs. ϵ (b) RS2: Def. Exp. Loss vs. δ vs. ϵ

Fig. 3. Simulation results of SPECTRE R: Defender Expected Loss vs. resource imbalance

y axis. In all of the plots, *for each epsilon value*, there is a δ value (indicated with gray vertical lines) at which collusion breaks and also a δ^* value (which corresponds to an optimal resource imbalance θ^*) at which collusion is broken and defender loss is minimized (indicated with solid black vertical lines). The higher the benefit of collusion, the larger the loss of the defender. Note that before collusion is broken, imposing a resource imbalance sometimes increases the defender's loss (see plots for $\epsilon = 3$) because the defender deviates from the optimal coverage probabilities for a traditional SSG without reaping the benefit of reduced cooperation. Similarly, note that defender loss increases for $\delta > \delta^*$ since cooperation is already broken, so the defender only suffers by further reducing coverage on the advantaged player. This emphasizes the importance of precision in modeling and recognizing the optimal δ for allocating resources in real-world settings.

6 Human Behavioral Approach

6.1 COSG Model for Bounded Rational Adversaries

While for perfectly rational adversaries the calculations shown in Fig. 3 would hold, our observations from human subject experiments did not match this expectation; the probability of collusion varied continuously with the level of asymmetry in the adversary's' situations. To address this problem, we propose a two layered model which is able to predict (i) the probability of collusion between the adversaries and (ii) the probability of attack over each target for each type of adversary. These layers account for ways in which human behavior experimentally differed from perfect rationality. We then use this model to generate the corresponding optimal patrol schedule.

Probability of attack over targets: We use a separate set of SUQR parameters for each adversary introduced in Sect. 3.1 to reflect differences in decision making. A generalized form of subjective expected utility is defined in Eq. 22 which is a linear function of the modified defender coverage, $\hat{c}_{t_i}$ at target t_i, the uncovered payoff of the attacker, $U^u_{\Psi_i}(t_i)$, the bonus for collusion ϵ and the covered payoff of the attacker $U^c_{\Psi_i}(t_i)$. β is the attackers' decision variable about collusion. A vector of $\omega^\beta_i = (\omega^\beta_{i,1}, \omega^\beta_{i,2}, \omega^\beta_{i,3})$ is assigned to each adversary. Each component of ω^β_i indicates the relative weights that the adversary gives to each feature.

$$\hat{U}_{\Psi_i}(t_i, \beta) = \omega^\beta_{i,1}.\hat{c}_{t_i} + \omega^\beta_{i,2}.(U^u_{\Psi_i}(t_i) + \beta.\epsilon) + \omega^\beta_{i,3}.U^c_{\Psi_i}(t_i) \tag{22}$$

The modified coverage probability, $\hat{c}_{t_i}$, is defined based on Prospect Theory mentioned in Sect. 2 and is related to the actual probability, c_{t_i}, via Eq. 23, where γ and η determine the elevation and curvature of the S-shaped function [6], respectively. These functions are plotted in Sect. 7.3.

$$\hat{c}_{t_i} = \frac{\eta c^\gamma_{t_i}}{\eta c^\gamma_{t_i} + (1 - c_{t_i})^\gamma} \tag{23}$$

By the SUQR model mentioned in Sect. 2, the probability (conditioned on the decision about collusion) that the adversary, i, will attack target t_i is given by:

$$q_{t_i}(\hat{C} \mid \beta) = \frac{e^{\hat{U}_{\Psi_i}(t_i, \hat{C}, \beta)}}{\displaystyle\sum_{t_i \in T_i} e^{\hat{U}_{\Psi_i}(t_i, \hat{C}, \beta)}} \tag{24}$$

For each attacker, the SUQR weight vector ω^β_i, and the probability perception parameters γ^β_i and η^β_i are estimated via maximum likelihood (MLE) using data collected from the human subject experiments. This resembles obtaining past data on poaching as mentioned in Sect. 3.2 to learn these parameters.

Probability of offering to collude: We propose a model which is intuitively based on SUQR to predict the probability of offering collusion by each adversary from a behavioral perspective. Different from the rational behavior model (see Fig. 3) where collusion is deterministic, this model assumes that the attackers make stochastic decisions concerning collusion.

The probability of collusion for each adversary is calculated using Eq. 25. Here, $\bar{U}^c_{\Psi_i} = \sum_{i \in N} \sum_{t_i \in T_i} \hat{U}_{\Psi_i}(t_i, \beta = 1)/(N.|T_i|)$ is the average adversary utility over all targets for a collusive attack and $\bar{U}^{nc}_{\Psi_i} = \sum_{t_i \in T_i} \hat{U}_{\Psi_i}(t_i, \beta = 0)/|T_i|$ is the average adversary utility over all targets for an individual attack.

$$q_i(\beta = 1) = \frac{e^{\bar{U}^c_{\Psi_i}}}{e^{\bar{U}^c_{\Psi_i}} + e^{\bar{U}^{nc}_{\Psi_i}}} \tag{25}$$

The coefficients in ω^β_i are learned for advantaged and disadvantaged attackers and $\beta = 0, 1$ using MLE and data collected from human subject experiments.

6.2 SPECTRE-BR: Optimal Defender Strategy for Bounded Rational Adversaries

The two above mentioned models are incorporated in SPECTRE-BR (Strategic Patrolling to Extinguish Collaborative ThREats from Boundedly Rational adversaries) to generate the defender optimal strategy by maximizing the expected utility of the defender given in Eq. 26 where the defender expected utility is computed as $U_\Theta(t_i, C, \beta) = c_{t_i} \cdot U_\Theta^c + (1 - c_{t_i})(U_\Theta^u + \beta\epsilon)$ for target t_i, mixed strategy C and the collusion variable β. In this equation, $\mathscr{C}$ represents the set of all possible coverage vectors. We define $q(\beta=1) = min(q_1(\beta=1), q_2(\beta=1))$ and $q(\beta=0)=1-q(\beta=1)$. This assumption is supported by the fact that collusive attacks happen only when both parties are sufficiently inclined to collude, and the advantaged player will always be less inclined to offer collusion.

$$\max_{C \in \mathscr{C}} \left(\sum_{i=1}^{N} \sum_{t_i \in T_i} \sum_{\beta=0}^{1} U_\Theta(t_i, C, \beta) q_{t_i}(C \mid \beta) q(\beta) \right) \tag{26}$$

7   Human Subject Experiments

To determine how the behavior of human players differs from perfect rationality, we recruited participants from Amazon Mechanical Turk to play the game described in Sect. 3. Each experiment used 50 participants. Here we report on the results.

7.1   Resource Imbalance Effect on Collusion

HYPOTHESIS A. *There exists a switch-over δ^* value, at which it is not rational for the adversaries to collude. Consequently, collusion will be broken completely.*

METHOD A. Given the intuition from the rational adversary model, the defender achieves higher expected utility by breaking collusion between the two adversaries. The main idea for preventing collusion was to place one adversary in the advantaged condition so he will avoid collusion. The corresponding optimal strategy results in an asymmetry between the maximum expected utilities on both sides which we referred to as δ. This δ is correlated with the difference between aggregated defender coverage on both sides, θ which is defined in Observation 2. Figure 4(a) illustrates this relationship by plotting δ on the x axis against the total resource imbalance on the y axis for $RS2$. As δ increases, the resource imbalance also increases. To see how deviating from balanced resource allocation affects human adversaries' decisions about collusion, we ran human subjects experiments on AMT for various δ values for two reward structures $RS1$ and $RS2$. Figures 4(b) and (c) illustrate two sample mixed strategy (defender coverage over targets) that we deployed on AMT for $RS2$. In Fig. 4(b), resources are distributed symmetrically, while in Fig. 4(c) δ was set equal to 1 and one side is covered more than the other. Next, as shown in Fig. 5(a), for each reward structure, we tested 4 different coverage distribution i.e., $\delta \in \{0, 1, 2, 3\}$. For each defender strategy we recruited 50 AMT workers. It is worth noting that the models introduced in this paper are valid for both symmetric and asymmetric payoff structures; however, we show the simulation results and experiments

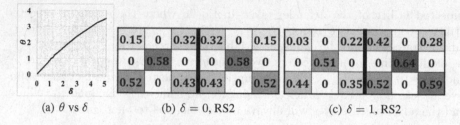

(a) θ vs δ (b) $\delta = 0$, RS2   (c) $\delta = 1$, RS2

Fig. 4. Defender strategy deployed on AMT and resource imbalance

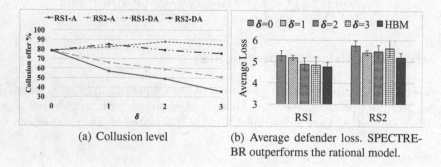

(a) Collusion level

(b) Average defender loss. SPECTRE-BR outperforms the rational model.

Fig. 5. Collusion level and average defender loss

for the symmetric case to hold the effect of other variables constant and focus mostly on the distribution of security resources.

OBSERVATION A. The experiments showed that for human adversaries, there is no switch-over point or sharp change in behavior as predicted in Fig. 3 when assuming rational adversaries. Rather, the probability of offering collusion decreased smoothly as δ increased for both $RS1$ and $RS2$. This completely contradicts the results assuming a rational adversary as seen in Fig. 3. These results are shown in Fig. 5(a). δ varies on the x axis while the y axis shows the probability of collusion. For advantaged attackers (denoted RS1-A and RS2-A in Fig. 5(a)), we observe a smooth decline in collusion as δ increases. However, for disadvantaged attackers (RS1-DA and RS2-DA), we did not observe a significant change in the level of collusion; the disadvantaged attacker always offered to collude with high probability.

ANALYSIS A. The previous observation has several implications: (i) for small values of δ there were a considerable number of human players in advantaged situations who refused to collude despite the fact that collusion was rational. (ii) For large values of δ, there were a considerable number of human players in advantaged situations who chose to collude despite the fact that collusion was an irrational decision in that situation. This behavior might indicate that the bounded rationality model might be a better fit than the model assuming full rationality when modeling collusive adversaries.

7.2 SPECTRE-BR Outperforms Model Assuming Perfectly Rational Adversaries

HYPOTHESIS B. *A lower probability of collusion decreases defender loss.*

METHOD B. See method A.

OBSERVATION B. Figure 5(b) shows the average defender loss obtained by different strategies for both reward structures, $RS1$ and $RS2$. Strategies generated based on the human behavior model (SPECTRE-BR) are labeled "HBM", while the other bars represent strategies generated by the MILP from Sect. 4 using the specified δ. Figure 5(b) shows the empirical utility obtained by each strategy. We calculated the average loss from human players who were in the advantaged and disadvantaged position and who decided to collude and not collude. Figure 5(b) plots the average of these losses weighted according to the frequencies with which players decided to collude, observed in the experiments. We see that the human behavior model obtains uniformly lower loss than the perfect rationality model. In nearly all populations, the difference in utility between the strategies generated by the human behavioral model and those generated by the MILP is statistically significant ($p < 0.05$). Table 2 gives t-test results from comparing the utility obtained by the human behavioral model against each other strategy.

Table 2. Statistical significance (t-Test p values for SPECTRE-BR and rational strategies)

RS	Rational Strategies (δ)			
	$\delta = 0$	$\delta = 1$	$\delta = 2$	$\delta = 3$
1	3.8×10^{-2}	6.6×10^{-4}	4.0×10^{-3}	4.6×10^{-3}
2	3.5×10^{-6}	1.9×10^{-3}	2.6×10^{-1}	5.1×10^{-2}

ANALYSIS B. Importantly, Fig. 5(b) shows that breaking collusion does not always decrease defender loss. For example, in $RS2$, defender loss is lower at $\delta = 2$ compared to $\delta = 3$; however, the chance of collusion (as seen in Fig. 5a) is higher for $\delta = 2$. Hence, simply decreasing the level of collusion (which is correlated with an increase in δ per OBSERVATION A) may not always be optimal for the defender.

7.3 Defender Coverage Perception

HYPOTHESIS C. *Human adversaries' probability weightings follow S-shaped curves independent of their decision about collusion.*

METHOD C. Parameters of S-curves, γ and η in Eq. 23 are learned for the data sets described in METHOD A using the techniques presented in Sect. 6.

OBSERVATION C. Figures 6(a) and (b) show the probability weighting functions learned for the disadvantaged and advantaged adversaries for both groups who are colluding and not colluding for $RS1$. In these figures the defender coverage varies along the x axis, and the attackers' perceptions of defender coverage are shown along the y axis. Figures 6(c) and (d) show the same for $RS2$.

ANALYSIS C. There are two main points in these results: (i) probability weightings followed S-shaped curves, contradicting prospect theory [9,25], i.e., low probabilities are underweighted and high probabilities are overweighted. (ii) Probability perceptions differed between those who decided to collude and not to collude. This analysis supports the use of SPECTRE-BR because humans' probability weightings are indeed nonlinear.

7.4 Individualism vs. Collectivism

HYPOTHESIS D. *Human adversaries who are collectivists are more likely to collude than individualists in nearly all cases.*

METHOD D. All of the participants in our experiments were presented with a survey after playing the game. Eight questions were selected from the 16-item individualism-collectivism scale. Questions with the highest factor loading were selected because prior research shows that these are the most accurate indicators of individualism vs. collectivism [21]. Players responded on a scale from 1 (strongly disagree) to 7 (strongly agree). These responses were used to create a player's OI:OC (overall individualism to overall collectivism) ratio as follows. First, the sum of a player's collectivism responses, c, from collectivism-oriented questions, q_j and individualistic responses, i, from individualism-oriented questions, m_k were calculated as $c = \sum_{j=1}^{4} q_j, \{q_j \in \mathbb{R}^+ : 1 \leq q_j \leq 7\}$ and $i = \sum_{k=1}^{4} m_k, \{m_k \in \mathbb{R}^+ : 1 \leq m_k \leq 7\}$. A player's OI:OC ratio is simply i/c. A player is called an individualist if his OI:OC ratio falls above the median OI:OC score for all players, otherwise he is called a collectivist. We next explore how decisions differ between the two groups. Also please note that the order effect on individualism vs. collectivism analysis is discussed in the online appendix[3] due to space consideration.

OBSERVATION D. The data confirmed that regardless of setting, collectivists are more likely to collude than individualists. This principle was applicable regardless of a player's reward structure, the game's δ value, and whether a player was predetermined to play in an advantaged or disadvantaged state. Figure 7 shows the chance of collusion on the y axis versus δ on the x axis for our two reward structures and in situations where the human is in the advantaged and then disadvantaged situations; we see that the chance of offering collusion for collectivists is always higher than individualists. There is one exception in Fig. 7(c), $\delta = 2$, where the chance of collusion for collectivists and individualists is approximately the same (a difference of less than 0.1 is observed). This single case can be considered an exception to the general rule.

---

[3] https://www.dropbox.com/s/uk9wqrdfq85vhk9/ICAppendix.pdf?dl=0.

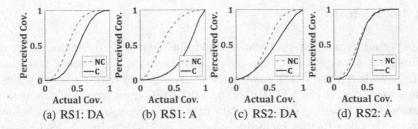

Fig. 6. Probability perception curves learned based on PT

ANALYSIS D. Due to factors like morality, social systems, cultural patterns, personality, etc. collectivists may prefer working with a fellow player [24] regardless of reward structure and delta value. However, the fact that collusion decreases as delta value increases has valuable implications. In security games, this means that adopting more rigorous defender strategies has the effect of dissolving collusion amongst attacker groups regardless of their OI:OC ratio. However, it is important to notice that if attackers have a relatively high OI:OC ratio (meaning they are individualists), the defender strategies given here are even more effective at preventing collusion. Please see the appendix for more individualism/collectivism analysis.

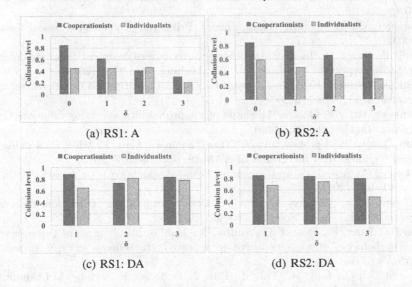

Fig. 7. Cooperation level for collectivists and individualists. RS1 and RS2 indicate the reward structure, while A and DA indicate that a player was on the advantaged or disadvantaged side.

8　Conclusion

This paper addresses the problem of collusion between adversaries in security domains from a game-theoretic and human behavioral perspective. Our contributions include: (i) the COSG model for security games with potential collusion among adversaries, (ii) SPECTRE-R to solve COSGs and break collusion assuming rational adversaries, (iii) observations and analyses of adversary behavior and the underlying factors including bounded rationality, imbalanced-resource-allocation effect, coverage perception, and individualism/collectivism attitudes within COSGs with data from 700 human subjects, (iv) a human behavioral model learned from the data which incorporates these underlying factors, and (v) SPECTRE-BR to optimize against the learned behavior model to provide better defender strategies against human subjects than SPECTRE-R.

Acknowledgement. This research is supported by MURI grant W911NF-11-1-0332.

References

1. Bartilow, H.A., Eom, K.: Free traders and drug smugglers: the effects of trade openness on states' ability to combat drug trafficking. Lat. Am. Polit. Soc. **51**(2), 117–145 (2009)
2. Berg, N.: Behavioral economics. 21st century economics: A reference handbook (2010)
3. Camerer, C.: Behavioral Game Theory. Princeton University Press, Princeton (2003)
4. Fang, F., Stone, P., Tambe, M.: When security games go green: designing defender strategies to prevent poaching and illegal fishing. In: IJCAI (2015)
5. Fehr, E., Schmidt, K.M.: A theory of fairness, competition, and cooperation. Q. J. Econ. **114**, 817–868 (1999)
6. Gonzalez, R., Wu, G.: On the shape of the probability weighting function. Cogn. Psychol. **38**(1), 129–166 (1999)
7. Guo, Q., An, B., Vorobeychik, Y., Tran-Thanh, L., Gan, J., Miao, C.: Coalitional security games. In: Proceedings of AAMAS, pp. 159–167 (2016)
8. Johnson, C.: America's first consumer financial watchdog is on a leash. Cath. UL Rev. **61**, 381 (2011)
9. Kahneman, D., Tversky, A.: Prospect theory: an analysis of decision under risk. Econometrica J. Econ. Soc. **47**, 263–291 (1979)
10. Kar, D., Fang, F., Fave, F.D., Sintov, N., Tambe, M.: A game of thrones: when human behavior models compete in repeated stackelberg security games. In: AAMAS (2015)
11. Kiekintveld, C., Jain, M., Tsai, J., Pita, J., Ordóñez, F., Tambe, M.: Computing optimal randomized resource allocations for massive security games. In: AAMAS (2009)
12. Korzhyk, D., Conitzer, V., Parr, R.: Complexity of computing optimal stackelberg strategies in security resource allocation games. In: AAAI (2010)
13. Korzhyk, D., Conitzer, V., Parr, R.: Security games with multiple attacker resources. In: IJCAI Proceedings, vol. 22, p. 273 (2011)

14. McFadden, D.L.: Quantal choice analaysis: a survey. Ann. Econ. Soc. Measur. 5(4), 363–390 (1976). NBER
15. McKelvey, R.D., Palfrey, T.R.: Quantal response equilibria for normal form games. Games Econ. Behav. 10(1), 6–38 (1995)
16. Narrod, C., Tiongco, M., Scott, R.: Current and predicted trends in the production, consumption and trade of live animals and their products. Rev. Sci. Tech. Off. Int. Epiz. 30(1), 31–49 (2011)
17. Nguyen, T.H., Kar, D., Brown, M., Sinha, A., Tambe, M., Jiang, A.X.: Towards a science of security games. New Frontiers of Multi-Disciplinary Research in STEAM-H (2015)
18. Nguyen, T.H., Sinha, A., Gholami, S., Plumptre, A., Joppa, L., Tambe, M., Driciru, M., Wanyama, F., Rwetsiba, A., Critchlow, R., Beale, C.: Capture: a new predictive anti-poaching tool for wildlife protection. In: AAMAS (2016)
19. Nguyen, T.H., Yang, R., Azaria, A., Kraus, S., Tambe, M.: Analyzing the effectiveness of adversary modeling in security games. In: AAAI (2013)
20. Restrepo, A.L., Guizado, Á.C.: From smugglers to warlords: twentieth century Colombian drug traffickers. Can. J. Lat. Am. Caribb. Stud. 28(55–56), 249–275 (2003)
21. Singelis, T.M., Triandis, H.C., Bhawuk, D.P., Gelfand, M.J.: Horizontal and vertical dimensions of individualism and collectivism: a theoretical and measurement refinement. Cross Cult. Res. 29(3), 240–275 (1995)
22. Sivadas, E., Bruvold, N.T., Nelson, M.R.: A reduced version of the horizontal and vertical individualism and collectivism scale. J. Bus. Res. 61(1), 201 (2008)
23. Tambe, M.: Security and Game Theory: Algorithms, Deployed Systems, Lessons Learned. Cambridge University Press, New York (2011)
24. Triandis, H.C., Gelfand, M.J.: Converging measurement of horizontal and vertical individualism and collectivism. J. Pers. Soc. Psychol. 74(1), 118. (1998)
25. Tversky, A., Kahneman, D.: Advances in prospect theory: cumulative representation of uncertainty. J. Risk Uncertainty 5(4), 297–323 (1992)
26. Warchol, G.L., Zupan, L.L., Clack, W.: Transnational criminality: an analysis of the illegal wildlife market in Southern Africa. Int. Crim. Justice Rev. 13(1), 1–27 (2003)
27. Wyler, L.S., Sheikh, P.A.: International illegal trade in wildlife. DTIC Document (2008)
28. Yang, R.: Human adversaries in security games: integrating models of bounded rationality and fast algorithms. Ph.D. thesis, University of Southern California (2014)
29. Yin, Z., Korzhyk, D., Kiekintveld, C., Conitzer, V., Tambe, M.: Stackelberg vs. nash in security games: interchangeability, equivalence, and uniqueness. In: AAMAS (2010)

A Game-Theoretic Approach to Respond to Attacker Lateral Movement

Mohammad A. Noureddine[1,3]([email]), Ahmed Fawaz[2,3], William H. Sanders[2,3], and Tamer Başar[2,3]

[1] Department of Computer Science, University of Illinois at Urbana-Champaign, 1308 W. Main Street, Urbana, IL 61801, USA
nouredd2@illinois.edu
[2] Department of Electrical and Computer Engineering, University of Illinois at Urbana-Champaign, 1308 W. Main Street, Urbana, IL 61801, USA
[3] Coordinated Science Laboratory, University of Illinois at Urbana-Champaign, 1308 W. Main Street, Urbana, IL 61801, USA
{afawaz2,whs,basar1}@illinois.edu

Abstract. In the wake of an increasing number in targeted and complex attacks on enterprise networks, there is a growing need for timely, efficient and strategic network response. Intrusion detection systems provide network administrators with a plethora of monitoring information, but that information must often be processed manually to enable decisions on response actions and thwart attacks. This gap between detection time and response time, which may be months long, may allow attackers to move freely in the network and achieve their goals. In this paper, we present a game-theoretic approach for automatic network response to an attacker that is moving laterally in an enterprise network. To do so, we first model the system as a network services graph and use monitoring information to label the graph with possible attacker lateral movement communications. We then build a defense-based zero-sum game in which we aim to prevent the attacker from reaching a sensitive node in the network. Solving the matrix game for saddle-point strategies provides us with an effective way to select appropriate response actions. We use simulations to show that our engine can efficiently delay an attacker that is moving laterally in the network from reaching the sensitive target, thus giving network administrators enough time to analyze the monitoring data and deploy effective actions to neutralize any impending threats.

1 Introduction

In the wake of the increasing number of targeted and complex network attacks, namely Advanced Persistent Threats (APTs), organizations need to build more resilient systems. *Resiliency* is a system's ability to maintain an acceptable level of operation in light of abnormal, and possibly malicious, activities. The key feature of resilient systems is their ability to react quickly and effectively to different types of activities. There has been an ever-increasing amount of work on detecting network intrusions; Intrusion Detection Systems (IDSs) are widely

© Springer International Publishing AG 2016
Q. Zhu et al. (Eds.): GameSec 2016, LNCS 9996, pp. 294–313, 2016.
DOI: 10.1007/978-3-319-47413-7_17

deployed as the first layer of defense against malicious opponents [10]. However, once alarms have been raised, it may take a network administrator anywhere from weeks to months to effectively analyze and respond to them. This delay creates a gap between the intrusion detection time and the intrusion response time, thus allowing attackers a sometimes large time gap in which they can move freely around the network and inflict higher levels of damage.

An important phase of the life cycle of an APT is lateral movement, in which attackers attempt to move laterally through the network, escalating their privileges and gaining deeper access to different zones or subnets [2]. As today's networks are segregated by levels of sensitivity, lateral movement is a crucial part of any successful targeted attack. An attacker's lateral movement is typically characterized by a set of causally related chains of communications between hosts and components in the network. This creates a challenge for detection mechanisms since attacker lateral movement is usually indistinguishable from administrator tasks. It is up to the network administrator to decide whether a suspicious chain of communication is malicious or benign. This gap between the detection of a suspicious chain and the administrator's decision and response allows attackers to move deeper into the network and thus inflict more damage. It is therefore essential to design response modules that can quickly respond to suspicious communication chains, giving network administrators enough time to make appropriate decisions.

Intrusion Response Systems (IRSs) combine intrusion detection with network response. They aim to reduce the dangerous time gap between detection time and response time. Static rule-based IRSs choose response actions by matching detected attack steps with a set of rules. Adaptive IRSs attempt to dynamically improve their performance using success/failure evaluation of their previous response actions, as well as IDS confidence metrics [21,23]. However, faced with the sophisticated nature of APTs, IRSs are still unable to prevent network attacks effectively. Rule-based systems can be easily overcome by adaptive attackers. Adaptive systems are still not mature enough to catch up with the increased complexity of APTs.

In this paper, we present a game-theoretic network response engine that takes effective actions in response to an attacker that is moving laterally in an enterprise network. The engine receives monitoring information from IDSs in the form of a network services graph, which is a graph data structure representing vulnerable services running between hosts, augmented with a labeling function that highlights services that are likely to have been compromised. We formulate the decision-making problem as a defense-based zero-sum matrix game that the engine analyzes to select appropriate response actions by solving for saddle-point strategies. Given the response engine's knowledge of the network and the location of sensitive components (e.g., database servers), its goal is to keep the suspicious actors as far away from the sensitive components as possible. The engine is not guaranteed to neutralize threats, if any, but can provide network administrators with enough time to analyze suspicious movement and take appropriate neutralization actions. The decision engine will make use of the monitoring information to decide which nodes' disconnection from the network would slow down the attacker's movements and allow administrators to take neutralizing actions.

An important feature of our approach is that, unlike most IRSs, it makes very few pre-game assumptions about the attacker's strategy; we only place a bound on the number of actions that an attacker can make within a time period, thus allowing us to model the problem as a zero-sum game. By not assuming an attacker model beforehand, our engine can avoid cases in which the attacker deviates from the model and uses its knowledge to trick the engine and cancel the effectiveness of its actions. We show that our engine is effectively able to increase the number of attack steps needed by an attacker to compromise a sensitive part of the network by at least 50 %. Additionally, in most cases, the engine was able to deny the attacker access to the sensitive nodes for the entire period of the simulation.

The rest of this paper is organized as follows. We describe the motivation behind our work in Sect. 2. We then present an overview of our approach and threat model in Sect. 3. Section 4 formally presents the response engine and the algorithms we use. We discuss implementation and results in Sect. 5. We review past literature in Sect. 6, which is followed by presentation of challenges and future directions in Sect. 7. We conclude in Sect. 8.

2   Motivation

The life cycle of an APT consists of the following steps [2,7,11]. The first is intelligence gathering and reconnaissance, which is followed by the establishment of an entry point into the target system. Subsequently, the attacker establishes a connection to one or more command and control (C&C) servers, and uses these connections to control the remainder of the operation. Following C&C establishment is *lateral movement*, wherein the attacker gathers user credential and authentication information and moves laterally in the network in order to reach a designated target. The last step includes performance of specific actions on the targets, such as data exfiltration or even physical damage [13].

Lateral movement allows attackers to achieve persistence in the target network and gain higher privileges by using different tools and techniques [2]. In a number of recent security breaches, the examination of network logs has shown that attackers were able to persist and move laterally in the victim network, staying undetected for long periods of time. For example, in the attack against the Saudi Arabian Oil Company, the attackers were able to spread the malware to infect 30,000 personal machines on the company's network through the use of available file-sharing services [8]. In the Ukraine power grid breach, attackers used stolen credentials to move laterally through the network and gain access to Supervisory Control and Data Acquisition (SCADA) dispatch workstations and servers. The attackers had enough privileges to cause more damage to the public power grid infrastructure [14]. Furthermore, through the use of USB sticks and exploitation of zero-day vulnerabilities in the Windows operating system, the Stuxnet malware was able to move between different workstations in an Iranian nuclear facility until it reached the target centrifuge controllers [13].

Early detection of lateral movement is an essential step towards thwarting APTs. However, without timely response, attackers can use the time gap between detection and administrator response to exfiltrate large amounts of data or inflict severe damage to the victim's infrastructure. It took network administrators two weeks to effectively neutralize threats and restore full operation to the Saudi Arabian Oil Company's network [8]. Furthermore, attackers attempt to hide their lateral movement through the use of legal network services such as file sharing (mainly Windows SMB), remote desktop tools, secure shell (SSH) and administrator utilities (such as the Windows Management Instrumentation) [2]. This stealthy approach makes it harder for network administrators to decide whether the traffic they are observing is malicious lateral movement or benign user or administrative traffic.

In this work, we present a game-theoretic approach for autonomous network response to potentially malicious lateral movement. The response actions taken by our engine aim to protect sensitive network infrastructure by keeping the attacker away from it for as long as possible, thus giving network administrators enough time to assess the observed alerts and take effective corrective actions to neutralize the threats.

3 Overview

We assume, in our framework, the presence of network level IDSs (such as Snort [20] and Bro [1]) that can provide the response engine with the necessary monitoring information. The response engine maintains the state of the network in the form of a network services graph, a graph data structure that represents the active services between nodes in the network. It then uses IDS information to define a labeling function over the graph that marks suspicious nodes and communications used for a possible compromise. Using the labels, the engine observes chains of communications between likely compromised nodes. Such chains are considered suspicious and require the engine to take immediate response actions. The engine considers all suspicious chains as hostile; its goal is to prevent any attackers from reaching specified sensitive nodes in the network, typically database servers or physical controllers.

From the observed states, the response engine can identify compromised nodes and possible target nodes for the attacker. It will take response actions that disconnect services from target nodes so that it prevents the attacker from reaching the sensitive node. This step can provide the network administrators with enough time to assess the IDS alerts and take appropriate actions to neutralize any threats. Figures 1 and 2 illustrate high-level diagrams of our response engine and a sample observed network state with 10 nodes, respectively.

Our threat model allows for the presence of a sophisticated attacker that has already established an entry point in an enterprise network, typically using spear phishing and social engineering, and aims to move laterally deeper into the network. Starting from a compromised node, the attacker identifies a set of possible target nodes for the next move. We assume that the attacker compromises one

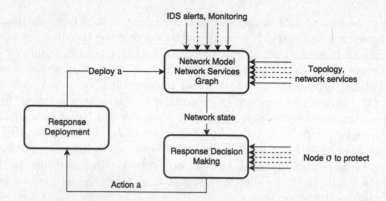

Fig. 1. Our defender model. The defense module uses IDS alerts and monitoring data along with observed attacker steps to build a network model. Trying to protect a sensitive node σ, it builds a zero-sum game and solves for the saddle-point strategies in order to select an appropriate response action a. The *Response Deployment* module is then responsible for the implementation of a in the network.

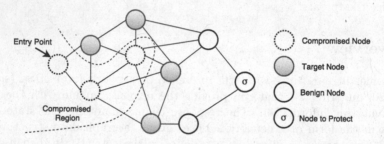

Fig. 2. An illustration of our game model. The attacker has compromised 3 nodes in the network, and has four potential targets to compromise next. The defender, seeing the three compromised nodes, has to decide where the attacker is going to move next and disconnect services from the node, thus slowing down the attack.

node at a time in order to avoid detection. We argue that this assumption is reasonable since attackers typically want to use legitimate administrator tools to hide their lateral movement activities [24]. Therefore, unlike computer worms that propagate widely and rapidly [26], lateral movement tends to be targeted, slow and careful. We will explore more sophisticated types of attackers with multi-move abilities in our future work.

Figure 2 illustrates an example network services graph with ten nodes, where an attacker has established a point of entry and already compromised three nodes. We highlight the target nodes that the attacker can choose to compromise next. We assume no prior knowledge of the strategy by which the attacker will choose the next node to compromise. Building our response engine on the assumption of like-minded attackers would lead to a false sense of security, since attackers with different motives would be able to overcome the responses of our engine, or possibly use them to their own advantage. Therefore, we formulate a defense-based game that attempts to protect a sensitive node in the network, regardless of the goals that the attacker is trying to achieve.

4   The Response Engine

In this section, we formally introduce our response decision-making problem and its formulation as a zero-sum game. We provide formal definitions for the network state, attack and response actions, and attack and response strategies, and then present how we build and solve the matrix game. We formulate the response engine's decision-making process as a complete information zero-sum game, in which the players are the engine and a potentially malicious attacker. We assume that both players take actions simultaneously, i.e., no player observes the action of the other before making its own move. In what follows, without loss of generality, we use the term attacker to refer to a suspicious chain of lateral movement communications. The response engine treats all communication chains as malicious and takes response actions accordingly. We use the terms *defender* and *response engine* interchangeably.

4.1   Definitions

Definition 1 (Network services graph). *A network services graph (NSG) is an undirected graph $G = <V, E>$ where V is the set of physical or logical nodes (workstations, printers, virtual machines, etc.) in the network and $E = V \times V$ is a set of edges.*

An edge $e = (v_1, v_2) \in E$ represents the existence of an active network service, such as file sharing, SSH, or remote desktop connectivity, between nodes v_1 and v_2 in the network.

For any $v \in V$, we define a $\texttt{neighborhood}(v)$ as the set

$$\texttt{neighborhood}(v) = \{u \in V | \exists (u, v) \in E\} \tag{1}$$

Definition 2 (Alert labeling function). *Given an NSG $G = <V, E>$, we define an Alert Labeling Function (ALF) as a labeling function ℓ over the nodes V and edges E of G such that*

$$For\ v \in V,\ \ell(v) = \begin{cases} \textit{True} & \textit{iff } v \textit{ is deemed compromised,} \\ \textit{False} & \textit{otherwise.} \end{cases} \tag{2}$$

$$For\ e = (u, v) \in E,\ \ell(e) = \begin{cases} \textit{True} & \textit{iff } \ell(u) = \textit{True} \wedge \ell(v) = \textit{True,} \\ \textit{False} & \textit{otherwise.} \end{cases} \tag{3}$$

A *suspicious chain* is then a sequence of nodes $\{v_1, v_2, \ldots, v_k\}$ such that

$$\begin{cases} v_1, v_2, \ldots, v_k \in V, \\ (v_i, v_{i+1}) \in E \quad \forall i \in \{1, \ldots, k-1\}, \text{ and} \\ \ell(v_i) = \texttt{True} \quad \forall i \in \{1, \ldots, k\} \end{cases}$$

We assume that an ALF is obtained from monitoring information provided by IDSs such as Snort [20] and Bro [1]. A suspicious chain can be either a malicious attacker moving laterally in the network, or a benign legal administrative task. The goal of our response engine is to slow the spread of the chain and keep it away from the sensitive infrastructure of the network, thus giving network administrators enough time to assess whether the chain is suspicious or not, and take appropriate corrective actions when needed.

Definition 3 (Network state). *We define the state of the network as a tuple $s = (G_s = <V_s, E_s>, \ell_s)$ where G_s is an NSG and ℓ_s is its corresponding ALF. We use S to refer to the set of all possible network states.*

For a given network state s, we define the set of vulnerable nodes $\mathcal{V}_s$ as

$$\mathcal{V}_s = \left\{ u \mid \left(u \in \bigcup_{v \in V_s \wedge \ell_s(v) = \texttt{True}} \texttt{neighborhood}(v) \right) \wedge \ell_s(u) = \texttt{False} \right\} \quad (4)$$

Definition 4 (Attack action). *Given a network state $s \in S$, an attack action a_e is a function over the ALFs, in which a player uses the service provided by edge $e = (v, v')$ such that $\ell_s(v) = \texttt{True}$ and $v' \in \mathcal{V}_s$, in order to compromise node v'. Formally we write*

$$a_e(\ell_s) = \ell' \text{ such that } \ell'(v') = \textit{True} \wedge \ell'(e) = \textit{True} \quad (5)$$

For a network state s, the set of possible attack actions $\mathcal{A}_s$ is defined as

$$\mathcal{A}_s = \{a_e \mid e = (u, v) \in E_s \wedge \ell_s(u) = \texttt{True} \wedge v \in \mathcal{V}_s\} \quad (6)$$

Definition 5 (Response action). *Given a network state s, a response action d_v is a function over the NSG edges, in which a player selects a node $v \in \mathcal{V}_s$, and disconnects available services on all edges $e = (u, v) \in E_s$ such that $\ell_s(u) = \textit{True}$. Formally, we write*

$$d_v(E_s) = E' \text{ such that } E' = E_s \backslash \{(u, v) \in E_s \mid \ell_s(u) = \textit{True}\} \quad (7)$$

For a network state s, we define the set of all possible response actions $\mathcal{D}_s$ as

$$\mathcal{D}_s = \{d_v \mid v \in \mathcal{V}_s\} \quad (8)$$

Definition 6 (Response strategy). *Given a network state s with a set of response actions $\mathcal{D}_s$, a strategy $\mathbf{p}_r : \mathcal{D}_s \longrightarrow [0, 1]^{|\mathcal{D}_s|}$ where $\sum_{d_v \in \mathcal{D}_s} \mathbf{P}_r(d_v) = 1$ is a probability distribution over the space of available response actions.*

A response strategy $\mathbf{p}_r$ is a *pure response strategy* iff

$$\exists\, d_v \in \mathcal{D}_s \text{ such that } \mathbf{p}_r(d_v) = 1 \land (\forall d_{v'} \neq d_v,\ \mathbf{p}_r(d_{v'}) = 0) \tag{9}$$

A response strategy that is not pure is a *mixed response strategy*. Given a network state s, after solving a zero sum game, the response engine samples its response action according to the computed response strategy.

Definition 7 (Attack strategy). *Given a network state s and a set of attack actions $\mathcal{A}_s$, an attack strategy $\mathbf{p}_a : \mathcal{A}_s \longrightarrow [0,1]^{|\mathcal{A}|}$ where $\sum_{a_e \in \mathcal{A}_s} \mathbf{p}_a(a_e) = 1$ is a probability distribution over the space of available attack actions $\mathcal{A}_s$.*

Definition 8 (Network next state). *Given a network state s, a response action $d_v \in \mathcal{D}_s$ for $v \in V_s$, and an attack action $a_e \in \mathcal{A}_s$ for $e = (u, w) \in E_s$, using Eqs. (5) and (7), we define the network next state (nns) as a function $\mathcal{S} \times \mathcal{D}_s \times \mathcal{A}_s \longrightarrow \mathcal{S}$ where*

$$nns(s, d_v, a_e) = s' \text{ where } \begin{cases} (G_{s'} = <V_s, d_v(E_s)>, \ell_s) & iff\ v = w, \\ (G_{s'} = <V_s, d_v(E_s)>, a_e(\ell_s)) & otherwise \end{cases} \tag{10}$$

4.2   Formulation as a Zero-Sum Game

The goal of our response engine is to keep an attacker, if any, as far away from a network's sensitive node (database server, SCADA controller, etc.) as possible. In the following, we assume that the engine is configured to keep the attacker at least `threshold` nodes away from a database server σ containing sensitive company data. The choices of `threshold` and σ are determined by the network administrators prior to the launch of the response engine.

Figure 3 shows the steps taken by our response engine at each time epoch $t_0 < t_1 < t_2 < \ldots < t$. In every step, the defender constructs a zero-sum defense-based matrix game and solves it for the saddle-point response strategy from which it samples an action to deploy. Assume that in a network state s, the response engine chooses to deploy action $d_v \in \mathcal{D}_s$ for $v \in V_s$, and the

---

```
1: for each time epoch t₀ < t₁ < t₂ < ... do
2:     (1) Obtain network state s = (Gₛ, ℓₛ).
3:     (2) Compute the sets of possible attack and response actions 𝒜ₛ and 𝒟ₛ
4:     (3) Compute the payoff matrix Mₛ = BUILD_GAME(𝒜ₛ, 𝒟ₛ, threshold, σ)
5:     (4) Compute the equilibrium response strategy p̂ᵣ
6:     (6) Sample response action dᵥ ∈ 𝒟ₛ from p̂ᵣ
7: end for
```

Fig. 3. The steps taken by our response engine at each time epoch. The engine first obtains the state of the network from the available monitors, and uses it to compute the sets of possible attack and response actions $\mathcal{A}_s$ and $\mathcal{D}_s$. It then builds the zero-sum game matrix M_s using Algorithm 1, and solves for the equilibrium response strategy $\hat{p}_s$. It finally samples a response action d_v from $\hat{p}_s$ that it deploys in the network.

attacker chooses to deploy action $a_e \in \mathcal{A}_s$ for $e = (u, w) \in E_s$. In other words, the defender disconnects services from node v in the network while the attacker compromises node w starting from the already compromised node u. If $v = w$, then the attacker's efforts were in vain and the response engine was able to guess correctly where the attacker would move next. However, when $v \neq w$, the attacker would have successfully compromised the node w. Note that this is not necessarily a loss, since by disconnecting services from certain nodes on the path, the response engine might be redirecting the attacker away from the target server σ. Furthermore, by carefully selecting nodes to disconnect, the engine can redirect the attacker into parts of the network where the attacker can no longer reach the target server σ, and thus cannot win the game. The attacker wins the game when it is able to reach a node within one hop of target server σ. The game ends when (1) the attacker reaches σ; (2) either player runs out of moves to play; or (3) the attacker can no longer reach σ.

Let $\mathtt{sp}(u, \sigma)$ be the length of the shortest path (in number of edges) in G_s from node u to the target server σ. We define the payoffs for the defender in terms of how far the compromised nodes are from the target server σ. A positive payoff indicates that the attacker is more than $\mathtt{threshold}$ edges away from σ. A negative payoff indicates that the attacker is getting closer to σ, an undesirable situation for our engine. Therefore, we define the payoff for the defender when the attacker compromises node w as $\mathtt{sp}(w, \sigma) - \mathtt{threshold}$. If $\mathtt{sp}(w, \sigma) > \mathtt{threshold}$ then the attacker is at least $\mathtt{sp}(w, \sigma) - \mathtt{threshold}$ edges away from the defender's predefined dangerous zone. Otherwise, attacker is $\mathtt{threshold} - \mathtt{sp}(w, \sigma)$ edges into the defender's dangerous zone. Moreover, when the defender disconnects a node w that the attacker wanted to compromise, two cases might arise. First, if $\mathtt{sp}(w, \sigma) = \infty$, i.e., w cannot reach σ, then it is desirable for the defender to lead the attacker into w, and thus the engine assigns d_w a payoff of 0 so that it wouldn't consider disconnecting w. Otherwise, when $\mathtt{sp}(w, \sigma) < \infty$, by disconnecting the services of w, the defender would have canceled the effect of the attacker's action, and thus considers it a win with payoff $\mathtt{sp}(w, \sigma) < \infty$.

Algorithm 1 illustrates how our response engine builds the zero-sum matrix game. For each network state s, the algorithm takes as input the set of response actions $\mathcal{D}_s$, the set of attack actions $\mathcal{A}_s$, the defender's $\mathtt{threshold}$, and the target server to protect σ. The algorithm then proceeds by iterating over all possible combinations of attack and response actions and computes the defender's payoffs according to Eq. (11). It then returns the computed game payoff matrix M_s with dimensions $|\mathcal{D}_s| \times |\mathcal{A}_s|$.

Formally, for player actions $d_v \in \mathcal{D}_s$ and $a_e \in \mathcal{A}_s$ where $v \in V_s$ and $e = (u, w) \in E_s$, we define the response engine's utility as

$$
u_d(d_v, a_e) = \begin{cases} 0 & \text{iff } v = w \wedge \mathtt{sp}(w, \sigma) = \infty \\ \mathtt{sp}(w, \sigma) & \text{iff } v = w \wedge \mathtt{sp}(w, \sigma) < \infty \\ \mathtt{sp}(w, \sigma) - \mathtt{threshold} & \text{iff } v \neq w \end{cases} \quad (11)
$$

Since the game is zero-sum, the utility of the attacker is $u_a(a_e, d_v) = -u_d(d_v, a_e)$.

Algorithm 1. Algorithm M_s = BUILD_GAME $(\mathcal{D}_s, \mathcal{A}_s, \texttt{threshold}, \sigma)$

1: **Inputs:** $\mathcal{D}_s, \mathcal{A}_s, \texttt{threshold}, \sigma$
2: **Outputs:** Zero-sum game payoff matrix M_s
3: **for each** response action $d_v \in \mathcal{D}_s$ **do**
4:    **for each** attack action $a_e \in \mathcal{A}_s$ **do**
5:      let $e \leftarrow (u, w)$
6:      **if** $v = w$ **then**
7:        **if** $\texttt{sp}(w, \sigma) = \infty$ **then**
8:          $M_s(v, w) \leftarrow 0$
9:        **else**
10:          $M_s(v, w) \leftarrow \texttt{sp}(w, \sigma)$
11:        **end if**
12:      **else**
13:        $M_s(v, w) \leftarrow \texttt{sp}(w, \sigma) - \texttt{threshold}$
14:      **end if**
15:    **end for**
16: **end for**

For a response strategy $\mathbf{p}_r$ over $\mathcal{D}_s$ and an attack strategy $\mathbf{p}_a$ over $\mathcal{A}_s$, the response engine's expected utility is defined as

$$U_d(\mathbf{p}_r, \mathbf{p}_a) = \sum_{d_v \in \mathcal{D}_s} \sum_{a_e \in \mathcal{A}_s} \mathbf{p}_r(d_v) u_d(d_v, a_e) \mathbf{p}_a(a_e) \tag{12}$$

Similarly, the attacker's expected payoff is $U_a(\mathbf{p}_a, \mathbf{p}_r) = -U_d(\mathbf{p}_r, \mathbf{p}_a)$.

In step 4 of Fig. 3, the response engine computes the saddle-point response strategy $\hat{\mathbf{p}}_r$ from which it samples the response action to deploy. $\hat{\mathbf{p}}_r$ is the best response strategy that the engine could adopt for the worst-case attacker. Formally, for saddle-point strategies $\hat{\mathbf{p}}_r$ and $\hat{\mathbf{p}}_a$,

$$U_d(\hat{\mathbf{p}}_r, \hat{\mathbf{p}}_a) \geq U_d(\mathbf{p}_r, \hat{\mathbf{p}}_a) \text{ for all } \mathbf{p}_r, \text{ and}$$
$$U_a(\hat{\mathbf{p}}_a, \hat{\mathbf{p}}_r) \leq U_a(\mathbf{p}_a, \hat{\mathbf{p}}_r) \text{ for all } \mathbf{p}_a \tag{13}$$

Finally, the engine chooses an action $d_v \in \mathcal{D}_s$ according to the distribution $\hat{\mathbf{p}}_r$ and deploys it in the network. In this paper, we assume that response actions are deployed instantaneously and successfully at all times; response action deployment challenges are beyond the scope of this paper.

5   Implementation and Results

We implemented a custom Python simulator in order to evaluate the performance of our proposed response engine. We use Python iGraph [9] to represent NSGs, and implement ALFs as features on the graphs' vertices. Since the payoffs for the response engine's actions are highly dependent on the structure of the NSG, we use three different graph topology generation algorithms to generate the initial graphs. The Waxman [25] and Albert-Barabási [3] algorithms are widely used to model interconnected networks, especially for the evaluation of different routing approaches. In addition, we generate random geometric graphs, as they are widely used for modeling social networks as well as studying the spread of epidemics and computer worms [17,19]. Because of the lack of publicly available data sets capturing lateral movement, we assume that the Waxman and Albert-Barabási models provide us with an appropriate representation of the structural characteristics of interconnected networks.

We use the geometric graph models in order to evaluate the performance of our engine in highly connected networks. We pick the initial attacker point of entry in the graph ω and the location of the database server σ such that $sp(\omega, \sigma) = d$, where d is the diameter of the computed graph. This is a reasonable assumption, since in APTs, attackers usually gain initial access to the target network by targeting employees with limited technical knowledge (such as customer service representatives) through social engineering campaigns, and then escalate their privileges while moving laterally in the network.

We implement our response engine as a direct translation of Fig. 3 and Algorithm 1, and we use the Gambit [16] Python game theory API in order to solve for the saddle-point strategies at each step of the simulation. We use the NumPy [12] Python API to sample response and attack actions from the computed saddle-point distributions. As stated earlier, we assume that attack and response actions are instantaneous and always successful, and thus implement the actions and their effects on the network as described in the network next-state function in Eq. (10).

We evaluate the performance of our response engine by computing the average percentage increase in the number of attack steps (i.e., compromises) needed by an adversary to reach the target server σ. We compute the average increase with respect to the shortest path that the attacker could have adopted in the absence of the response engine. Formally, let k be the number of attack steps needed to reach σ and d be the diameter of the NSG; then, the percentage increase in attack steps is $\frac{k-d}{d} \times 100$. If the attacker is unable to reach the target server, we set the number of attack steps k to the maximum allowed number of rounds of play in the simulation, which is 40 in our simulations.

In addition, we report on the average attacker distance from the server σ as well as the minimum distance that the attacker was able to reach. As discussed earlier, we measure the distance in terms of the number of attack steps needed to compromise the server. A minimum distance of 1 means that the attacker was able to successfully reach σ. We also report and compare the average achieved payoff for the defender while playing the game. We ran our simulations on a Macbook Pro laptop running OSX El Capitan, with 2.2 GHz Intel Core i7 processors and 16 GB of RAM. We start by describing our results for various defender `threshold` values for an NSG with 100 nodes, and then fix the `threshold` value and vary the number of nodes in the NSG. Finally, we report on performance metrics in terms of the time needed to perform the computation for various NSG sizes.

5.1   Evaluation of threshold Values

We start by evaluating the performance of our response engine for various values of the threshold above which we would like to keep the attacker away from the sensitive node σ. We used each graph generation algorithm to generate 10 random NSGs, simulated the game for threshold $\in \{1, 2, 3, 4, 5, 6\}$, and then computed the average values of the metrics over the ten runs.

Table 1 shows the structural characteristics in terms of the number of vertices, average number of edges, diameter, and maximum degree of the graphs generated by each algorithm. All of the graphs we generated are connected, with the geometric graphs showing the largest levels of edge connectivity, giving attackers more space to move in the network. The Waxman and Barabási generators have lower levels of edge connectivity, making them more representative of network services topologies than the geometric graphs are.

Table 1. Characteristics of generated NSGs (averages)

NSG Generator	$|V|$	$|E|$	Diameter	Max Degree
Barabási	100	294	4	50.2
Waxman	100	336.6	4.9	13.7
Geometric	100	1059.8	5.2	34.5

Figure 4a shows the average percentage increase in attacker steps needed to reach the target (or reach the simulation limit) for the various values of threshold. The results show that in all cases, our engine was able to increase the number of steps needed by the attacker by at least 50 %. Considering only the Waxman and Barabási graphs, the engine was able to increase the number of steps needed by the attacker by at least 600 %. This is a promising result that shows the effectiveness of our engine, especially in enterprise networks. Further, the results show that smaller values for threshold achieve a greater average increase in attacker steps. This is further confirmed by the average defender payoff curves shown in Fig. 4b, in which smaller values of threshold achieve greater payoffs. In fact, this result is a direct implication of our definition of the payoff matrix values in Eq. (11). The smaller the values of threshold, the more the engine has room to take actions that have a high payoff, and the more effective its strategies are in keeping the attacker away from the server.

Figures 4c and d show the average distance between the attacker and the server, and the minimum distance reached by the attacker, respectively. For the Waxman and Barabási graphs, the results show that our engine keeps the attacker, on average, at a distance close to the graph's diameter, thus keeping the attacker from penetrating deeper into the network. For both types of graphs, Fig. 4d confirms that the attacker was unable to reach the target server (average minimum distance ≥ 1).

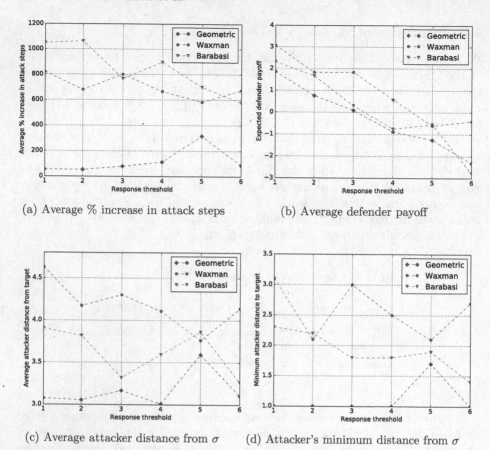

(a) Average % increase in attack steps   (b) Average defender payoff

(c) Average attacker distance from σ (d) Attacker's minimum distance from σ

Fig. 4. Performance evaluation of our response engine with varying threshold values. a shows that our engine was able to increase the number of compromises needed by the attacker by at least 55 %. b illustrates that the zero-sum game's payoff for the defender decreases almost linearly as the `threshold` increases. c shows that the average attacker's distance from σ is very close to the NSG's diameter, while d shows that, with the exception of the geometric NSG, our engine was able to keep that attacker from reaching the target data server σ. It was able, however, in the geometric NSG case, to increase the number of compromises needed to reach σ by at least 55 %.

In the case of the geometric graphs, Fig. 4d shows that the attacker was almost always able to reach the target server. We attribute this attacker success to the high edge connectivity in the geometric graphs. Although our engine is able to delay attackers, because of the high connectivity of the graph, they may find alternative ways to reach the server. Nevertheless, our response engine was always able to cause at least a 50 % increase in the number of attack steps needed to reach the server.

In summary, the results show that our response engine is able to effectively delay, and on average prevent, an attacker that is moving laterally in the network from reaching the target database server. It was effectively able to increase the number of attack steps needed by the adversary by at least 600 % for the graphs that are representative of real-world network topologies. In addition, even when the graphs were highly connected, our engine was still able to increase the attacker's required amount of attack steps by at least 50 %.

5.2   Scalability

Next, we measured the scalability of our response engine as the network grew in size. We varied the number of nodes in the network from 100 to 300 in steps of 50 and measured the average percentage increase in attack steps as well as the attacker's average distance from the target σ. Figure 5 shows our results for averages measured over five random NSGs generated by each of the NSG generation algorithms. We set the defender's `threshold` values to those that achieved a maximum average increase in attack steps as shown in Fig. 4a, which are 5 for geometric NSGs, 2 for Barabási NSGs, and 3 for Waxman NSGs.

As shown in Fig. 5a, our response engine can scale well as the size of the network increases, providing average percentage increases in attack steps between 550 % and 700 % for Waxman NSGs, 750 % and 1150 % for Barabási NSGs, and 50 % and 220 % for geometric NSGs. These results show that as the number of nodes, and thus the number of connection edges, increases in the network, our engine is able to maintain high-performance levels and delay possible attackers, even when they have more room to evade the engine's responses and move laterally in the network. This is further confirmed by the results shown in Figs. 5b and c. For the Waxman and Barabási NSGs, the response engine is always capable of keeping the attacker at an average distance from the target server equal to the diameter of the graph. For the geometric NSGs, the attacker is always capable of getting close to and reaching the target server, regardless of the diameter of the graph. Our engine, however, is always capable of increasing the number of attack steps required by at least 50 %, even for larger networks.

5.3   Computational Performance

Finally, we evaluated the computational performance of our game engine as the scale of the network increased from 100 to 300 nodes. We used the same values for `threshold` as in the previous subsection, and measured the average time to solve for the saddle-point strategies as well as the average size of the matrix game generated during the simulation. Since all of the payoff matrices we generated are square, we report on the number of rows in the matrix games. The rows correspond to the number of available attack or response actions for the players (i.e., for a state s, we report on $|\mathcal{A}_s| = |\mathcal{D}_s|$). Our engine makes use of the `ExternalLogitSolver` solver from the Gambit software framework [16] to solve for the saddle-point strategies at each step of the simulation. In computing our metrics, we averaged the computation time and matrix size over 10 random graphs from each algorithm, and we limited the number of steps in the simulation (i.e., the number of game turns) to 10.

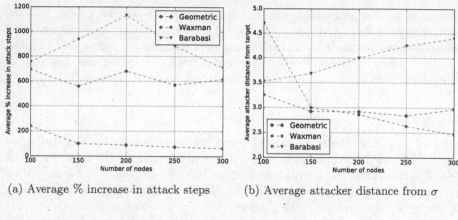

(a) Average % increase in attack steps   (b) Average attacker distance from σ

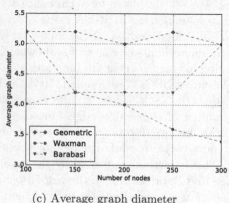

(c) Average graph diameter

Fig. 5. Performance evaluation of our response engine with increasing number of nodes in the network. a shows that our engine maintains high levels of performance even when the network grows larger. The engine is also capable of keeping the attacker at an average distance close to the graph's diameter in the cases of the Waxman and Barabási NSGs, as shown in b and c.

Figure 6b shows that for all NSG-generation algorithms, the size of the payoff matrices for the generated zero-sum game increases almost linearly with the increase in the size of the nodes in the network. In other words, the average number of available actions for each player increases linearly with the size of the network. Consequently, Fig. 6a shows that the computational time needed to obtain the saddle-point strategies scales very efficiently with the increase in the size of the network; the engine was able to solve 50 × 50 matrix games in 15 s, on the average. The short time is a promising result compared to the time needed by an administrator to analyze the observed alerts and deploy strategic response actions.

In summary, our results clearly show the merits of our game engine in slowing down the advance of an attacker that is moving laterally within an enterprise

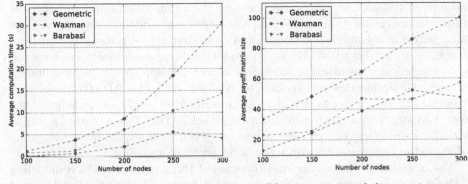

(a) Average time (s) to solve matrix game   (b) Average size of the matrix game

Fig. 6. Computational performance evaluation of the engine for larger networks. Our response engine scales well with the increase in the size of the network.

network, and its ability to protect a sensitive database server effectively from compromise. For all of the NSG-generation algorithms, our engine was able to increase the number of attack steps needed by an attacker to reach the sensitive server by at least 50 %, with the value increasing to 600 % for the Waxman and Barabási NSG-generation algorithms. The results also show that our engine is able to maintain proper performance as networks grow in size. Further, the computational resources required for obtaining the saddle-point strategies increased linearly with the number of the nodes in the network.

6 Related Work

Several researchers have tackled the problem of selecting cyber actions as a response to intrusions. The space can be divided into three parts; automated response through rule-based methods, cost-sensitive methods, and security games.

In rule-based intrusion response, each kind of intrusion alert is tagged with a suitable response. The static nature of rule-based intrusion response makes it predictable and limits its ability to adapt to different attacker strategies. Researchers have extended rule-based intrusion response systems to become cost-sensitive; cost models range from manual assessment of costs to use of dependency graphs on the system components to compute a response action's cost. In all of those cases, the process of selecting a response minimizes the cost of response actions over a set of predefined actions that are considered suitable for tackling a perceived threat. Stakhanova surveyed this class of systems in [26]. While cost-sensitive intrusion response systems minimize the cost of responding, they are still predictable by attackers, and a large effort is required in order to construct the cost models.

Bloem *et al.* [6] tackled the problem of intrusion response as a resource allocation problem. Their goal was to manage the administrator's time, a critical and limited resource, by alternating between the administrator and an imperfect automated intrusion response system. The problem is modeled as a nonzero-sum game between automated responses and administrator responses, in which an attacker gain (utility) function is required. Obtaining such functions, however, is hard in practice, as attacker incentives are not known. The problem of finding attacker-centric metrics was tackled by ADAPT [22]. The ADAPT developers attempted to find a taxonomy of attack metrics that require knowledge of the cost of an attack and the benefit from the attack. ADAPT has created a framework for computing the metrics needed to set up games; however, assigning values to the parameters is still more of an art than a science.

Use of security games improved the state of IRSs, as they enabled modeling of the interaction between the attacker and defender, are less predictable, and can learn from previous attacker behavior [5,15]. In [4], the authors model the security game as a two-player game between an attacker and a defender; the attacker has two actions (to attack or not attack), and the defender has two actions (to monitor or not monitor). The authors consider the interaction as a repeated game and find an equilibrium strategy. Nguyen *et al.* [18] used fictitious play to address the issue of hidden payoff matrices. While this game setup is important on a high level and can be useful as a design guideline for IDSs, it does not help in low-level online response selection during a cyber attack.

To address the issue of high level abstraction in network security games, Zonouz [28] designed the Response and Recovery Engine (RRE), an online response engine modeled as a Stackelberg game between an attacker and a defender. Similar to work by Zhu and Başar [27], the authors model the system with an attack response tree (ART); the tree is then used to construct a competitive Markov decision process to find an optimal response. The state of the decision process is a vector of the probabilities of compromise of all the components in the system. The authors compute the minimax equilibrium to find an optimal response. The strategy is evaluated for both finite and infinite horizons. Scalability issues are tackled using finite lookahead. The game, however, has several limitations: (1) the model is sensitive to the assigned costs; (2) the model required *a priori* information on attacks and monitoring (conditional probabilities) which is not available; and (3) the system uses a hard-to-design ART to construct the game.

7   Discussion and Future Work

The goals of our response engine are to provide networked systems with the ability to maintain acceptable levels of operation in the presence of potentially malicious actors in the network, and to give administrators enough time to analyze security alerts and neutralize any threats, if present. Our results show that the engine is able to delay, and often prevent, an attacker from reaching a sensitive database server in an enterprise network. However, the response actions that our engine deploys can have negative impacts on the system's provided services and overall performance. For example, disconnecting certain nodes as part of our engine's response to an attacker can compromise other nodes' ability to reach the database service. This can have severe impacts on the system's resiliency, especially if it is part of a service provider's infrastructure. In the future, we plan to augment our engine with response action cost metrics that reflect their impact on the network's performance and resiliency. We plan to add support for a resiliency budget that the engine should always meet when making response action decisions. In addition, we will investigate deployment challenges for the response actions. We envision that with the adoption of Software Defined Networks (SDNs), the deployment of such actions will become easier. Our engine can be implemented as part of the SDN controller, and can make use of an SDN control protocols to deploy its response actions.

In the context of APTs, attackers are often well-skilled, stealthy, and highly adaptive actors that can adapt to the changes in the network, including the response actions deployed by our engine. We will investigate more sophisticated models of attackers, specifically ones that can compromise more than one node in each attack step, and can adapt in response to our engine's deployed actions. In addition, knowledge of the attacker's strategies and goals would provide our response engine with the ability to make more informed strategic decisions about which response actions to deploy. Therefore, we plan to investigate online learning techniques that our engine can employ in order to predict, with high accuracy, an attacker's strategies and goals. However, the main challenge that we face in our framework's design and implementation is the lack of publicly available datasets that contain traces of attackers' lateral movements in large-scale enterprise networks. In addition to simulations, we will investigate alternative methods with which we can evaluate our response engine and the learning techniques that we devise. Such methods can include implementation in a real-world, large-scale testbed.

8 Conclusion

Detection of and timely response to network intrusions go hand-in-hand when secure and resilient systems are being built. Without timely response, IDSs are of little value in the face of APTs; the time delay between the sounding of IDS alarms and the manual response by network administrators allows attackers to move freely in the network. We have presented an efficient and scalable game-theoretic response engine that responds to an attacker's lateral movement in an enterprise network, and effectively protects a sensitive network node from compromise. Our response engine observes the network state as a network services graph that captures the different services running between the nodes in the network, augmented with a labeling function that captures the IDS alerts concerning suspicious lateral movements. It then selects an appropriate response action by solving for the saddle-point strategies of a defense-based zero-sum game, in which payoffs correspond to the differences between the shortest path from the attacker to a sensitive target node, and an acceptable engine safety distance threshold. We have implemented our response engine in a custom simulator and evaluated it for three different network graph generation algorithms. The results have shown that our engine is able to effectively delay, and often stop, an attacker from reaching a sensitive node in the network. The engine scales well with the size of the network, maintaining proper operation and efficiently managing computational resources. Our results show that the response engine constitutes a significant first step towards building secure and resilient systems that can detect, respond to, and eventually recover from malicious actors.

Acknowledgment. This work was supported in part by the Office of Naval Research (ONR) MURI grant N00014-16-1-2710. The authors would like to thank Jenny Applequist for her editorial comments.

References

1. The Bro network security monitor (2014). https://www.bro.org/
2. Lateral movement: How do threat actors move deeper into your network. Technical report, Trend Micro (2003)
3. Albert, R., Barabási, A.: Statistical mechanics of complex networks. Rev. Mod. Phys. **74**, 47–97 (2002)
4. Alpcan, T., Başar, T.: A game theoretic approach to decision and analysis in network intrusion detection. In: Proceedings of the 42nd IEEE Conference on Decision and Control, vol. 3, pp. 2595–2600, December 2003
5. Alpcan, T., Başar, T.: Network Security: A Decision and Game-Theoretic Approach. Cambridge University Press, New York (2010)
6. Bloem, M., Alpcan, T., Başar, T.: Intrusion response as a resource allocation problem. In: Proceedings of the 45th IEEE Conference on Decision and Control, pp. 6283–6288, December 2006
7. Brewer, R.: Advanced persistent threats: minimizing the damage. Netw. Secur. **2014**(4), 5–9 (2014)
8. Bronk, C., Tikk-Rangas, E.: Hack or attack? Shamoon and the evolution of cyber conflict, February 2013. http://ssrn.com/abstract=2270860
9. Csardi, G., Nepusz, T.: The iGraph software package for complex network research. InterJ. Complex Syst. **1695**(5), 1–9 (2006)
10. Di Pietro, R., Mancini, L.V. (eds.): Intrusion Detection Systems. Springer, New York (2008)
11. Hutchins, E.M., Cloppert, M.J., Amin, R.M.: Intelligence-driven computer network defense informed by analysis of adversary campaigns and intrusion kill chains. In: Leading Issues in Information Warfare and Security Research, vol. 1, p. 80 (2011)
12. Jones, E., Oliphant, T., Peterson, P.: SciPy: open source scientific tools for Python (2001). http://www.scipy.org/. Accessed 16 June 2016
13. Langner, R.: Stuxnet: dissecting a cyberwarfare weapon. IEEE Secur. Priv. **9**(3), 49–51 (2011)
14. Lee, R.M., Assante, M.J., Conway, T.: Analysis of the cyber attack on the Ukrainian power grid. SANS Industrial Control Systems (2016)
15. Manshaei, M.H., Zhu, Q., Alpcan, T., Başar, T., Hubaux, J.: Game theory meets network security, privacy. ACM Comput. Surv. **45**(3), 25:1–25:39 (2013)
16. McKelvey, R.D., McLennan, A.M., Turocy, T.L.: Gambit: software tools for game theory. Technical report, Version 15.1.0 (2016)

17. Nekovee, M.: Worm epidemics in wireless ad hoc networks. New J. Phys. 9(6), 189 (2007)
18. Nguyen, K.C., Alpcan, T., Başar, T.: Fictitious play with time-invariant frequency update for network security. In: Proceedings of the IEEE International Conference on Control Applications, pp. 65–70, September 2010
19. Penrose, M.: Random Geometric Graphs, vol. 5. Oxford University Press, Oxford (2003)
20. Roesch, M.: Snort: lightweight intrusion detection for networks. In: Proceedings of USENIX, LISA 1999, pp. 229–238 (1999)
21. Shameli-Sendi, A., Ezzati-Jivan, N., Jabbarifar, M., Dagenais, M.: Intrusion response systems: survey and taxonomy. Int. J. Comput. Sci. Netw. Secur 12(1), 1–14 (2012)
22. Simmons, C.B., Shiva, S.G., Bedi, H.S., Shandilya, V.: ADAPT: a game inspired attack-defense and performance metric taxonomy. In: Janczewski, L.J., Wolfe, H.B., Shenoi, S. (eds.) SEC 2013. IAICT, vol. 405, pp. 344–365. Springer, Heidelberg (2013). doi:10.1007/978-3-642-39218-4_26
23. Stakhanova, N., Basu, S., Wong, J.: A taxonomy of intrusion response systems. Int. J. Inf. Comput. Secur. 1(1–2), 169–184 (2007)
24. Trend Micro: Understanding targeted attacks: six components oftargeted attacks, November 2015. http://www.trendmicro.com/vinfo/us/security/news/cyber-attacks/targeted-attacks-six-components. Accessed 06 May 2016
25. Waxman, B.M.: Routing of multipoint connections. IEEE J. Sel. Areas Commun. 6(9), 1617–1622 (1988)
26. Weaver, N., Paxson, V., Staniford, S., Cunningham, R.: A taxonomy of computer worms. In: Proceedings of the 2003 ACM Workshop on Rapid Malcode, pp. 11–18. ACM, New York (2003)
27. Zhu, Q., Başar, T.: Dynamic policy-based IDS configuration. In: Proceedings of the 48th IEEE Conference on Decision and Control, pp. 8600–8605, December 2009
28. Zonouz, S.A., Khurana, H., Sanders, W.H., Yardley, T.M.: RRE: a game-theoretic intrusion response and recovery engine. IEEE Trans. Parallel Distrib. Syst. 25(2), 395–406 (2014)

GADAPT: A Sequential Game-Theoretic Framework for Designing Defense-in-Depth Strategies Against Advanced Persistent Threats

Stefan Rass[1] and Quanyan Zhu[2(✉)]

[1] Institute of Applied Informatics, System Security Group,
Universität Klagenfurt, Klagenfurt, Austria
stefan.rass@aau.at
[2] Department of Electrical and Computer Engineering,
Tandon School of Engineering,
New York University, Brooklyn, NY 11201, USA
qz494@nyu.edu

Abstract. We present a dynamic game framework to model and design defense strategies for advanced persistent threats (APTs). The model is based on a sequence of nested finite two-person zero-sum games, in which the APT is modeled as the attempt to get through multiple protective shells of a system towards conquering the target located in the center of the infrastructure. In each stage, a sub-game captures the attack and defense interactions between two players, and its outcome determines the security level and the resilience against penetrations as well as the structure of the game in the next stage. By construction, interdependencies between protections at multiple stages are automatically accounted for by the dynamic game. The game model provides an analysis and design framework to develop effective protective layers and strategic defense-in-depth strategies against APTs. We discuss a few closed form solutions of our sequential APT-games, upon which design problems can be formulated to optimize the quality of security (QoS) across several layers. Numerical experiments are conducted in this work to corroborate our results.

1 Introduction

The recent advances in the information and communications technologies (ICTs) have witnessed a gradual migration of many critical infrastructures such as electric power grid, gas/oil plants and waste water treatment into open public networks to increase its real-time situational awareness and the operational efficiency. However, this paradigm shift has also inherited existing vulnerabilities of ICTs and posed many challenges for providing information assurance to the legacy systems. For example, the recent computer worm, Stuxnet, have been spread to target Siemens Supervisory Control And Data Acquisition (SCADA) systems that are configured to control and monitor specific industrial processes.

© Springer International Publishing AG 2016
Q. Zhu et al. (Eds.): GameSec 2016, LNCS 9996, pp. 314–326, 2016.
DOI: 10.1007/978-3-319-47413-7_18

Cyber security mechanisms need to be built into multiple layers of the system to protect critical assets against security threats.

Traditional design of security mechanisms relies heavily on cryptographic techniques and the secrecy of cryptographic keys or system states. However, the landscape of system security has recently evolved considerably. The attacks have become more sophisticated, persistent and organized over the years. The attackers can use a wide array of tools such as social engineering and side channel information to steal the full cryptographic keys, which violates the key secrecy assumption in cryptographic primitives. This type of attacks is often referred to as Advanced Persistent Threats (APTs), which can persist in a system for a long period of time, advance stealthy and slowly to maintain a small footprint and reduce detection risks.

In this work, we present a dynamic game framework to capture the distinct feature of APTs in control systems. The objective of using game theory for APT is a paradigm shift from designing perfect security to prevent attacks to strategic planning and design of security mechanisms that allow systems to adapt and mitigate its loss over time. The interactions between an stealthy attacker and the system can be modeled through a sequence of nested zero-sum games in which the attacker can advance or stay at each stage of the game, while the system designer aims to detect and thwart the attack from reaching the target or the most critical assets located at the center of the infrastructure.

The nested feature of the game integrates multiple layers of the infrastructure together. At each layer, a sub-game captures the local attack and defense interactions, and its outcome determines the security level and the resilience against APT penetrations at the current stage, as well as the structure of the game in the next layer. The nested structure enables a holistic integration of multiple layers of the infrastructure, which often composed of cyber-layer communications and networking protocol and the physical-layer control and automation algorithms. The nested structure can also capture different domains within one layer of the infrastructure. For example, an APT can advance from the domain of enterprise Intranet to the domain of utility networks at the cyber layer of an infrastructure.

Another distinct feature of the model is to capture the dynamic behaviors of APT and its dynamic interactions with different layers of the systems at distinct stages. The dynamic game framework allows the system to adapt to the real-time observations and information collected at each stage and implement an automated policy that will enable the system to adjust its security response at different stages. The game model also provides a computational and design framework to develop effective protective layers and strategic defense-in-depth strategies against APTs. We discuss a few closed form solutions of our sequential APT-games, upon which design problems can be formulated to optimize the quality of security (QoS) across several layers.

Below, we present related work, and Sect. 2 introduces the dynamic nested sequential game model for APT and presents the analytical results of the equilibrium analysis. Section 3 focuses on the design of security mechanism enabled by the framework. The paper is concluded in Sect. 4.

Related Work: Game-theoretic methods have been widely used in modeling attacker-defender interactions in communication networks [1–4] and cyber-physical systems [5–8]. The application of game theory to APT has been recently studied in [9,10]. Game-theoretic techniques provide a natural framework to capture the dynamic and strategic conflicting objectives between an APT who aims to inflict maximum damage on the network and a defender who aims to maximize its utility while minimizing his risk [11]. In [9], FlipIt game is proposed as the framework for "Stealthy Takeover," in which players compete to control a shared resource. In [10], a game-of-games structure is proposed to compose the FlipIt game together with a signaling game to capture the stealthy behaviors between an attacker and a cloud user. Our work is also related to the recent literature on proactive cyber defense mechanisms to defend against intelligent and adaptive adversaries by reducing cyber-system signatures observed by the adversaries and increasing their cost and difficulty to attack. Different types of proactive mechanisms have been investigated including moving target defense [12,13], randomization techniques [14–17], deception [18–20], and software diversity [21–23]. Game theory provides a scientific framework to address the security mechanism design of proactive cyber security.

2   APTs as Inspection Games

The game is defined as a walk on a directed acyclic graph $G = (V, E)$. This graph is created from the infrastructure directly, but may also be derived from an attack graph (related to the infrastructure). Let $v_0 \in V$ be the smallest node w.r.t. the topological ordering of G. For the APT game, we think of v_0 as the *target node* that the adversary seeks to reach against the actions of the defender. In G, define the k-th level set L_k as the set of nodes that are equidistantly separated from v_0 by k edges. Formally, we can think of L_k as a "concentric circle" in the graph around v_0, with $L_0 = \{v_0\}$. The game starts with the adversary located at the outermost level L_k, in which a local game is played towards reaching the next level L_{k-1}.

Within each level, the concrete game structure is determined by the particular nodes located at distance k and their vulnerabilities (whose exploits are the strategies for the opponent player). At any such (fixed) stage, the strategies for both players depend on which physical parts of the system (computers, routers, etc.) are located at distance k, which vulnerabilities can be exploited for these, and which counteractions can be adopted to prevent attacks.

This view defines a sequence of k games, $G_k, G_{k-1}, \ldots, G_1$, where in every game G_i, the attacker has two possible actions, which are either to *move onwards to the next level*, or to *stay at the current level*, e.g., to gather necessary information for the next steps and/or to remain undetected upon being idle (and thus not noticeable; like a "dropper" malware). The payoffs in each case depend on the chances to successfully enter the next game G_{i-1} in case of a "move", or to become detected in case of a "stay". Let the attacker be player 2 in our inspection game.

The defender, acting as player 1, can choose any component $v \in V$ in the graph G for inspection, which results in one out of two results: (1) it can (even unknowingly) close an existing backdoor that the adversary was using (this would be a "catch" event), or (2) it can have chosen the wrong spot to check, so the adversary's outside connection up to its current position in the G remains intact (this would be a "miss" event). It is important to stress the stealthiness of the situation here, as the spot inspections by player 1 may neither indicate the current nor past presence of the attacker at a node. That is, the defender's (player 1's) actual move in the game comprises two steps: it randomly chooses a node to inspect, and then resets it to a valid reference state (e.g., patch it, update its configuration, completely reinstall it, etc.). If there was a backdoor at a node being active, the defender's action may have closed it, even though the defender itself never noticed this success.

The adversary, once having successfully entered game G_i, maintains a path in the graph from the outermost level L_k up to the current level L_i (with $1 \le i < k$) in G. Call a particular such path $P_i \subseteq V$, and define it to be a set of consecutive nodes in G. Whichever node player 1 chooses to inspect, the attacker looses the game at this stage (but not the overall game) if any node in P_i is inspected (as the backdoor is closed by then), no matter if it decided to stay or move. The likelihood for this event to happen is determined by the randomized rule by which nodes for inspections are being selected, which is the behavior that player 1 seeks to optimize using game theory.

If the attacker decides to stay and remains uncaught, this leaves his current profit unchanged, since it took it no closer to its goal. If the adversary decides to move and remains uncaught, this adds to his revenue, since it now is in game G_{i-1}. If the backdoor is closed by an inspection (the path P_i is broken), then the game ends, returning the so-far collected payoff for the adversary, and the respective negative value for the defender (zero-sum). The zero-sum assumption is convenient here for providing a valid worst-case assessment without requiring a payoff (incentive) model for the attacker. It thus simplifies the modeling at the cost of possibly giving a pessimistic security assessment (if the attacker is less malicious than presumed). In any case, the game automatically ends once G_1 is won, returning the maximal revenue/loss to the attacker/defender.

To illustrate the modeling, consider the generic SCADA infrastructure as depicted in Fig. 1. The distance (level) in the game is determined by the *number of access controls* (e.g., firewalls) between the attacker and the underlying utility controller nodes (e.g., valves in the water supply, which are shown as ⋈ in Fig. 1). To define the games on each stage, a topological vulnerability analysis may be used to dig up possible exploits related to each component, so as to define the opponent's action set and the respective countermeasures. An example output of such an analysis could look like shown in Fig. 1, with the most important part for our game modeling being the vulnerability assessment. This can, for example, be done using a CVSS scoring. Although the score is only for comparative purposes and is devoid of a physical meaning as such, it can be taken as a qualitative indication of severity, on which an "informed guess" for the probability of an

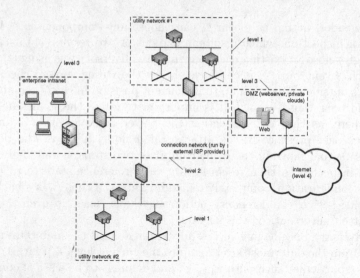

Fig. 1. Example SCADA infrastructure for APT modeling

exploit to happen can be based. This will become handy when the inspection game model is defined.

Let $1 \leq n \leq k$ be the stage of the game, where k is the maximal distance between the attacker and its goal (i.e., the number of stages in the gameplay). Our model is essentially a version of a sequential inspection game (with reversed roles) in the sense that

- Up to k "inspections" are possible,
- but an "inspection" here means a penetration attempt for the attacker, seeking to get to game G_{n-1} from G_n.
- the defender (player 1) then takes any (reasonable) number of random checks on the infrastructure towards maximizing security.

Let $I(n)$ denote the equilibrium payoff in the game at stage n, then – for simplicity – let us think of only two actions for each player, which are to defend or not to defend (spot checking by player 1), or to penetrate or stay (player 2). Obviously, if player 1 does not defend (inspect), then the attacker will successfully get from G_n to G_{n-1} upon a penetration attempt (outcome $I(n - 1)$), or will stay where it is (outcome $I(n)$). In both cases, we assume an investment of c for an attack and a cost of z to remain idle. Likewise, if player 1 defends, then an attack will succeed with a certain likelihood p, and an inactive ("staying") attacker will go undetected with likelihood q. Upon detection, the attacker looses all that has been accomplished so far (revenue $-I(n)$), or retains $I(n)$ since the attacker stays where it is. The payoff structure can be described by the matrix displayed in Fig. 2. Both parameters p and q may explicitly depend on (and can be tailored to) the stage n.

It is reasonable to assume a cyclic structure in this game, since:

pl. 1 \ pl. 2	penetrate	stay
defend	$p(n) \cdot I(n-1) + (1-p(n)) \cdot I(n) - c$	$q(n) \cdot I(n) + (1-q(n)) \cdot (-I(n)) - z$
do not defend	$I(n-1) - c$	$I(n) - z$

Fig. 2. Sequential 2-player game model for advanced persistent threats

- if "stay" is a pure strategy equilibrium for the attacker, then there is nothing to actively defend, since the attacker has no incentive to get to the center.
- if "penetrate" is a pure strategy equilibrium for the attacker, then "defend" is the rational choice for player 1. In that case, we get a recurrence equation $I(n) = p(n) \cdot I(n-1) + (1-p(n)) \cdot I(n) - c$ with the closed form solution $I(n) = -c \cdot n + I(0)$.
- obviously, "do not defend" cannot be a pure strategy equilibrium, since this would defeat the whole purpose of security.
- if "defend" is a pure strategy equilibrium for the defender, then our goal would be designing the system such that the attacker has an incentive to refrain from attacking continuously. As in the first case, a characteristic property of APTs is their stealthiness, so that an attacker will not expose her/himself to the risk of getting detected upon too much activity.

Under this assumption, we find $I(n)$ to be given by the recursion

$$I(n) = \frac{\sqrt{(2c(1-q) + 2(q-1)I(n-1) - pz + z)^2 + 8(p-1)(q-1)zI(n-1)}}{4(q-1)}$$
$$+ \frac{-2cq + 2c + 2qI(n-1) - 2I(n-1) - pz + z}{4(q-1)}. \tag{1}$$

Technically, we can directly obtain $I(n)$ under the assumption of a unique equilibrium in mixed strategies, but this assumption needs to be verified. Thus, take $I(n)$ as "given", and let us take the reverse route of starting from (1) as a mere definition for $I(n)$, and verify it to be a valid equilibrium of the sequential game in Fig. 2. To this end, we need to assure that $I(n)$, upon substituting it into the payoff matrix, induces a circular structure and hence a unique mixed equilibrium. This equilibrium can then (independently) be calculated by well-known closed-form formulas, whose result will match our definition (1). To materialize this plan, we need a preliminary result:

Lemma 1. *Assume $I(n)$ to be defined by (1), and take $I(0) > 0$, as well as $0 < p, q < 1$ and $c, z > 0$ all being constant. Then, $I(n)$ is monotonously decreasing, and $I(n) \leq 0$ for all $n \geq 1$.*

Proof. (sketch) First, we can show that $I(1) < 0$ whenever $I(0) > 0$ (under the stated hypotheses on p, q, c, z), and then induct on n, while using the implication $[I(n) \leq I(n-1)] \rightarrow [I(n+1) \leq I(n)]$ for $n \geq 1$. $\square$

The conclusion made by the lemma is indeed practically meaningful, considering that the sequential game is played "backwards" from n to stage $n-1$ to stage $n-2$, etc. To reach the center (payoff $I(0)$), the attacker has to *invest* something, hoping to get refunded with the value $I(0)$ upon conquering the goal. Thus, the sign of $I(n) \leq 0$ for $n \geq 1$ indicates the a-priori imprest before the reward is gained when the game ends.

By refining the hypothesis of Lemma 1, we obtain a sufficient condition for the payoff structure induced by $I(n)$ to have a unique mixed equilibrium:

Proposition 1. *Let p, q, c, z in Fig. 2 be constants, and assume $0 < p < 1, 0 < q < 1/2, c > 0, z > 0$ as well as $c \cdot q + \frac{p}{2} \cdot z < c$. Then, $I(n)$ as defined by (1) with the initial condition $0 < I(0) < \frac{c}{p} + \frac{z}{2(q-1)}$ has a unique equilibrium in mixed strategies for all $n \geq 1$ in the sequential game as defined by Fig. 2.*

Proof. (sketch) Let the payoff structure be $A = \begin{pmatrix} a & b \\ c & d \end{pmatrix}$ and define the predicate $Q(A) := (a < b) \wedge (c > d) \wedge (a < c) \wedge (b > d)$ as an indicator for a circular preference structure. It is a matter of easy yet messy algebra to show that $Q(A)$ holds under the stated assumptions, together with the upper bound $I(n) \leq 0$ implied by Proposition 1 for $n \geq 1$. Hence, by backsubstituting (1) into the payoff matrix (Fig. 2), we have the circular structure being guaranteed, which then implies the existence of only one equilibrium in mixed strategies.   □

Corollary 1. *Under the conditions of Proposition 1, $I(n)$ as given by (1) gives the unique equilibrium value in the n-th stage of the game, with the respective equilibrium strategies obtained from the resulting payoff structure (with $I(n)$ and $I(n-1)$ being substituted).*

Proof. This immediately follows by computing the equilibrium value from the payoff structure using the closed form formula, which is valid for matrices without a saddle-point in mixed strategies. Specifically, using the notation as in the proof of Proposition 1, and $Q(A)$ then its saddle-point value is given by $v(A) = \det(A)/N$, with $N = a - b - c + d$. The equilibrium strategies are found as $(p^*, 1 - p^*)$ and $(q^*, 1 - q^*)$ with $p^* = (d - c)/N$ and $q^* = (d - b)/N$. The corollary then follows by writing down $v(A)$ as a quadratic equation in $I(n)$ and $I(n-1)$, solving for $I(n)$, and observing that one of the two solutions matches (1). In fact, using the parameter configuration as given in Example 1, we get an immediate counter-example showing that the second solution of the quadratic equation defining $I(n)$ does not yield a payoff matrix with a circular preference structure.   □

Note that the likelihood p may indeed depend on the stage n in reality, and is determined by the equilibrium payoff in the n-th stage. Formally, we may think of this value to depend on n via the game G_n being defined with indicator-valued loss functions. That is, the game G_n is defined with payoffs from $\{0, 1\}$ to indicate either a successful penetration (outcome 1) or a successful defense (outcome 0), so that the (long-run) average revenue is the probability to successfully get from the n-th stage to stage $n-1$. The shape of the game G_n depends on the number r_n of possible exploits and the number s_n of countermeasures at level n. Since both action sets are finite, and if the success or failure of a countermeasure can be meaningfully determined, the game $G_n \in \{0,1\}^{s_n \times r_n}$ is actually a matrix game over $\{0,1\}$. Then, we can set $p(n) := val(G_n)$, when val denotes the saddle-point value of G_n. The parameter c shown above captures costs associated with the penetration attempt. Likewise, the parameter $q(n)$ and the cost z are specified in an analogous way. They describe the likelihood of being detected during an idle phase of information gathering, and the cost for the attacker in playing the strategy "stay".

Example 1. In many infrastructures, one (not exclusive) purpose of firewalls is to concentrate traffic at a single entry or exit point. So, if the firewall separates stage n from stage $n-1$ (cf. Fig. 1), the "local" game G_n is played at the intersection point between the networks. The defending player 1 is herein the totality of countermeasures against unauthorized traffic, say, packet filters (the firewall directly), but also intrusion detection mechanisms, access control, or similar. Likewise, player 2 is the intruder having various options to penetrate the barrier between stage n and stage $n-1$, such as forged emails, conquering a local computer, etc. Table 1 lists some of the particular scenarios that may possibly define game G_n. In practice, we recommend resorting to specific catalogues of vulnerabilities and respective countermeasures, such as are provided by the ISO27000 norm [24] or related.

Table 1. Lists of defense and attack actions, as two separate lists (same rows thus do not reflect any correspondence between defender's and attacker's actions).

Defender action	Attacker action
d_1 : Inspect packets (intrusion detection)	a_1 : Use open ports
d_2 : Check firewall filter rules	a_2 : Zero-day exploits
d_3 : Update firewall firmware	a_3 : Drop sleeping trojan
d_4 : Local malware scans	a_4 : Use shared network drive being accessible from stage $n-1$ and stage n
d_5 : Reinstall computer	a_5 : Email spoofing
$\vdots$	$\vdots$

The gameplay itself would be defined as a 0-1-valued matrix in which each scenario is assigned an outcome of either "success" or "failure". However, many of these actions are inherently probabilistic in the sense that there is no 100 % detection rate of the intrusion detection, malware scan, or similar. Other measures like reinstalling a computer from a reference image, however, may indeed have a guarantee to wipe out all malware (unless a virulent email stored elsewhere is re-opened and re-infect the machines). If the defense strategy is a local

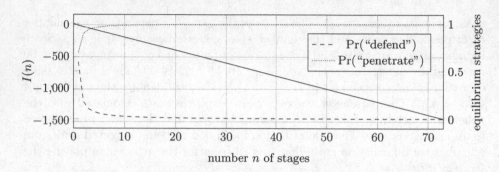

Fig. 3. Example sequential game equilibrium $I(n)$

malware scan (strategy d_4), then we may have likelihood p_{43} to succeed in finding a sleeping trojan (attack strategy a_5). Similarly, reinstalling the computer afresh removes the trojan, i.e., has success rate 1 in the scenario (d_5, a_4). Checking a firewall rule from time to time (defense action d_2) would in turn be effective against exploits of open ports (attack action a_1), provided that those are not in permanent legitimate use. The outcome of this scenario can then be taken as a probability p_{22}, with its value assessed upon expert ratings of this threat/vulnerability combination. The resulting matrix defining the game G_n would then end up as the labeled matrix $A \in [0,1]^{n \times n}$ with values ranging over the entire unit interval.

Likewise, if the attacker decides to remain stealthy, it pays less $z \leq c$ that upon trying to penetrate, but plays a different game G'_n (with its own strategies and outcomes, depending on how the defender acts). Its saddle-point value $q(n) :=$ $val(G'_n)$ then tells the likelihood of being detected. If the attacker tried to remain stealthy and is detected, the game terminates with the path P_n being closed, so that the full lot of $I(n)$ is lost. It is an easy matter of solving this equation numerically for computing the value $I(n)$ at several stages, starting from $I(0)$.

Example 2. Figure 3 displays a solution to $I(n)$ for the parameter set $p = 0.7, q = 0.1, c = 20, z = 10$ and $I(0) = 15$ (note that the parameters satisfy the hypothesis of Proposition 1). The second ordinate (on the right) refers to the equilibrium strategies by specifying the likelihoods to play "defend" (for player 1) and "penetrate" for player 2. As expected, the defense is becoming more intense in the proximity of the center. Likewise, the attacker is best advised to behave more aggressively when the goal (stage 0) is near.

3   Design Problems

Practically, altering the infrastructure towards enhancing security amounts to changing one or more individual stage games A_i, A'_i, say by adding a firewall, reconfiguring a surveillance or intrusion detection system, etc. Towards automating this process, we may think of the possible changes being modeled by decision variables that influence the parameters in the respective stage-game matrix; cf. Example 1. Call their entirety a vector $\boldsymbol{\theta}$. The result is an update to the parameters p, q, c and z depending on the particular values of the decision variables, either by making the detection probability higher (increase p, q), or increase the cost to penetrate (raise c) or stay undetected (raise z).

Essentially, we thus have a design problem to optimize the parameters p, q, c, z in an expectedly resource-constrained manner. To formalize the matter, recall that $p(n), q(n)$ have been defined as the values of the games G_n, G'_n played within the n-th stage of the protection. Each value is thus the optimum of the linearly constrained program $p(n) = \max_{\boldsymbol{x}}(v)$ subject to $\boldsymbol{A}_n \cdot \boldsymbol{x}_n < \boldsymbol{b}_n$, where the matrix $\boldsymbol{A}_n$ and vector $\boldsymbol{b}_n$ are defined to resemble the well-known constraints on the variables $(v, \boldsymbol{x})$, where $\boldsymbol{x} = (x_1, \ldots, x_n)$ ranges over all mixed strategies on the strategy space of player 1 (the defender in our game G_n). Likewise, we can abstractly write $q(n) = \max_{\boldsymbol{y}}(u)$ subject to $\boldsymbol{A}'_n \cdot \boldsymbol{x}'_n \leq \boldsymbol{b}'_n$, with $(u, \boldsymbol{x}')$ determining the value u and optimal mixed strategy $\boldsymbol{x}'$ in game G'_n.

The problem of optimizing the value of the sequential game I can come in different flavours. The difference is in the goal expression, which can be:

- $I(k)$, e.g., the value of the game at the outermost stage: this is a static design problem and refers to optimizing the protection from an external perspective.
- $I(k)$ for some large but fixed k: this is also a static optimization problem and attempts to optimize an approximation to the limit that the sequence $I(n)$ approaches when n tends to infinity (i.e., the limit then measures the overall strength of protection irrespectively of the number of stages and games).
- I as a function of $n \in \{1, \ldots, k\}$: Optimizing $I(n)$ over all stages defines a dynamic optimization problem. Note that we exclude $I(0)$ here as this measures the value of the innermost asset, and thus may be fixed a priori.

We will leave the particular details and issues of solving these various kinds of optimizations as an interesting route of future research. Here, let us complete our discussion by adding the constraints that the above optimization problems are subject to. The decision variables over which the optimization is done are only implicitly available here and primarily define the values of the inner sub-games $G_1, G'_1, G_2, G'_2, \ldots, G_k, G'_k, G_k, G'_k$. Let us assume all of them to be

- finite (as there are clearly not infinitely many attack and defense strategies available),
- and zero-sum (for the sake of the sequential game becoming a worst-case model across the entire infrastructure).

The vector of decision variables $\boldsymbol{\theta}$ defines a sequence of game-matrices $\boldsymbol{A}_1(\boldsymbol{\theta})$, $\boldsymbol{A}'_1(\boldsymbol{\theta})$, $\boldsymbol{A}_2(\boldsymbol{\theta})$, $\boldsymbol{A}'_2(\boldsymbol{\theta})$, $\ldots$, $\boldsymbol{A}_k(\boldsymbol{\theta})$, $\boldsymbol{A}'_k(\boldsymbol{\theta})$. The n-th such pair of matrices $A_n(\boldsymbol{\theta})$, $\boldsymbol{A}'_n(\boldsymbol{\theta})$ give rise to two linear optimization problems with constraint matrices $\mathbf{B}_n, \mathbf{B}'_n$, that again depend on the (not necessarily all) decision variables $\boldsymbol{\theta}$. We omit this dependence hereafter to simplify our notation.

The constraints to the infrastructure design problem are found by gluing together the constraints for the stage-games, resulting in a (large) block matrix

$B = \mathrm{diag}(B_1, B_1', B_2, B_2', \ldots, B_k, B_k')$. The overall optimization is then over the vector θ and subject to the constraint $B(\theta) \cdot x \leq b$, with the right-hand side vector b collecting the constraints from all the optimization problems (including the variables $p(n), q(n)$ for all stages), defining the respective games. The goal function in this problem is the solution to the sequential game model. This solution can be worked out numerically (which we assume as feasible, since there are not too many stages to be expected in real life).

4   Conclusion and Outlook

Modeling advanced persistent threats by game theory is a so far largely open issue, and the inherent nature of an APT to be stealthy and highly tailored to the particular infrastructure makes accurate modeling into a challenge. To account for this, we designed a simple 2×2-game on top of individual games within an infrastructure, so that the sub-games define the overall APT sequential game model. In this way, we can accurately model the infrastructure at hand, while retaining an analytically and numerically feasible game-theoretic model. As our experiments indicated, the model, despite its simplicity, provides a quite rich dynamics, which under slight alterations even exhibits interesting phenomena like the convergence of the equilibrium values as the number of stages in the game increases (e.g., such as is observed when some of the cost parameters are allowed with negative values to reflect a gain in some situations). An analytic treatment of this is currently in progress, and will be reported in companion work. As a byproduct, the model allows to define design-problems to optimize security investments to mitigate APT risks. This route of usage is particularly interesting for practitioners, seeking to improve the resilience of an IT-infrastructure.

Acknowledgments. This work is partially supported by the grant CNS-1544782 from National Science Foundation, as well as by the European Commission's Project No. 608090, HyRiM (Hybrid Risk Management for Utility Networks) under the 7th Framework Programme (FP7-SEC-2013-1).

References

1. Zhu, Q., Saad, W., Han, Z., Poor, H.V., Başar, T.: Eavesdropping and jamming in next-generation wireless networks: a game-theoretic approach. In: MILCOM 2011 Military Communications Conference, pp. 119–124 (2011)
2. Conti, M., Di Pietro, R., Mancini, L.V., Mei, A.: Emergent properties: detection of the node-capture attack in mobile wireless sensor networks. In: Proceedings of WiSec 2008, pp. 214–219. ACM (2008)
3. Zhu, Q., Bushnell, L., Başar, T.: Game-theoretic analysis of node capture and cloning attack with multiple attackers in wireless sensor networks. In: Proceedings of IEEE CDC (2012)
4. Shree, R., Khan, R.: Wormhole attack in wireless sensor network. Int. J. Comput. Netw. Commun. Secur. **2**(1), 22–26 (2014)

5. Xu, Z., Zhu, Q.: Secure and resilient control design for cloud enabled networked control systems. In: Proceedings of CPS-SPC 2015, pp. 31–42. ACM, New York (2015)
6. Zhu, Q., Başar, T.: Game-theoretic methods for robustness, security, and resilience of cyberphysical control systems: games-in-games principle for optimal cross-layer resilient control systems. IEEE Control Syst. **35**(1), 46–65 (2015)
7. Miao, F., Zhu, Q.: A moving-horizon hybrid stochastic game for secure control of cyber-physical systems. In: Proceedings of IEEE CDC, pp. 517–522, December 2014
8. Zhu, Q., Bushnell, L., Başar, T.: Resilient distributed control of multi-agent cyber-physical systems. In: Tarraf, C.D. (ed.) Control of Cyber-Physical Systems. LNCS, vol. 449, pp. 301–316. Springer, Heidelberg (2013)
9. Dijk, M., Juels, A., Oprea, A., Rivest, R.L.: Flipit: The game of "stealthy takeover". J. Cryptol. **26**(4), 655–713 (2013)
10. Pawlick, J., Farhang, S., Zhu, Q.: Flip the cloud: cyber-physical signaling games in the presence of advanced persistent threats. In: Khouzani, M.H.R., Panaousis, E., Theodorakopoulos, G. (eds.) GameSec 2015. LNCS, vol. 9406, pp. 289–308. Springer, Heidelberg (2015). doi:10.1007/978-3-319-25594-1_16
11. Manshaei, M.H., Zhu, Q., Alpcan, T., Bacşar, T., Hubaux, J.P.: Game theory meets network security and privacy. ACM Comput. Surv. **45**(3), 25 (2013)
12. Jajodia, S., Ghosh, A.K., Swarup, V., Wang, C., Wang, X.S. (eds.): oving Target Defense - Creating Asymmetric Uncertainty for Cyber Threats. Advances in Information Security. Springer, New York (2011)
13. Jajodia, S., Ghosh, A.K., Subrahmanian, V.S., Swarup, V., Wang, C., Wang, X.S.: Moving Target Defense II - Application of Game Theory and Adversarial Modeling. Advances in Information Security, vol. 100. Springer, New York (2013)
14. Kc, G.S., Keromytis, A.D., Prevelakis, V.: Countering code-injection attacks with instruction-set randomization. In: Proceedings of the 10th ACM Conference on Computer and Communications Security, CCS 2003, pp. 272–280. ACM, New York (2003)
15. Al-Shaer, E.: Toward network configuration randomization for moving target defense. In: Jajodia, S., Ghosh, K.A., Swarup, V., Wang, C., Wang, S.X. (eds.) Moving Target Defense: Creating Asymmetric Uncertainty for Cyber Threats. Advances in Information Security, vol. 54, pp. 153–159. Springer, New York (2011)
16. Jafarian, J.H., Al-Shaer, E., Duan, Q.: Openflow random host mutation: transparent moving target defense using software defined networking. In: Proceedings of the First Workshop on Hot Topics in Software Defined Networks, HotSDN 2012, pp. 127–132. ACM, New York (2012)
17. Al-Shaer, E., Duan, Q., Jafarian, J.H.: Random host mutation for moving target defense. In: Keromytis, A.D., Pietro, R. (eds.) SecureComm 2012. LNICSSITE, vol. 106, pp. 310–327. Springer, Heidelberg (2013). doi:10.1007/978-3-642-36883-7_19
18. McQueen, M.A., Boyer, W.F.: Deception used for cyber defense of control systems. In: 2nd Conference on Human System Interactions, pp. 624–631, May 2009
19. Zhuang, J., Bier, V.M., Alagoz, O.: Modeling secrecy and deception in a multiple-period attackerdefender signaling game. Eur. J. Oper. Res. **203**(2), 409–418 (2010)
20. Pawlick, J., Zhu, Q.: Deception by design: evidence-based signaling games for network defense. CoRR abs/1503.05458 (2015)
21. Ammann, P.E., Knight, J.C.: Data diversity: an approach to software fault tolerance. IEEE Trans. Comput. **37**(4), 418–425 (1988)

22. Dalton, M., Kannan, H., Kozyrakis, C.: Raksha: a flexible information flow architecture for software security. SIGARCH Comput. Archit. News **35**(2), 482–493 (2007)
23. Chen, P., Kataria, G., Krishnan, R.: Software diversity for information security. In: WEIS (2005)
24. International Standards Organisation (ISO): ISO/IEC 27001 - Information technology - Security techniques - Information security management systems - Requirements (2013). http://www.iso.org/iso/iso27001. Accessed 11 Apr 2016

Incentives and Cybersecurity Mechanisms

Optimal Contract Design Under Asymmetric Information for Cloud-Enabled Internet of Controlled Things

Juntao Chen[✉] and Quanyan Zhu

Department of Electrical and Computer Engineering,
Tandon School of Engineering, New York University, Brooklyn, NY 11201, USA
{jc6412,qz494}@nyu.edu

Abstract. The development of advanced wireless communication technologies and smart embedded control devices makes everything connected, leading to an emerging paradigm of the *Internet of Controlled Things* (IoCT). IoCT consists of two layers of systems: cyber layer and physical layer. This work aims to establish a holistic framework that integrates the cyber-physical layers of the IoCT through the lens of *contract theory*. For the cyber layer, we use a `FlipIt` game to capture the cloud security. We focus on two types of cloud, high-type and low-type, in terms of their provided quality of service (QoS). The cloud's type is of private information which is unknown to the contract maker. Therefore, the control system administrator (CSA) at the physical layer needs to design a menu of two contracts for each type of service provider (SP) due to this asymmetric information structure. According to the received contract, SP decides his cyber defense strategy in the `FlipIt` game of which the *Nash equilibrium* determines the QoS of the cloud, and further influences the physical system performance. The objective of CSA is to minimize the payment to the cloud SP and the control cost jointly by designing optimal contracts. Due to the interdependence between the cyber and physical layers in the cloud-enabled IoCT, we need to address the cloud security and contract design problems in an integrative manner. We find that CSA always requires the *best* QoS from two types of cloud. In addition, under the optimal contracts, the utilities of both SPs are constants. Furthermore, no contracts will be offered to the cloud if the resulting service cannot stabilize the physical system.

1 Introduction

Driven by the advances in sensing, processing, storage and cloud technologies, sensor deployments are pervasive, and thus an increasing amount of information is available to devices in the Internet of Things (IoT) [1]. With the emerging of smart home and smart cities in which physical systems play a critical role, the sensing, actuation and control of devices in the IoT have given rise to an

This work is partially supported by the grant CNS-1544782, EFRI-1441140 and SES-1541164 from National Science Foundation.

Q. Zhu et al. (Eds.): GameSec 2016, LNCS 9996, pp. 329–348, 2016.
DOI: 10.1007/978-3-319-47413-7_19

expanded term: *Internet of Controlled Things* (IoCT). A cloud-enabled IoCT allows heterogeneous components to provide services in an integrated system. For an instance, cloud resources can provide data aggregation, storage and processing for the physical systems.

Figure 1 shows a framework of the cloud-enabled IoCT. The sensors associated with devices can send data to the remote controllers through up-links, and the control commands can be sent back to the actuator via down-links. Both up-link and down-link are enabled by the cloud. This framework provides an efficient approach to remote control of systems due to the integration of cloud. Basically, the cloud-enabled IoCT can be divided into two parts including the cyber layer and physical layer. The cloud at the cyber layer faces cyber threats. Malicious attackers may steal or infer keys used to authenticate sensors or controllers in the cloud-enabled IoCT. These types of attacks are categorized as *advanced persistent threats* (APTs) [2], since they lead to complete compromise of the cloud without the detection of network administrators. To address APTs in the cloud-enabled IoCT, we use a `FlipIt` game-theoretic framework in which the system administrator reclaims the control of the cloud by renewing the password [9]. Therefore, the cloud security level corresponds to the renewing frequency of the cloud defender at cyber layer.

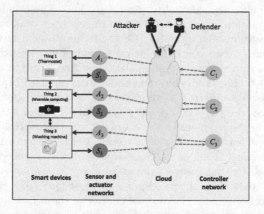

Fig. 1. Illustration of the cloud-enabled Internet of controlled things. The sensors associated with physical devices can send data to the remote controllers, and the control commands can be sent back to the actuators. Both directions of communications are enabled by the cloud in the middle layer. The cloud faces cyber threats, and we use a `FlipIt` game-theoretic model to capture its security. The security level of cloud impacts the communication quality and therefore influences the control system performance.

At the physical layer, the control systems use optimal control to minimize the control cost. As shown in Fig. 1, the performance of the physical system is closely related to the security of cloud. To use the cloud resources, the control system administrator (CSA) should make a contract with the cloud owner or service provider (SP). Various cloud companies have different quality of services

(QoS) in terms of the cloud security. In this paper, we focus on two types of cloud including high-type and low-type. The CSA has no knowledge about the type of cloud that nature picks which is private information of the cyber layer. Therefore, CSA needs to design two contracts for each type of cloud SP. We find the optimal contracts by formulating a *mechanism design* problem. At the cyber layer, based on the received contract, cloud SP designs cyber defense strategy, and the resulting *Nash equilibrium* of FlipIt game determines the QoS. The optimal controller of physical system is designed according to the received communication quality. In addition, the control system performance and the payment to the cloud SP together guide the contract design of CSA. The interdependencies between the cyber and physical layers make the contract design and cloud protection coupled, and thus should be addressed jointly.

The applications of the proposed cloud-enabled IoCT framework and the adopted contract-based approach are massive, such as remote surgery, control of mobile robotic networks, remote 3D printing, smart home automation, and networking in smart grids [4–8].

The main contributions of this paper are summarized as follows:

1. We propose a cloud-enabled IoCT framework which includes cyber-physical layers, and it is suitable for remote control.
2. We use contract theory to study the strategic behaviors of control system administrator and cloud service provider based on the emerging trend of *everything as a service* in IoCT.
3. The cloud security is modeled by a FlipIt game of which the equilibrium is determined by accepted contracts. The designed contracts of CSA take the control system performance and cloud service payment into account jointly.
4. In terms of the type of cloud SP, we obtain the optimal contract for CSA under asymmetric information through mechanism design.

1.1   Related Work

Cloud-enabled networked control systems are becoming popular and have been investigated in a number of fields including 3D printers [4], robotics [5] and smart grid [8]. The security issues in the networked control systems have been studied in [3,4] through using a multi-layer game-theoretic framework. In addition, in terms of the cyber security, the authors in [9] have proposed a FlipIt game model to capture the stealthy and complete compromise of attackers via advanced persistent threats.

Contract design has a rich literature in economics and operation research which has been widely applied to financial markets [10–12], insurances [13,14] and supply chains [15,16]. Contract-based approach has been adopted in various application domains including cyber-physical control systems [17,18], power and vehicle networks [19], environmental services [20] and smart cities [21]. To develop reliable software, engineers in the area of system and software engineering have taken contract-based method to enable their design [22,23].

One major class of contract design problems have the asymmetric information structure between two entities. For example, in the risk management of supply chains [15,16,24], the suppliers have a number of types which is an unknown information to the buyer. Thus, the buyer needs to design a menu of contracts in terms of the type of the supplier. Our work captures the asymmetric information structure between the cloud SP and CSA, and designs optimal contracts in the cloud-enabled Internet of controlled things.

1.2   Organization of the Paper

The rest of this paper is organized as follows. Section 2 introduces the optimal control of the physical layer and formulates the contract design problem. Analysis of the FlipIt game in the cloud and the optimal control of physical systems are given in Sect. 3. Section 4 designs the optimal contracts for the control system administrator under asymmetric information. Case studies are presented in Sect. 5, and Sect. 6 concludes the paper.

2   Problem Formulation

In this section, we first present a bi-level framework that captures the features of the contract design for cloud-enabled IoCT. Then, we formulate the problem which includes the optimal control of physical systems, a two-person nonzero-sum FlipIt game for the cloud protection and the contract between the control system administrator and the cloud service provider.

2.1   Bi-level Framework

The proposed bi-level framework including the cyber and physical layers is shown in Fig. 2. In the cyber layer, a FlipIt game-theoretic framework [9] is adopted to capture the interactions between the defender and attacker and hence model the cloud security. The outcome of the FlipIt game determines the communication quality provided to the control system. In terms of the quality of service that cyber layer offers, the cloud can be divided into two types including high-type (H-type) and low-type (L-type).

At the physical layer, the systems use the optimal control based on the received communication service. In addition, since CSA has no information about the type of the cloud that nature picks, he needs to design two types of contracts: $(\bar{p}_H, p_H, q_H, v_H)$ and $(\bar{p}_L, p_L, q_L, v_L)$, where $\bar{p}_i$ is the transfer payment from the physical system; p_i is the unit payment of service; q_i is the targeted communication quality; and v_i is a positive parameter corresponding to the penalty of degraded service, for $i \in \{H, L\}$. Detailed contract design is introduced in Sect. 2.4.

The offered contracts to the cloud layer influence the strategy of the cyber defender in the FlipIt game, and this feedback structure makes the decision-makings of the cloud SP and the CSA interdependent.

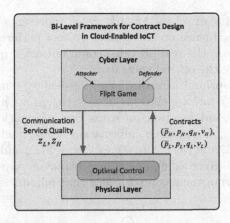

Fig. 2. Bi-level framework for the optimal contract design in the cloud-enabled IoCT. The cyber layer uses a `FlipIt` game to capture the interactions between the defender and attacker, and then determines the communication quality provided to the control system based on the offered contract from the physical layer. The physical systems in the lower layer adopt optimal control based on the received communication quality. The loop structure makes the decision-makings of two layers correlated.

2.2 Optimal Control of the Physical System

The control system with discrete-time dynamics under unreliable communication links can be captured by

$$x_{k+1} = Ax_k + \alpha_k Bu_k + w_k, \tag{1}$$
$$y_k = \beta_k Cx_k, \tag{2}$$

for $k = 0, 1, ...$, where $x_k \in \mathbb{R}^n$ is the state; $u_k \in \mathbb{R}^m$ is the control input; $w_k \in \mathbb{R}^n$ is the exogenous disturbance with mean zero; $y_k \in \mathbb{R}^l$ is the sensor output; and A, B, C are time-invariant matrices with appropriate dimensions. Note that w_k, $\forall k$, are independent. The stochastic process $\{\alpha_k\}$ models the unreliable nature of the communication link from the controller to the actuator, and $\{\beta_k\}$ captures the vulnerability of the link from the sensor to controller.

Without loss of generality, we assume that, in the cloud, the uplink and downlink are facing the *same probability* of cyber threats, i.e., they are of the same quality. Therefore, α_k and β_k are Bernoulli random variables with the same probability mass function (PMF). The provided cloud service is divided into two types including H-type and L-type. Specifically, each type of service has

$$\alpha_k^i = \begin{cases} 1, & \text{with probability } \rho_i, \\ 0, & \text{with probability } 1 - \rho_i, \end{cases} \tag{3}$$

for $i \in \{H, L\}$. In addition, we have $1 \geq \rho_H > \rho_L > 0$ to distinguish two types of clouds.

Remark: The value of ρ_i, $i \in \{H, L\}$, which represents the communication quality has a direct influence on the control system performance given by (4). In the IoCT framework, the real provided communication quality by the cyber layer is determined by the offered contracts.

We consider the optimal control of the physical system in a infinite horizon, and define the control policy as $\Pi = \{\mu_0, \mu_1, ..., \mu_{N-1}\}$, where N is the decision horizon, and function μ_k maps the information I_k to some control space, i.e., $u_k = \mu_k(I_k)$. The information set I_k includes $(\alpha_0, ..., \alpha_{k-1})$, $(\beta_0, ..., \beta_k)$, $(y_0, ..., y_k)$, and $(u_0, ..., u_{k-1})$, for $k = 1, 2, ...$, and specially for $k = 0$, $I_0 = (y_0, \beta_0)$. With a given communication parameter ρ_i, the physical control system aims to find an optimal control policy that minimizes the quadratic cost function

$$J(\Pi^*|\rho_i) = \limsup_{N \to \infty} \frac{1}{N} \mathbb{E}\left\{ \sum_{k=0}^{N-1} (x_k^T Q x_k + \alpha_k^i u_k^T R u_k) \right\}, \tag{4}$$

while considering the system dynamics (1) and (2), where $i \in \{H, L\}$, $R \succ 0$, and $Q \succeq 0$. Note that R and Q are two matrices that capture the cost of state deviation and control effort, respectively.

For notational brevity, we drop the type index H, L when necessary according to the context.

2.3 FlipIt Game for the Cloud Security

We use a FlipIt game to model the interactions between the defender ($\mathcal{D}$) and attacker ($\mathcal{A}$) over a communication channel. Specifically, the strategies of $\mathcal{D}$ and $\mathcal{A}$ are to choose f and g, the renewal and attacking frequencies with which they claim to control of the communication channel. Note that f and g are chosen by prior commitment such that neither $\mathcal{D}$ nor $\mathcal{A}$ know the opponent's action when making choices. In addition, we focus the FlipIt game analysis on periodic strategies, in which the moves of $\mathcal{D}$ and $\mathcal{A}$ are both spaced equally apart, and their phases are chosen randomly from a uniform distribution [9].

Based on f and g, we can compute the expected proportions of time that $\mathcal{D}$ and $\mathcal{A}$ control the communication channel. Denote the proportions by z and $1 - z$ for $\mathcal{D}$ and $\mathcal{A}$, respectively, and we obtain

$$z = \begin{cases} 1, & \text{if } g = 0, \\ \frac{f}{2g}, & \text{if } g > f \geq 0, \\ 1 - \frac{g}{2f}, & \text{if } f \geq g > 0. \end{cases} \tag{5}$$

Notice that when $g > f \geq 0$, i.e., the attacking frequency of $\mathcal{A}$ is larger than the renewal frequency of $\mathcal{D}$, then the proportion of time that the cloud is secure is $z < \frac{1}{2}$; and when $f \geq g > 0$, we obtain $z \geq \frac{1}{2}$.

Remark: When the defender controls the cloud, then the signals are successfully transmitted over the communication channel through the cloud. In addition, in the FlipIt game, when the interval between consecutive moves is small which results in high play rates of $\mathcal{D}$ and $\mathcal{A}$, then z can be interpreted as the probability of random variable α_k being 1 in the control system. Therefore, the FlipIt game outcome z determines the communication reliability measure ρ in (3). We use z to represent the provided service of the cloud in the following.

Then, the optimization problem for the cloud SP under the H-type contract $(\bar{p}_H, p_H, q_H, v_H)$ can be formulated as $\pi_H(\bar{p}_H, p_H, q_H, v_H) = \max_{f_H} \{\bar{p}_H + \min\{p_H z_H, p_H q_H\} - C_H(f_H) - V_H(q_H, z_H, v_H)\}$, where $\bar{p}_H, p_H, q_H$, and v_H have been introduced in Sect. 2.1; f_H is the defense strategy of the cloud SP; and z_H is obtained through (5) based on f_H and g_H, where g_H denotes the attacker's strategy. C_H and V_H are defense cost and penalty functions which have mappings $C_H : \mathbb{R}^+ \to \mathbb{R}^+$ and $V_H : (0,1] \times (0,1] \times \mathbb{R}^+ \to \mathbb{R}^+$, respectively. The term $\min\{p_H z_H, p_H q_H\}$ indicates that the physical system will not pay more to the cloud for receiving better communication service as requested in the contract. Similarly, for the L-type contract $(\bar{p}_L, p_L, q_L, v_L)$, the problem becomes

$$\pi_L(\bar{p}_L, p_L, q_L, v_L) = \max_{f_L} \{\bar{p}_L + \min\{p_L z_L, p_L q_L\} - C_L(f_L) - V_L(q_L, z_L, v_L)\}.$$

The attacker and defender in the FlipIt game determine an equilibrium strategy (f^*, g^*) which has a unique mapping to z^* through (5).

2.4   Contract Design for the Physical Layer

The type of the cloud is private information that the cloud SP can take advantage of. Then, the CSA designs a menu of two contracts, $(\bar{p}_H, p_H, q_H, v_H)$ and $(\bar{p}_L, p_L, q_L, v_L)$, based on the prior probability σ of the cloud being H-type.

Due to this asymmetric information structure, we find the optimal contracts by formulating it as a mechanism design problem. Specifically, by using the *revelation principle* [25], we address the contract design by focusing on the incentive compatible and direct revelation mechanisms. The contract design problem (CDP) for the physical CSA is as follows:

$$\min_{\substack{(\bar{p}_H, p_H, q_H, v_H) \\ (\bar{p}_L, p_L, q_L, v_L)}} \sigma \left(\bar{p}_H + p_H z_H^* + \phi_H \frac{U^o - U(z_H^*)}{U^o} - V_H(q_H, z_H^*, v_H) \right)$$

$$+ (1 - \sigma) \left(\bar{p}_L + p_L z_L^* + \phi_L \frac{U^o - U(z_L^*)}{U^o} - V_L(q_L, z_L^*, v_L) \right) \tag{6}$$

$$\text{s.t.} \quad \pi_H(\bar{p}_H, p_H, q_H, v_H) \geq \pi_H(\bar{p}_L, p_L, q_L, v_L), \tag{7a}$$

$$\pi_L(\bar{p}_L, p_L, q_L, v_L) \geq \pi_L(\bar{p}_H, p_H, q_H, v_H), \tag{7b}$$

$$\pi_H(\bar{p}_H, p_H, q_H, v_H) \geq \epsilon_H, \tag{7c}$$

$$\pi_L(\bar{p}_L, p_L, q_L, v_L) \geq \epsilon_L, \tag{7d}$$

$$p_H > 0, \ p_L > 0, \tag{7e}$$

$$q_{H,\min} \leq q_H \leq q_{H,\max} < 1, \tag{7f}$$

$$q_{L,\min} \leq q_L \leq q_{L,\max} < 1, \tag{7g}$$

where ϕ_H and ϕ_L are positive weighting parameters, $U : (0,1] \rightarrow \mathbb{R}^+$ is the utility of the control system, and U^o denotes the optimal utility of control system under $z = 1$ which is a known constant to the CSA. ϵ_H and ϵ_L are the minimum required profits of H-type and L-type clouds, respectively. $q_{H,\min}$, $q_{H,\max}$, $q_{L,\min}$ and $q_{L,\max}$ are bounds of the required communication quality in each contract.

Note that (7a) and (7b) are *incentive compatibility* (IC) constraints which ensure that a cloud does not benefit from lying about its type to the CSA. In addition, (7c) and (7d) are called *individual rationality* (IR) constraints which indicate that a cloud accepts the contract only when its minimum profit is met.

Timing of contract design: The formulated problem in this section can be divided into two main stages including the contracting and execution. First, the nature reveals the type to the cloud but not to the CSA which introduces the *asymmetric information* structure. Then, the CSA designs and offers two contracts in terms of the type of cloud SP. The cloud picks one of the contracts which completes the contracting stage. In the execution stage, based on the accepted contract, the cloud owner makes a defending strategy against the cloud attacker to achieve a certain level of security of the cloud resources. Then the remote control of the physical system is enabled by the resulting cloud service. If the provided communication quality does not meet the required one in the contract, then the cloud SP pays a penalty.

3   Analysis of the Cloud Security and Physical Control Systems

In this section, we first analyze the FlipIt game that captures the cloud layer security, and then present the optimal control results of the physical systems. In addition, we discuss the impact of cloud layer strategies on the performance of physical control systems.

3.1   Security Analysis of the Cloud Layer

In order to design the strategy of the cloud defender, first, we need to analyze the FlipIt game under each type of contract.

The cost function $C_\mathcal{M}(f_\mathcal{M})$ can be defined as

$$C_\mathcal{M}(f_\mathcal{M}) := \psi_\mathcal{D}^\mathcal{M} f_\mathcal{M} \tag{8}$$

which comes from the FlipIt game, where $\mathcal{M} \in \{H, L\}$, and $\psi_\mathcal{D}^\mathcal{M} > 0$ is unit defending cost of the cloud. The penalty function $V_\mathcal{M}(q_\mathcal{M}, z_\mathcal{M}, v_\mathcal{M})$ needs to be defined. The cloud provides different types of services in terms of the package drop rate which is captured by $1 - z$. The objective function of the cloud does not include the terms related to the dynamics at the physical layer.

The penalty function $V_\mathcal{M}$ admits the form

$$V_\mathcal{M}(q_\mathcal{M}, z_\mathcal{M}, v_\mathcal{M}) = v_\mathcal{M}(q_\mathcal{M} - z_\mathcal{M})\mathbf{1}_{\{q_\mathcal{M} > z_\mathcal{M}\}}, \tag{9}$$

where $z_\mathcal{M}$ is obtained through (5), and $\mathbf{1}_{\{\bullet\}}$ is an indicator function.

Some natural assumptions on the parameters are as follows.

Assumption 1. *For the penalty and payment parameters, we have*

$$0 < p_L < p_H, \tag{10}$$
$$0 < v_L < v_H, \tag{11}$$
$$p_H < v_H \le v_{H,\max}, \tag{12}$$
$$p_L < v_L \le v_{L,\max}, \tag{13}$$

where $v_{H,\max}$ and $v_{L,\max}$ are the maximum unit penalty in the H-type and L-type contracts.

Note that the inequalities (10) and (11) differentiate the unit payment and penalty in the H-type and L-type contracts. In addition, (12) and (13) indicate that the unit penalty in both contracts is larger than the unit payment and bounded above.

Based on (5), we need to discuss two cases of the `FlipIt` game. Specifically, when $f_H \ge g_H > 0$, the `FlipIt` game for the cloud under the H-type contract can be formulated as

$$F_\mathcal{D}^H(f_H, g_H | \bar{p}_H, p_H, q_H, v_H) = \max_{f_H} \left\{ \bar{p}_H + \min \left\{ p_H \left(1 - \frac{g_H}{2f_H} \right), p_H q_H \right\} \right.$$
$$\left. - \psi_\mathcal{D}^H f_H - v_H \left(q_H + \frac{g_H}{2f_H} - 1 \right) \mathbf{1}_{\left\{ q_H > 1 - \frac{g_H}{2f_H} \right\}} \right\},$$

$$F_\mathcal{A}^H(f_H, g_H | \bar{p}_H, p_H, q_H, v_H) = \min_{g_H} \left\{ \psi_\mathcal{A}^H g_H - u_\mathcal{A}^H \frac{g_H}{2f_H} \right\},$$

where $F_\mathcal{D}^H : \mathbb{R}^+ \times \mathbb{R}^+ \to \mathbb{R}$ and $F_\mathcal{A}^H : \mathbb{R}^+ \times \mathbb{R}^+ \to \mathbb{R}$ are objective functions of the cloud defender and attacker, respectively, and $\psi_\mathcal{A}^H > 0$ and $u_\mathcal{A}^H > 0$ are unit cost of attacking the cloud and unit payoff of controlling the cloud. For $f_L \ge g_L > 0$, the `FlipIt` game under the L-type contract can be formulated similarly.

Another nontrivial case in the `FlipIt` game is when $g_\mathcal{M} > f_\mathcal{M} > 0$, for $\mathcal{M} \in \{H, L\}$. The defender's and attacker's problems are

$$F_\mathcal{D}^\mathcal{M}(f_\mathcal{M}, g_\mathcal{M} | \bar{p}_\mathcal{M}, p_\mathcal{M}, q_\mathcal{M}, v_\mathcal{M}) = \max_{f_\mathcal{M}} \left\{ \bar{p}_\mathcal{M} + \min \left\{ p_\mathcal{M} \frac{f_\mathcal{M}}{2g_\mathcal{M}}, p_\mathcal{M} q_\mathcal{M} \right\} \right.$$
$$\left. - \psi_\mathcal{D}^\mathcal{M} f_\mathcal{M} - v_\mathcal{M} \left(q_\mathcal{M} - \frac{f_\mathcal{M}}{2g_\mathcal{M}} \right) \mathbf{1}_{\left\{ q_\mathcal{M} > \frac{f_\mathcal{M}}{2g_\mathcal{M}} \right\}} \right\},$$

$$F_\mathcal{A}^\mathcal{M}(f_\mathcal{M}, g_\mathcal{M} | \bar{p}_\mathcal{M}, p_\mathcal{M}, q_\mathcal{M}, v_\mathcal{M}) = \min_{g_\mathcal{M}} \left\{ \psi_\mathcal{A}^\mathcal{M} g_\mathcal{M} - u_\mathcal{A}^\mathcal{M} \left(1 - \frac{f_\mathcal{M}}{2g_\mathcal{M}} \right) \right\}.$$

Definition 1 (Nash Equilibrium). *A Nash equilibrium of the `FlipIt` game is a strategy profile $(f_\mathcal{M}^*, g_\mathcal{M}^*)$ such that*

$$f_\mathcal{M}^* \in \arg\max_{f_\mathcal{M} \in \mathbb{R}_+} F_\mathcal{D}^\mathcal{M}(f_\mathcal{M}, g_\mathcal{M}^* | \bar{p}_\mathcal{M}, p_\mathcal{M}, q_\mathcal{M}, v_\mathcal{M}), \tag{14}$$

$$g_{\mathcal{M}}^* \in \arg\min_{g_{\mathcal{M}} \in \mathbb{R}_+} F_{\mathcal{A}}^{\mathcal{M}}\left(f_{\mathcal{M}}^*, g_{\mathcal{M}} | \bar{p}_{\mathcal{M}}, p_{\mathcal{M}}, q_{\mathcal{M}}, v_{\mathcal{M}}\right), \tag{15}$$

for $\mathcal{M} \in \{H, L\}$.

Based on the Nash equilibrium of the FlipIt game, we obtain

$$\pi_{\mathcal{M}}(\bar{p}_{\mathcal{M}}, p_{\mathcal{M}}, q_{\mathcal{M}}, v_{\mathcal{M}}) = F_{\mathcal{D}}^{\mathcal{M}}\left(f_{\mathcal{M}}^*, g_{\mathcal{M}}^* | \bar{p}_{\mathcal{M}}, p_{\mathcal{M}}, q_{\mathcal{M}}, v_{\mathcal{M}}\right). \tag{16}$$

For example, under the H-type contract and when $f_H \geq g_H > 0$, we obtain $\pi_H(\bar{p}_H, p_H, q_H, v_H) = \max_{f_H} \left\{ \bar{p}_H + \min\{p_H(1 - \frac{g_H}{2f_H}), p_H q_H\} - \psi_{\mathcal{D}}^H f_H - v_H(q_H + \frac{g_H}{2f_H} - 1)\mathbf{1}_{\{q_H > 1 - \frac{g_H}{2f_H}\}} \right\}$. Next, we analyze the cloud's optimal strategy for a given contract $(\bar{p}, p, q, v)$. We first have the following proposition.

Proposition 1. *Given a contract $(\bar{p}, p, q, v)$, the Nash equilibrium strategy of the FlipIt game leads to $q + \frac{g^*}{2f^*} - 1 \geq 0$ for $f^* \geq g^* > 0$, and $q - \frac{f^*}{2g^*} \geq 0$ for $g^* > f^* > 0$.*

Proof. In the region of $f^* \geq g^* > 0$, assume that $q + \frac{g^*}{2f^*} - 1 < 0$, then $p(1 - \frac{g^*}{2f^*}) > pq$ and $\min\{p(1 - \frac{g^*}{2f^*}), pq\} = pq$. By focusing on $F_{\mathcal{D}}^{\mathcal{M}}$ in the FlipIt game, there exists at least one pair (f', g') such that $f' < f^*$ and $1 - \frac{g'}{2f'} = q$. Then, $F_{\mathcal{D}}^{\mathcal{M}}(f', g' | \bar{p}, p, q, v) < F_{\mathcal{D}}^{\mathcal{M}}(f^*, g^* | \bar{p}, p, q, v)$ which indicates that (f^*, g^*) is not a FlipIt game equilibrium. Hence, $q + \frac{g^*}{2f^*} - 1 < 0$ does not hold. Similar analysis applies to the region $g^* > f^* > 0$, and we can obtain $q - \frac{f^*}{2g^*} \geq 0$. $\square$

The Proposition 1 indicates that the cloud will not provide better communication quality according to the contract. Otherwise, it can spend less protection effort on cyber defense to achieve more profits. Therefore, based on Proposition 1, we can simplify the FlipIt game according to $\min\{p_{\mathcal{M}}(1 - \frac{g_{\mathcal{M}}^*}{2f_{\mathcal{M}}^*}), p_{\mathcal{M}} q_{\mathcal{M}}\} = p_{\mathcal{M}}(1 - \frac{g_{\mathcal{M}}^*}{2f_{\mathcal{M}}^*})$, $v_{\mathcal{M}}(q_{\mathcal{M}} + \frac{g_{\mathcal{M}}^*}{2f_{\mathcal{M}}^*} - 1)\mathbf{1}_{\{q_{\mathcal{M}} > 1 - \frac{g_{\mathcal{M}}^*}{2f_{\mathcal{M}}^*}\}} = v_{\mathcal{M}}(q_{\mathcal{M}} + \frac{g_{\mathcal{M}}^*}{2f_{\mathcal{M}}^*} - 1)$, $\min\{p_{\mathcal{M}} \frac{f_{\mathcal{M}}^*}{2g_{\mathcal{M}}^*}, p_{\mathcal{M}} q_{\mathcal{M}}\} = p_{\mathcal{M}} \frac{f_{\mathcal{M}}^*}{2g_{\mathcal{M}}^*}$, and $v_{\mathcal{M}}(q_{\mathcal{M}} - \frac{f_{\mathcal{M}}^*}{2g_{\mathcal{M}}^*})\mathbf{1}_{\{q_{\mathcal{M}} > \frac{f_{\mathcal{M}}^*}{2g_{\mathcal{M}}^*}\}} = v_{\mathcal{M}}(q_{\mathcal{M}} - \frac{f_{\mathcal{M}}^*}{2g_{\mathcal{M}}^*})$. Then, we can obtain the Nash equilibrium of the FlipIt game as follows.

Theorem 1. *The Nash equilibria of the FlipIt game are summarized as follows:*

(i) *when $\frac{\psi_{\mathcal{D}}}{v+p} < \frac{\psi_{\mathcal{A}}}{u_{\mathcal{A}}}$, then $f^* = \frac{u_{\mathcal{A}}}{2\psi_{\mathcal{A}}}$, and $g^* = \frac{\psi_{\mathcal{D}}}{2\psi_{\mathcal{A}}^2} \cdot \frac{u_{\mathcal{A}}^2}{v+p}$;*

(ii) *when $\frac{\psi_{\mathcal{D}}}{v+p} > \frac{\psi_{\mathcal{A}}}{u_{\mathcal{A}}}$, then $f^* = \frac{\psi_{\mathcal{A}}}{2\psi_{\mathcal{D}}^2} \cdot \frac{(v+p)^2}{u_{\mathcal{A}}}$, and $g^* = \frac{v+p}{2\psi_{\mathcal{D}}}$;*

(iii) *when $\frac{\psi_{\mathcal{D}}}{v+p} = \frac{\psi_{\mathcal{A}}}{u_{\mathcal{A}}}$, then $f^* = \frac{u_{\mathcal{A}}}{2\psi_{\mathcal{A}}}$, and $g^* = \frac{v+p}{2\psi_{\mathcal{D}}}$.*

Proof. For $f \geq g > 0$, the cloud defender's problem is $\max_f \{\bar{p} + p(1 - \frac{g}{2f}) - \psi_D f - v(q + \frac{g}{2f} - 1)\}$, which can be rewritten as $\max_f \{\bar{p} + (p + v)(1 - \frac{g}{2f}) - \psi_D f - vq\}$. In addition, the attacker is solving $\min_g \{\psi_A g - u_A \frac{g}{2f}\}$. Then, the above FlipIt game is strategically equivalent to the game that the defender solves $\max_f \{(1 - \frac{g}{2f}) - \frac{\psi_D}{p+v} f\}$, and the attacker solves $\max_g \{\frac{g}{2f} - \frac{\psi_A}{u_A} g\}$, which reduces the FlipIt game to the form in [9] and can be solved accordingly. The analysis is similar for the case $g > f > 0$. □

Under the Nash equilibrium presented in Theorem 1, the utility of the cloud can be expressed as

$$\pi(\bar{p}, p, q, v) = \bar{p} + p - \frac{u_A \psi_D}{\psi_A} - v(q - 1), \text{ for } \frac{\psi_D}{v + p} < \frac{\psi_A}{u_A}, \tag{17}$$

$$\pi(\bar{p}, p, q, v) = \bar{p} - vq, \text{ for } \frac{\psi_D}{v + p} \geq \frac{\psi_A}{u_A}. \tag{18}$$

Remark: Under $\frac{\psi_D}{v+p} \geq \frac{\psi_A}{u_A}$, the provided communication quality satisfies $z \leq \frac{1}{2}$ which means that more than half of the total packages are lost during transmission, and generally this is detrimental to the stability of the physical control system. Therefore, we mainly focus on the region of $\frac{\psi_D}{v+p} < \frac{\psi_A}{u_A}$ in the following paper, and the equilibrium strategy of the FlipIt game is $f^* = \frac{u_A}{2\psi_A}$, and $g^* = \frac{\psi_D}{2\psi_A^2} \cdot \frac{u_A^2}{v+p}$. Then,

$$z^* = 1 - \frac{g^*}{2f^*} = 1 - \frac{\psi_D u_A}{2\psi_A(v + p)}, \tag{19}$$

and we further have $q_{H,\min} > \frac{1}{2}$, $q_{L,\min} > \frac{1}{2}$.

The difference between the H-type and L-type cloud's profit under a certain contract is critical in designing the optimal contracts in Sect. 4, and we define it as follows.

Definition 2. *For a given contract* $(\bar{p}, p, q, v)$, *the benefit of being a H-type cloud over the L-type cloud is defined as* $\delta := \pi_H(\bar{p}, p, q, v) - \pi_L(\bar{p}, p, q, v)$.

Note that δ is not a function of contract terms, since $\bar{p}, p, q, v$ are not coupled with other parameters of the attacker and defender as seen from Eqs. (17) and (18). Furthermore, due to the IR constraints, we obtain

$$\delta \geq \epsilon_H - \epsilon_L. \tag{20}$$

To facilitate the optimal contract design, without loss of generality, we make the following assumption of parameters at the cyber layer.

Assumption 2. *The parameters satisfy* $\frac{\psi_D^H}{\psi_A^H} < \frac{\psi_D^L}{\psi_A^L}$, *and* $u_A^H = u_A^L$.

In Assumption 2, the inequality $\frac{\psi_D^H}{\psi_A^H} < \frac{\psi_D^L}{\psi_A^L}$ indicates that the H-type cloud is more resistant to malicious attacks, and the equality $u_A^H = u_A^L$ represents that the unit payoff of compromising two types of cloud are the same. Note that Assumption 2 is not strict, and we use it to determine the sign of δ below.

Based on the Assumption 2 and (17), we obtain

$$\delta = \frac{u_A^L \psi_D^L}{\psi_A^L} - \frac{u_A^H \psi_D^H}{\psi_A^H} > 0. \tag{21}$$

Thus, the profit of the cloud is larger of being H-type than L-type for a given contract $(\bar{p}, p, q, v)$.

Remark: The parameter δ is not necessary positive, and Assumption 2 is not strict. Without loss of generality, we choose δ to be positive. The results obtained in this section can be easily extended to the case with negative values of δ.

3.2 Physical Control System Analysis

The cloud defense strategy at the cyber layer and the contract design of CSA are interdependent. At the physical layer, one critical problem is the stability of the control system. First, we present the following theorem.

Theorem 2 ([26]). *Let $(A, \sqrt{Q})$ be observable. Then, under the communication quality $z_{\mathcal{M}}$, the condition ensuring the mean-square stability of the physical control system is*

$$\zeta := \max |\lambda(A)| < \frac{1}{\sqrt{1 - z_{\mathcal{M}}}}, \tag{22}$$

where ζ and $\lambda(A)$ denote the spectral radius and the eigenvalue of system matrix A, respectively.

Then, for $\mathcal{M} \in \{H, L\}$, we define the utility function of the control system as

$$U(z_{\mathcal{M}}) = -J(\Pi^* | z_{\mathcal{M}}), \tag{23}$$

where $J(\Pi^* | z_{\mathcal{M}})$ denotes the optimal control cost under $z_{\mathcal{M}}$.

Remark: Based on (19), if $\frac{1}{\sqrt{\frac{g_{\mathcal{M}}}{2f_{\mathcal{M}}}}} \leq \zeta$, then the control system is unstable, and $U(1 - \frac{g_{\mathcal{M}}}{2f_{\mathcal{M}}}) \to -\infty$ under which the contract design problem is not feasible. Hence, the contract should be designed in a way such that if it is picked by the cloud SP, the provided communication QoS stabilizes the physical system.

Another problem of the control system is to obtain the optimal control cost $J(\Pi^* | z_{\mathcal{M}})$. We state the solution to the optimal controller over unreliable communication channels as follows.

Theorem 3 ([26]). *The optimal control law is*

$$u_k^* = G_k \hat{x}_k, \tag{24}$$

where the matrix $G_k = -(R+B^T K_{k+1}B)^{-1}B^T K_{k+1}A$, *with* K_k *recursively given by the Riccati equation*

$$P_k = z_\mathcal{M} A^T B (R + B^T K_{k+1}B)^{-1}B^T K_{k+1}A,$$
$$K_k = A^T K_{k+1}A - P_k + Q. \tag{25}$$

The estimator $\hat{x}_k$ *takes the form*

$$\hat{x}_k = \begin{cases} A\hat{x}_{k-1} + \alpha_{k-1}Bu_{k-1}, & \beta_k = 0, \\ x_k, & \beta_k = 1. \end{cases} \tag{26}$$

In addition, when $k \to \infty$, $\lim_{k\to\infty} G_k = G$, *and the controller takes*

$$G = -(R + B^T K B)^{-1}B^T K A,$$
$$K = A^T K A + Q - z_\mathcal{M} A^T K B (R + B^T K B)^{-1}B^T K A. \tag{27}$$

Note that the control parameter K in Theorem 3 corresponds to the communication reliability $z_\mathcal{M}$. In addition, at the steady state, we have

$$J(\Pi^*|z_\mathcal{M}) = \mathbb{E}\{x_{N-1}^T Q x_{N-1} + z_\mathcal{M} u_{N-1}^T R u_{N-1}\}$$
$$= \mathbb{E}\{x_{N-1}^T Q x_{N-1} + z_\mathcal{M} \hat{x}_{N-1}^T G^T R G \hat{x}_{N-1}\}, \tag{28}$$

where gain G is given in Theorem 3.

The relationship between $z_\mathcal{M}$ and $J(\Pi^*|z_\mathcal{M})$ is critical during the contract design stage. Specifically, we have the following lemma.

Lemma 1. *Under* $\zeta < \frac{1}{\sqrt{1-z_\mathcal{M}}}$, *the control cost* $J(\Pi^*|z_\mathcal{M})$ *of the physical system is monotonically decreasing with the increase of* $z_\mathcal{M}$.

Remark: The monotonicity of $J(\Pi^*|z_\mathcal{M})$ with respect to $z_\mathcal{M}$ is verified by case studies in Sect. 5. The interpretation is as follows. With smaller $z_\mathcal{M}$, i.e., the package drop rate over communication channels is huge, then the physical system state and control input will encounter large deviations from nominal ones frequently, and therefore the control cost $J(\Pi^*|z_\mathcal{M})$ increases.

4   Optimal Contracts Design Under Asymmetric Information

We have analyzed the FlipIt game at the cyber layer and the optimal control of the physical system over unreliable communication links in Sect. 3. In this section, we design the optimal H-type and L-type contracts for the CSA under the asymmetric information structure. First, we simplify the constrained contract design problem formulated in Sect. 2.4 as follows.

Proposition 2. *The contract design problem in Sect. 2.4 is equivalent to*

$$\text{CDP}': \quad \min_{(\bar{p}_H, p_H, q_H, v_H)} \left\{ \sigma \left(\delta + \epsilon_L + \psi_D^H f_H^* + \phi_H \frac{U^o - U(z_H^*)}{U^o} \right) \right\}$$

$$+ \min_{(\bar{p}_L, p_L, q_L, v_L)} \left\{ (1 - \sigma) \left(\epsilon_L + \psi_D^L f_L^* + \phi_L \frac{U^o - U(z_L^*)}{U^o} \right) \right\}$$

$$\text{s.t.} \quad p_H > 0, \; p_L > 0,$$

$$q_{H,\min} \leq q_H \leq q_{H,\max} < 1,$$

$$q_{L,\min} \leq q_L \leq q_{L,\max} < 1. \tag{29}$$

Proof. Note that under $\frac{\psi_D^M}{v_M + p_M} < \frac{\psi_A^M}{u_A^M}$, we have $\pi_M(\bar{p}_M, p_M, q_M, v_M) = \bar{p}_M + p_M(1 - \frac{g_M^*}{2f_M^*}) - \psi_D^M f_M^* - v_M(q_M + \frac{g_M^*}{2f_M^*} - 1)\mathbf{1}_{\{q_M > 1 - \frac{g_M^*}{2f_M^*}\}}$, which yields $\bar{p}_M + p_M(1 - \frac{g_M^*}{2f_M^*}) - v_M(q_M + \frac{g_M^*}{2f_M^*} - 1)\mathbf{1}_{\{q_M > 1 - \frac{g_M^*}{2f_M^*}\}} = \pi_M(\bar{p}_M, p_M, q_M, v_M) + \psi_D^M f_M^*$. Thus, the objective function (6) in CDP can be rewritten as

$$\min_{\substack{(\bar{p}_H, p_H, q_H, v_H) \\ (\bar{p}_L, p_L, q_L, v_L)}} \sigma \left(\pi_H(\bar{p}_H, p_H, q_H, v_H) + \psi_D^H f_H^* + \phi_H \frac{U^o - U(z_H^*)}{U^o} \right)$$

$$+ (1 - \sigma) \left(\pi_L(\bar{p}_L, p_L, q_L, v_L) + \psi_D^L f_L^* + \phi_L \frac{U^o - U(z_L^*)}{U^o} \right). \tag{30}$$

Furthermore, based on the Definition 2, we have

$$\pi_H(\bar{p}_L, p_L, q_L, v_L) = \pi_L(\bar{p}_L, p_L, q_L, v_L) + \delta,$$

$$\pi_L(\bar{p}_H, p_H, q_H, v_H) = \pi_H(\bar{p}_H, p_H, q_H, v_H) - \delta. \tag{31}$$

Then, plugging (31) into the IC constraints (7a) and (7b) yields

$$\delta \geq \pi_H(\bar{p}_H, p_H, q_H, v_H) - \pi_L(\bar{p}_L, p_L, q_L, v_L) \geq \delta,$$

$$\Rightarrow \pi_H(\bar{p}_H, p_H, q_H, v_H) - \pi_L(\bar{p}_L, p_L, q_L, v_L) = \delta. \tag{32}$$

The constraints (7a)–(7d) can be equivalently captured by (32) together with $\pi_L(\bar{p}_L, p_L, q_L, v_L) \geq \epsilon_L$ since $\delta > 0$.

On the other hand, notice that for given p_M and v_M, the objective function (30) is minimized if $\pi_M(\bar{p}_M, p_M, q_M, v_M)$ achieves the minimum. The underlying interpretation is that lower utility of the cloud leads to higher quality of the communication which is beneficial for the physical systems. Therefore, based on (32) and the IR constraint $\pi_L(\bar{p}_L, p_L, q_L, v_L) \geq \epsilon_L$, we obtain $\pi_H(\bar{p}_H, p_H, q_H, v_H) - \pi_L(p_L, q_L, v_L) = \delta$, and $\pi_L(\bar{p}_L, p_L, q_L, v_L) = \epsilon_L$. Therefore, the constraints (7a)–(7d) further become

$$\pi_H(\bar{p}_H, p_H, q_H, v_H) = \delta + \epsilon_L, \tag{33}$$

$$\pi_L(\bar{p}_L, p_L, q_L, v_L) = \epsilon_L, \tag{34}$$

which result in CDP'. $\qquad \square$

Remark: Note that in CDP′, IC and IR constraints are incorporated into the objective function. In addition, two separate minimization terms in CDP′ are *decoupled* through the decision variables, and thus can be solved independently which simplifies the optimal contracts design.

First, we focus on $\min_{(\bar{p}_H, p_H, q_H, v_H)} \left\{ \sigma \left(\delta + \epsilon_L + \psi_D^H f_H^* + \phi_H \frac{U^o - U(z_H^*)}{U^o} \right) \right\}$, where $f_H^* = \frac{u_A^H}{2\psi_A^H}$, and $z_H^* = 1 - \frac{\psi_D^H u_A^H}{2\psi_A^H(v_H + p_H)}$ based on (19). Three underlying constraints are $\frac{\psi_D^H}{v_H + p_H} < \frac{\psi_A^H}{u_A^H}$, $q_H \geq z_H^*$, and $\zeta < \sqrt{\frac{2\psi_A^H(v_H + p_H)}{\psi_D^H u_A^H}}$. Then, we obtain Lemma 2.

Lemma 2. *The contract design is only dependent on the physical control system performance, and larger value of $v_H + p_H$ is desirable.*

Proof. Notice that $\underset{(\bar{p}_H, p_H, q_H, v_H)}{\arg\min} \left\{ \sigma \left(\delta + \epsilon_L + \psi_D^H f_H^* + \phi_H \frac{U^o - U(z_H^*)}{U^o} \right) \right\}$ is equivalent to $\underset{(\bar{p}_H, p_H, q_H, v_H)}{\arg\max} \ U(z_H^*)$, since $f_H^* = \frac{u_A^H}{2\psi_A^H}$ is irrelevant to the contract parameters. Thus, the contract design only relates to the control system performance. Through Lemma 1 together with z_H^* in (19), we obtain the result. $\square$

Next, through analyzing the impact of contract on the physical systems, we can obtain the optimal H-type contract as follows.

Theorem 4. *The optimal H-type contract $(\bar{p}_H, p_H, q_H, v_H)$ is designed as*

$$\bar{p}_H = 0, \tag{35}$$

$$q_H = q_{H,\max}, \tag{36}$$

$$v_H = \min\left\{ v_{H,\max}, \frac{\psi_D^H u_A^H}{2\psi_A^H(1 - q_{H,\max})} - p_H \right\}, \tag{37}$$

$$p_H = \frac{u_A^L \psi_D^L}{\psi_A^L} + (q_{H,\max} - 1)v_H + \epsilon_L. \tag{38}$$

Proof. From (33), we obtain $\pi_H(\bar{p}_H, p_H, q_H, v_H) = \bar{p}_H + p_H - \frac{u_A^H \psi_D^H}{\psi_A^H} - v_H(q_H - 1) = \delta + \epsilon_L = \frac{u_A^L \psi_D^L}{\psi_A^L} - \frac{u_A^H \psi_D^H}{\psi_A^H} + \epsilon_L$, which yields $\bar{p}_H + p_H - v_H(q_H - 1) = \frac{u_A^L \psi_D^L}{\psi_A^L} + \epsilon_L$. Therefore, $p_H + v_H = \frac{u_A^L \psi_D^L}{\psi_A^L} + v_H q_H - \bar{p}_H + \epsilon_L$. To maximize $p_H + v_H$, we obtain $\bar{p}_H = 0$, $q_H = q_{H,\max}$. We can also verify that $q_H = q_{H,\max}$ from the constraint $q_H \geq z_H^*$. In addition, $q_H \geq 1 - \frac{\psi_D^H u_A^H}{2\psi_A^H(v_H + p_H)}$ yields $v_H \leq \frac{\psi_D^H u_A^H}{2\psi_A^H(1 - q_H)} - p_H$. Therefore, together with the bound, the penalty parameter v_H takes the value $v_H = \min\{v_{H,\max}, \frac{\psi_D^H u_A^H}{2\psi_A^H(1 - q_{H,\max})} - p_H\}$, and then the unit payment p_H is equal to $p_H = \frac{u_A^L \psi_D^L}{\psi_A^L} + (q_{H,\max} - 1)v_H + \epsilon_L$. $\square$

Remark: When $v_{H,\max} > \frac{\psi_D^H u_A^H}{2\psi_A^H(1 - q_{H,\max})} - p_H$, we can obtain the contract terms v_H and p_H by solving Eqs. (37) and (38) jointly. In addition, if $v_H + p_H \leq$

$\frac{\zeta^2 \psi_D^H u_A^H}{2\psi_A^H}$, then *no contract* is placed to the H-type cloud, since the provided communication service cannot stabilize the control system.

For the second part $\min_{(\bar{p}_L, p_L, q_L, v_L)} \left\{ (1-\sigma)\left(\epsilon_L + \psi_D^L f_L^* + \phi_L \frac{U^o - U(z_L^*)}{U^o}\right) \right\}$ in the objective function of CDP′, we can equivalently solve the optimization problem $\max_{(\bar{p}_L, p_L, q_L, v_L)} U(z_L^*)$. Remind that based on (34), $\pi_L(\bar{p}_L, p_L, q_L, v_L) = \bar{p}_L + p_L - \frac{u_A^L \psi_D^L}{\psi_A^L} - v_L(q_L - 1) = \epsilon_L$, which gives $\bar{p}_L + p_L - v_L(q_L - 1) = \frac{u_A^L \psi_D^L}{\psi_A^L} + \epsilon_L$, and thus $p_L + v_L = \frac{u_A^L \psi_D^L}{\psi_A^L} + v_L q_L - \bar{p}_L + \epsilon_L$. Then, the L-type contract $(\bar{p}_L, p_L, q_L, v_L)$ immediately follows the similar analysis in Theorem 4.

Theorem 5. *The optimal L-type contract $(\bar{p}_L, p_L, q_L, v_L)$ is given by*

$$\bar{p}_L = 0, \quad q_L = q_{L,\max},$$
$$v_L = \min\left\{ v_{L,\max}, \frac{\psi_D^L u_A^L}{2\psi_A^L(1 - q_{L,\max})} - p_L \right\},$$
$$p_L = \frac{u_A^L \psi_D^L}{\psi_A^L} + (q_{L,\max} - 1)v_L + \epsilon_L. \tag{39}$$

Similarly, when the optimal contract satisfies $v_L + p_L \leq \frac{\zeta^2 \psi_D^L u_A^L}{2\psi_A^L}$, then the CSA will not place the contract due to the instability of the control system.

We have obtained the optimal contracts for CSA. Several characteristics in both the optimal H-type and L-type contracts are summarized as follows:

1. In the focused region of $\frac{\psi_D}{v+p} < \frac{\psi_A}{u_A}$, i.e., the provided service $z > \frac{1}{2}$, the contract design problem is simply *reduced to* the minimization of the control system cost.
2. *No contract* will be offered to the cloud SP if the resulting communication cannot stabilize the physical system. One reason accounting for this situation is that the ratio between the unit defending cost and the unit utility of SP is relatively large, and thus the cloud defender is reluctant to spend too much effort on protecting the cloud resources.
3. In both H-type and L-type contracts, the transfer payment is equal to zero, and the requested communication quality achieves the upper bound.
4. Under the accepted optimal contracts, the payoffs of the H-type and L-type clouds are equal to $\delta + \epsilon_L$ and ϵ_L, respectively, which are both *constants*.

5   Case Studies

In this section, we illustrate the designed optimal contracts via case studies.

In the physical layer, the control system is described by Eqs. (1) and (2), and the system matrices are given by $A = \begin{bmatrix} 1 & -1 \\ 3 & 1 \end{bmatrix}$, $B = C = R = Q = \begin{bmatrix} 1 & 0 \\ 0 & 1 \end{bmatrix}$. Then the spectral radius of matrix A is equal to $\zeta = 2$. Based on Theorem 2, to stabilize the control system, the minimum communication quality is $z = 0.75$. In addition, the exogenous disturbance to the system is with zero mean and unit variance.

Several parameters in the contract design are summarized as follows: $\psi_{\mathcal{A}}^H = 8$, $\psi_{\mathcal{A}}^L = 4$, $u_{\mathcal{A}}^H = u_{\mathcal{A}}^L = 20$, $v_{H,\max} = 100$, $v_{L,\max} = 90$, $\epsilon_L = 10$, $\epsilon_H = 20$, $q_{L,\min} = 0.75$, $q_{L,\max} = 0.82$, $q_{H,\min} = 0.86$, and $q_{H,\max} = 0.93$.

First, we illustrate the case of H-type contract design. Specifically, the unit defending cost of the L-type cloud is chosen as $\psi_{\mathcal{D}}^L = 8$, and we design the H-type contract in the reasonable region of $\psi_{\mathcal{D}}^H$ that satisfies the conditions in Assumption 2. The corresponding results are shown in Fig. 3. In the contractable region of Fig. 3(a), with the increasing of $\psi_{\mathcal{D}}^H$, the unit payment p_H decreases first and then keeps as a constant. In contract, in Fig. 3(b), the unit penalty v_H increases first and then becomes unchanged. The unchanging region of $\psi_{\mathcal{D}}^H$ is due to the fact that v_H achieves the maximum. The utility of the H-type cloud is decreasing as the defending cost becomes larger as depicted in Fig. 3(c), and this property can be verified by Eq. (33). Note that when $\psi_{\mathcal{D}}^H/\psi_{\mathcal{A}}^H \geq 1.59$, no H-type contract is accepted by the cloud since the cloud's minimum utility ϵ_H cannot be met by providing the service. The required and real provided communication quality are presented in Fig. 3(d). As shown in Proposition 1, the provided z_H will never be greater than the required q_H. In addition, in the middle region, z_H is decreasing as the increase of the defending cost. The reason is that the penalty v_H and the payment p_H are constant, and then the cloud can earn more profit by spending less effort in protecting the cloud resources which results in worse communication service. Figure 3(e) shows the control cost of the physical system. With larger z_H, the control cost is lower which corroborates Lemma 1.

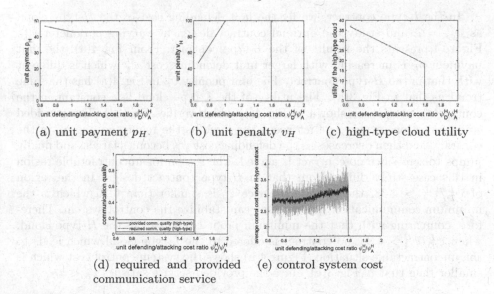

(a) unit payment p_H (b) unit penalty v_H (c) high-type cloud utility

(d) required and provided communication service   (e) control system cost

Fig. 3. (a) and (b) show the unit payment p_H and the unit penalty v_H in the designed H-type contract. (c), (d), and (e) represent the utility of the H-type cloud, the provided communication quality z_H, and the control system performance under the corresponding contract, respectively.

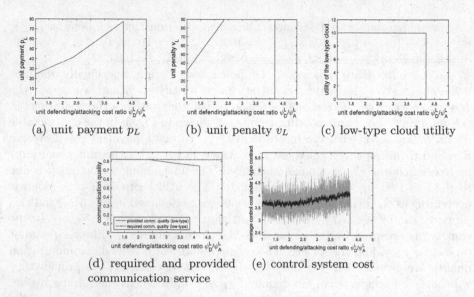

(a) unit payment p_L (b) unit penalty v_L (c) low-type cloud utility

(d) required and provided   (e) control system cost
communication service

Fig. 4. (a) and (b) show the unit payment p_L and the unit penalty v_L in the designed L-type contract. (c), (d), and (e) represent the utility of the L-type cloud, the provided communication quality z_L, and the average control system cost under the corresponding contract, respectively.

In the L-type contract, we fix the unit defending cost of the H-type cloud as $\psi_{\mathcal{D}}^H = 7$, and study the optimal contract design by varying parameter $\psi_{\mathcal{D}}^L$. Figure 4 presents the results of the L-type contract. From Fig. 4(a), the unit payment p_L is increasing with larger unit defending cost $\psi_{\mathcal{D}}^L$ which is different with that in the H-type contract. The unit penalty v_L in Fig. 4(a) has the same trend as that in Fig. 3(a). The utility of the L-type cloud is a constant in the contractable region as shown in Fig. 4(c) which verifies Eq. (34). The provided communication service z_L first keeps the same as the requested one q_L in the contract, and then decreases as the defending cost $\psi_{\mathcal{D}}^L$ becomes larger, and finally jumps to zero since no contract is agreed. The reason for uncontractable region in this case study differs from that in H-type contract design. In the region of $\psi_{\mathcal{D}}^L/\psi_{\mathcal{A}}^L > 4.38$, the provided service z_L is smaller than 0.75 which is the minimum communication quality that can stabilize the control system. Therefore, comparing with that the minimum profit is not met in the H-type cloud, when $\psi_{\mathcal{D}}^L/\psi_{\mathcal{A}}^L > 4.38$, no contract is offered to the L-type cloud which leads to the uncontractable situation. Figure 4(e) shows the system control cost which is smaller than that in Fig. 3(e), since the provided service satisfies $z_L < z_H$.

6   Conclusion

We have studied the optimal contract design for the cloud-enabled Internet of Controlled Things (IoCT) under the asymmetric information structure. In the proposed bi-level cyber-physical framework, we have used a `FlipIt` game to model the interactions between the cloud defender and attacker at the upper cyber layer. The cloud defense strategy is influenced by the offered contracts by the physical system administrator. At the lower physical layer, the devices have adopted the optimal control based on the received communication quality from the cyber layer. We have designed two optimal contracts in terms of the type of the cloud. Case studies have been provided to corroborate the obtained results. The future work would be quantifying the value of information to the cloud service provider and extending the contract analysis to the entire region.

References

1. Jayavardhana, G., Buyya, R., Marusic, S., Palaniswami, M.: Internet of Things (IoT): a vision, architectural elements, and future directions. Future Gener. Comput.Syst. **29**(7), 1645–1660 (2013)
2. Colin, T.: Advanced persistent threats and how to monitor and deter them. Netw. Secur. **2011**(8), 16–19 (2011)
3. Pawlick, J., Farhang, S., Zhu, Q.: Flip the cloud: cyber-physical signaling games in the presence of advanced persistent threats. In: Khouzani, M.H.R., Panaousis, E., Theodorakopoulos, G. (eds.) GameSec 2015. LNCS, vol. 9406, pp. 289–308. Springer, Heidelberg (2015). doi:10.1007/978-3-319-25594-1_16
4. Xu, Z., Zhu, Q.: Cross-layer secure cyber-physical control system design for networked 3D printers. In: Proceedings of American Control Conference (ACC), pp. 1191–1196 (2016)
5. Hu, G., Tay, W.P., Wen, Y.: Cloud robotics: architecture, challenges and applications. IEEE Netw. **26**(3), 21–28 (2012)
6. Chen, J., Zhu, Q.: Resilient and decentralized control of multi-level cooperative mobile networks to maintain connectivity under adversarial environment. arXiv preprint arXiv:1505.07158 (2016)
7. Krutz, R.L., Vines, R.D.: Cloud security: a comprehensive guide to secure cloud computing. Wiley Publishing, Indianapolis (2010)
8. Fang, X., Misra, S., Xue, G., Yang, D.: Managing smart grid information in the cloud: opportunities, model, and applications. IEEE Netw. **26**(4), 32–38 (2012)
9. van Dijk, M., Juels, A., Oprea, A., Rivest, R.L.: FlipIt: the game of stealthy takeover. J. Cryptology **26**(4), 655–713 (2013)
10. Triantis, G.G.: Financial contract design in the world of venture capital. University of Chicago Law Review 68 (2001)
11. Fehr, E., Klein, A., Schmidt, M.: Fairness and contract design. Econometrica **75**(1), 121–154 (2007)
12. DellaVigna, S., Malmendier, U.: Contract design and self-control: theory and evidence. Q. J. Econ. **CXIX**(2), 353–402 (2004)
13. Doherty, N.A., Dionne, G.: Insurance with undiversifiable risk: contract structure and organizational form of insurance firms. J. Risk Uncertainty **6**(2), 187–203 (1993)
14. Skees, J.R., Black, J.R., Barnett, B.J.: Designing and rating an area yield crop insurance contract. Am. J. Agric. Econ. **79**(2), 430–438 (1997)
15. Yang, Z., Aydin, G., Babich, V., Beil, D.R.: Supply disruptions, asymmetric information, and a backup production option. Manage. Sci. **55**(2), 192–209 (2009)

16. Corbett, C.J., de Groote, X.: A supplier's optimal quantity discount policy under asymmetric information. Manage. Sci. **46**(3), 444–450 (2000)
17. Pant, Y.V., Abbas, H., Mohta, K., Nghiem, T.X., Devietti, J., Mangharam, R.: Co-design of anytime computation and robust control. In: Proceedings of Real-Time Systems Symposium, pp. 43–52 (2015)
18. Azab, M., Eltoweissy, M.: Defense as a service cloud for cyber-physical systems. In: Proceedings of the Conference on Collaborative Computing: Networking, Applications and Worksharing, pp. 392–401 (2011)
19. Gao, Y., Chen, Y., Wang, C.Y., Liu, K.R.: A contract-based approach for ancillary services in V2G networks: optimality and learning. In: Proceedings of IEEE International Conference on Computer Communications (INFOCOM), pp. 1151–1159 (2013)
20. Ferraro, P.J.: Asymmetric information and contract design for payments for environmental services. Ecol. Econ. **65**(4), 810–821 (2008)
21. Perera, C., Zaslavsky, A., Christen, P., Georgakopoulos, D.: Sensing as a service model for smart cities supported by internet of things. Trans. Emerg. Telecommun. Technol. **25**(1), 81–93 (2014)
22. Jezequel, J.-M., Train, M., Mingins, C.: Design Patterns with Contracts. Addison-Wesley Longman Publishing Co., Boston (1999)
23. Heckel, R., Lohmann, M.: Towards contract-based testing of web services. Electr. Notes Theoret. Comput. Sci. **116**, 145–156 (2005)
24. Sharpe, S.A.: Asymmetric information, bank lending, and implicit contracts: a stylized model of customer relationships. J. Financ. **45**(4), 1069–1087 (1990)
25. Myerson, R.B.: Incentive compatibility and the bargaining problem. Econometrica **47**(1), 61–74 (1979)
26. Imer, O.C., Yuksel, S., Baar, T.: Optimal control of LTI systems over unreliable communication links. Automatica **42**(9), 1429–1439 (2006)

Becoming Cybercriminals: Incentives in Networks with Interdependent Security

Aron Laszka$^{(\boxtimes)}$ and Galina Schwartz

University of California, Berkeley, Berkeley, USA
laszka@berkeley.edu

Abstract. We study users' incentives to become cybercriminals when network security is interdependent. We present a game-theoretic model in which each player (i.e., network user) decides his type, honest or malicious. Honest users represent law-abiding network users, while malicious users represent cybercriminals. After deciding on their types, the users make their security choices. We will follow [29], where breach probabilities for large-scale networks are obtained from a standard interdependent security (IDS) setup. In large-scale IDS networks, the breach probability of each player becomes a function of two variables: the player's own security action and network security, which is an aggregate characteristic of the network; network security is computed from the security actions of the individual nodes that comprise the network. This allows us to quantify user security choices in networks with IDS even when users have only very limited, aggregate information about security choices of other users of the network.

Keywords: Interdependent security · Cybercrime · Security economics · Game theory · Nash equilibrium · Security investments

1 Introduction

Due to technological reasons, network security features multiple layers of interdependencies. Interdependent security has been extensively studied, see [20] for a recent survey; however, most of the existing literature does not address the strategic reasons of the losses; i.e., there is no explicit modeling of attackers' incentives to become engaged in cybercrime. In this paper, we look at users' incentives for becoming attackers (malicious users), and study how users' security choices and utilities are affected by the number of attackers.

Another distinctive feature of our setup, which is non-standard for the IDS literature, is that our model can deal with large-scale IDS networks. In many cases, the IDS papers do not emphasize the effects of large-scale games. Notable exceptions closely related to our work are [24] and [1]. In the latter, the authors consider a model with multiple IDS players similar to our setup, and in the former, large-scale networks with different topologies are studied. Ideas from [24] were further developed in [29], whose setup we expand to study incentives for becoming a cybercriminal.

© Springer International Publishing AG 2016
Q. Zhu et al. (Eds.): GameSec 2016, LNCS 9996, pp. 349–369, 2016.
DOI: 10.1007/978-3-319-47413-7_20

We consider a large-scale IDS network with strategic players (i.e., network users or nodes), who choose to which type they will belong, honest or malicious; the players also make choices of their security investments; we allow continuous security choices.

A common trend in numerous papers approaching economic aspects of cybercrime is inquiry into the "production technology" of cybercrime.[1] Our approach is complementary: we give virtually no details about the implementation side of cybercrime. We take a large-scale, macro perspective, and reduce the problem to the following base level parameters: risk aversion, loss size, degree of IDS, and costs of improving security. In this paper, we consider a more aggregate perspective. We build on the framework of risk assessment for large-scale IDS networks, developed by [29], and model users' incentives to become cybercriminals. While at present our model is minimalistic and stylized, it could be extended to include more parameters, such as different costs of attacking, and attacks with different IDS features.

Following a seminal contribution of Tullock [31], we approach incentives for cybercrime in the perspective of rent seeking. The core idea of rent seeking was originally coined by Tullock to study any non-productive wealth redistribution. Rent seeking was demonstrated to be useful methodology for the analysis of diverse subjects, ranging from monopolist's (over)pricing and losses from imposition of tariffs to corruption, fraud, theft, and other criminal endeavors. The distinguished feature of rent seeking is its wasteful and oftentimes openly coercive nature. The propensity of rent-seeking activities depends on institutions and enforcement capabilities. The prevalence of inefficient, corrupt institutions results in higher rent-seeking activities, and it is associated with poor economic performance and growth.

In [26,27], Olson connected an increase of rent-seeking activities with increased severity of the problem(s) of collective action. In the cybersecurity economics literature, this problem is studied under the name of free riding. The problem arises when individually and socially optimal actions differ, and a large number of dispersed players is present, with each player's gains or losses being trivial in size. In such cases, mechanisms to align individually and socially optimal actions are hard to find. Investments in cybersecurity are well known to have a marked presence of free riding [2,3,32], and thus, in general, suboptimal. Proliferation of rent seeking (in our case, cybercrime) negatively affects growth, as it shifts resources away from productive activities.

Consider for example the papers modeling one of the most widespread cybercrimes – phishing. The modeling literature originated by [7] looks at specific costs (number of targets, strength of the attack, probability of being caught, and the size of the fine) and the benefits (revenues resulting from the losses of the targets, such as stolen bank account information). The authors discuss the difficulties of designing effective countermeasures. From their analysis, increased penalties have limited impact. They advocate that improving the controls to prevent trading of stolen data will be more impactful. Followup papers introduce

---

[1] For example in [4,18,22], cybercrime is approached from value-chain perspective.

additional considerations and tools, such as risk simulation approach [17]. At the same time, the literature acknowledges practical complications: while preventing trading will be highly effective, it is questionable that this recommendation can be achieved in practice: it requires global enforcement institutions with novel legal rights and technological capabilities.

In the world with global connectivity, crime is becoming global as well due to the increased proliferation of the cybercrime. The global world is facing new threats, with meager existing institutions to counteract them. This situation requires developing novel tools to reduce user incentives for becoming malicious. Designing new economic institutions to be charged with mitigating rent-seeking incentives to engage in cybercrime is socially desirable as only such institutions will preclude the formation and syndication of organized international cybercrime. Our work permits quantifiable assessment and comparative analysis of various policy tools and institutions.

1.1 Applications

Our analysis can be applied to address robustness of large-scale cyber-physical systems (CPS). In [16], Knowles et al. present a comprehensive review of security approaches for CPS, and survey methodologies and research for measuring and managing cyber-risks in industrial control systems.

Since modern CPS are increasingly networked, achieving robust performance requires addressing the problem of interdependencies (see Sect. 6.2.3 of [16]). The authors identify the importance of system-wide risk assessment for CPS, and discuss three difficulties: (i) scant data availability, (ii) lack of established framework for defining and computing risk metrics, and (iii) lack of reliable performance evaluation of security measures. The focus of our paper is (ii). We use IDS framework, and demonstrate how system security evolves when the attacker choices are endogenous.

For example, the perpetrators of the Energetic Bear (a.k.a. Dragonfly) cyberespionage campaign exploited interdependence between energy companies and industrial control system (ICS) manufacturers [30]. In order to penetrate highly-secure targets (e.g., energy grid operators, major electricity generation firms, petroleum pipeline operators in the U.S., Germany, Turkey, etc.), the attackers compromised ICS manufacturers and inserted malware into software updates distributed by these manufacturers, which were downloaded and applied by the targets, leading to their compromise.

While Knowles et al. discuss the problem of interdependencies, they also express skepticism about the realistic options of improving the current state of cybercrime reality [16]. In fact, the authors expect slow progress due to lack of incentives for private entities to share information about risks. Our setup allows circumventing the problem of data limitations as our analysis relies on aggregate information about network security only.

The remainder of this paper is organized as follows. In Sect. 2, we discuss related work on interdependent security. In Sect. 3, we introduce our model of interdependent security and incentives for malicious behavior. In Sect. 4, we study the Nash equilibria of our model. In Sect. 5, we present numerical illustrations for our theoretical results. Finally, in Sect. 6, we offer concluding remarks and outline future work.

2   Related Work

In this section, we provide a brief overview of the most related papers from the interdependent security literature. For a more detailed review of the relevant literature, we refer the interested reader to [20].

The interdependent security problem was originally introduced in the seminal paper of Kunreuther and Heal, who initially formulated an IDS model for airline security. They extended their model to cover a broad range of applications, including cybersecurity, fire protection, and vaccinations [19]. They study the Nash equilibria of the model, and examine various approaches for incentivizing individuals to invest in security by internalizing externalities, such as insurance, fines, and regulations. In follow-up work, they extend their analysis to study tipping (i.e., when inducing some individuals to invest in security results in others investing as well) and coalitions of individuals that can induce tipping [10,11]. Other authors have also used this model to study various phenomena, including uncertainty and systematic risks [13,14,21].

Öğüt et al. introduce an interdependence model for cybersecurity, which they use to study the effects of interdependence on security investments and cyber-insurance [24]. Similar to the model of Kunreuther and Heal, the model of Öğüt et al. is based on the probabilistic propagation of security compromises from one entity to the other. In follow-up work, the authors extend their analysis by considering subsidies provided by a social planner, and find that subsidies for security investments can induce socially optimal investments, but subsidies for insurance do not provide a similar inducement [25].

Varian introduces and studies three prototypical interdependence models for system reliability: total effort, weakest link, and best shot [32]. In these models, the overall level of reliability depends respectively on the sum of efforts exerted by the individuals, the minimum effort, and the maximum effort. Later, these models have been widely used for studying security interdependence.

For example, Grossklags et al. compare Nash equilibrium and social optimum security investments in the total effort, weakest link, and best shot models [8]. In another example, Honeyman et al. address investment suboptimalities when users cannot distinguish between security failures (weakest link), and reliability failures (total effort) [12].

Khouzani et al. consider security interdependence between autonomous systems (AS), and study the effect of regulations that penalize outbound threat activities [15]. The authors find that free-riding may render regulations ineffective when the fraction of AS over which the regulator has authority is lower than a certain threshold, and show how a regulator may use information regarding the heterogeneity of AS for more effective regulation.

In most interdependent security models, adversaries are not strategic decision-makers. Nonetheless, there are a few research efforts that do consider strategic adversaries. Hausken models adversaries as a single, strategic player, who considers the users' strategies and substitutes into the most optimal attack allocation [9]. This substitution effect creates negative externalities between the users' security investments, which are fundamentally different from the positive externalities considered in our model. Moscibroda et al. consider malicious users [23] in the inoculation game, which was introduced originally by Aspnes et al. [5,6]. In the model of Moscibroda et al., malicious users are byzantine: they appear to be non-malicious users who invest in security, but they are actually not secure at all. Furthermore, the set of malicious users is assumed to be exogenous to the model. Grossklags et al. introduce an interdependence model, called weakest target, in which an attacker targets and always compromises the user with lowest security effort [8].

In another related paper, Acemoglu et al. focus on security investments of interconnected agents, and study contagion due to the possibility of cascading failures [1]. They analyze how individual and social optima behave in the presence of endogenous attacks. The authors formulate the sufficient conditions for underinvestment in security, and demonstrate that overinvestment occurs in some cases. Interestingly, in contrast to our results, overinvestment in security may intensify when attacks are endogenous in [1]. In our paper, the imposition of fast growing security costs guarantees that underinvestment occurs.

3 Model

Here, we introduce our model of non-malicious and malicious users, their incentives, and the security interdependence between them. A list of symbols used in this paper can be found in Table 1.

We assume that the number of users is fixed and denoted by N. Each user chooses his type, malicious or honest (i.e., attacker or defender). We will denote the number of malicious users and honest users by M and $N - M$, respectively. Each user's objective is to maximize his expected payoff (i.e., utility) u

$$u_i = u_i(\mathbf{t}, \mathbf{s}) = \max_{t_i, s_i} \{u_i, v_i\},$$

where $v_i = v_i(\mathbf{t}, \mathbf{s})$ and $u_i = u_i(\mathbf{t}, \mathbf{s})$ denote respective utilities of malicious and honest users, and $\mathbf{s} = (s_1, \ldots, s_N)$ is a vector of the players' security choices, and $\mathbf{t} = (t_1, \ldots, t_N)$ is a vector of user types, with $t_i = 1/0$, for malicious/honest user respectively, which allows us to express the number of malicious users M as:

$$M := \sum_{i=1}^{N} t_i. \tag{1}$$

Each honest user i objective is to maximize his expected utility $u_i = u_i(\mathbf{t}, \mathbf{s})$

$$u_i = [1 - B_i(\mathbf{s})] U(W) + B_i(\mathbf{s})U(W - L) - h(s_i), \tag{2}$$

where $B_i(\mathbf{s}) = B_i(s_i, s_{-i})$ is the probability that user i suffers a security breach, $U(w)$ is the utility with wealth w, W is the initial user wealth, and L is the

Table 1. List of symbols

Symbol	Description
Constants	
N	Number of users
W	Initial wealth of a user
L	Loss of a user in case of a security breach
μ	Probability of a malicious user getting caught
q_∞	Defined as $\lim_{N \to \infty} q(N)N$
Functions	
$q(N)$	Strength of interdependence between N users
$h(s)$	Cost of security level s
$B_i(s_1, \ldots, s_N)$	Security breach probability of user i
$G_i(M, s_1, \ldots, s_N)$	Financial gain of malicious user i
$U(\ldots)$	Utility function of a user
Variables	
s_i	Security level of user i
M	Number of malicious users
$\hat{s}$	Equilibrium security level of honest users

loss in case of a security breach. We assume that $L \in (0, W)$. The function $h(s)$ is security cost function, with $s \in [0, 1)$ denoting the security level of the user. While we view h as the "cost" of attaining a given security level, we model these costs as separable from U because security costs are often non-monetary (e.g., inconvenience and effort).

We assume $h'(s) > 0$ and $h''(s) > 0$ for $s_i \in (0, 1)$ for every $s \in [0, 1)$, $h(0) = h'(0) = 0$, and $h(1) = \infty$.[2] In addition, we will impose $h'''(s) > 0$ to simplify the exposition. Intuitively, with these assumptions, the marginal productivity of investing in security is decreasing rapidly, and the cost of attaining perfect security is prohibitively high. We assume that the users are risk-averse, that is, the function U is concave at any wealth $w \geq 0$: $U'(w) > 0$ and $U''(w) < 0$; also we let $U(0) = 0$.

Each malicious user j maximizes $v_j = v_j(\mathbf{t}, \mathbf{s})$

$$v_j = (1 - \mu)U(G_j(\mathbf{t}, \mathbf{s})) + \mu U(0) - h(s_j), \tag{3}$$

where μ is the probability of a malicious user being caught and punished (e.g., by law enforcement), and G_j is the gain of user j from engaging in cyber-crime. We assume that honest users' losses are distributed evenly between the malicious users:

$$G_j(\mathbf{t}, \mathbf{s}) = \frac{\sum_{i \in \text{honest users}} B_i(\mathbf{s})L}{M}, \tag{4}$$

---

[2] In other words, the Inada conditions hold.

and M is given by Eq. (1).

In our model, each user has two strategic actions: (i) user decides on his type (malicious or honest), and on his security level s (and thus, cost $h(s)$). In the next section (Sect. 4), we will study the Nash equilibria of our model, which are defined as follows.

Definition 1 (Nash Equilibrium). *A strategy profile* $(\mathbf{t}, \mathbf{s})$ *is a Nash equilibrium if*

- *being malicious is a best response for every malicious user and*
- *being non-malicious and investing in security level s_i is a best response for every non-malicious user i.*

3.1 Interdependent Security Model

For breach probabilities B_i, we will assume interdependent security (IDS). Our model builds on well-known interdependent security model of Kunreuther and Heal [19].

In this model, a user can be compromised (i.e., breached) in two ways: (i) directly and (ii) indirectly. The probability of a *direct breach* reflects the probability that an honest user is breached directly by an adversary. For each user i, the probability of being compromised directly is modeled as Bernoulli random process, with the failure probability equal to $(1 - s_i)$ when security investment is $h(s_i)$. This means that the probability of user i being safe from direct attacks is equal to that user's security level s_i, and does not depend on other users' security choices. We assume that for any two users, the probabilities of direct compromise are independent Bernoulli random processes.

Indirect breach probability reflects the presence of IDS – the users are interdependent. More specifically, we assume that in addition to direct compromise, the user can be breached indirectly – i.e., via a connection to another user, who was compromised directly. The assumption of indirect compromise reflects the connectivity and trust between the users. Let $q_{ij}(N)$ denote the conditional probability that user i is compromised indirectly by user j in the network with N users, given that user j is directly compromised. To simplify, for now we will assume that $q_{ij}(N)$ is a constant (independent of i and j): $q_{ij}(N) = q(N)$. Then, the probability of user i to be breached indirectly can be expressed as

$$\Pr[\text{compromised indirectly}]$$
$$= 1 - \Pr[\text{not compromised indirectly}] \tag{5}$$
$$= 1 - \prod_{j \neq i} \Pr[\text{no indirect compromise from user } j] \tag{6}$$
$$= 1 - \prod_{j \neq i} (1 - \Pr[\text{user } j \text{ is directly compromised}] \Pr[\text{successful propagation}]) \tag{7}$$
$$= 1 - \prod_{j \neq i} (1 - (1 - s_j)q(N)). \tag{8}$$

Next, let $B_i = B_i(\mathbf{s})$ denote the probability that user i is compromised (i.e., breached) either directly or indirectly:

$$B_i = 1 - \Pr[\text{not compromised}] \tag{9}$$

$$= 1 - \Pr[\text{not compromised directly}]\Pr[\text{not compromised indirectly}] \tag{10}$$

$$= 1 - s_i \prod_{j \neq i}(1 - (1 - s_j)q(N)). \tag{11}$$

In practical scenarios, $q(N)$ must decrease with N (the number of network users). As it is standard in aggregative games, we let the limit of $q(N)$ equal to zero as N approaches infinity.

4 Analysis

Next, we present theoretical results on our model of interdependent security and incentives for malicious behavior. First, in Sect. 4.1, we consider breach probabilities in large-scale networks. We show that the IDS model allows approximating a user's breach probability using the user's own security level and the average security level of the network. Second, in Sect. 4.2, we study equilibrium security choices for a game with a fixed number of malicious users. Finally, in Sect. 4.3, we study the equilibrium of the game where the number of malicious users is endogenous: it is determined by user choices.

4.1 Large-Scale Networks

We begin our analysis by studying the honest users' breach probabilities in large-scale networks (i.e., when the number of users N is high). Our goal here is to express the breach probabilities in a simpler form, which will facilitate the subsequent analysis of the users' equilibrium choices.

First, recall that in practical scenarios, $q(N)$ approaches zero as N grows (i.e., $\lim_{N \to \infty} q(N) = 0$). Hence, we can discard the terms with $q(N)^2, q(N)^3, \ldots$, and obtain the following approximation for large-scale networks:

$$B_i(\mathbf{s}) = 1 - s_i \prod_{j \neq i}(1 - (1 - s_j)q(N)) \tag{12}$$

$$\approx 1 - s_i \left(1 - \sum_{j \neq i}(1 - s_j)q(N)\right) \tag{13}$$

$$\approx 1 - s_i \left[1 - q(N)N\left(1 - \frac{\sum_{j \neq i} s_j}{N}\right)\right]. \tag{14}$$

Let $\bar{s}$ denote the average of the security levels taken over all users; formally, let $\bar{s} = \frac{\sum_j s_j}{N}$. Next, we use that the fraction $\frac{\sum_{j \neq i} s_j}{N}$ approaches the average security level $\bar{s}$ as N grows, and obtain:

$$1 - s_i \left[1 - q(N)N\left(1 - \frac{\sum_{j \neq i} s_j}{N}\right)\right] \approx 1 - s_i\left(1 - q(N)N(1 - \bar{s})\right). \tag{15}$$

Finally, we assume that $q(N)N$ has a limit as N approaches infinity, and this limit is less than 1. Then, we let $q_\infty = \lim_{N\to\infty} q(N)N$, which gives us:

$$1 - s_i\left(1 - q(N)N(1 - \bar{s})\right) \approx 1 - s_i(1 - q_\infty(1 - \bar{s})) \tag{16}$$
$$= 1 - s_i(1 - q_\infty) - s_i q_\infty \bar{s}. \tag{17}$$

Thus, for large-scale networks, breach probability B_i is a function of user security s_i and the average security $\bar{s}$:

$$B_i(s_i, s_{-i}) = 1 - s_i(1 - q_\infty) - s_i q_\infty \bar{s}. \tag{18}$$

In the remainder of the paper, we use (18) for breach probability B_i of user i.

4.2 Game with Exogenous Number of Malicious Users

Next, let us consider a game with a fixed number M of malicious users, that is, a game in which the strategic choice of every user i is limited to selecting security s_i. From Eq. (3), malicious users incur no losses, thus, they will not invest in network security (see Sect. 3.1). Hence, in any equilibrium, $s_j = 0$ for every malicious user j.

Let $\bar{s}_H$ denote the average security level of honest users:

$$\bar{s}_H = \frac{\sum_{j\in\text{honest users}} s_j}{N - M}. \tag{19}$$

Recall that malicious users contribute zero towards the security of the network, that is, $s_j = 0$ for every malicious user j. Hence, the breach probability of an honest user i can be expressed as

$$B_i(s_i, \bar{s}_H) = 1 - s_i(1 - q_\infty) - s_i q_\infty \bar{s} \tag{20}$$
$$= 1 - s_i(1 - q_\infty) - s_i q_\infty \frac{N - M}{N} \bar{s}_H. \tag{21}$$

Using $B_i(s_i, \bar{s}_H)$, the expected utility of user i can be expressed as

$$u = [1 - B_i(s_i, \bar{s}_H)]\, U(W) + B_i(s_i, \bar{s}_H)U(W - L) - h(s_i) \tag{22}$$
$$= U(W - L) + [1 - B_i(s_i, \bar{s}_H)]\, \Delta_0 - h(s_i), \tag{23}$$

where

$$\Delta_0 = U(W) - U(W - L). \tag{24}$$

Our goal is to characterize the equilibrium security levels when user types are given. Thus, in the game $\Gamma(M)$ we assume that the users' types are fixed and their strategic choices are restricted to selecting security levels, and we study the Nash equilibrium of this game.

Definition 2 (Nash Equilibrium with Fixed M). *Consider the game $\Gamma(M)$ in which the number of malicious users M is given. A strategy profile $(s_1, \ldots, s_N)$ is a Nash equilibrium if security level s_i is a best response for every user.*

Lemma 1. *In any equilibrium of the game $\Gamma(M)$, for each user type, security choices are identical.*

Proof. First, we notice that for any M, malicious users do not invest in security. From the definition of malicious user utilities (3), they have no losses, and thus have no incentive to invest in security: thus, for any M, it is optimal to choose $s_j^*(M) = 0$ for every malicious user j.

Second, we show that every honest user has a unique best response, and this best response is independent of user identity, which means that any equilibrium is symmetric. Consider some $\mathbf{s} = (\cdot, s_{-i})$. To find user i's optimal security (i.e., the utility maximizing security s_i), we take the first derivative of (2) with respect to s_i (user i FOC):

$$\frac{d}{ds_i} u_i = -\frac{d}{ds_i} B_i(s_i, s_{-i}) \Delta_0 - h'(s_i) = 0, \tag{25}$$

where we use B_i given by (14)

$$\frac{d}{ds_i} B_i(s_i, s_{-i}) = \frac{d}{ds_i} \left(1 - s_i \left[1 - q(N)N \left(1 - \frac{\sum_{j \neq i} s_j}{N} \right) \right] \right) \tag{26}$$

$$= -\left[1 - q(N)N \left(1 - \frac{\sum_{j \neq i} s_j}{N} \right) \right]. \tag{27}$$

Since the second order condition (SOC) is negative:

$$\frac{d^2}{ds_i^2} u_i = -h''(s_i) < 0,$$

there exists a unique optimal response s_i^* to any $s_i^* = s_i^*(M, s_{-i})$, and it is given by the solution of FOC (25).

For large N, we have:

$$\frac{d}{ds_i} u_i = -\frac{d}{ds_i} B_i(s_i, s_{-i}) \Delta_0 - h'(s_i) \tag{28}$$

$$= \underbrace{\left[1 - \underbrace{q_\infty \left(1 - \frac{N - M}{N} \bar{s}_H \right)}_{<1} \right] \Delta_0}_{>0} - h'(s_i). \tag{29}$$

Since $h'(s_i)$ is increasing in s_i, the derivative u' is a decreasing function of s_i. Furthermore, since the first term is positive and $h'(0) = 0$, the derivative u' is positive at $s_i = 0$. Consequently, user i best response s_i^* is interior (because $s_i = 1$ cannot be optimal as it is unaffordable), and it is given by:

$$u' = 0 \tag{30}$$

$$\left[1 - q_\infty \left(1 - \frac{N - M}{N} \bar{s}_H \right) \right] \Delta_0 - h'(s_i) = 0 \tag{31}$$

$$\Delta_0 = \frac{h'(s_i)}{1 - q_\infty \left(1 - \frac{N-M}{N} \bar{s}_H \right)}. \tag{32}$$

Finally, since the solution of (32) is independent of user identity, we infer that best responses are identical for all honest users.   □

From Lemma 1, we infer that the honest users' security levels are identical in an equilibrium. The following theorem shows that the equilibrium security level always exists, and is unique. This implies that there is a unique Nash equilibrium of the game $\Gamma(M)$.

Theorem 1. *For a given M, the honest users' equilibrium security $s^*(M)$ is unique.*

Proof. By definition, identical security level s is an equilibrium if and only if security level s is a best response for every honest user. Consequently, it follows from the proof of Lemma 1 that an identical security level s is an equilibrium if and only if

$$\Delta_0 = R(s, M), \tag{33}$$

where

$$R(s, M) = \frac{h'(s)}{1 - q_\infty + q_\infty \frac{N-M}{N} s}. \tag{34}$$

In order to prove the claim of the theorem, we have to show that Eq. (33) has a unique solution.

First, notice that

$$R(0, M) = 0 \tag{35}$$

since $h'(0) = 0$, and

$$R(1, M) = \infty \tag{36}$$

since $h(s)$ grows without bound as s approaches 1. Therefore, there must exist a value s^* between 0 and 1 for which $R(s^*, M) = \Delta_0$ as $R(s, M)$ is a continuous function on $[0, 1)$.

To prove that this s^* exists uniquely, it suffices to show that $\frac{d}{ds} R(s, M) > 0$ on $(0, 1)$. The first derivative of $R(s, M)$ with respect to s is

$$\frac{d}{ds} R(s, M) = \frac{h''(s) \left[1 - q_\infty + q_\infty \frac{N-M}{N} s\right] - h'(s) q_\infty \frac{N-M}{N}}{\left[1 - q_\infty + q_\infty \frac{N-M}{N} s\right]^2}. \tag{37}$$

Since the denominator is always positive, we only have to show that the numerator is positive on $(0, 1)$. First, observe that the numerator is non-negative at $s = 0$, since

$$\underbrace{h''(0)}_{\geq 0} \underbrace{\left[1 - q_\infty + q_\infty \frac{N-M}{N} s\right]}_{>0} - \underbrace{h'(0)}_{=0} q_\infty \frac{N-M}{N} \geq 0. \tag{38}$$

Finally, we prove that the numerator is strictly increasing on $[0,1)$ by showing that its first derivative with respect to s is positive:

$$\frac{d}{ds}\left(h''(s)\left[1 - q_\infty + q_\infty \frac{N-M}{N}s\right] - h'(s)q_\infty\frac{N-M}{N}\right)$$

$$= h'''(s)\left[1 - q_\infty + q_\infty\frac{N-M}{N}s\right] + h''(s)q_\infty\frac{N-M}{N}$$

$$- h''(s)q_\infty\frac{N-M}{N} \tag{39}$$

$$= \underbrace{h'''(s)}_{>0}\underbrace{\left[1 - q_\infty + q_\infty\frac{N-M}{N}s\right]}_{>0} \tag{40}$$

$$>0. \tag{41}$$

Since the numerator is non-negative at $s = 0$ and it is strictly increasing in s on $[0,1)$, it must be positive for any $s \in (0,1)$. Therefore, the first derivative of $R(s,M)$ is also positive, which proves that the solution s^* exists uniquely for a given number of malicious users M.   □

Equilibrium in the game $\Gamma(M)$ exists and is unique. This allows us to define the equilibrium security level as a function $s^*(M)$ of M.

Theorem 2. *As the number of malicious users M increases, the honest users' equilibrium security $s^*(M)$ decreases.*

Proof. Since $\Delta_0 = R(s^*(M), M)$ must hold for every pair $(s^*(M), M)$ (see Eq. (33)), we have

$$0 = \frac{d}{dM}R(s^*(M), M) \tag{42}$$

$$0 = \frac{h''(s^*(M))s^{*\prime}(M)\left(1 - q_\infty + q_\infty\frac{N-M}{N}s^*(M)\right)}{\left(1 - q_\infty + q_\infty\frac{N-M}{N}s^*(M)\right)^2}$$

$$- \frac{h'(s^*(M))q_\infty\left(\frac{-1}{N}s^*(M) + \frac{N-M}{N}s^{*\prime}(M)\right)}{\left(1 - q_\infty + q_\infty\frac{N-M}{N}s^*(M)\right)^2} \tag{43}$$

$$- h'(s^*(M))q_\infty\left(\frac{-1}{N}s^*(M) + \frac{N-M}{N}s^{*\prime}(M)\right) \tag{44}$$

$$0 = s^{*\prime}(M)\left[h''(\hat{s}(M))\left(1 - q_\infty + q_\infty\frac{N-M}{N}s^*(M)\right) - h'(s^*(M))q_\infty\frac{N-M}{N}\right]$$

$$- h'(s^*(M))q_\infty\frac{-1}{N}s^*(M) \tag{45}$$

$$s^{*\prime}(M) = \frac{h'(s^*(M))q_\infty\frac{1}{N}s^*(M)}{h'(s^*(M))q_\infty\frac{N-M}{N} - h''(s^*(M))\left(1 - q_\infty + q_\infty\frac{N-M}{N}s^*(M)\right)}. \tag{46}$$

Notice that the denominator of the above fraction is the inverse of the numerator of the right-hand side of Eq. (37). Since we have shown in the proof of Theorem 1 that the numerator of the right-hand side of Eq. (37) is positive, we have that the denominator of the above fraction is negative. Further, the numerator of the above fraction is obviously positive since it consists of only positive factors. Hence, $s^{*'}(M)$ is negative, which proves that the honest users' equilibrium security decreases as the number of malicious users increases.   □

Unfortunately, $s^*(M)$ cannot be expressed in closed form. Nonetheless, we can easily find $s^*(M)$ numerically for any M. On the other hand, we can express the number of malicious users as a function $M(s^*)$ of the equilibrium security level s^* in closed form:

$$\Delta_0 = \frac{h'(s^*)}{1 - q_\infty + q_\infty \frac{N-M}{N} s^*} \tag{47}$$

$$q_\infty \frac{N-M}{N} s^* = \frac{h'(s^*)}{\Delta_0} + q_\infty - 1 \tag{48}$$

$$M(s^*) = N \left[1 - \frac{\frac{h'(s^*)}{\Delta_0} + q_\infty - 1}{q_\infty s^*} \right]. \tag{49}$$

The value of $M(s^*)$ can be interpreted as the number of malicious users which induces the honest users to choose security $s^*(M)$. Note that from Theorem 2, we readily have that $M(s^*)$ is a decreasing function of s^*.

4.3   Incentives for Becoming Malicious

In the previous subsection, we studied a restricted version of our game $\Gamma(M)$, in which the number of malicious users was exogenously given. We found the equilibrium of the game $\Gamma(M)$ as the solution of (34), from which the honest users' equilibrium security levels can be found.

Next, we will study the game Γ, in which users choose their types (honest or malicious). We will solve the game Γ by building on the results of the previous subsection.

First, Theorem 1 provides the honest users' equilibrium security level $s^*(M)$. Thus, we can express a malicious user's gain as a function $G_i(M)$ of the number of malicious users M:

$$G_i(M) = \frac{\sum_{j \in \text{honest users}} B_j(s_j, \bar{s}_H) L}{M} \tag{50}$$

$$= \frac{(N - M)\left(1 - s^*(1 - q_\infty) - s^{*2} \frac{N-M}{N} q_\infty\right) L}{M}. \tag{51}$$

From Theorem 1, honest users choose $s^*(M)$ in an equilibrium. Next, we will find an equilibrium number of malicious users of M. For this purpose, we have to determine the combinations of M and $s^*(M)$ that form a strategy profile such that being malicious is a best response for malicious users and being honest is a best response for honest users.

Finally, now we are ready to prove that there always exists an equilibrium of the game in which users self-select their types (honest or malicious). Effectively, for each equilibrium number of malicious users M, the equilibrium security choices will be identical to equilibrium security $s^*(M)$ in the game $\Gamma(M)$ with that same fixed number of malicious users M.

Theorem 3. *There exists at least one Nash equilibrium.*

Proof. Assume the reverse. Then, at any $M \in [0, N-1]$ there exists (i) malicious or (ii) honest user, for whom a deviation to the opposite user type is profitable:

$$v(M, s^*(M)) < u|_{M-1, s_{-i}=s^*(M)} := \max_{s_i} u_i(M-1, s_i, s_{-i}), \text{ (i)} \qquad (52)$$

or

$$u(M, s^*(M)) < v(M+1, s^*(M)), \text{ (ii)} \qquad (53)$$

where $v(M, s^*(M))$ and $u(M, s^*(M))$ denote, respectively, the malicious and honest users' utility with M malicious users and all honest users choosing security $s^*(M)$, and $u_i(M, s_i, s_{-i})$ denotes honest user i's utility given that he chooses security s_i and all other honest users choose s_{-i}. From Lemma 1, the honest users' best response to M and $s_{-i} = s^*(M)$ is $s^*(M)$, which gives:

$$u|_{M, s^*(M)} \leq u(M, s^*(M)). \qquad (54)$$

From Theorem 2, $s^*(M)$ decreases in M, which gives:

$$v(M+1, s^*(M)) < v(M+1, s^*(M+1)), \qquad (55)$$

because ceteris paribus, lower security benefits malicious users. Similarly, we have from Theorem 2 and (54) that:

$$u|_{\tilde{M}, s^*(\tilde{M}+1)} < u|_{\tilde{M}, s^*(\tilde{M})} \leq u(M, s^*(M)) \qquad (56)$$

because ceteris paribus, higher security benefits honest users.

Let (52) hold[3] for any $M > \tilde{M}$, but not for $\tilde{M}$. Hence, at $\tilde{M}+1$ we have:

$$v(\tilde{M}+1, s^*(\tilde{M}+1)) < u|_{\tilde{M}, s^*(\tilde{M}+1)}. \qquad (57)$$

Then, if $\tilde{M}$ is not an equilibrium, (53) must hold:

$$u(\tilde{M}, s^*(\tilde{M})) < v(\tilde{M}+1, s^*(\tilde{M})). \qquad (58)$$

Combining (58) and (57) with (55) and (56) provides:

$$u(\tilde{M}, s^*(\tilde{M})) < v(\tilde{M}+1, s^*(\tilde{M})) < v(\tilde{M}+1, s^*(\tilde{M}+1)) \qquad (59)$$

$$v(\tilde{M}+1, s^*(\tilde{M}+1)) < u|_{\tilde{M}, s^*(\tilde{M}+1)} \leq u(\tilde{M}, s^*(\tilde{M})), \qquad (60)$$

which contradict each other. Thus, Theorem 3 is proven. □

---

[3] If (52) holds for all $M \in [1, N-1]$, we let $\tilde{M} = 0$.

5   Numerical Illustrations

Here, we present present numerical results showcasing our model and illustrating our theoretical findings. First, we instantiate our model using the following parameter values:

- number of users $N = 500$,
- initial wealth $W = 100$,
- potential loss $L = 30$,
- security interdependence $q_\infty = 0.5$,
- probability of a malicious user getting caught $\mu = 0.2$,

and we use the following security-cost function (see Fig. 1):

$$h(s) = 10\frac{s^2}{\sqrt{1-s}} \tag{61}$$

and the following utility function:

$$U(x) = x^{0.9}. \tag{62}$$

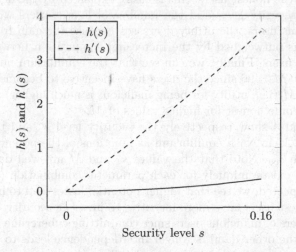

Fig. 1. The security-cost function $h(s)$ and its first derivative $h'(s)$ used in the numerical illustrations.

Figure 2 shows the honest users' equilibrium security level $s^*(M)$ as a function of M. Furthermore, it also shows the honest and malicious users' utilities u and v for these equilibrium security levels (i.e., utilities given that there are M malicious users and the honest users choose $s^*(M)$). We see that – as established by Theorem 2 – the equilibrium security level is a strictly decreasing function of the number of malicious users. Moreover, we see that the utilities are also strictly

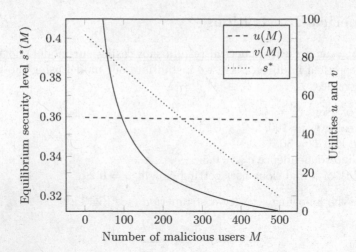

Fig. 2. The equilibrium security level s^* and the resulting utilities u and v for honest and malicious users as functions of the number of malicious users M. Please note the different scalings of the vertical axes.

decreasing. For the honest users, this is easily explained by the decrease in both the individual security level and the number of honest users who contribute. For the malicious users, the utility decreases because the gain from decreasing security levels is outweighed by the increasing competition between more and more malicious users. Finally, we can see that the equilibrium number of malicious users is at $M = 96$ since the users have incentive to become malicious for lower values of M (i.e., utility for being malicious is much higher) and they have incentive to become honest for higher values of M.

Figures 3 and 4 show respectively the security level s^* and the number of malicious users $\hat{M}$ in Nash equilibrium as functions of the potential loss L and interdependence q_∞. Note that the values s^* and $\hat{M}$ are well defined because the equilibrium exists uniquely for each parameter combination (q_∞, L) in this example. As expected, we see that higher potential losses lead to higher security levels since honest users have more incentive to invest in security, and they lead to higher numbers of malicious users since committing cybercrime becomes more profitable. On the other hand, stronger interdependence leads to lower security levels since the honest users' breach probabilities becomes less dependent on their own security levels, which disincentivizes investing or making an effort. Conversely, stronger interdependence leads to higher numbers of malicious users since propagating security breaches becomes easier, which makes cybercrime more profitable.

Figure 5 shows the security level s^* and the number of malicious users $\hat{M}$ in Nash equilibrium as functions of the probability μ of a malicious user getting caught. Note that the values s^* and $\hat{M}$ are again well defined because the equilibrium exists uniquely for each parameter value μ in this example. As expected,

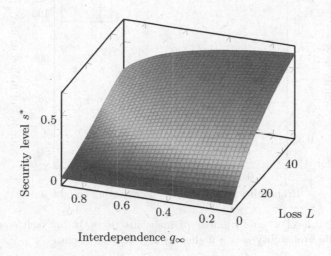

Fig. 3. Security level s^* in Nash equilibrium as a function of potential loss L and interdependence q_∞.

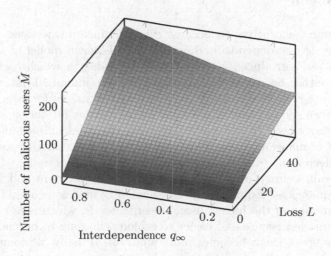

Fig. 4. Number of malicious users $\hat{M}$ in Nash equilibrium as a function of potential loss L and interdependence q_∞.

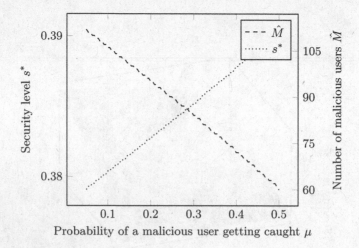

Fig. 5. Security level s^* and number of malicious users $\hat{M}$ in Nash equilibrium as functions of the probability μ of a malicious user getting caught.

we see that a higher probability of getting caught disincentivizes users from engaging in cybercrime and reduces the number of malicious users. On the other hand, the probability of getting caught has an almost negligible effect on the honest users security level.

6   Conclusion

We studied users' incentives to become cybercriminals in networks where the users' security is interdependent. Based on a well-known model of interdependent security, we introduced a game-theoretic model, in which each user can choose to be either honest or malicious (i.e., cybercriminal). First, we showed how to compute security-breach probabilities in this model for large-scale networks. Then, we showed that if users are homogeneous, all honest users select the same security level in an equilibrium, and this level exists uniquely for a fixed number of malicious users. Furthermore, we found that this security level is a strictly decreasing function of the number of malicious users, which means that the overall security of a network drops rapidly as more and more users choose to be malicious. Equivalently, the number of malicious users is a strictly decreasing function of the honest users' security levels, which is not surprising: as users become less secure and easier to exploit, choosing to be malicious and taking advantage of them becomes more profitable. Finally, we found that the game always has a Nash equilibrium.

There are multiple directions for extending our current work. Firstly, we plan to study heterogeneous users, who may have different initial wealth, probability of getting caught, etc. While our current model, which assumes homogeneous users, is very useful for studying how the users' choices are affected by changing various parameters, a heterogeneous-user model will enable us to study the differences between individual users' choices. We conjecture that even though users may choose different security levels, their equilibrium security levels will decrease as the number of malicious users increases. Secondly, we plan to extend our current model by considering cyber-insurance, that is, by allowing users to purchase cyber-insurance policies in addition to investing in security. In practice, the adoption of cyber-insurance is growing rapidly as the market size is estimated to increase from $2.5 billion in 2015 to $7.5 billion in 2020 [28]. Consequently, users' security choices are increasingly affected by the availability of cyber-insurance. We conjecture that increasing the number of malicious users will have an opposite effect on cyber-insurance as compared to security investments: decreasing security levels will result in increasing adoption of cyber-insurance.

Acknowledgment. This work was supported in part by FORCES (Foundations Of Resilient CybEr-Physical Systems), which receives support from the National Science Foundation (NSF award numbers CNS-1238959, CNS-1238962, CNS-1239054, CNS-1239166).

References

1. Acemoglu, D., Malekian, A., Ozdaglar, A.: Network security and contagion. Working Paper 19174, National Bureau of Economic Research. http://www.nber.org/papers/w19174
2. Anderson, R., Moore, T.: The economics of information security. Science **314**(5799), 610–613 (2006)
3. Anderson, R., Barton, C., Böhme, R., Clayton, R., Van Eeten, M.J., Levi, M., Moore, T., Savage, S.: Measuring the cost of cybercrime. In: Böhme, R. (ed.) The Economics of Information Security and Privacy, pp. 265–300. Springer, Heidelberg (2013)
4. Asghari, H., Van Eeten, M., Arnbak, A., Van Eijk, N.: Security economics in the HTTPS value chain. In: 12th Workshop on the Economics of Information Security (WEIS) (2013)
5. Aspnes, J., Chang, K., Yampolskiy, A.: Inoculation strategies for victims of viruses and the sum-of-squares partition problem. In: Proceedings of the 16th Annual ACM-SIAM Symposium on Discrete Algorithms (SODA), pp. 43–52. SIAM (2005)
6. Aspnes, J., Chang, K., Yampolskiy, A.: Inoculation strategies for victims of viruses and the sum-of-squares partition problem. J. Comput. Syst. Sci. **72**(6), 1077–1093 (2006)
7. Fultz, N., Grossklags, J.: Blue versus red: towards a model of distributed security attacks. In: Dingledine, R., Golle, P. (eds.) FC 2009. LNCS, vol. 5628, pp. 167–183. Springer, Heidelberg (2009). doi:10.1007/978-3-642-03549-4_10
8. Grossklags, J., Christin, N., Chuang, J.: Secure or insure?: a game-theoretic analysis of information security games. In: Proceedings of the 17th International Conference on World Wide Web (WWW), pp. 209–218. ACM (2008)
9. Hausken, K.: Income, interdependence, and substitution effects affecting incentives for security investment. J. Account. Public Policy **25**(6), 629–665 (2006)
10. Heal, G., Kunreuther, H.: Interdependent security: a general model. Technical report, Working Paper 10706, National Bureau of Economic Research (2004)

11. Heal, G., Kunreuther, H.: Modeling interdependent risks. Risk Anal. **27**(3), 621–634 (2007)
12. Honeyman, P., Schwartz, G., Assche, A.V.: Interdependence of reliability and security. In: 6th Workshop on the Economics of Information Security (WEIS) (2007)
13. Johnson, B., Grossklags, J., Christin, N., Chuang, J.: Uncertainty in interdependent security games. In: Alpcan, T., Buttyán, L., Baras, J.S. (eds.) GameSec 2010. LNCS, vol. 6442, pp. 234–244. Springer, Heidelberg (2010). doi:10.1007/978-3-642-17197-0_16
14. Johnson, B., Laszka, A., Grossklags, J.: The complexity of estimating systematic risk in networks. In: Proceedings of the 27th IEEE Computer Security Foundations Symposium (CSF), pp. 325–336 (2014)
15. Khouzani, M.R., Sen, S., Shroff, N.B.: An economic analysis of regulating security investments in the internet. In: Proceedings of the 32nd IEEE International Conference on Computer Communications (INFOCOM), pp. 818–826. IEEE (2013)
16. Knowles, W., Prince, D., Hutchison, D., Disso, J.F.P., Jones, K.: A survey of cyber security management in industrial control systems. Int. J. Crit. Infrastruct. Prot. **9**, 52–80 (2015)
17. Konradt, C., Schilling, A., Werners, B.: Phishing: an economic analysis of cybercrime perpetrators. Comput. Secur. **58**, 39–46 (2016). http://www.sciencedirect.com/science/article/pii/s0167404815001844
18. Kraemer-Mbula, E., Tang, P., Rush, H.: The cybercrime ecosystem: online innovation in the shadows? Technol. Forecast. Soc. Change **80**(3), 541–555 (2013). Future-Oriented Technology Analysis. http://www.sciencedirect.com/science/article/pii/S0040162512001710
19. Kunreuther, H., Heal, G.: Interdependent security. J. Risk Uncertain. **26**(2–3), 231–249 (2003)
20. Laszka, A., Felegyhazi, M., Buttyan, L.: A survey of interdependent information security games. ACM Comput. Surv. **47**(2), 1–38 (2014)
21. Laszka, A., Johnson, B., Grossklags, J., Felegyhazi, M.: Estimating systematic risk in real-world networks. In: Proceedings of the 18th International Conference on Financial Cryptography and Data Security (FC), pp. 417–435 (2014)
22. Levchenko, K., Pitsillidis, A., Chachra, N., Enright, B., Félegyházi, M., Grier, C., Halvorson, T., Kanich, C., Kreibich, C., Liu, H., et al.: Click trajectories: end-to-end analysis of the spam value chain. In: Proceedings of the 32nd IEEE Symposium on Security and Privacy (S&P), pp. 431–446. IEEE (2011)
23. Moscibroda, T., Schmid, S., Wattenhofer, R.: When selfish meets evil: Byzantine players in a virus inoculation game. In: Proceedings of the 25th Annual ACM Symposium on Principles of Distributed Computing (PODC), pp. 35–44. ACM (2006)
24. Öğüt, H., Menon, N., Raghunathan, S.: Cyber insurance and IT security investment: impact of interdependence risk. In: 4th Workshop on the Economics of Information Security (WEIS) (2005)
25. Öğüt, H., Raghunathan, S., Menon, N.: Cyber security risk management: public policy implications of correlated risk, imperfect ability to prove loss, and observability of self-protection. Risk Anal. **31**(3), 497–512 (2011)
26. Olson, M.: The Rise and Decline of Nations: Economic Growth, Stagflation, and Social Rigidities. Yale University Press, New Haven (2008)
27. Olson, M.: The logic of Collective Action, vol. 124. Harvard University Press, Cambridge (2009)
28. PricewaterhouseCoopers: Insurance 2020 & beyond: reaping the dividends of cyber resilience (2015). http://www.pwc.com/insurance. Accessed 16 June 2016

29. Schwartz, G.A., Sastry, S.S.: Cyber-insurance framework for large scale interdependent networks. In: Proceedings of the 3rd International Conference on High Confidence Networked Systems (HiCoNS), pp. 145–154. ACM (2014)
30. Symantec: Emerging threat: Dragonfly/Energetic Bear - APT group. Symantec Connect. http://www.symantec.com/connect/blogs/emerging-threat-dragonfly-energetic-bear-apt-group. Accessed 16 Feb 2016
31. Tullock, G.: The welfare costs of tariffs, monopolies, and theft. Econ. Inq. **5**(3), 224–232 (1967)
32. Varian, H.: System reliability and free riding. In: Camp, L.J., Lewis, S. (eds.) Economics of Information Security, pp. 1–15. Springer, New York (2004)

A Logic for the Compliance Budget

Gabrielle Anderson[1], Guy McCusker[2(✉)], and David Pym[1]

[1] University College London, London, UK
gabrielle.anderson@cantab.net, d.pym@ucl.ac.uk
[2] University of Bath, Bath, UK
g.a.mccusker@bath.ac.uk

Abstract. Security breaches often arise as a result of users' failure to comply with security policies. Such failures to comply may simply be innocent mistakes. However, there is evidence that, in some circumstances, users choose not to comply because they perceive that the security benefit of compliance is outweighed by the cost that is the impact of compliance on their abilities to complete their operational tasks. That is, they perceive security compliance as hindering their work. The 'compliance budget' is a concept in information security that describes how the users of an organization's systems determine the extent to which they comply with the specified security policy. The purpose of this paper is to initiate a qualitative logical analysis of, and so provide reasoning tools for, this important concept in security economics for which quantitative analysis is difficult to establish. We set up a simple temporal logic of preferences, with a semantics given in terms of histories and sets of preferences, and explain how to use it to model and reason about the compliance budget. The key ingredients are preference update, to account for behavioural change in response to policy change, and an ability to handle uncertainty, to account for the lack of quantitative measures.

1 Introduction

The security of systems is not simply a technical problem. While encryption, robust protocols, verified code, and network defences are critical aspects of system security, the behaviour of system managers and users, and the policies that are intended to manage their behaviour, are also of critical importance.

Many security breaches are the result of users' failure to comply with security policies. Failure to comply may simply be the result of a mistake, because of a misunderstanding, or derive from users' being required to form an effectively impossible task.

In recent years, many effective tools for analysing security behaviour and investments have been provided by economics, beginning with significant work by Anderson and Moore [2,3], explaining the relevance of economics to information security, and Gordon and Loeb [10,11], considering optimal investment in information security. Since then, there has been a vast development in security economics, too extensive to survey in this short paper. Game theory and decision theory have been significant parts of this development; see, for example, [1,20],

© Springer International Publishing AG 2016
Q. Zhu et al. (Eds.): GameSec 2016, LNCS 9996, pp. 370–381, 2016.
DOI: 10.1007/978-3-319-47413-7_21

and much more besides. Some of us have contributed to the use of methods from economics to assess the role of public policy in the management of information security [16] and in system management policy [14,15].

A key aspect of the management of system security policies concerns the relationship between the human users of systems and the security policies with which they are expected to comply. This relationship has been explored, in the context of security economics, by Beautement et al. [5,6] through the concept of the *compliance budget*. The idea here is that users have a limited appetite for engagement in the behaviour that is required in order to ensure compliance with policy if that behaviour detracts from their primary operational tasks.

In Sect. 2, we explain the concept of the compliance budget as introduced in [5,6], building on earlier work in [7]. In Sect. 3, we introduce a simple temporal logic with a semantics that is based on histories of events and agents' preferences. In Sect. 4, we consider an example of how agents' behaviour can be understood in terms of the compliance budget and reasoned about logically. In Sect. 5, we consider our model of the compliance behaviour in the context of incomplete information, and briefly set out a programme of further work.

This paper is intended to be conceptual rather than technical in nature and, to this end, we deliberately employ a slightly informal style. Its purpose is to initiate a qualitative logical analysis of an important concept in security economics for which quantitative analysis is difficult to establish. We are not aware of any related work on logical analyses of the compliance budget or similar concepts.

This work was supported by UK EPSRC EP/K033042/1 and EP/K033247/1.

2   The Compliance Budget

Organizations' security policies are enforced using tools of different kinds, ranging from simple instructions from managers through to complex combinations of hardware, software, and tokens. For example, access control via 'something you are, something you have, and something you know'. In situations in which non-compliance with the policy is possible, most of an organization's employees will nevertheless comply provided compliance does not require additional effort.

If extra effort is required, individuals will weigh this extra effort, and the opportunity cost that it implies in terms of their production task, against the benefits they obtain from compliance. If there is good alignment (i.e., of incentives) between the individual's goals as an employee and the organization's goals, then there will be little or no conflict as the behaviour required from the individual for compliance causes no friction.

However, most individuals will tend not to choose to comply with the security behaviour required by an organization if that behaviour conflicts with the behaviour that they perceive to be required in order to achieve their own goals. In such a situation, goals are less likely to be met and effort is likely to be wasted. This benefit-cost analysis is illustrated in Fig. 1.

Alternative rates of compliance expenditure are also shown for comparison. Once the compliance threshold is crossed security effectiveness drops sharply as

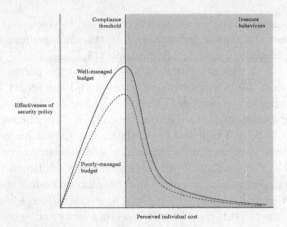

Fig. 1. The relationship between the perceived cost to an individual of compliance and the effectiveness of a security policy [5,6]. More effective policies achieve greater effectiveness at a given cost to an individual.

employees elect to complete tasks that benefit them more directly as individuals rather than security tasks that more benefit the organization as a whole. A well-managed budget will spend perceived effort at a slower rate, so that more security policies can be accommodated before the compliance threshold is reached, resulting in a higher level of achieved security. If the limit is exceeded, security policies are less likely to be followed; achieved levels of security will decline.

Following [5,6], we remark that, in the absence of quantitative data, the precise shape of the graph in Fig. 1 cannot be plotted precisely. Moreover, there will be variations from individual to individual, although the same core features will occur. These behaviours have been investigated in extensive empirical studies [5–7], supporting the formulation of the concept of *the compliance budget*.

3   A Logic for the Compliance Budget

In this section, we introduce a (multi-modal) temporal logic of preferences with which we can reason about the compliance budget. For convenience, we name the logic CBL, for 'compliance budget logic'.

The logic includes temporal modalities, modalities for agents' preferences, and a modality for preference update [4]. Each modality has a specific role in our modelling of reasoning about the compliance budget.

- The temporal modalities, $\bigcirc$ (next time step) and $\mathcal{U}$ (until) are familiar from temporal logic [19] (see [13] for a tutorial exposition) and are used to reason about how properties of the system change over time.
- The modality $\Diamond_i$, together with its dual $\Box_i$, is used to reason about the preferences of the agents (or principals, or players) i in a system. It denotes

decision-making capability between outcomes with different 'worth' to the agents in the system.

– The modality $[!\Phi]$, for a finite set of formulae Φ, is reminiscent of the key modality in public announcement logic (see, e.g., [9] for a convenient summary); it is used to reason about how the preferences of agents in the system are changed by the imposition of policies by those who manage the system.

The semantics of the logic is specified in terms of history-based structures [18], and is explained below. Histories (sequences of events) can be used to represent the trace semantics of complex systems models, and can be seen as a simple version of the process-theoretic models explored in, for example, [8].

Definition 1 (Syntax of CBL). *Given a set of propositional variables* $\mathbf{P}$*, with elements* p, q*, etc., the syntax of the logic CBL is defined as follows:*

$$\phi ::= \mathrm{p} \mid \perp \mid \top \mid \neg\phi \mid \phi \vee \phi \mid \phi \wedge \phi \mid \phi \rightarrow \phi \quad \textit{classical propositionals}$$
$$\mid \bigcirc\phi \mid \phi\, \mathcal{U}\, \phi \qquad\qquad\qquad\qquad \textit{temporal modalities}$$
$$\mid \Diamond_i\phi \mid \Box_i\phi \mid [!\Phi]\phi \qquad\qquad\quad \textit{preference modalities.}$$

We write formulae as ϕ, ψ, etc., and finite sets of formulae as Φ, Ψ, etc. The existential preference modal operator for agent i is $\Diamond_i\phi$. The dual universal preference modal operator for agent i is $\Box_i\phi$. When there is only a single agent in the system, we sometimes drop the agent annotation. The temporal 'next-time' operator is $\bigcirc\phi$. The temporal 'until' operator is $\phi\,\mathcal{U}\,\psi$.

The preference update modality—which updates agents' preferences—is written $[!\Phi]\phi$ and denotes that ϕ holds when the model is updated to disregard preferences between pairs of histories that (respectively) do and do not satisfy some formula in Φ. We refer to a formula as *update-free* if it contains no uses of the preference update modality.

The compliance budget [5,6] is qualitative rather than quantitative concept, and accepts that accurate measures of the effort taken to follow a given policy, and the effort that an employee has to expend, can generally not be practically obtained. As a result, a preference update consists of a set of formulae according to which the preferences are updated without any formal notion of likelihood or probability between the different facts; that is, it is a qualitative update to preferences rather than a quantitative update. The preference-update modality $[!\Phi]$ will be used to give a logical account of the behavioural changes brought about by the implementation of a policy. The set of formulae Φ represents the impact on agents' decision-making under a new policy; we allow Φ to be a finite set rather than a single formula in order to incorporate uncertainty in this decision-making impact. This set-up will be essential to our logical description of the compliance budget.

First, we need some notation. Let $\mathcal{E}$ be the set of events (the 'global event set of a model') and $\mathcal{A}$ be the set of agents of a history-based model.

A history H over a set of events $\mathcal{E}$ is a possibly infinite sequence of events drawn from the set $\mathcal{E}$. ϵ denotes the empty history.

If a history H is of at least length $m \in \mathbb{N}$, then let $H(m)$ be the mth element of the sequence, H_m be the m-length prefix of the history. We emphasize that a history is finite using lower case. Let $h; H$ denote the concatenation of a finite history h with a (possibly infinite) history H. In this case, we say that h is a prefix of $h; H$. (Note that $\epsilon; H = H$.)

A *protocol* $\mathcal{H}$ is a set of histories closed under finite prefix.

Definition 2. *A preference relation $\prec$ is a strict partial order on a protocol.*

In a system with multiple agents, we use a different preference relation for each agent, to describe their separate interests. Such a collection of preferences is specified as a preference structure.

Definition 3 (Preference structure). *A preference structure for agents $\mathcal{A}$ over histories $\mathcal{H}$ is given by a tuple $(\prec_1, \ldots, \prec_n)$, where $\mathcal{A} = \{1, \ldots, n\}$, and, for all $i \in \mathcal{A}$, $\prec_i$ is a preference relation on the protocol $\mathcal{H}$.*

We write preference structures π, π', etc., and sets of preference structures Π, Π', etc.

$$H, t \models_\mathcal{M} \mathrm{p} \text{ iff } H_t \text{ is defined and } H_t \in \mathcal{V}(\mathrm{p})$$
$$H, t \models_\mathcal{M} \bot \text{ never} \quad H, t \models_\mathcal{M} \top \text{ always} \quad H, t \models_\mathcal{M} \neg \phi \text{ iff } H, t \not\models_\mathcal{M} \phi$$
$$H, t \models_\mathcal{M} \phi \vee \psi \text{ iff } H, t \models_\mathcal{M} \phi \text{ or } H, t \models_\mathcal{M} \psi \quad H, t \models_\mathcal{M} \phi \wedge \psi \text{ iff } H, t \models_\mathcal{M} \phi \text{ and } H, t \models_\mathcal{M} \psi$$

$$H, t \models_\mathcal{M} \bigcirc \phi \quad \text{iff } H, t+1 \models_\mathcal{M} \phi$$
$$H, t \models_\mathcal{M} \phi \, \mathcal{U} \, \psi \text{ iff there exists } k \in \mathbb{N} \text{ such that } t \leq k, \, H, k \models_\mathcal{M} \psi$$
$$\text{and, for all } l \in \mathbb{N}, \, t \leq l < k \text{ implies } H, l \models_\mathcal{M} \phi$$

$$H, t \models_\mathcal{M} \Diamond_i \phi \quad \text{iff there exist } H' \in \mathcal{H} \text{ and } \pi \in \Pi$$
$$H \pi_i H', \text{ and, } H', t \models_\mathcal{M} \phi$$
$$H, t \models_\mathcal{M} \Box_i \phi \quad \text{iff for all } H' \in \mathcal{H} \text{ and all } \pi \in \Pi,$$
$$H \pi_i H' \text{ implies } H', t \models_\mathcal{M} \phi$$
$$H, t \models_\mathcal{M} [!\Phi] \phi \text{ iff } H, t \models_{\mathcal{M}[!\Phi, t]} \phi$$

Fig. 2. Satisfaction relation

We can now define models of CBL. Satisfaction and model update are defined mutually inductively.

Definition 4 (History-based preference models). *A tuple $(\mathcal{E}, \mathcal{A}, \mathcal{H}, \mathcal{V}, \Pi)$ is a history-based preference model (HBPM), or a* history-based model *for short, if $\mathcal{E}$ is a set of events, $\mathcal{A} = \{1, \ldots, n\}$ is a set of agents, $\mathcal{H}$ is a protocol, $\mathcal{V}$ is a valuation function from propositions to subsets of $\mathcal{H}$ containing only finite histories, and Π is a set of preference structures for agents $\mathcal{A}$ over $\mathcal{H}$.*

Models are denoted $\mathcal{M}$, $\mathcal{M}'$, etc. The interpretation of the connectives and modalities is given in Fig. 2, where satisfaction of a formula ϕ in history H at time t in a model M is written $H, t \vDash_{\mathcal{M}} \phi$. Note that the semantics of preference update depends on the definition of preference-based model update, which is explained below. The necessary model update, as defined in Definition 6, requires only the strictly smaller formula ϕ.

The modality $\Diamond_i \phi$ denotes the existence of a history (trace) that is preferred by agent i in some possible preference relation and in which ϕ holds. The modality $\bigcirc \phi$ denotes that ϕ holds at the next time point. The modality $\phi \mathcal{U} \psi$ denotes that ϕ holds until some time point, at which ψ holds.

In order to reason about the impact of a policy, it is helpful to be able to modify the preferences of the principals in the logic. This can be modelled using preference updates, which can remove (but cannot add) preferences between pairs of histories. A preference update is performed using a *distinguishing formula*, ϕ. Given two histories H, H', if $H, t \vDash_{\mathcal{M}} \phi$ but $H', t \nvDash_{\mathcal{M}} \phi$, then we call ϕ a 'distinguishing formula' for (H, t) and (H', t). In this case, preference update for agent i leads to a new preference relation $\prec'_i$ such that $H \nprec'_i H'$. The notion of preference update in history-based models that we use in this paper was introduced in [4].

Definition 5 (Preference relation update). *Let $\prec$ be a preference relation and $\mathcal{M} = (\mathcal{E}, \mathcal{A}, \mathcal{H}, \mathcal{V}, \Pi)$ be a history-based model. The preference relation updated according to a formula ϕ at time t, $\prec^{\phi, \mathcal{M}, t}$, is defined as*

$$\prec^{\phi, \mathcal{M}, t} := \prec \setminus \{(H, H') \mid H, t \vDash_{\mathcal{M}} \phi \text{ and } H', t \nvDash_{\mathcal{M}} \phi\},$$

Lemma 1. *If $\mathcal{M} = (\mathcal{E}, \mathcal{A}, \mathcal{H}, \mathcal{V}, \Pi)$ is a history-based model, $\prec$ is a preference relation over histories $\mathcal{H}$, ϕ is a formula, and t is a time-point, then $\prec^{\phi, \mathcal{M}, t}$ is a preference relation over histories $\mathcal{H}$.*

Proof. Establishing this amounts to checking that the given relation is transitive. Suppose $H \prec^{\phi, \mathcal{M}, t} H' \prec^{\phi, \mathcal{M}, t} H''$. If $H, t \vDash_{\mathcal{M}} \phi$, then $H', t \vDash_{\mathcal{M}} \phi$, so that $H'', t \vDash_{\mathcal{M}} \phi$. Therefore $H \prec^{\phi, \mathcal{M}, t} H''$.

We extend updates of preference relations pointwise to updates of preference structures. We can then use preference relation update to update a model using a finite set of distinguishing formulae.

Definition 6 (Preference-based model update). *Let $\mathcal{M} = (\mathcal{E}, \mathcal{A}, \mathcal{H}, \mathcal{V}, \Pi)$ be a history-based preference model. The updated preference model $\mathcal{M}[!\Phi, t]$ (with respect to a finite set of distinguishing non-updating formulae Φ and time-point t) is defined as $\mathcal{M}[!\Phi, t] = (\mathcal{E}, \mathcal{A}, \mathcal{H}, \mathcal{V}, \{\pi^{\phi, \mathcal{M}, t} \mid \pi \in \Pi \text{ and } \phi \in \Phi\})$.*

We represent a preference update within the logic via the $[!\Phi]$ modality. Given a model $\mathcal{M}$ and a finite set of distinguishing non-updating formulae Φ, a preference update modality is satisfied by history H at time t in model $\mathcal{M}$ (i.e., $H, t \vDash_{\mathcal{M}} [!\Phi]\psi$), if and only if ψ holds in the model updated by Φ at time-point t (i.e., $H, t \vDash_{\mathcal{M}[!\Phi, t]} \psi$).

Proposition 1. *The logic CBL as defined in Fig. 2, together with the supporting definitions, is a conservative extension of the temporal fragment (classical
propositionals and temporal modalities) without the preference fragment (preference modalities, $\Diamond_i\phi$, $\Box_i\phi$, $[!\Phi]\phi$).*

Proof. Consider that all of the satisfaction clauses for the temporal modalities
are independent of the structures required to define the preference modalities.

Example 1. Suppose a set of events $\mathcal{E} = \{c, d\}$, denoting compliance and deviation from management-desired behaviour, and a set of histories of all traces over
the events of at most length two, that is $\mathcal{H} = \{\epsilon, c, d, (c; c), (c; d), (d; c), (d; d)\}$.

We consider only one agent in the system, an employee (that is, $\mathcal{A} = \{1\}$).
The employee always prefers to deviate; that is, $\pi = (\prec)$ is given by the transitive
closure of

$$c \prec d \qquad c; c \prec c; d \qquad c; c \prec d; c \qquad c; d \prec d; d \qquad d; c \prec d; d.$$

Let p_c be a proposition that holds for a history when the last event in its
sequence is c; that is, $h \in \mathcal{V}(p_c)$ if and only if h is of the form $h'; c$. Let p_d be
defined similarly.

Let $\mathcal{M} = (\mathcal{E}, \mathcal{A}, \mathcal{H}, \mathcal{V}, \{\pi\})$. We can use the logic CBL to say that the
employee prefers to deviate from the behaviour desired by the manager at the
first opportunity; that is, $(c; c), 0 \vDash_{\mathcal{M}} \Diamond_1 \bigcirc p_d$.

Suppose the policy-maker introduces incentives to encourage greater compliance with policy. In CBL, this is modelled as a preference update with the
formula $\phi = \bigcirc p_c$. Updating the preferences using this formula results in $\prec^{\phi,\mathcal{M},0}$,
consisting in just

$$c; c \prec^{\phi,\mathcal{M},0} c; d \qquad d; c \prec^{\phi,\mathcal{M},0} d; d.$$

This update removes the employee's preference to deviate at the first opportu-
nity, but not at later opportunities; formally, $(c; c), 0 \vDash_{\mathcal{M}} [!\{\phi\}]\neg\Diamond_1 \bigcirc p_d$.

To deal with the second opportunity to deviate from policy, let $\psi = \bigcirc \bigcirc p_c$.
Updating the original preferences using this formula results in $\prec^{\psi,\mathcal{M},0}$, given by

$$c \prec^{\psi,\mathcal{M},0} d \qquad c; c \prec^{\psi,\mathcal{M},0} d; c \qquad c; d \prec^{\psi,\mathcal{M},0} d; d.$$

This update removes the employee's preference to deviate at the second oppor-
tunity, but not at other opportunities; formally, $(c; c), 0 \vDash_{\mathcal{M}} [!\{\psi\}]\neg\Diamond_1 \bigcirc \bigcirc p_d$.

In some situations, the policy-maker may have less fine-grained control over
the employees. For example, they can prevent one deviation, but have no control
over which deviation is prevented. This is represented by updating the preferences using the set of formulae $\Phi = \{\phi, \psi\}$, resulting in the two possible pref-
erence relations above; that is, $\mathcal{M}_{[!\Phi,0]} = (\mathcal{E}, \mathcal{A}, \mathcal{H}, \mathcal{V}, \{\prec^{\phi,\mathcal{M},0}, \prec^{\psi,\mathcal{M},0}\})$. This
update removes the employee's preference to deviate twice. However, there is
now uncertainty about the preferences, and properties that hold for updates
according to ϕ and ψ no longer hold. Indeed,

$$(c; c), 0 \vDash_{\mathcal{M}} [!\Phi]\Diamond_1 \bigcirc p_d \qquad \text{and} \qquad (c; c), 0 \vDash_{\mathcal{M}} [!\Phi]\Diamond_1 \bigcirc \bigcirc p_d.$$

We do, however, have the weaker property, that the employee does not prefer to deviate at both opportunities; formally,

$$(c; c), 0 \models_{\mathcal{M}} [!\Phi] \neg \Diamond_1((\bigcirc p_d) \wedge (\bigcirc \bigcirc p_d)).$$

To see this, note that the only histories preferable to c; c are c; d, from the update for ϕ, and d; c, from the update for ψ, and $(c; d), 0 \not\models_{\mathcal{M}_{[!\Phi,0]}} ((\bigcirc p_d) \wedge (\bigcirc \bigcirc p_d))$ and $(d; c), 0 \not\models_{\mathcal{M}_{[!\Phi,0]}} ((\bigcirc p_d) \wedge (\bigcirc \bigcirc p_d))$. □

Building on this set-up, we now introduce a logical model of the compliance budget. To this end, we let $\mathsf{load}_n(\phi, i)$ denote that agent i has at least n distinct situations in which it would prefer to violate the policy ϕ.

Definition 7. *Let i be an agent and ϕ be an update-free formula. The load formulae of these parameters are defined by*

$$\mathsf{load}_0(\phi, i) \triangleq \top$$
$$\mathsf{load}_{n+1}(\phi, i) \triangleq (\phi \wedge \Box_i \phi) \, \mathcal{U} \, (\phi \wedge \Diamond_i \neg \phi \wedge \bigcirc \mathsf{load}_n(\phi, i)).$$

Given a load formula $\mathsf{load}_n(\phi, i)$, we refer to n as the load value, and ϕ as the policy. The intuition for this is that if $\mathsf{load}_n(\phi, i)$ holds, agent i has complied with policy ϕ, but would have preferred to deviate on n occasions, so expending compliance budget.

If we perform a preference update according to the formula $\mathsf{load}_n(\phi, i)$, we will remove the preference to deviate from the policy ϕ at the first n opportunities. We can represent a bound on uncertainty on an agent's compliance budget— that is, uncertainty on how much more the agent will comply with policies—by updating according to a set of load formulae with a range of load values:

$$\mathsf{load}_m(\phi, i), \mathsf{load}_{m+1}(\phi, i), \ldots, \mathsf{load}_{n-1}(\phi, i), \mathsf{load}_n(\phi, i).$$

4   An Access Control Example

We illustrate the facility to model ideas from the compliance budget using an example concerning remote access policy. We suppose a business setting, with an internal (local) network that contains some core systems resources (for example, databases, workflow tools, or email servers). This core system can be divided into high security and low security components, where high security components are more valuable or more vulnerable.

Principals using devices on the local network can access the entire core system, both high and low security. Principals using devices on remote networks can access the core system, but with certain restrictions based on the type of connection that is used to access the resources.

The system is technologically configured so that a connection using a virtual private network (VPN) can access the whole core system, but a connection using a secure shell (SSH) can only access the low-security component of the core system. This is an attempted implementation of the (currently informal) policy that the high-security component of the core system should only be remotely accessed via VPN. Principals can, however, use SSH to connect to locally owned machines, and then connect directly to the high component of the core system. Hence, the policy can be circumvented by a determined principal.

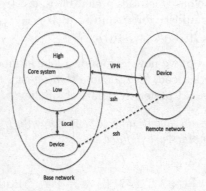

Fig. 3. Access control from a remote network

This scenario is depicted graphically in Fig. 3. To model this, we consider the policy-maker M, and the employee (principal) P. We assume a set of events $\mathcal{E}$, comprising: cc_{loc}, cc_V, and cc_S, connecting to the core system via the local network, VPN, and SSH; dc_{loc}, dc_V, and dc_S, disconnecting from the core system via the local network, VPN, and SSH; cd_S and dd_S, connecting to and disconnecting from an employee-controlled device on the local network via SSH; and a_L and a_H, accessing the low- and high-security component of the core system.

The technological configuration places various constraints on the behaviour of the system, represented by the set of histories that we consider, $\mathcal{H}$. An access event e occurs within the scope of a connection event e' within a history h if and only if there exist histories h_1, h_2, h_3 such that $h = h_1; e'; h_2; e; h_3$ and h_2 does not contain any connect or disconnect events. For example, the a_L event does occur within the scope of the cc_S event in the history $cc_S; a_L$, but does not occur within the scope of cc_S event in the history $cc_S; dc_S; a_L$.

The set of histories contains all finite sequences of the events, except for those where some a_L event does not occur within the scope of a connection cc_{loc}, cc_V, or cc_S, and those where some a_H event does not occur within the scope of a connection cc_{loc} or cc_V. For example, the history $cc_S; a_H$ is *not* included in $\mathcal{H}$, but the histories $cc_V; a_H$ and $cd_S; cc_{loc}; a_H$ *are* included in $\mathcal{H}$. The history $cc_V; a_H$ conforms to the informal policy that the high security component of the core system should only be remotely accessed via VPN. The history $cd_S; cc_{loc}; a_H$, however, does not conform to the informal policy, but it is included in our set of histories $\mathcal{H}$ as the a_H event *does occur* within the scope of a connection cc_{loc}.

There are various costs and benefits to the employee for their different actions/events that they can choose. Working locally (on site) gives direct, secure access, but comes with the possibility, for example, of being interrupted by a colleague. Working remotely removes the possibility of being interrupted, but requires accessing the core system via some additional security mechanism. Using a VPN to connect remotely to the core system gives full, secure access, but has the costs that the VPN is harder (than SSH) to operate, is more faulty (than SSH), and removes the ability, for example, to use a printer on the remote network. Using SSH to connect remotely to the core system gives secure access and is easier (than the VPN) to operate, is less faulty (than the VPN), and retains the ability to use a printer on the remote network, but has the cost that it has limited access only to the low security component of the core system.

We demonstrate how to model the imposition of a policy that explicitly guides against using SSH to access the core system. In the remainder of this section, we overload our syntax slightly and use an event e as a proposition which is satisfied by a history if and only if the last event in the history is the given event. The histories that comply with this policy are those that satisfy

$$\phi \triangleq (\mathsf{cd}_S \to ((\neg\mathsf{cc}_{loc})\,\mathcal{U}\,\mathsf{dd}_S)$$

at every time step. Note that a finer-grained policy that only prohibits the use of such connections to access high-security resources could be described similarly.

Consider a user working at a remote site, engaged in a task which requires two accesses to the high-security resources at base, with intervening access to remote-site resources that are not available when connected via VPN. As described above, we endow the user with a preference relation favouring SSH connections above VPN: $\mathsf{cc}_V; \mathsf{a}_H; \mathsf{dc}_V \prec \mathsf{cd}_S; \mathsf{cc}_{loc}; \mathsf{a}_H; \mathsf{dc}_{loc}; \mathsf{dd}_S$ (and similarly for longer histories containing these as subsequences). The user may complete the task with any of the following three histories:

$$\mathsf{cc}_V; \mathsf{a}_H; \mathsf{dc}_V; \mathsf{cc}_V; \mathsf{a}_H; \mathsf{dc}_V \tag{1}$$

$$\prec \mathsf{cc}_V; \mathsf{a}_H; \mathsf{dc}_V; \mathsf{cd}_S; \mathsf{cc}_{loc}; \mathsf{a}_H; \mathsf{dc}_{loc}; \mathsf{dd}_S \tag{2}$$

$$\prec \mathsf{cd}_S; \mathsf{cc}_{loc}; \mathsf{a}_H; \mathsf{a}_H; \mathsf{dc}_{loc}; \mathsf{dd}_S. \tag{3}$$

Consider a model $\mathcal{M}$ that embodies this scenario. To model the imposition of the access control policy, we perform preference update according to the set of formulae $\Phi \triangleq \{\mathsf{load}_i(\phi) \mid i = 1, 2\}$, arriving at model $\mathcal{M}' \triangleq \mathcal{M}[!\Phi]$ This update reflects the policy-maker's inevitable uncertainty in the compliance budget of the user. Because of the user's prior preference, compliance with policy ϕ comes at a cost. Some of this cost is directly observable: witness the disconnections and reconnections in history (1), the least-preferred, most-compliant behaviour. However, other costs may not be observed, for instance the possible failure of attempts to access resources at the remote site while connected via VPN. Not only is the policy-maker unable to judge the effort available for compliance, but also the effort required to comply is uncertain. In our model, updating preference with $\mathsf{load}_n(\phi)$ reflects the willingness of a user to deviate from prior preference in favour of compliance up to n times. Model $\mathcal{M}'$ contains a preference structure for each n, that is, for each possible value of the (unmeasurable) compliance budget. A highly compliant user ($n = 2$ in this example) becomes indifferent between histories 1–3. A user with low compliance budget ($n = 1$) retains just the preference for history 2 over the fully-compliant 1. Thus for the highly compliant user, preference and policy are aligned, so there is no reason to violate the policy. For the less compliant user, after the first VPN connection the budget is exhausted and the user prefers to revert to SSH, contravening the policy.

The scenario we have modelled ascribes at least some propensity for compliance to the user: we do not include $load_0(\phi)$ in the set of preference update formulae. As a result, we are able to draw some conclusions about the preferences of the user under the policy ϕ. For instance, each of the histories 1–3 satisfies

$$H, 0 \models_{\mathcal{M}} [!\Phi]\Box(cd_S \rightarrow \neg(\neg dd_S \; \mathcal{U} \; a_H \land \bigcirc(\neg dd_S \; \mathcal{U} \; a_H)));$$

that is, the user would never prefer to adopt a behaviour incorporating two accesses to high-security resources via SSH.

5   Further Work: Incomplete Information Reasoning

Our model of the compliance budget has been designed to account for the fact that the 'budget' is not a quantifiable value, and the rate at which it is depleted is unknown, as explained in [5,6]. This has led to a model in which we have, for each agent, a set of *possible* preference relations over histories. That is, our model incorporates *uncertainty* about the preferences of the agents: we know that eventually the compliance budget will be exhausted, but we do not know how long that will take. The impact of imposing a new policy ϕ is modelled by updating the agents' preferences with $load_n(\phi)$ for an uncertain value of n.

Uncertainty over preferences is the qualitative analogue of uncertainty over payoffs. Harsanyi [12] demonstrates that *all* uncertainty over the structure of a game can be reduced to uncertainty over payoffs. Our model is therefore a simple qualitative setting in which to study situations of incomplete information. Security policy decisions are typically incomplete information situations because of uncertainty over the compliance budget of agents. As Harsanyi's reduction suggests, this uncertainty subsumes lack of knowledge of the consequences of compliance on productivity. In the VPN example, the policy-maker insisting on VPN for remote access is not aware of the implications for individual agents, who may have difficulty accessing local resources (e.g., network printers) while connected to a VPN. Such issues may or may not be the reason that compliance is reduced, but, in our model, it does not matter: uncertainty in the compliance budget accounts for uncertainty over the details of agent behaviour, allowing us to model behaviour at an appropriate level of abstraction.

Much work remains, including: the metatheory of the logic and the theory of load formulae (e.g., for the interaction of multiple policies); other logics, such as epistemic variants to internalize uncertainty (note that history-based semantics supports epistemic constructions [17]); decision- and game-theoretic considerations such as optimality and equilibria; consideration of richer and larger models in order to explore the value of the approach for security policy-makers.

References

1. Alpcan, T., Başar, T., Security, N.: Decision and Game-Theoretic Approach. Cambridge University Press, Cambridge (2010)
2. Anderson, R.: Why information security is hard: an economic perspective. In: Proceedings of the 17th Annual Computer Security Applications Conference, pp. 358–265. IEEE (2001)
3. Anderson, R., Moore, T.: The economics of information security. Science **314**, 610–613 (2006)
4. Baskent, C., McCusker, G.: Preferences and equilibria in history based models. In: Proceedings of the 12th Conference on Logic and the Foundations of Game and Decision Theory (2016). http://loft.epicenter.name
5. Beautement, A., Sasse, A., Wonham, M.: The compliance budget. In: Proceedings of the New Security Paradigms Workshop (NSPW 2008), pp. 47–55. ACM (2008) doi:10.1145/1595676.1595684
6. Beautement, A., Sasse, A.: The economics of user effort in information security. Comput. Fraud Secur. **10**, 8–12 (2009). doi:10.1016/S1361-3723(09)70127-7
7. Beautement, A., Coles, R., Griffin, J., Ioannidis, C., Monahan, B., Pym, D., Sasse, A., Wonham, M.: Modelling the human and technological costs and benefits of USB memory stick security. In: Johnson, M.E. (ed.) Managing Information Risk and the Economics of Security, pp. 141–163. Springer, New York (2009)
8. Collinson, M., Monahan, B., Pym, D.: A Discipline of Mathematical Systems Modelling. College Publications, London (2012)
9. van Ditmarsch, H., Halpern, J., van der Hoek, W., Kooi, B. (eds.): Handbook of Epistemic Logic. College Publications, London (2015)
10. Gordon, L.A., Loeb, M.P.: The economics of information security investment. ACM Trans. Inf. Syst. Secur. **5**(4), 438–457 (2002)
11. Gordon, L.A., Loeb, M.P., Resources, M.C.: A Cost-Benefit Analysis. McGraw Hill, New York (2006)
12. Harsanyi, J.: Games with incomplete information played by 'Bayesian' players, Part III. Manag. Sci. **14**(7), 486–502 (1968)
13. Huth, M., Ryan, M.: Logic in Computer Science: Modelling and Reasoning About Systems. Cambridge University Press, Cambridge (2004)
14. Ioannidis, C., Pym, D., Williams, J.: Investments and trade-offs in the economics of information security. In: Proceedings of the Financial Cryptography, Data Security, pp. 148–162 (2009)
15. Ioannidis, C., Pym, D., Williams, J.: Information security trade-offs and optimal patching policies. Eur. J. Oper. Res. **216**(2), 434–444 (2012). doi:10.1016/j.ejor.2011.05.050
16. Ioannidis, C., Pym, D., Williams, J.: Is public co-ordination of investment in information security desirable? J. Inf. Secur. **7**, 60–80 (2016). http://dx.doi.org/10.4236/jis.2016.72005
17. Pacuit, E.: Some comments on history based structures. J. Appl. Logic **5**(4), 613–624 (2007)
18. Parikh, R., Ramanujam, R.: A knowledge-based semantics of messages. J. Logic Lang. Inf. **12**(4), 453–467 (2003)
19. A. Pnueli. The temporal logic of programs. In: Proceedings of the 18th Annual Symposium on Foundations of Computer Science (FOCS), pp. 46–57 (1977). doi:10.1109/SFCS.1977.32
20. Tambe, M.: Security and Game Theory: Algorithms, Deployed Systems, Lessons Learned. Cambridge University Press, Cambridge (2011)

A Game-Theoretic Analysis of Deception over Social Networks Using Fake Avatars

Amin Mohammadi[1], Mohammad Hossein Manshaei[1(✉)],
Monireh Mohebbi Moghaddam[1], and Quanyan Zhu[2]

[1] Department of Electrical and Computer Engineering,
Isfahan University of Technology, 84156-83111 Isfahan, Iran
{amin.mohammadi,monireh.mohebbi}@ec.iut.ac.ir
[2] Department of Electrical and Computer Engineering,
New York University, New York, USA
manshaei@cc.iut.ac.ir, quanyan.zhu@nyu.edu

Abstract. In this paper, we formulate a deception game in networks in which the defender deploys a fake avatar for identification of the compromised internal user. We utilize signaling game to study the strategy of the deployed fake avatar when she interacts with external users. We consider a situation where the fake avatar as the defender is uncertain about the type of a connected external user, which can be a normal user or an attacker. We attempt to help the defender in selecting her best strategy, which is alerting to the system for detecting an attack or not alert. For this purpose, we analyze the game for finding the Perfect Bayesian Nash equilibria. Our analysis determines for which probability of the external user being an attacker, the defender should launch a defending mechanism.

Keywords: Network security · Deception · Fake avatar · Social network · Game theory · Signaling game

1 Introduction

As the number and complexity of cyber-attacks has been increasing in the last years, security becomes an essential requirement for any network [1]. Generally, two categories of mechanisms used for guaranteeing the network security which are prevention-based and detection-based techniques. The former technologies that provide confidentiality, integrity and authentication security requirements usually include cryptography, key management and so on [2]. This class of techniques will fail against sophisticated attackers, and it is essential to utilize the detection measures in some situations. One of the powerful detection techniques, which has attracted the attention of many researchers recently, is *deception*. Deception refers to the actions used by the defender to persuade an adversary

Q. Zhu—This work is partially supported by the grant CNS-1544782 from National Science Foundation.

Q. Zhu et al. (Eds.): GameSec 2016, LNCS 9996, pp. 382–394, 2016.
DOI: 10.1007/978-3-319-47413-7_22

to believe that false information they were given was actually true [3]. The use of deception techniques increases the possibility of detecting attacks in the early stage of the attack life-cycle. Deception has proved that has a positive effect for the defenders, and conversely a negative impact on the attackers [4]. Various deception techniques have been proposed in the literatures, such as honeypots and fake avatars.

Avatar deployment is one of the deception techniques that can be used by a defender for identifying malicious activity in various networks, especially in the social networks. Avatars should appear realistic to the people from both inside and outside the organization and has the positions that are likely to be interesting to the attackers. In addition, such avatars should have closely monitored accounts in the organization (e.g., active directory accounts), as well as valid email addresses. Interaction with avatars should be regularly monitored by the network administrator [1]. In this study, we consider this deception approach and aim to model the defender-attacker interaction using game theory. In our model, the attacker attempts to mislead the defender and subsequently, we try to help the defender in making the best decision. Specifically, the scenario we examine in this paper is described as follows.

A defender deploys fake avatars in her network. External users (e.g., interested in applying for a position in this organization) may contact the human avatar. However, since internal users know the correct contact details, communication between an internal user and the avatar can be considered suspicious. Such interactions could be a sign that the user's account has been compromised by an attacker [1]. With this assumption, the fake avatar can easily detect the interaction from the malicious insider users. Moreover, as these avatars look like a real entity for external users, they communicate her to gain some information. Suppose that we can divide all communications between the external users and fake avatars to two categories: *suspicious* or *non-suspicious*. *Suspicious communication* refers to the communication type which is potentially risky and the attacker can obtain a great success in her attack. However, a normal user can also communicate with avatar through a suspicious connection and earn more information and save money and time by using these communication type. Given the above description, a fake avatar potentially receives suspicious or non-suspicious signals that have been generated by a normal user or an attacker. Upon receiving these signals, she should decide to whether provide an alert for the system administrator or not. This decision is a bit hard as she is not sure about the sender's type. Given this scenario, we make a game theoretical model to investigate the involved parties interaction.

We use the *signaling game* to capture and analyze this interaction. Signaling game refers to a class of two-player dynamic games in which one player (called the Sender) is informed and the other (called the Receiver) is not. In other words, the sender has the private information (i.e., its type) while the receiver has a common information about the type distribution of the sender. The sender's strategy set consists of sending messages depends on its type, while the receiver's strategy set consists of actions contingent on the sender's choices [2].

In the scenario under study, the user moves first by choosing whether or not to send a suspicious signal to the avatar, after which, the fake avatar decides whether to alert or not. The incomplete information arises from the fake avatar's uncertainty of the user type. We aim to help the fake avatar to make the best decision given the signal received from the user. To this end, we determine and characterize the Perfect Bayesian Equilibria of the defined signaling game.

Several previous works focus on the deception as an effective defending technique [1,5–11]. In [1], Virvilis et al. proposed the use of multiple deception techniques to defend against *advanced persistent threats* (APT) and malicious insider attacks. In [8], Costarella et al. focused on the honeynet-aware botnet attacks and attempted to improve the depth of deception levels by this approach. In [9], authors proposed an system which can be interfaced with social networking sites for creating deceptive honeybots and leveraging them for receiving information from botnets. Our work is mainly different from the above work as we model the mentioned deception technique by a mathematical tool. However, our ideas partly derive from [1], but have significant differences. As authors in this paper just introduced the fake avatar as a deception technique in the social networks, but we formulate this type of the deception by a theoretical framework, which is game theory. Additionally, we model the interactions between a fake avatar and an external user, not an internal one, with a signaling game in the deception scenario. In addition, we seek the pure-strategy BNE. These equilibriums determine when and how the avatar takes a defense action.

There is a number of work in which authors used the signaling game for modeling different issues in different networks, for example [13] studied power control management problem in wireless network. Since in this paper we focus on the network security, we mainly review the researches in which this game model is employed for formulating the security scenarios in computer networks, such as [2,4,12,14–18]. In [17], authors model the interaction between a service provider and his client in the presence of the attacker. Similarly, [16] modeled the same situation, but in the cloud systems. [15] formulated the interaction between the fingerprinter and the defender as a signaling game. In [19], authors employ this game model for analyzing the intrusion detection in mobile ad hoc networks. They model the interaction between an attacker and an intrusion detection system. In a similar way, authors in [20] present a signaling game model to analyze intrusion detection in wireless sensor network, with focusing on the dropping packets attacks. In [21], authors propose a Bayesian signaling game formulation for intrusion detection system in wireless ad hoc networks. Multi-step attack-defense scenario on confidentiality has been modeled in [22] by repeated signaling game. Among these work, [4,18] formulated the deception scenarios by using this game model, but they mainly focused on the honeypot as the deception technique. None of these mentioned works utilized the deception as the defending strategy.

In summary, compared to all previous work, we focus on the *fake avatar* as a *deception technique* that defender used. Moreover, we utilize a *signaling game* model to decide *the strategy of fake avatar* in response to the received signal

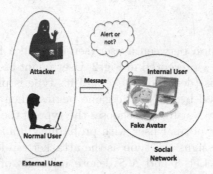

Fig. 1. System model: defender deploys a fake avatar in her network. Internal users knows the details and the address of the fake avatar, but external user does not know. The external user can be a normal user or an attacker. She can send a suspicious signal or non-suspicious one. The avatar should decide to generate an alert or not under the uncertainty about the user's type.

from *the external users*. Note that in our model, the defender is a second mover and has the incomplete information about the user type, while in the former ones such as the honeypot scenario [8], the defender appears as the sender with complete information.

The remainder of this paper is structured as follows. In Sect. 2, we describe our system model. Then, we discuss our game model in Sect. 3. In Sect. 4, we analyze the proposed signaling game and derive the set of equilibria. Finally, Sect. 5 summarizes the conclusion.

2   System Model

In this paper, we focus on a system model depicted in Fig. 1. This model consists of three main entities: a defender, a normal user and an attacker. The defender protects her network by deploying some fake avatars[1]. Users from inside and outside the network may connect to these avatars. Both the normal user and the attacker are external users who convey messages to fake avatars to reach their goals through receiving the necessary information. They send their requests to the avatar. The fake avatar needs to deal with the attacker and the normal user simultaneously, while she does not know explicitly about the sender type. The fake avatar should decide whether this communication is coming from a normal user or not, and subsequently alert to the system or not. She can not detect the received signal is from which type of uses, as both of them may send a suspicious signal. Hence, we need to model the uncertainty of a given fake avatar about the user type as well as the dynamic nature of the user and the avatar movements. We model this deception in the next section by a signaling game.

---

[1] As fake avatars have been deployed by defender in our model, we use these words interchangeably.

3   Game Model

We model the interaction between a user (sender) and a fake avatar (receiver) as a signaling game $\mathcal{G}_{\mathcal{FA}}$ presented in Fig. 2. User might be an attacker (denoted by A) or a normal user (denoted by N) and its type is private information to the fake avatar (denoted by FR). The game begins when the user attempts to contact the fake avatar. Nature (C) chooses the type of the user. Nature chooses either type attacker or normal user with probability θ and $1 - \theta$, respectively. We interpret Nature assigns the type as the attacker randomly selecting a user to contact the avatar. Let $\mathcal{S}$ and $\mathcal{NS}$ denote messages of suspicious and non-suspicious, respectively. Therefore, the action space of the user, both normal one and the attacker, is $\{\mathcal{S}, \mathcal{NS}\}$. The avatar receives the signal and then chooses an action, which is either alert ($\mathcal{A}$), or not ($\mathcal{NA}$) to the system. Obviously, if the defender knows the opponent, chooses her action given the type of the user, $\mathcal{A}$ in facing an attacker and $\mathcal{NA}$ in response to a normal user.

If the avatar makes an alert upon receiving a signal from the normal user, imposed the cost υ to the system, because of keeping busy the network resources, as well as the cost ι for generating the false positive alert. Cost υ is incurred to system even when the avatar receives a signal from the attacker while she earns gain Φ for the attacker detection, result in correct alerting. This action has a negative effect on the attacker and causes the cost κ to her. All other used notations in this game as well as their description is summarized in Table 1.

We assume that users (i.e., both the attacker and the normal user) prefer to send a suspicious signal if they know the avatar does not make any alert to the system by receiving their signals. Given this assumption, the cost of sending the suspicious signal than the received gain from choosing this action is negligible ($\mu \geq c_a$ and $\nu \geq c_n$). Furthermore, since compromising the system by the attacker imposed a significant cost to the system, the following conditions could be likely: $\upsilon \leq \Phi$ and $\iota \leq \Phi$. We will analyze our model in the next section by considering these assumptions.

4   Game Analysis

In this section, first we briefly define the Perfect Bayesian Nash Equilibria (PBNE). Then, we analyze the deception signaling game for finding the pure strategy PBNE. Finally, we present a case study and discuss our results.

4.1   Equilibrium Analysis

We now review the concept of Perfect Bayesian Equilibrium, the extension of sub-game perfection to games of incomplete information. Basically, in non-Bayesian games, a strategy profile is a Nash equilibrium (NE) if every strategy in that profile is a best response to every other strategy in the profile. Resulting from this definition, NE results in a steady state. But, in Bayesian games, players are seeking to maximize their expected payoffs, given their beliefs about the

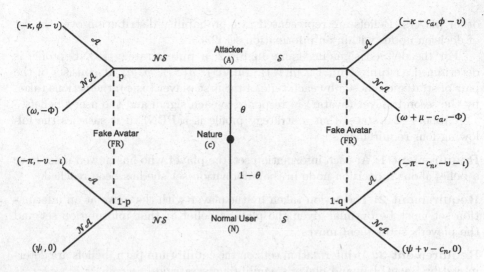

Fig. 2. The representation of the fake avatar deception as a signaling game. The players are a user and a fake avatar. The belief of the fake avatar about the sender type is modeled by θ. The fake avatar observes the actions of the user, i.e., Suspicious and Non-Suspicious signal. The actions of the fake avatar are Alert and Not-Alert.

Table 1. Deception game notations

Notation	Description
$\mathcal{S}$	Specious signal
$\mathcal{NS}$	Non-suspicious signal
$\mathcal{A}$	Alert to the system
$\mathcal{NA}$	Not alert to the system
κ	Cost of the attack detection that the attacker incurs
ω	Gain of the attacker for successfully sending the signal
c_a	Difference between the cost of sending specious and non-specious signal imposed on the attacker
c_n	Difference between the cost of sending specious and non-specious signal imposed on the normal user
μ	Difference between the gain from successfully sending specious and non-specious signal earned by the attacker
ν	Difference between the gain from successfully sending specious and non-specious signal earned by the normal user
Φ	Gain of the attacker identification earned by the defender
ϕ	Cost of not detection of the attacker imposed on the defender
υ	The defender's cost for making the alert
ι	The defender's cost for generating the false positive alert
π	Cost of alerting imposed on the normal user
ψ	Benefit earned by the normal user from receiving the desired response

other players. Beliefs are represented as a probability distribution over the set of decision nodes within an information set [23].

For the defined signaling game in Fig. 2, a pure strategy PBNE profile is determined as tuple $((m(A), m(N)), (a(\mathcal{NS}), a(\mathcal{S})), p, q)$. It consists of the pair of strategies chosen by each type of the first player (user), the actions taken by the second player (avatar) in response to each signal and the user's beliefs.

Gibbons [24] states that a strategy profile is a PBNE if it satisfies the following four requirements:

Requirement 1: At each information set, the player who has moved must have a belief about a decision node in the information set she has been reached.

Requirement 2: The action taken by the player with the move at an information set must be optimal given the player's belief at that information set and the player's subsequent moves.

Requirement 3: At information sets on the equilibrium path, beliefs are determined by Bayes' law and player's equilibrium strategies.

Requirement 4: At information sets off the equilibrium path, beliefs are determined by Bayes' law and the players' equilibrium strategies where possible.

Note that on and off equilibrium path imply to information sets which are reached respectively with positive and zero probability, when the equilibrium strategies are played. In each signaling game, two types of PBNE are possible, called pooling and separating, which are defined as follows:

Definition 1 (Pooling equilibrium): A PBNE is a pooling equilibrium if the first player sends the same signal regardless of his type [23].

Definition 2 (Separating equilibrium): A PBNE is a separating equilibrium if the first player sends different signals, depending on his type [23].

For example, in our defined signaling game, if $m(A) = m(N)$ at a given PBNE, it will represent a pooling equilibrium. A PBNE in which $m(A) \neq m(N)$ is a separating equilibrium.

4.2 Deception Signaling Game Equilibria

We now find the pure-strategy separating and pooling equilibria of the fake avatar deception game, according to the above definitions.

Theorem 1. *There is a pooling equilibrium on the strategy* $(\mathcal{NS}, \mathcal{NS})$ *of the sender for every* θ *in* $\mathcal{G}_{\mathcal{FA}}$.

Proof. Suppose that the sender strategy is $(\mathcal{NS}, \mathcal{NS})$. In this case and according to the rule of belief updating on the equilibrium path based on the Bayes' law, we have $p = \theta$. Given the sender strategy, if FR chooses action $\mathcal{A}$ in response to the strategy $\mathcal{NS}$, gains payoff $\theta \times (\phi - v) + (1 - \theta) \times (-v - \iota)$. On the other side, if she plays action $\mathcal{NA}$, obtains a gain equal to $\theta \times (-\Phi) + (1 - \theta) \times (0)$. As $v \leq \Phi$, we have $0 < \frac{v+\iota}{\phi+\Phi+\iota} < 1$. Hence, for each θ, there are the following possible cases:

1. $\theta \geq \frac{v+\iota}{\phi+\Phi+\iota}$: Given the condition on the θ we have:

$$\theta \times (\phi - v) + (1 - \theta) \times (-v - \iota) \geq \theta \times (-\Phi) \Longleftrightarrow p = \theta \geq F := \frac{v + \iota}{\phi + \Phi + \iota}$$

As a result, action $\mathcal{A}$ is the receiver's best response to the sender strategy $\mathcal{NS}$. If q, which is the receiver's belief off equilibrium path, is greater than $\frac{v+\iota}{\times} \phi + \Phi + \iota$, the action $\mathcal{A}$ will be the best response to the strategy $\mathcal{S}$. Because $q \times (\phi-v)+(1-q)\times(-v-\iota) \geq q\times(-\Phi)+(1-q)\times(0) \Longleftrightarrow q \geq F = \frac{v+\iota}{\phi+\Phi+\iota}$, where $q \times (\phi-v)+(1-q)\times(-v-\iota)$ and $q\times(-\Phi)+(1-q)\times(0)$ are the expected payoff of the receiver (avatar) from playing $\mathcal{A}$ and $\mathcal{NA}$, respectively. If the sender sends signal $\mathcal{S}$ instead of $\mathcal{NS}$, it obtains payoff $-\kappa - c_a$ which is less than $-\kappa$, which is not a profitable deviation. Similarly, if the normal user deviates from strategy $\mathcal{NS}$, it earns the benefit equal to $-\pi - c_n$ instead of payoff $-\pi$ for playing strategy $\mathcal{S}$. Therefore, $\{(\mathcal{NS},\mathcal{NS}),(\mathcal{A},\mathcal{A}),q,p = \theta\}$ is a PBNE for $q \geq F$.

2. $\theta < \frac{v+\iota}{\phi+\Phi+\iota}$: Given the condition on the θ, we have:

$$\theta \times (\phi - v) + (1 - \theta) \times (-v - \iota) \leq \theta \times (-\Phi) \Longleftrightarrow p = \theta \leq F = \frac{v + \iota}{\phi + \Phi + \iota}$$

Therefore, action $\mathcal{NA}$ is the best response to the sender's strategy $\mathcal{NS}$. If $q \leq \frac{v+\iota}{\phi+\Phi+\iota}$, then the strategy $\mathcal{A}$ is the best response to the $\mathcal{S}$. This is because of $q \times (\phi-v)+(1-q)\times(-v-\iota) \geq q\times(-\Phi)+(1-q\times(0) \Longleftrightarrow q \geq F = \frac{v+\iota}{\phi+\Phi+\iota}$, where $q \times (\phi - v) + (1 - q) \times (-v - \iota)$ is the expected payoff of the receiver for playing $\mathcal{A}$ and $q \times (-\Phi) + (1 - q) \times (0)$ is the receiver's expected payoff for choosing the strategy $\mathcal{NA}$. If the attacker picks up the strategy $\mathcal{S}$ instead of the $\mathcal{NS}$, she earns less payoff $(-\kappa - c_a < -\kappa)$. This is true for the second type of the sender. As a result, $\{(\mathcal{NS},\mathcal{NS}),(\mathcal{NA},\mathcal{A}),q,p = \theta\}$ is a PBNE for $q \geq F$.

Theorem 1 states if the selected strategies of both user types (the attacker and the normal user) are $\mathcal{NS}$, then there is an equilibrium. In this case, depending on the θ's value, the defender selects one of the strategies $\mathcal{A}$ or $\mathcal{NA}$. In other words, if the probability of user being an attacker (θ) is greater than $\frac{v+\iota}{\phi+\Phi+\iota}$, she must alert to the system. Otherwise, her best response is $\mathcal{NA}$. In both cases and for $q \geq F$, if the external user selects $\mathcal{S}$, her best response is $\mathcal{A}$.

Theorem 2. *There exists a pooling equilibrium on the strategy $(\mathcal{S},\mathcal{S})$ if $\theta \leq F$.*

Proof. Assume that $(\mathcal{S},\mathcal{S})$ is the sender's strategy. In this case, we have: $q = \theta$ based on the Bayes' rule. Given the sender's strategy $(\mathcal{S})$, the fake avatar gains payoff $\theta \times (\phi - v) + (1 - \theta) \times (-v - \iota)$ if she plays the strategy $\mathcal{A}$, and payoff $\theta \times (-\Phi) + (1 - \theta) \times (0)$ if she takes the action $\mathcal{NA}$ in response to the strategy $\mathcal{NS}$. There are two possible cases for each value of θ as the following:

1. $\theta \geq \frac{v+\iota}{\phi+\Phi+\iota}$: Given the value of θ, we have:

$$\theta \times (\phi - v) + (1 - \theta) \times (-v - \iota) \geq \theta \times (-\Phi) \Longleftrightarrow q = \theta \geq F := \frac{v + \iota}{\phi + \Phi + \iota}$$

Accordingly, the fake avatar should play the action $\mathcal{A}$ in response to the $\mathcal{S}$, as it is the best response. But, if the attacker violates from the sending signal $\mathcal{S}$, obtains payoff $-\kappa$ or ω, which is more than $-\kappa - c_a$, the payoff earned by the attacker for choosing strategy $\mathcal{S}$. Therefore, for $\theta \geq \frac{v+\iota}{\phi+\Phi+\iota}$, the sender strategy is not steady and subsequently there is no PBNE on $\mathcal{S}$ for this value of θ.

2. $\theta \leq \frac{v+\iota}{\phi+\Phi+\iota}$: According to the θ values, we have:

$$\theta \times (\phi - v) + (1 - \theta) \times (-v - \iota) \leq \theta \times (-\Phi) \Longleftrightarrow q = \theta \leq F = \frac{v + \iota}{\phi + \Phi + \iota}$$

Consequently, $\mathcal{NA}$ is the best response to the $\mathcal{S}$. If the condition $p \leq \frac{v+\iota}{\phi+\Phi+\iota}$ being held, strategy $\mathcal{NA}$ will be the best response to the $\mathcal{NS}$, due to the following: $p \times (\phi - v) + (1 - p) \times (-v - \iota) \leq p \times (-\Phi) + (1 - p) \times (0) \Longleftrightarrow p \leq F = \frac{v+\iota}{\phi+\Phi+\iota}$. The attacker gets payoff ω instead of $\omega + \mu - c_a$ if she sends the signal $\mathcal{NS}$ rather $\mathcal{S}$. The former payoff value is less than the latter as $\mu \geq c_a$. The similar result will obtain for the normal user. Therefore, the strategy profile $\{(\mathcal{S}, \mathcal{S}), (\mathcal{NA}, \mathcal{NA}), p, q = \theta\}$ is an PBNE for $p \leq F$.

Theorem 2 states that if the value of θ is less than $\frac{v+\iota}{\phi+\Phi+\iota}$, playing the strategy $\mathcal{S}$ by both user type is an equilibrium. In summary, this means if the external user knows that the belief of the defender in being the external user as an attacker is less than the mentioned value, the user should take the strategy $\mathcal{S}$. Otherwise, there is no equilibrium. In this case (i.e., $\theta \leq F$) and by observing the strategy $\mathcal{S}$, the fake avatar's best response is not alerting to the system.

Theorem 3. *There is no separating equilibria in the game $\mathcal{G}_{\mathcal{FA}}$.*

Proof. Consider $(\mathcal{NS}, \mathcal{S})$ represents the selected strategy by each type of the sender, in which the attacker chooses strategy $\mathcal{NS}$ and the normal user plays the strategy $\mathcal{S}$. As a result of the Bayes rule, we have: $p = 1$ and $q = 0$. Subsequently, given the payoff values for the fake avatar, the receiver best response will be $\mathcal{A}$ to the strategy $\mathcal{NS}$ as well as $\mathcal{NA}$ to $\mathcal{S}$. In more details, the fake avatar gains payoff $\phi - v$ for playing $\mathcal{A}$ in response to the sender's strategy $\mathcal{NS}$, which is greater than the earned payoff for playing $\mathcal{NA}$ (which is $-\Phi$). Therefore, she should select $\mathcal{A}$. In other hand, when the avatar receives the signal $\mathcal{S}$ from the normal user, she obtains payoff zero by playing $\mathcal{NA}$ which is greater than the negative value $-v - \iota$ for choosing strategy $\mathcal{A}$. Therefore, she will select $\mathcal{NA}$ in response to $\mathcal{S}$. Now, according to the fake avatar's best responses, we should check whether the deviation of the selected strategy by the sender is beneficial for her or not. As shown in the Fig. 2, the attacker gains payoff $-\kappa$ by playing $\mathcal{NS}$, as the fake avatar responds to this signal by $\mathcal{A}$. If the attacker deviates and chooses strategy $\mathcal{S}$, she earns a payoff equal to $\omega + \mu - c_a$, as the avatar

Table 2. Perfect bayesian equilibria and their conditions

Condition on θ	Possible PBNE profiles	Preferred PBNE profile
$\theta \geq \frac{\upsilon+\iota}{\phi+\Phi+\iota}$	$\mathcal{PBNE}_1 = \{(\mathcal{NS},\mathcal{NS}),(\mathcal{A},\mathcal{A}),q,p=\theta\}$	$\mathcal{PBNE}_1$
$\theta \leq \frac{\upsilon+\iota}{\phi+\Phi+\iota}$	$\mathcal{PBNE}_2 = \{(\mathcal{S},\mathcal{S}),(\mathcal{NA},\mathcal{NA}),p,q=\theta\}$	$\mathcal{PBNE}_2$
	$\mathcal{PBNE}_3 = \{(\mathcal{NS},\mathcal{NS}),(\mathcal{NA},\mathcal{A}),q,p=\theta\}$	

plays $\mathcal{NA}$ in response to $\mathcal{S}$. This value is more than $-\kappa$. Hence, sending signal $\mathcal{NS}$ instead $\mathcal{S}$ results in the more payoff for the attacker. As a result, there is no PBNE on the $(\mathcal{NS},\mathcal{S})$. With the same analysis, we can conclude that there is no PBNE on the $(\mathcal{S},\mathcal{NS})$, too.

The last theorem explains the situation in which the selected strategy by the different type of the sender, i.e., the attacker and the normal user, is one of the following strategy pair: $(\mathcal{NS},\mathcal{S})$ or $(\mathcal{S},\mathcal{NS})$. In this case and according to the above theorem, there is no equilibrium. In other words and according to the explained theorems, for being in the equilibrium, both type of the external users should play the same strategy. Theorems 1, 2 and 3 characterize the Perfect Bayesian Nash equilibria. But, as there is more than one equilibrium point for some values of θ, the question is how the defender can use the above analysis in the best way for choosing among these equilibria. We address this issue for different θ values in the following:

- $\theta \geq \frac{\upsilon+\iota}{\phi+\Phi+\iota}$: According to the Theorem 1, the only PBNE is the strategy profile $\{(\mathcal{NS},\mathcal{NS}),(\mathcal{A},\mathcal{A}),q,p=\theta\}$. We call it $\mathcal{PBNE}_1$.
- $\theta \leq \frac{\upsilon+\iota}{\phi+\Phi+\iota}$: Given the Theorems 1 and 2, there are two equilibria. We call them $\mathcal{PBNE}_2$ and $\mathcal{PBNE}_3$ and are presents as the follow:

$$\mathcal{PBNE}_2 = \{(\mathcal{S},\mathcal{S}),(\mathcal{NA},\mathcal{NA}),p,q=\theta\}$$
$$\mathcal{PBNE}_3 = \{(\mathcal{NS},\mathcal{NS}),(\mathcal{NA},\mathcal{A}),q,p=\theta\}$$

The attacker and the normal user payoffs in the strategy profile $\mathcal{PBNE}_2$ are $\omega + \mu - c_a$ and $\psi + \nu - c_n$, respectively, while they gain payoffs ω and ψ for playing in $\mathcal{PBNE}_3$, respectively. As $\nu \geq c_n$ and $\mu \geq c_a$ are held, the sender prefers to play $\mathcal{PBNE}_2$ rather $\mathcal{PBNE}_3$, because of the more payoff she obtains.

We summarized the above explanation in Table 2 as well as in the following corollary.

Corollary 1. *In the signaling game $\mathcal{G_{FA}}$, the best response for the attacker, normal user and defender for $\theta \leq F$ and $\theta \geq F$ are $\{(\mathcal{S},\mathcal{S}),(\mathcal{NA},\mathcal{NA}),p,q=\theta\}$ and $\{(\mathcal{NS},\mathcal{NS}),(\mathcal{A},\mathcal{A}),q,p=\theta\}$, respectively.*

The above corollary specifies the best strategy for players in the game $\mathcal{G_{FA}}$. In summery, it recommends to the attacker as well as the normal user to select the strategy $\mathcal{S}$ when the value of θ is small enough, otherwise play the strategy $\mathcal{NS}$. The best response of the defender can be summarized as follow.

1. If $\theta \leq F$, the best response for the avatar is $(\mathcal{NA}, \mathcal{NA})$.
2. If $\theta \geq F$, the avatar' best strategy is $(\mathcal{A}, \mathcal{A})$.

4.3 Case Study and Discussion

Following the discussion of the previous subsection, we investigate a case study to illustrate the decision process resulting from Corollary 1. Consider the following values for system parameters: $\kappa = 2$, $\phi = 4$, $\Phi = 5$, $\omega = 5$, $\upsilon = .5$, $c_a = 1$, $c_n = 1$, $\mu = 2.5$, $\nu = 4$, $\pi = 1$, $\psi = 2$ and $\iota = 0.5$. These values states that the gain of the fake avatar for attacker identification (Φ) as well as the gain of the attacker for successfully achieving to her goal (ω) is high in respect to the other parameters, such as the defender's cost for generating false positive alert or making an alert. Additionally, the cost imposed on the defender if she does not detect the attacks (ϕ) is considered high. Assigned values to the other parameters are chosen with the same reasoning.

Given these values, and by calculating the value of F as well as the expected payoff of the sender and receiver for pooling and separating equilibria for different values of θ, we can see in Fig. 3 that the expected payoff for the sender in the pooling strategy profile on the $\mathcal{S}$ is greater than $\mathcal{NS}$ for $\theta = F = 0.105$. Hence, selecting the strategy pair $(\mathcal{S}, \mathcal{S})$ is more beneficial for the sender. For $\theta = F > 0.105$, there is just one equilibrium which is pooling on the $\mathcal{NS}$. Hence, she plays this strategy profile. Following the sender decision, the avatar should plays the strategy profile $(\mathcal{NA}, \mathcal{NA})$ for $\theta < 0.105$ and $(\mathcal{A}, \mathcal{A})$ for $\theta \geq 0.105$.

Typically, the following relations in the sensitive network environment are valid: $\upsilon \ll \Phi$ and $\iota \ll \Phi$. Therefore, $\theta = F = \frac{\upsilon + \iota}{\phi + \Phi + \iota}$ has small value, as we

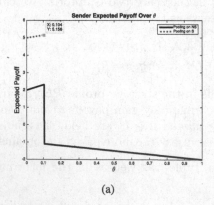
(a)

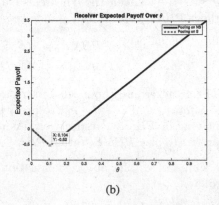
(b)

Fig. 3. Expected payoff of the (a) sender, (b) receiver in the case study. For $\theta < 0.105$, the sender's expected payoff in the pooling strategy profile on the $\mathcal{S}$ is greater than $\mathcal{NS}$. Therefore, she must select $(\mathcal{S}, \mathcal{S})$. The receiver's expected payoff for these θ values are the same in both pooling equilibria. But, according to the sender strategy, the rational fake avatar should select $(\mathcal{NA}, \mathcal{NA})$. After this point, there is no PBNE on the $(\mathcal{S}, \mathcal{S})$ and the only PBNE is on $(\mathcal{NS}, \mathcal{NS})$. Hence, the sender's best decision is $(\mathcal{NS}, \mathcal{NS})$ and the avatar's best response is playing $(\mathcal{A}, \mathcal{A})$.

showed in the case study. In other words, because of the huge cost that the network administrator should pay in the case of information thief by the attacker (value of Φ), the cost of generating the alert as well as the false positive alerting are neglectable. Consequently, the value of the $\theta = F$ converges to zero. Hence, according to the little value of the θ and given the Corollary 1, if the probability of the user being the attacker becomes more than F, the fake avatar should alert to the system. Particularly, just when the environment is enough safe, the avatar does not generate any alert. As soon as the risk of presence of the attacker increases ($\theta > F$), the fake avatar should take the action $\mathcal{A}$. This outcome is consistent with the reality because of the high imposed risk to the system in the case of the existence of the attacker and not alerting. But the result of the signaling game analysis exactly specifies the border of this decision.

5   Conclusion

In this paper, we investigated a deception game in which the fake avatar deployed by the defender interacts with an external user. As the fake avatar is unaware from the type of the external user, we used the signaling game to capture and analyze this interaction. We examined the game and determined its potential equilibria. We then show how the defender can use this analysis to better chooses her strategy in different situations. We illustrated our finding by a case study. As results of our analyses, in Subsect. 4.3, we concluded that if the probability of the user being the attacker becomes more than calculated F, the fake avatar should alert to the system.

References

1. Virvilis, N., Serrano, O.S., Vanautgaerden, B.: Changing the game: the art of deceiving sophisticated attackers. In: 6th International Conference On Cyber Conflict (CyCon 2014), pp. 87–97. IEEE (2014)
2. Shen, S., Li, Y., Xu, H., Cao, Q.: Signaling game based strategy of intrusion detection in wireless sensor networks. Comput. Math. Appl. **62**(6), 2404–2416 (2011)
3. Ahmad, A., Maynard, S.B., Park, S.: Information security strategies: towards an organizational multi-strategy perspective. J. Intell. Manuf. **25**(2), 357–370 (2014)
4. Carroll, T.E., Grosu, D.: A game theoretic investigation of deception in network security. Secur. Commun. Netw. **4**(10), 1162–1172 (2011)
5. Almeshekah, M.H., Spafford, E.H.: Planning and integrating deception into computer security defenses. In: Proceedings of the 2014 workshop on New Security Paradigms Workshop, pp. 127–138. ACM (2014)
6. Zarras, A.: The art of false alarms in the game of deception: leveraging fake honeypots for enhanced security. In: 2014 International Carnahan Conference onSecurity Technology (ICCST), pp. 1–6. IEEE (2014)
7. Wang, W., Bickford, J., Murynets, I., Subbaraman, R., Forte, A.G., Singaraju, G., et al.: Detecting targeted attacks by multilayer deception. J. Cyber Secur. Mob. **2**(2), 175–199 (2013)
8. Costarella, C., Chung, S., Endicott-Popovsky, B., Dittrich, D.: Hardening Honeynets against Honeypot-aware Botnet Attacks. University of Washington, US (2013)

9. Zhu, Q., Clark, A., Poovendran, R., Basar, T.: Deployment and exploitation of deceptive honeybots in social networks. In: Conference on Decision and Control. IEEE (2013)

10. Clark, A., Zhu, Q., Poovendran, R., Başar, T.: Deceptive routing in relay networks. In: Grossklags, J., Walrand, J. (eds.) GameSec 2012. LNCS, vol. 7638, pp. 171–185. Springer, Heidelberg (2012). doi:10.1007/978-3-642-34266-0_10

11. Zhu, Q., Clark, A., Poovendran, R., Basar, T.: Deceptive routing games. In: IEEE 51st Conference on Decision and Control (CDC), pp. 2704–2711. IEEE (2012)

12. L'Huillier, G., Weber, R., Figueroa, N.: Online phishing classification using adversarial data mining and signaling games. In: Proceedings of the ACM SIGKDD Workshop on CyberSecurity and Intelligence Informatics, pp. 33–42. ACM (2009)

13. Ibrahimi, K., Altman, E., Haddad, M.: Signaling game-based approach to power control management in wireless networks. In: Proceedings of Performance monitoring and measurement of heterogeneous wireless and wired networks, pp. 139–144. ACM (2013)

14. Casey, W., Morales, J.A., Nguyen, T., Spring, J., Weaver, R., Wright, E., Metcalf, L., Mishra, B.: Cyber security via signaling games: toward a science of cyber security. In: Natarajan, R. (ed.) ICDCIT 2014. LNCS, vol. 8337, pp. 34–42. Springer, Heidelberg (2014). doi:10.1007/978-3-319-04483-5_4

15. Rahman, M.A., Manshaei, M.H., Al-Shaer, E.: A game-theoretic approach for deceiving remote operating system fingerprinting. In: 2013 IEEE Conference on Communications and Network Security (CNS), pp. 73–81. IEEE (2013)

16. Pawlick, J., Farhang, S., Zhu, Q.: Flip the cloud: cyber-physical signaling games in the presence of advanced persistent threats. In: Khouzani, M.H.R., Panaousis, E., Theodorakopoulos, G. (eds.) GameSec 2015. LNCS, vol. 9406, pp. 289–308. Springer, Heidelberg (2015). doi:10.1007/978-3-319-25594-1_16

17. Mohebbi Moghaddam, M., Manshaei, M.H., Zhu, Q.: To trust or not: a security signaling game between service provider and client. In: Khouzani, M.H.R., Panaousis, E., Theodorakopoulos, G. (eds.) GameSec 2015. LNCS, vol. 9406, pp. 322–333. Springer, Heidelberg (2015). doi:10.1007/978-3-319-25594-1_18

18. Pawlick, J., Zhu, Q.: Deception by design: evidence-based signaling games for network defense. arXiv preprint arXiv:1503.05458 (2015)

19. Patcha, A., Park, J.M.: A game theoretic formulation for intrusion detection in mobile ad hoc networks. IJ Netw. Secur. 2(2), 131–137 (2006)

20. Estiri, M., Khademzadeh, A.: A theoretical signaling game model for intrusion detection in wireless sensor networks. In: 2010 14th International Telecommunications Network Strategy and Planning Symposium (NETWORKS), pp. 1–6. IEEE (2010)

21. Liu, Y., Comaniciu, C., Man, H.: A bayesian game approach for intrusion detection in wireless ad hoc networks. In: Workshop on Game theory for communications and networks. ACM (2006)

22. Lin, J., Liu, P., Jing, J.: Using signaling games to model the multi-step attack-defense scenarios on confidentiality. In: Grossklags, J., Walrand, J. (eds.) GameSec 2012. LNCS, vol. 7638, pp. 118–137. Springer, Heidelberg (2012). doi:10.1007/978-3-642-34266-0_7

23. Shoham, Y., Leyton-Brown, K.: Multiagent Systems: Algorithmic, Game-theoretic, and Logical Foundations. Cambridge University Press, Cambridge (2008)

24. Gibbons, R.: Game Theory for Applied Economists. Princeton University Press, Princeton (1992)

Intrusion Detection and Information Limitations in Security

Network Elicitation in Adversarial Environment

Marcin Dziubiński$^{(\boxtimes)}$, Piotr Sankowski, and Qiang Zhang

Institute of Informatics, University of Warsaw, Banacha 2, 02-097 Warsaw, Poland
{m.dziubinski,sank,qzhang}@mimuw.edu.pl

Abstract. We study a problem of a defender who wants to protect a network against contagious attack by an intelligent adversary. The defender could only protect a fixed number of nodes and does not know the network. Each of the nodes in the network does not know the network either, but knows his/her neighbours only. We propose an incentive compatible mechanism allowing the defender to elicit information about the whole network. The mechanism is efficient in the sense that under truthful reports it assigns the protection optimally.

1 Introduction

The problem of protecting networks against external threats is of great practical importance and has attracted interest from many research communities, from graph theory, operations research and computer science, to physics and economics. Examples of applications include telecommunication networks, power grids, computer networks, social and financial networks. An interesting aspect is protection against intentional, targeted, disruptions caused by intelligent adversaries aiming to exploit networks weakest spots. This is particularly relevant in modern era where terrorist or cyber-terrorist attacks pose a realistic threat and computer aided security support systems become increasingly important [13,21]. However, all previous works on this topic share one important weakness—they assume perfect knowledge about the state and topology of the network. It seems that in the modern distributed systems this assumption rarely can be justified.

This paper aims to study a novel model of network protection where the defending agent does not have a full knowledge about the connections in the network and needs to elicit it from nodes present in the system. We introduce three stages in our model. In the first stage, the defender ask all the nodes in the network to report their ties. In the second stage, the defender decides which nodes to protect. In the last, third stage, the adversary attacks one of the nodes. The attack is infectious and the infection spreads across the network through

Marcin Dziubiński was supported by the Strategic Resilience of Networks project realized within the Homing Plus programme of the Foundation for Polish Science, co-financed by the European Union from the Regional Development Fund within Operational Programme Innovative Economy ("Grants for Innovation"). Piotr Sankowski was supported by ERC project PAAl-POC 680912, EU FET project MULTIPLEX 317532 and polish funds for years 2012-2016 for co-financed international projects.

© Springer International Publishing AG 2016
Q. Zhu et al. (Eds.): GameSec 2016, LNCS 9996, pp. 397–414, 2016.
DOI: 10.1007/978-3-319-47413-7_23

unprotected nodes. The main challenge in this model is to design a truthful mechanism: can we incentivize the autonoumous and self-interested nodes to reveal their true ties in the network via the protection mechanism only, and ensuring, at the same time, that the network is well protected? We not only show that this is possible but we present optimal truthful mechanism as well: when nodes report truthfully, the mechanism assigns protection optimally. In other words, we show that incentive compatibility does not entail any additional costs for the defender.

One of the earliest papers addressing the adversary-defender model of network protection is [5], where a problem of network attack and defence is studied. The players move sequentially. First, the designer chooses a network and strengths of links. Then the adversary distributes his attacking resources across all the links. A link is removed if the amount of assigned attacking resources is at least equal to its strength. It is shown that the complexity of computing optimal attack and optimal defence is strongly polynomial. The next paper [15] studies the network attack and defence problem for a particular class of networks – complete multilayered networks. The network is known to both the adversary and the designer and they simultaneously choose how to distribute their attack and defence resources across the nodes. A node is removed if the strength of attack exceeds the strength of defence assigned to it. In more recent papers [9–11,16,17] models of network defence are introduced where the defender chooses a spanning tree of a network, while the attacker chooses a link to remove. The defender and the adversary also move simultaneously. The attack is successful if the chosen link belongs to the chosen spanning trees. Polynomial time algorithms for computing optimal attack and defence strategies are provided for several variants of this game.

A number of papers consider a model which is closely related to the model introduced in this paper: attacks by an adversary are contagious and defence is perfect. Aspnes *et al.* [2] present a model with random attacks, where infection spreads from an infected node to all unprotected nodes with probability one. The authors focus on computing Nash equilibria and show that the problem is NP-hard. They provide approximation algorithms for finding the equilibria. A number of follow up papers extend this model by: introducing malicious nodes that cooperate with the adversary [20], considering link protection [14], considering links and nodes protection [12]. On the other hand, the paper [1] uses random networks setting to study incentives to protect when nodes care about their own survival only. Other related works [18,19] use techniques based on mean field analysis to study the problem of incentives and externalities in network security on random networks. Lastly, some papers study the problem of network defence in combination with problem of network design. For example, a problem of choosing both the network and protection of the nodes prior to subsequent contagious attack by an adversary is considered in [8]. A similar problem under non-contagious attacks was analysed by [7]. Paper [4] proposes a problem of network design prior to decentralized protection decision by the nodes, followed by an attack of an adversary. They show how inefficiencies due to decentralization can be mitigated by choosing right network topologies.

We stress that all the studies listed above assume that the network is known to the defender (or the nodes, if defence decisions are decentralized) and focus on finding optimal or stable protection. In this paper we relax the assumption of full information about the network and consider a mechanism design problem with the aim to assign protection optimally in situations where neither the defender nor the nodes know the entire network. We propose a mechanism where truth telling (revealing all the private information about the network each node has) is a weakly dominant strategy. In other words, we propose an incentive compatible mechanism. The mechanism is also efficient: it assigns defence optimally when the network is reported truthfully.

Notice that designing a mechanism for defence allocation that would be incentive compatible is not difficult: we could simply ignore the reports and assign the defence in some predefined way. Having a mechanism that is incentive compatible and, at the same time, efficient is the challenge. The literature is full of impossibility results for more or less general settings. We show that in our setting it is theoretically feasible and, in addition, the defender is able to learn the whole network topology, as a byproduct. To our knowledge this is the first paper to consider this aspect of the problem and to propose a mechanism allowing for network elicitation.

A mechanism presented in this paper may be useful in situations where the defender has no full knowledge of the network and where discovering the network topology is either infeasible or very costly. This might be an issue in the case of ad hoc, content sharing software networks (like content sharing P2P networks, networks of e-mail contacts, or social networks) as well as in the case of open multiagent systems. Even if discovering the full topology of such networks was feasible, it may be costly or restricted by the law. In such cases the mechanism that we show in the paper offers an alternative method of gathering information about the network, which might be less costly than other methods.

The rest of the paper is organized as follows. In Sect. 2 we provide a simple example illustrating the problem studied in the paper. In Sect. 3 we define the model and introduce all the graph theoretic and game theoretic notions used in the paper. Section 4 contains the analysis of the problem and our results. We conclude in Sect. 5.

2 Example

To see potential problems faced by the defender with incomplete information about the network, consider the following example. Suppose that the actual network is a star over 4 nodes, $1, 2, 3, 4$, with node 1 being the centre of the star (c.f. Fig. 1(a)). Suppose that defender's budget allows him to protect at most one node in the network. The protected network is going to be attacked by an adversary, who will infect one node in the network and the infection will spread through all unprotected nodes reachable from the attacked node. The defender assumes that the adversary aims to infect as many nodes as possible. Moreover, each node assumes the worst case scenario: it will get infected unless it is defended.

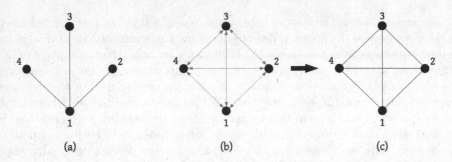

Fig. 1. (a) The actual network; (b) Neighbours reported by the nodes; (c) The network reconstructed from reports.

If the defender knew the network, it would be optimal for him to protect the central node, 1, and one of the periphery nodes would be removed by the attack (since the nodes are pessimistic, each of the assumes to be the one to get infected, hence each periphery node wants to receive protection). Suppose that the defender uses the following, naïve, method of learning the topology of the network. All the nodes are asked to report their neighbours. The defender reconstructs the network by adding to it every link reported by at least one node. The defender promises to protect one of the optimal nodes in thus reconstructed network. Every periphery node has incentive to report all other nodes as its neighbours (c.f. Fig. 1(b)). In effect the reconstructed graph would be a clique (c.f. Fig. 1(c)) and each periphery node would receive protection with probability 1/4, instead of 0. Thus the simple method provides incentives for the nodes to submit untruthful reports. Is there a method that would allow the defender to allocate protection optimally when information about the network topology is lacking? As we show in the paper, such a method exists. It removes incentives to misreport from the nodes and it allows the defender to elicit the actual network topology.

The important feature of the simple example above is that there are no externalities, in the sense that every node wants to receive protection for itself and no node can benefit from any other node receiving protection. This is a key feature of the general model defined and studied in this paper.

3   The Model

Let $N = \{1, \ldots, n\}$ be a set of nodes (e.g. computers + their users in a computer network). The nodes are connected forming an undirected graph $G = (N, E)$, where E is the set of links between the nodes from N. The set of all graphs over a set of nodes $M \subseteq N$ is denoted by $\mathcal{G}(M)$. The set of all graphs that can be formed over the set of nodes N or any of its subsets is $\mathcal{G} = \bigcup_{M \subseteq N} \mathcal{G}(M)$. The nodes are under a threat of contagious attack by an adversary, who infects a chosen node in the network. To lower the damage an attack can cause to the network, some nodes of the network can be protected. Protection is perfect and a protected node cannot be infected.[1] The infection spreads from the attacked node to any node in the network that can be reached from it by a path consisting of unprotected nodes.

Before defining the problem of a defender who wants to allocate protection to selected nodes in the network, we provide formal description of network attack. We start with graph theoretic terminology that will be used. *Undirected graph G over a set of nodes N* is a pair (N, E) such that $E \subseteq \{ij : i \neq j \text{ and } i, j \in N\}$ is the set of edges (or links) between nodes from N (where ij is an abbreviation for $\{i, j\}$). We will use $E(G)$ to denote the set of links of G. Given graph $G = (N, E)$ and a set of nodes $M \subseteq N$, $G[M] = (M, E|_M)$, $E|_M = \{ij \in E : i, j \in M\}$, denotes the *subgraph of G induced by M*. $G - M = G[N \backslash M]$ denotes the graph obtained from G by removing the set of nodes M from it. Two nodes $i, j \in N$ are *connected* in G if there exists a sequence of nodes $i_0, \ldots, i_m$, called a *path*, such that $i = i_0$, $j = i_m$, and for all $k \in \{1, \ldots, m\}$, $i_{k-1}i_k \in E(G)$. A *component* of graph G is a maximal set of nodes such that every two nodes from the set are connected in G. The set of all components of G is denoted by $\mathcal{C}(G)$. Given node i, component $C_i(G)$ is the component in $\mathcal{C}(G)$ containing i. As is common, we will abuse the terminology by using the term component to refer to a graph induced by component (set of nodes) C, $G[C]$. Graph G is connected if $|\mathcal{C}(G)| = 1$. Throughout the paper we restrict attention to connected networks only.

Let $G \in \mathcal{G}(N)$ be a connected graph over the set of nodes N and let $D \subseteq N$ be the set of protected nodes in the network. The graph $G - D$, obtained from G by removing all the protected nodes is called an *attack network* (of G and D). An attack on an unprotected node $x \in N \backslash D$, infects and removes the component of the attack networks containing x, $C_x(G - D)$. This leads to *residual network* $G - C_x(G - D)$. An attack on a protected node has no effect on the network. An example of attack on a defended network is presented in Fig. 2.

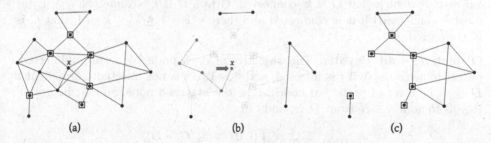

Fig. 2. (a) Network G with defended nodes, D, marked with square frames; (b) Attack network $G - D$ and infection of node x and its component $C_x(G - D)$ in $G - D$; (c) Residual network $G - C_x(G - D)$.

---

[1] The assumption of perfect defence is not a crucial one and is made for presentation simplicity. The mechanisms proposed in the paper would work even if protection could fail with some probability. Crucial is the fact that every node prefers to be protected to not being protected.

The defender wants to allocate perfect defence to selected nodes in the network (e.g. subsidise use of antivirus software). The defender has fixed number $b \geq 1$ of units of defence to allocate across the nodes. The defender knows the set of nodes, N, but she does not know the set of edges.

The nodes do not know the entire network. Each of them has information about its neighbourhood only. Given graph G over N and set of nodes $M \subseteq N$, $\partial_M(G) = \{j \in N \backslash M : i \in M$ and $ij \in E(G)\}$ is the *set of (external) neighbours of M in G*. In the case of M being a singleton set, $M = \{i\}$, we will write $\partial_i(G)$ rather than $\partial_{\{i\}}(G)$. Private information of each node i is represented by i's type, $\theta_i(G) = \partial_i(G)$, i.e. it is the set of nodes being i's neighbours in G. The vector $\boldsymbol{\theta}(G)$ is the vector of all nodes' types.

After the defence is allocated, one of the nodes in the network gets attacked, after which the infection spreads. All infected nodes obtain utility 0. All the remaining nodes obtain utility 1.

The above assumptions are reasonable in the settings where the nodes do not care about knowing the whole network (knowing the whole network and maintaining this information is usually costly). The assumption that the defender does not know the network is in line with the mechanism design approach to the problem. The objective is to propose a reusable mechanism that could be applied in many circumstances and that could be used to allocate the defence without the need to learn the network using other methods (which could require additional cost and time). The worst case scenario approach is common in the study of network vulnerability.[2]

Throughout the paper we will also use the following additional graph theoretic notation and terminology. A set of nodes $X \subseteq N$ is called a *k-cut* of G (for $k \geq 1$) if $|X| = k$, G is connected, and $G - X$ is not connected.[3] Graph G is *k-connected*, $k \geq 1$, if there exists a k-cut of G and for all $1 \leq k' < k$ there is no k'-cut of G. Graph G is k^+-*connected* (at least k-connected) if there exists $k' \geq k$ such that G is k'-connected. Graph G is k^--*connected*, $k \geq 1$, (at most k-connected) if it is connected and there exists $1 \leq k' \leq k$ such that G is k'-connected.

Objectives and Payoffs. The objective of each node is to stay uninfected: payoff to node i is 0, if i is infected, and it is 1, if i is not infected. Formally, let $D \subseteq N$ be a set of protected nodes, x be the attacked node, and G the graph. Payoff to node $i \in N$ from D, x, and G is

$$\Pi^i(D, x, G) = \begin{cases} 0, \text{ if } i \in C_x(G - D), \\ 1, \text{ otherwise.} \end{cases} \tag{1}$$

The objective of the defender is to maximise (utilitarian) social welfare from network defence., i.e. the sum of nodes utilities in the residual network. Formally,

---

[2] Another common approach is to consider average case scenario, which is common in the study of network reliability (c.f. [3], for example).

[3] This paper is concerned with vertex cuts. Therefore we will use a term 'cut' to refer to vertex cuts (as opposed to edge cuts, which are not in scope of this paper).

given set of protected nodes, D, the attacked node, x, and network G, social welfare from network defence is

$$W(D, x, G) = \sum_{i \in N} \Pi^i(D, x, G) = n - |C_x(G - D))|, \qquad (2)$$

that is it is equal to the number of uninfected nodes.

The defender wants to maximise social welfare in the worst case scenario. Therefore his objective is to maximise the pessimistic social welfare across all possible attacks. Pessimistic social welfare from a given set of protected nodes, D, and network, G, is

$$PW(D, G) = \min_{x \in N \setminus \{D\}} W(D, x, G). \qquad (3)$$

Nodes' approach is also pessimistic: every unprotected node assumes to get infected in result of the attack. Hence, given defence D, pessimistic payoff to node i is

$$\mathbf{P}\Pi^i(D) = \begin{cases} 1, \text{if } i \in D, \\ 0, \text{otherwise.} \end{cases} \qquad (4)$$

3.1   Mechanism

To allocate $b \geq 1$ units of defence, the defender will use a (randomised) mechanism. A mechanism $\mathcal{M}$ is a pair (M, α), where $M = (M_1, \ldots, M_n)$ are sets of messages of each player and $\alpha : \prod_{i=1}^{n} M_i \to \Delta(\bigcup_{j=0}^{b} \binom{N}{j})$ is a (randomised) *protection allocation function*: given a message profile $\mathbf{m} \in \prod_{i=1}^{n} M_i$, $\alpha(\mathbf{m})$ is a *randomized protection*, i.e. a probability distribution over the sets of nodes of cardinality at most b, including no protection ($\varnothing$). Given a finite set X, $\Delta(X)$ denotes the set of all lotteries over X. Given set X and $j \in [0..|X|]$, $\binom{X}{j}$ denotes the set of all j element subsets of X.

Given mechanism (M, α), a strategy of node i is a function $s_i : 2^N \to M$, mapping his types to the message set M. Thus a message of node i with neighbourhood θ_i is $s_i(\theta_i)$. In the case of *direct mechanisms*, the set of messages is $M = 2^N$ and a strategy of node i determines the set of reported neighbours as a function of the set of neighbours.

Given randomized protection allocation $\alpha \in \Delta(2^N)$, expected payoff to node i with set of neighbours θ_i, when the rest of the graph is unknown is

$$\mathbf{E}_\alpha \Pi^i = \sum_{D \in 2^N} \alpha(D) \mathbf{P}\Pi^i(D). \qquad (5)$$

Incentive Compatibility. Fix a (direct) mechanism $\mathcal{M} = (M, \alpha)$. A strategy $s_i \in M_i^{2^N}$ of node i is (weakly) dominant if for every strategy profile of the remaining nodes, $\mathbf{s}_{-i} \in \prod_{j \in N \setminus \{i\}} M_j^{2^N}$, every graph $G \in \mathcal{G}(N)$, and every other strategy $s_i' \in M_i^{2^N}$,

$$\mathbf{E}_{\alpha(s_i(\theta_i(G)), \mathbf{s}_{-i}(\boldsymbol{\theta}_{-i}(G)))} \Pi^i \geq \mathbf{E}_{\alpha(s_i'(\theta_i(G)), \mathbf{s}_{-i}(\boldsymbol{\theta}_{-i}(G)))} \Pi^i. \qquad (6)$$

Mechanism $\mathcal{M}$ is *incentive compatible* if truth telling is a dominant strategy for each node $i \in N$. A message profile, $\boldsymbol{\theta}$, where each node i reports its true type, θ_i, is called *truthful*.

4   Analysis

Assuming that the defender knows graph G, it is optimal to protect set of b nodes that maximizes defender's welfare from the residual graph. Given graph G and the number of defence resources, b, let

$$\mathcal{OPT}(G, b) = \arg \max_{D \in \binom{N}{b}} \min_{x \in N \setminus D} |N \setminus C_x(G - D)| = \arg \min_{D \in \binom{N}{b}} \max_{C \in \mathcal{C}(G - D)} |C| \quad (7)$$

denote the set of all optimal b-element sets of nodes to protect.

We start the analysis with an easier case of one unit of defence. This introduces the main ideas for the mechanism. The ideas for the case of more units of defence build on them but are a bit more involved.

4.1   Case of $b = 1$

We start with the case where the defender can protect at most one node, i.e. $b = 1$.

If G is 1-connected, then an optimal (singleton) set of nodes to protect is one of the 1-cuts of G. If G is 2^+-connected, then protecting any (singleton) set of nodes is equally good to the defender. The following lemma on the properties of optimal sets will be vital.

Lemma 1. *Let G be a connected graph over the set of nodes, N, $\{i\} \in \mathcal{OPT}(G, 1)$, $j \in N$, $i \neq j$, be two distinct nodes. Then the following hold*

(i). $\{i\} \in \mathcal{OPT}(G + ij, 1)$.
(ii). For all $l \in N$, if $\{l\} \in \mathcal{OPT}(G + ij, 1)$, then $\{l\} \in \mathcal{OPT}(G, 1)$.

Proof. Take any connected graph, G, and any two distinct nodes, $i, j \in N$, such that $i \in \mathcal{OPT}(G, 1)$, as stated in the lemma. If $ij \in G$, then the two points of the lemma are trivially satisfied. Suppose that $ij \notin G$. If G is 2^+-connected, then $G + ij$ is 2^+-connected as well. Consequently, $\mathcal{OPT}(G, 1) = \mathcal{OPT}(G + ij, 1) = \binom{N}{1}$, and points (i) and (ii) follow immediately.

Suppose that G is 1-connected, in which case $\{i\}$, and any element of $\mathcal{OPT}(G, 1)$, is a 1-cut in G. Let $H = G + ij$. Since $\{i\}$ is a 1-cut in G so $\{i\}$ is a 1-cut in H and $\mathcal{C}(G - \{i\}) = \mathcal{C}(H - \{i\})$. Thus

$$\max_{C \in \mathcal{C}(G - \{i\})} |C| = \max_{C \in \mathcal{C}(H - \{i\})} |C|. \quad (8)$$

Moreover, H is 1-connected and any element of $\mathcal{OPT}(H, 1)$ is a 1-cut of H.

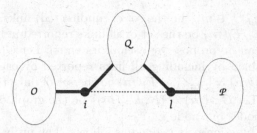

Fig. 3. Schematic presentation of graph G when i and l are 1-cuts. Oval regions represent sets of components, bold lines represent sets of edges between nodes i and l and the components in the oval regions. The dashed line between nodes i and j represent a possible link between the two nodes.

Suppose that $\{l\} \in \mathcal{OPT}(H, 1)$. It follows that

$$\max_{C \in \mathcal{C}(H-\{l\})} |C| \leq \max_{C \in \mathcal{C}(H-\{i\})} |C|. \tag{9}$$

If $l = i$, then the claim in the lemma follows automatically. Suppose that $l \neq i$.

The set of components of $\mathcal{C}(G - \{i, l\})$ can be partitioned into pairwise disjoint subsets $\mathcal{O}$, $\mathcal{P}$, and $\mathcal{Q}$ (some of them possibly empty) such that (c.f. Fig. 3):

$$\mathcal{O} = \{C \in \mathcal{C}(G - \{i, l\}) : \partial_C(G) = \{i\}\}$$
$$\mathcal{P} = \{C \in \mathcal{C}(G - \{i, l\}) : \partial_C(G) = \{l\}\}$$
$$\mathcal{Q} = \{C \in \mathcal{C}(G - \{i, l\}) : \partial_C(G) = \{i, l\}\}.$$

Since $\{l\}$ is a 1-cut in H so it must be that $j \in \bigcup \mathcal{O} \cup \bigcup \mathcal{Q} \cup \{l\}$. Hence $\mathcal{C}(G - \{l\}) = \mathcal{C}(H - \{l\})$ and

$$\max_{C \in \mathcal{C}(G-\{l\})} |C| = \max_{C \in \mathcal{C}(H-\{l\})} |C|. \tag{10}$$

Moreover, since $i \in \mathcal{OPT}(G, 1)$ so

$$\max_{C \in \mathcal{C}(G-\{i\})} |C| \leq \max_{C \in \mathcal{C}(G-\{l\})} |C|. \tag{11}$$

Combining this with (8–11), we get

$$\max_{C \in \mathcal{C}(G-\{i\})} |C| = \max_{C \in \mathcal{C}(G-\{l\})} |C| = \max_{C \in \mathcal{C}(H-\{l\})} |C| = \max_{C \in \mathcal{C}(H-\{i\})} |C|.$$

Thus $i \in \mathcal{OPT}(H, 1)$ and $l \in \mathcal{OPT}(G, 1)$, which proves points (i) and (ii). □

Mechanism. Consider the following (direct) mechanism $\mathcal{M} = (M, \alpha)$. The set of messages, M_i, of each node $i \in N$ is the same across nodes and $M_i = 2^N$. That is, each player submits a set of nodes. Given a message profile $\boldsymbol{m} = (m_1, \ldots, m_n)$ the defender allocates the unit of defence as follows.

Let $E_i(m) = \{ij : j \in m_i\}$ be the set of (undirected) links reported by node i. Let $E(m) = \bigcup_{l \in N} E_l(m)$ be the set of all links reported in message profile m (note that a link can be propose by each of its ends). Let $G(m) = (N, E(m))$ be the graph obtained by including all links reported by each node. For each node $i \in N$ let $E_{-i}(m) = \bigcup_{l \in N \setminus \{i\}} E_l(m)$ be the set of links reported by all the nodes but node i. Let $G_{-i}(m) = (N, E_{-i}(m))$ be the graph determined by the reports of all the nodes but node i.

The allocation function α is defined as follows. The probability that node i receives protection under message profile m depends on the graph $G_{-i}(m)$ only:

$$\alpha_{\{i\}}(m) = \begin{cases} \frac{1}{|\mathcal{OPT}(G_{-i}(m),1)|} & \text{if} \{i\} \in \mathcal{OPT}(G_{-i}(m),1) \\ & \text{and } G_{-i}(m) \text{ is connected,} \\ 0 & \text{otherwise.} \end{cases} \quad (12)$$

The probability that no node receives protection is

$$\alpha_{\varnothing}(m) = 1 - \sum_{i \in N} \alpha_{\{i\}}(m). \quad (13)$$

As we show below, mechanism $\mathcal{M}$ is valid and is incentive compatible.

Proposition 1. *Mechanism $\mathcal{M}$ is valid and is incentive compatible.*

Proof. To show that the mechanism is valid, we need to show that for all $m \in M$, $\alpha(m)$ is a valid probability distribution. Take any $m \in M$. It is clear that for all $i \in N, \alpha_{\{i\}}(m) \in [0, 1]$. Thus what remains to be shown is that $\sum_{i \in N} \alpha_{\{i\}}(m) \leq 1$. Let $\beta(m)$ be a probability distribution on $\binom{N}{1} \cup \{\varnothing\}$ mixing uniformly on the set $\mathcal{OPT}(G(m), 1)$ of optimal nodes to protect in $G(m)$, if $G(m)$ is connected, or assigning probability 1 to $\varnothing$, otherwise. Formally,

$$\beta_{\{i\}}(m) = \begin{cases} \frac{1}{|\mathcal{OPT}(G(m),1)|}, & \text{if} i \in \mathcal{OPT}(G(m), 1) \\ & \text{and } G(m) \text{ is connected} \\ 0, & \text{otherwise,} \end{cases} \quad (14)$$

and

$$\beta_{\varnothing}(m) = 1 - \sum_{i \in N} \beta_{\{i\}}(m). \quad (15)$$

Clearly $\beta(m)$ is a valid probability distribution. We will show that for all $i \in N$, $\alpha_{\{i\}}(m) \leq \beta_{\{i\}}(m)$, which implies that $\alpha(m)$ is a valid probability distribution. If $\{i\} \notin \mathcal{OPT}(G_{-i}(m), 1)$ or $G_{-i}(m)$ is not connected, then trivially $\alpha_{\{i\}}(m) = 0 \leq \beta_i(m)$. Suppose that $G_{-i}(m)$ is connected and $\{i\} \in \mathcal{OPT}(G_{-i}(m), 1)$. By Lemma 1, $\mathcal{OPT}(G(m), 1) \subseteq \mathcal{OPT}(G_{-i}(m), 1)$ and $\{i\} \in \mathcal{OPT}(G(m), 1)$. Hence $\alpha_{\{i\}}(m) = \frac{1}{|\mathcal{OPT}(G_{-i}(m),1)|} \leq \frac{1}{|\mathcal{OPT}(G(m),1)|} = \beta_{\{i\}}(m)$. This shows that the mechanism is valid.

Next we turn to incentive compatibility. Clearly, the utility of any node $i \in N$ is independent of the message of the node, m_i. Hence reporting $m_i = \theta_i$, i.e. revealing the true set of neighbours, is a best response to any message profile m_{-i}. $\square$

Since under truthful message profile, m, every link is reported twice (once by each end) so, for all $i \in N$, $G_{-i}(m) = G(m)$ and $\mathcal{OPT}(G_{-i}(m)) = \mathcal{OPT}(G(m))$. Hence the mechanism is optimal. This is stated in the following observation.

Observation 1. *Let G be a connected network and m be a truthful message profile. The allocation rule $\alpha(m)$ mixes uniformly on $\mathcal{OPT}(G(m, 1))$. Hence it implements optimal protection.*

4.2   Case of $b \geq 2$

The case of $b \geq n$ is trivial. An example of a direct incentive compatible mechanism is the one where each node reports its "neighbours" and the defender protects every node with probability 1. Similarly, the case of $b = n - 1$ is trivial, as in this case an optimal incentive compatible mechanism is the one where each node reports its "neighbours" and the defender picks one of the nodes with probability $\frac{1}{n}$, and protects all the remaining nodes. For the remaining part of this section we assume that $2 \leq b \leq n - 2$.

Analogously to the case of $b = 1$, if the network is $(b+1)^+$-connected, then protecting any b-element set of nodes is optimal for the defender. If the network is b^--connected, then an optimal set to protect must be a b-cut of G.

The following lemma, extending Lemma 1, will be crucial.

Lemma 2. *Let G be a connected graph over the set of nodes, N, $X \in \mathcal{OPT}(G, b)$, $i \in X$, $j \in N$, $i \neq j$, be two distinct nodes. Then the following hold*

(i). $X \in \mathcal{OPT}(G + ij, b)$.
(ii). For all $Y \in \binom{N}{b}$, if $Y \in \mathcal{OPT}(G + ij, b)$, then $Y \in \mathcal{OPT}(G, b)$.

Proof. Take any connected graph, G, any set of nodes $X \in \mathcal{OPT}(G, b)$ and any two distinct nodes, $i, j \in N$, such that $i \in X$, as stated in the lemma. If $ij \in G$, then the two points of the lemma are trivially satisfied. Suppose that $ij \notin G$. If G is $(b+1)^+$-connected, then $G + ij$ is $(b+1)^+$-connected as well. Consequently, $\mathcal{OPT}(G, b) = \mathcal{OPT}(G + ij, b) = \binom{N}{b}$, and points (i) and (ii) follow immediately.

Suppose that G is b^--connected, in which case X, and any element of $\mathcal{OPT}(G, b)$, is a b-cut in G. Let $H = G + ij$. Since X is a b-cut in G so X is a b-cut in H and $\mathcal{C}(G - X) = \mathcal{C}(H - X)$. Consequently,

$$\max_{C \in \mathcal{C}(G - X)} |C| = \max_{C \in \mathcal{C}(H - X)} |C|. \tag{16}$$

Moreover, H is b^--connected and any element of $\mathcal{OPT}(H, b)$ is a b-cut of H.
Suppose that $Y \in \mathcal{OPT}(H, b)$. It follows that

$$\max_{C \in \mathcal{C}(H - Y)} |C| \leq \max_{C \in \mathcal{C}(H - X)} |C|. \tag{17}$$

If $Y = X$, then the claim in the lemma follows immediately. Suppose that $Y \neq X$.

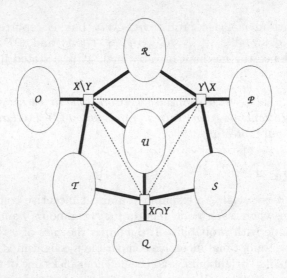

Fig. 4. Schematic presentation of graph G when X and Y are b-cuts. Squares represent subsets of the two cuts ($X\backslash Y$, $Y\backslash X$ and $X \cap Y$), oval regions represent sets of components, bold lines represent sets of edges between the sets of nodes $X\backslash Y$, $Y\backslash X$, and $X \cap Y$, and the components in the oval regions. The dashed lines between squares represent possible sets of links between nodes in $X\backslash Y$, $Y\backslash X$ and $X \cap Y$.

The set of components, $\mathcal{C}(G - (X \cup Y))$ can be partitioned into (pairwise disjoint) subsets $\mathcal{P}$, $\mathcal{Q}$, $\mathcal{R}$, $\mathcal{S}$, $\mathcal{T}$ and $\mathcal{U}$ (some of them may be empty), such that (c.f. Fig. 4):

$$\mathcal{O} = \{C \in \mathcal{C}(G - (X \cup Y)) : \partial_C(G) \subseteq X\backslash Y\}$$
$$\mathcal{P} = \{C \in \mathcal{C}(G - (X \cup Y)) : \partial_C(G) \subseteq Y\backslash X\}$$
$$\mathcal{Q} = \{C \in \mathcal{C}(G - (X \cup Y)) : \partial_C(G) \subseteq X \cap Y\}$$
$$\mathcal{R} = \{C \in \mathcal{C}(G - (X \cup Y)) : \partial_C(G) \subseteq (X \cup Y)\backslash(X \cap Y)$$
$$\text{and } \partial_C(G) \cap (Y\backslash X) \neq \varnothing \text{ and } \partial_C(G) \cap (X\backslash Y) \neq \varnothing\}$$
$$\mathcal{S} = \{C \in \mathcal{C}(G - (X \cup Y)) : \partial_C(G) \subseteq Y \text{ and }$$
$$\partial_C(G) \cap (Y\backslash X) \neq \varnothing \text{ and } \partial_C(G) \cap X \cap Y \neq \varnothing\}$$
$$\mathcal{T} = \{C \in \mathcal{C}(G - (X \cup Y)) : \partial_C(G) \subseteq X \text{ and }$$
$$\partial_C(G) \cap (X\backslash Y) \neq \varnothing \text{ and } \partial_C(G) \cap X \cap Y \neq \varnothing\}$$
$$\mathcal{U} = \{C \in \mathcal{C}(G - (X \cup Y)) : \partial_C(G) \subseteq X \cup Y \text{ and }$$
$$\partial_C(G) \cap (Y\backslash X) \neq \varnothing \text{ and } \partial_C(G) \cap (X\backslash Y) \neq \varnothing \text{ and } \partial_C(G) \cap X \cap Y \neq \varnothing\}.$$

Let $A = (X\backslash Y) \cup \bigcup(\mathcal{O} \cup \mathcal{R} \cup \mathcal{T} \cup \mathcal{U})$. Notice that a maximiser of $\{|C| : C \in \mathcal{C}(G - Y)\}$ is either A or an element of $\mathcal{Q}$, as otherwise it would hold that $\max_{C \in \mathcal{C}(G-X)} |C| > \max_{C \in \mathcal{C}(G-Y)} |C|$, a contradiction with the assumption that $X \in \mathcal{OPT}(G, b)$. Since Y is a separator in H so it must be that either $j \in A$, or $i \in X \cap Y$.

Suppose that $j \in A$. Since $i \in X$ so the new link, ij, is either added to A, or is between A and $X \cap Y$. Either way adding the link does not affect the sizes of components after cut Y is applied and so

$$\max_{C \in \mathcal{C}(H-Y)} |C| = \max_{C \in \mathcal{C}(G-Y)} |C|. \tag{18}$$

Equality (18) follows in the case of $i \in X \cap Y$, because then $\mathcal{C}(G-Y) = \mathcal{C}(H-Y)$. Since $X \in \mathcal{OPT}(G,b)$ so

$$\max_{C \in \mathcal{C}(G-X)} |C| \leq \max_{C \in \mathcal{C}(G-Y)} |C|. \tag{19}$$

Combining this with (16), (17) and (18), we get

$$\max_{C \in \mathcal{C}(H-Y)} |C| = \max_{C \in \mathcal{C}(H-X)} |C| = \max_{C \in \mathcal{C}(G-Y)} |C| = \max_{C \in \mathcal{C}(G-X)} |C|. \tag{20}$$

Thus $X \in \mathcal{OPT}(H,b)$ and $Y \in \mathcal{OPT}(G,b)$. This proves points (i) and (ii). $\square$

Mechanism. Consider the following (direct) mechanism (M, α). The set of messages for each node i is $M_i = 2^N$. The allocation function, α, is defined as follows. Given node $i \in N$, set of nodes $D \in \binom{N}{b}$, and message profile m, define

$$\xi_D^i(m) = \begin{cases} \frac{1}{|\mathcal{OPT}(G_{-i}(m),b)|}, & \text{if } D \in \mathcal{OPT}(G_{-i}(m), b), \\ & i \in D \text{ and } G_{-i}(m) \text{ is} \\ & \text{connected,} \\ 0, & \text{otherwise.} \end{cases} \tag{21}$$

Given set of nodes $D \in \binom{N}{b}$ and message profile m, let

$$\xi_D(m) = \max_{i \in N} \xi_D^i(m). \tag{22}$$

The intended interpretation of $\xi_D(m)$ is the "probability" of "considering" the set of nodes D for protection (of course it requires showing that this quantity is indeed a probability). After the defender chooses the set considered for protection, s/he protects a subset of this set according to the following procedure. Let $D = \{i_1, \ldots, i_b\}$ be such that $\xi_D^{i_1}(m) \geq \ldots \geq \xi_D^{i_b}(m)$. Conditioned on D being considered for protection, the probability that set of nodes $D(k) = \{i_1, \ldots, i_k\}$ is protected is

$$\zeta_{D(k)|D}(m) = \begin{cases} \frac{\xi_D^{i_k}(m)}{\xi_D(m)}, & \text{if } k = b \text{ and} \\ & \xi_D(m) \neq 0 \\ \frac{\xi_D^{i_k}(m) - \xi_D^{i_{k+1}}(m)}{\xi_D(m)}, & \text{if } k \in \{1, \ldots, b-1\} \\ & \text{and } \xi_D(m) \neq 0 \\ 0, & \text{if } \xi_D(m) = 0. \end{cases} \tag{23}$$

It will be convenient to extend the definition of $\zeta_{\cdot|D}(m)$ beyond the sets $D(k)$ by setting it to 0, i.e. given $T \subseteq N$ with $1 \leq |T| \leq b$,

$$\zeta_{T|D}(m) = 0, \text{if for all } k \in \{1, \ldots, b\}, T \neq D(k). \tag{24}$$

The allocation function α is defined as follows. The probability of protecting set of nodes $T \subseteq N$ with $1 \le |T| \le b$ is

$$\alpha_T(m) = \sum_{D \in \binom{N}{b}} \xi_D(m) \zeta_{T|D}(m). \tag{25}$$

The probability of protecting no node is

$$\alpha_\varnothing(m) = 1 - \sum_{i=1}^{b} \sum_{T \in \binom{N}{i}} \alpha_T(m). \tag{26}$$

The following theorem states the validity and incentive compatibility of the mechanism $\mathcal{M}$.

Theorem 1. *Mechanism $\mathcal{M}$ is valid and is incentive compatible.*

Proof. To show that the mechanism is valid, we need to show that for all $m \in M$, $\alpha(m)$ is a valid probability distribution. Take any $m \in M$ and consider $\xi_D(m)$, $D \in \binom{N}{b}$, first. Clearly, for all $D \in \binom{N}{b}$, $\xi_D(m) \in [0,1]$. We will show that

$$\sum_{D \in \binom{N}{b}} \xi_D(m) \le 1. \tag{27}$$

Let $\beta(m)$ be a probability distribution on $\binom{N}{b} \cup \{\varnothing\}$ mixing uniformly on the set $\mathcal{OPT}(G(m), b)$ of optimal nodes to protect in $G(m)$, if $G(m)$ is connected, and assigning probability 1 to $\varnothing$, otherwise. Formally,

$$\beta_D(m) = \begin{cases} \frac{1}{|\mathcal{OPT}(G(m),b)|}, & \text{if } D \in \mathcal{OPT}(G(m), b) \text{ and} \\ & \quad G(m) \text{ is connected} \\ 0, & \text{otherwise,} \end{cases} \tag{28}$$

and

$$\beta_\varnothing(m) = 1 - \sum_{D \in \binom{N}{b}} \beta_D(m). \tag{29}$$

Clearly $\beta(m)$ is a valid probability distribution. We will show that for all $D \in \binom{N}{b}$, $\xi_D(m) \le \beta_D(m)$. If $D \notin \mathcal{OPT}(G_{-i}(m), b)$ or $G_{-i}(m)$ is not connected, then trivially $\xi_D(m) = 0 \le \beta_D(m)$. Suppose that $G_{-i}(m)$ is connected and $D \in \mathcal{OPT}(G_{-i}(m), b)$. By Lemma 2, $\mathcal{OPT}(G(m), b) \subseteq \mathcal{OPT}(G_{-i}(m), b)$ and $D \in \mathcal{OPT}(G(m), b)$. Hence $0 \le \xi_D(m) = \frac{1}{|\mathcal{OPT}(G_{-i}(m),b)|} \le \frac{1}{|\mathcal{OPT}(G(m),b)|} = \beta_D(m)$. Combining this with the fact that

$$\sum_{D \in \binom{N}{b}} \beta_D(m) \le 1 \tag{30}$$

yields (27).

For any $D \in \binom{N}{b}$ consider $\zeta_{\cdot|D}(m)$. Notice that for all $T \subseteq N$ with $1 \leq |T| \leq b$ we have $\zeta_{T|D}(m) \in [0,1]$. Moreover

$$\sum_{k=1}^{b} \sum_{T \in \binom{N}{k}} \zeta_{T|D}(m) = \sum_{k=1}^{b} \zeta_{D(k)|D}(m) = \begin{cases} 1, & \text{if } \xi_D(m) > 0, \\ 0, & \text{if } \xi_D(m) = 0. \end{cases} \quad (31)$$

By (27) and (31),

$$\sum_{k=1}^{b} \sum_{T \in \binom{N}{k}} \alpha_T(m) = \sum_{k=1}^{b} \sum_{T \in \binom{N}{k}} \sum_{D \in \binom{N}{b}} \xi_D(m) \zeta_{T|D}(m) \quad (32)$$

$$= \sum_{D \in \binom{N}{b}} \xi_D(m) \sum_{k=1}^{b} \sum_{T \in \binom{N}{k}} \zeta_{T|D}(m) = \sum_{D \in \binom{N}{b}} \xi_D(m) \leq 1. \quad (33)$$

This and the fact that $\alpha_T(m) \geq 0$, for all $T \subseteq N$ with $0 \leq |T| \leq b$, implies that $\alpha(m)$ is a valid probability distribution.

For incentive compatibility, notice that the utility of any node $i \in N$ is independent of its message, m_i (the probability of i receiving protection is established merely on the basis of $G_{-i}(m)$). Hence reporting $m_i = \theta_i$, i.e. revealing the true set of neighbours, is a best response to any message profile m_{-i}.   □

Like in the case of one unit of defence, using the fact that under truthful message profile, m, every link is reported twice, we observe that for all $i \in N$, $G_{-i}(m) = G(m)$ and $\mathcal{OPT}(G_{-i}(m)) = \mathcal{OPT}(G(m))$. In other words, mechanism $\mathcal{M}$ is optimal. The following observation states this fact.

Observation 2. *Let G be a connected network and m be a truthful message profile. The allocation rule $\alpha(m)$ mixes uniformly on $\mathcal{OPT}(G(m,b))$. Hence it implements optimal protection.*

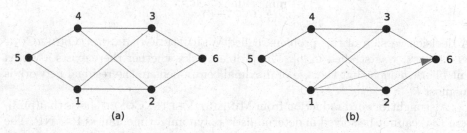

Fig. 5. (a) Network G; (b) Network extracted from reports where every node apart from 1 reports all his neighbours and 1 misreports by reporting the true neighbours and node 6 as a neighbour.

We end this section with an example illustrating how the mechanism works.

Example 1. Consider network G in Fig. 5(a) and suppose that the defender has $b = 2$ units of defence. There are 6 nodes, $N = \{1, \ldots, 6\}$, with types: $\theta_1(G) = \theta_4(G) = \{2, 3, 5\}$, $\theta_2(G) = \theta_2(G) = \{1, 4, 6\}$, $\theta_5(G) = \{1, 4\}$, and $\theta_6(G) = \{2, 3\}$.

Suppose first that every node i reports truthfully, i.e. submits its own type: $m_i = \theta_i(G)$, for all $i \in N$. In this case, since every link is reported by its both ends, the networks $G_{-i}(m) = G$, for all $i \in N$. The set of optimal 2-element sets of nodes to protect is $\mathcal{OPT}(G, 2) = \{\{1, 4\}, \{2, 3\}\}$. Hence $\xi^i_D(m) = \frac{1}{2}$, if $D \in \mathcal{OPT}(G, 2)$, and is 0, otherwise; consequently, $\xi_D(m) = \frac{1}{2}$, if $D = \{1, 4\}$, and $\xi_D(m) = 0$, otherwise. Furthermore $\zeta_{D(2)|D}(m) = 1$, if $D \in \mathcal{OPT}(G, 2)$, and is 0, otherwise. In effect $\alpha_D(m) = \frac{1}{2}$, if $D \in \mathcal{OPT}(G, 2)$, and is 0, otherwise. Thus the defender assigns protection optimally.

Now, suppose that $m_1 = \{2, 3, 5, 6\}$, i.e. in addition to the true neighbours 1 reports falsely that 6 is its neighbour. Suppose also that all the other nodes report truthfully (c.f. Fig. 5(b)). In this case for every $i \in N \backslash \{1\}$, the network $G_{-i}(m) = G'$ where G' is network G with an additional link, $\{1, 6\}$, and the network $G_{-1}(m) = G$. In result from adding link $\{1, 6\}$, $\{2, 3\}$ is no longer an optimal 2-element set of nodes to protect and $\mathcal{OPT}(G', 2) = \{1, 4\}$. Therefore $\xi^1_{\{1,4\}}(m) = \frac{1}{2}$, $\xi^4_{\{1,4\}}(m) = 1$, and $\xi^i_D(m) = 0$, otherwise. Consequently, $\xi_{\{1,4\}}(m) = 1$, and $\xi_D(m) = 0$, otherwise. Quantities $\zeta_{T|D}(m)$ are non-zero for $D = \{1, 4\}$ and $T \in \{\{4\}, \{1, 4\}\}$ only. Since $1 = \xi^4_{\{1,4\}}(m) \geq \xi^1_{\{1,4\}}(m) = \frac{1}{2}$ so $\{1, 4\}(1) = \{4\}$ and $\{1, 4\}(2) = \{1, 4\}$. Thus $\zeta_{\{1,4\}|\{1,4\}}(m) = \frac{1}{2}$ and $\zeta_{\{4\}|\{1,4\}}(m) = \frac{1 - \frac{1}{2}}{1} = \frac{1}{2}$. Hence $\alpha_{\{1,4\}}(m) = \frac{1}{2}$, $\alpha_{\{4\}}(m) = \frac{1}{2}$, and $\alpha_T(m) = 0$, for all remaining $T \subseteq N$ with $|T| \leq 2$. Thus reporting an additional neighbour, 6, does not improve the chances of 1 receiving protection and so 1 does not have incentive for this misreport.

4.3 A Note on Computational Complexity

Given the graph, G, over n nodes and b units of protection, the defender aiming to protect the network optimally has to solve the following optimization problem:

$$\min_{Z \in \binom{N}{b}} \max_{C \in \mathcal{C}(G-Z)} |C|. \tag{34}$$

A decision version of this problem, called WEIGHTEDCOMPONENTORDERCON-NECTIVITY, was studied in [6]. The problem asks whether there exists a vertex cut of size b such that the size of maximal component in the residual network is at most l.

A straightforward reduction from MINIMUMVERTEXCOVER shows that problem (34) cannot be solved in deterministic polynomial time, unless P = NP. The reduction involves solving the above optimization problem for subsequent values of b, starting from 1, until the maximum size of a component in the residual network is 1. Unfortunately, no approximation algorithms are known for the problem above. A similar problem, called SUMOFSQUARESPARTITION, asking for a vertex cut of size b that minimizes the sum of squares of component sizes in the residual network was proposed and studied in [2]. It is shown that optimal solution can be approximated to within a factor of $O(\log^{1.5}(n))$. This could be a suggestion that the problem (34) might have similar approximation guarantees.

A more promising approach could be by considering scenarios were defender's budget, b, is small (which, arguably, is a realistic assumption). It is shown in [6] that exact solution for WEIGHTEDCOMPONENTORDERCONNECTIVITY can be found in time $n \cdot 2^{O(b \log l)}$. This suggests that with low budget it may be possible to find optimal solutions for the optimization problem (34) in a reasonable time.

5   Conclusions

In this paper we considered a problem of protecting a network against contagious attacks by an adversary when neither the defender nor the nodes have full information about the network. We proposed a mechanism that allows the defender to allocate protection optimally by eliciting full information about the network from the nodes. To our knowledge this is the first paper that considers such a problem and proposes an incentive compatible mechanism that allows for eliciting the information about the network.

Interesting problems for future research would include finding mechanisms for network elicitation in adversarial domain where nodes form beliefs (in form of probability distribution) about the unknown aspects of the network, instead of using a pessimistic approach considered in this paper. Another interesting question would be to consider a mechanism for network elicitation in the settings where nodes care not only about their own survival but also about the value of the residual network (e.g. the number of neighbours that survive the attack or the size of their component after the attack). Lastly, algorithms for approximating set $\mathcal{OPT}(G, b)$ maintaining the monotonicity conditions of Lemma 2 could be sought in order to obtain an incentive compatible approximate version of the mechanism presented in the paper.

References

1. Acemoglu, D., Malekian, A., Ozdaglar, A.: Network Security and Contagion. MIT Mimeo, New York (2013)
2. Aspnes, J., Chang, K., Yampolskiy, A.: Inoculation strategies for victims of viruses and the sum-of-squares partition problem. J. Comput. Syst. Sci. **72**(6), 1077–1093 (2006)
3. Boesch, F., Satyanarayana, A., Suffel, C.: A survey of some network reliability analysis and synthesis results. Networks **54**(2), 99–107 (2009)
4. Cerdeiro, D., Dziubiński, M., Goyal, S.: Individual security and network design. In: Proceedings of the Fifteenth ACM Conference on Economics and Computation, EC 2014, pp. 205–206. ACM, New York (2014)
5. Cunningham, W.: Optimal attack and reinforcement of a network. J. ACM **32**(3), 549–61 (1985)
6. Drange, P.G., Dregi, M.S., Hof, P.: On the computational complexity of vertex integrity and component order connectivity. In: Ahn, H.-K., Shin, C.-S. (eds.) ISAAC 2014. LNCS, vol. 8889, pp. 285–297. Springer, Heidelberg (2014). doi:10.1007/978-3-319-13075-0_23

7. Dziubiński, M., Goyal, S.: Network design and defence. Games Econ. Behav. **79**(1), 30–43 (2013)
8. Goyal, S., Vigier, A.: Attack, defence, and contagion in networks. Rev. Econ. Stud. **81**(4), 1518–1542 (2014)
9. Gueye, A., Marbukh, V.: Towards a metric for communication network vulnerability to attacks: a game theoretic approach. In: Proceedings of the 3rd International ICTS Conference on Game Theory for Networks, GameNets 2012 (2012)
10. Gueye, A., Walrand, J.C., Anantharam, V.: Design of network topology in an adversarial environment. In: Alpcan, T., Buttyán, L., Baras, J.S. (eds.) GameSec 2010. LNCS, vol. 6442, pp. 1–20. Springer, Heidelberg (2010). doi:10.1007/978-3-642-17197-0_1
11. Gueye, A., Walrand, J., Anantharam, V.: A network topology design game: how to choose communication links in an adversarial environment. In: Proceedings of the 2nd International ICTS Conference on Game Theory for Networks, GameNets 2011 (2011)
12. He, J., Liang, H., Yuan, H.: Controlling infection by blocking nodes and links simultaneously. In: Chen, N., Elkind, E., Koutsoupias, E. (eds.) Internet and Network Economics. Lecture Notes in Computer Science, vol. 7090, pp. 206–217. Springer, Berlin Heidelberg (2011)
13. Jain, M., An, B., Tambe, M.: An overview of recent application trends at the AAMAS conference: security, sustainability and safety. AI Mag. **33**(3), 14–28 (2012)
14. Kimura, M., Saito, K., Motoda, H.: Blocking links to minimize contamination spread in a social network. ACM Trans. Knowl. Discovery Data **3**(2), 9: 1–9: 23 (2009)
15. Kovenock, D., Roberson, B.: The optimal defense of networks of targets. Purdue University Economics Working Papers 1251, Purdue University, Department of Economics (2010)
16. Laszka, A., Szeszlér, D., Buttyán, L.: Game-theoretic robustness of many-to-one networks. In: Proceedings of the 3rd International ICTS Conference on Game Theory for Networks, GameNets 2012 (2012)
17. Laszka, A., Szeszlér, D., Buttyán, L.: Linear loss function for the network blocking game: an efficient model for measuring network robustness and link criticality. In: Proceedings of the 3rd International Conference on Decision and Game Theory for Security, GameSec 2012 (2012)
18. Lelarge, M., Bolot, J.: A local mean field analysis of security investments in networks. In: Feigenbaum, J., Yang, Y.R. (eds.) NetEcon, pp. 25–30. ACM (2008)
19. Lelarge, M., Bolot, J.: Network externalities and the deployment of security features and protocols in the internet. ACM SIGMETRICS Perform. Eval. Rev. - SIGMETRICS **36**(1), 37–48 (2008)
20. Moscibroda, T., Schmid, S., Wattenhofer, R.: The price of malice: a game-theoretic framework for malicious behavior. Internet Math. **6**(2), 125–156 (2009)
21. Yang, R., Kiekintveld, C., Ordóñez, F., Tambe, M., John, R.: Improving resource allocation strategies against human adversaries in security games: an extended study. Artif. Intell. **195**, 440–469 (2013)

Optimal Thresholds for Anomaly-Based Intrusion Detection in Dynamical Environments

Amin Ghafouri[1]([⊠]), Waseem Abbas[1], Aron Laszka[2], Yevgeniy Vorobeychik[1], and Xenofon Koutsoukos[1]

[1] Institute for Software Integrated Systems, Vanderbilt University, Nashville, USA
{amin.ghafouri,waseem.abbas,yevgeniy.vorobeychik,
xenofon.koutsoukos}@vanderbilt.edu
[2] Department of Electrical Engineering and Computer Sciences,
University of California, Berkeley, USA
laszka@berkeley.edu

Abstract. In cyber-physical systems, malicious and resourceful attackers could penetrate a system through cyber means and cause significant physical damage. Consequently, early detection of such attacks becomes integral towards making these systems resilient to attacks. To achieve this objective, intrusion detection systems (IDS) that are able to detect malicious behavior early enough can be deployed. However, practical IDS are imperfect and sometimes they may produce false alarms even for normal system behavior. Since alarms need to be investigated for any potential damage, a large number of false alarms may increase the operational costs significantly. Thus, IDS need to be configured properly, as oversensitive IDS could detect attacks very early but at the cost of a higher number of false alarms. Similarly, IDS with very low sensitivity could reduce the false alarms while increasing the time to detect the attacks. The configuration of IDS to strike the right balance between time to detecting attacks and the rate of false positives is a challenging task, especially in dynamic environments, in which the damage caused by a successful attack is time-varying.

In this paper, using a game-theoretic setup, we study the problem of finding optimal detection thresholds for anomaly-based detectors implemented in dynamical systems in the face of strategic attacks. We formulate the problem as an attacker-defender security game, and determine thresholds for the detector to achieve an optimal trade-off between the detection delay and the false positive rates. In this direction, we first provide an algorithm that computes an optimal *fixed threshold* that remains fixed throughout. Second, we allow the detector's threshold to change with time to further minimize the defender's loss, and we provide a polynomial-time algorithm to compute time-varying thresholds, which we call *adaptive thresholds*. Finally, we numerically evaluate our results using a water-distribution network as a case study.

Keywords: Cyber-physical systems · Security · Game theory · Intrusion detection system

© Springer International Publishing AG 2016
Q. Zhu et al. (Eds.): GameSec 2016, LNCS 9996, pp. 415–434, 2016.
DOI: 10.1007/978-3-319-47413-7_24

1   Introduction

In recent years, we have seen an increasing trend of malicious intruders and attackers penetrating into various cyber-physical systems (CPS) through cyber means and causing severe physical damage. Examples of such incidents include the infamous Stuxnet worm [13], cyber attack on German steel plant [17], and Maroochy Shire water-services incident [1] to name a few. To maximize the damage, attackers often aim to remain covert and avoid getting detected for an extended duration of time. As a result, it becomes crucial for a defender to design and place efficient intrusion and attack detection mechanisms to minimize the damage. While attackers may be able to hide the specific information technology methods used to exploit and reprogram a CPS, they cannot hide their final intent: the need to cause an adverse effect on the CPS by sending malicious sensor or controller data that will not match the behavior expected by an anomaly-based detection system [7]. Anomaly-based detection systems incorporate knowledge of the physical system, in order to monitor the system for suspicious activities and cyber-attacks. An important design consideration in such detection systems is to carefully configure them in order to satisfy the expected monitoring goals.

A well-known method for anomaly-based detection is sequential change detection [11]. This method assumes a sequence of measurements that starts under the normal hypothesis and then, at some point in time, it changes to the attack hypothesis. Change detection algorithm attempts to detect this change as soon as possible. In a sequential change detection, there is a *detection delay*, that is, a time difference between when an attack occurs and when an alarm is raised. On the other hand, detection algorithms may induce *false positives*, that is, alarms raised for normal system behavior. In general, it is desirable to reduce detection delay as much as possible while maintaining an acceptable false positive rate. Nevertheless, there exists a trade-off between the detection delay and the rate of false positives, which can be controlled by changing the sensitivity of the detector. A typical way to control detector sensitivity is through a detection threshold: by decreasing (increasing) detection threshold, a defender can decrease (increase) detection delay and increase (decrease) false positive rate. Consequently, the detection threshold must be carefully selected, since a large value may result in excessive losses due to high detection delays, while a small value may result in wasting operational resources on investigating false alarms.

Finding an *optimal threshold*, that is, one that optimally balances the detection delay-false positive trade-off, is a challenging problem [14]. However, it becomes much more challenging when detectors are deployed in CPS with dynamic behavior, that is, when the expected damage incurred from undetected cyber-attacks depends on the system state and time. As a result, an attack on a CPS which is in a critical state is expected to cause more damage as compared to an attack in a less critical state. For example, in water distribution networks and electrical grids, disruptions at a high-demand time are more problematic than disruptions at a low-demand time. Hence, defenders need to incorporate time-dependent information in computing optimal detection thresholds when facing strategic attackers.

We study the problem of finding optimal detection thresholds for anomaly-based detectors implemented in dynamical systems in the face of strategic attacks. We model rational attacks against a system that is equipped with a detector as a two-player game between a defender and an attacker. We assume that an attacker can attack a system at any time. Considering that the damage is time-dependent, the attacker's objective is to choose the optimal time to launch an attack to maximize the damage incurred. On the other hand, the defender's objective is to select the detection thresholds to detect an attack with minimum delay while maintaining an acceptable rate of false positives. To this end, first we present an algorithm that selects an optimal threshold for the detector that is independent of time (i.e., *fixed*). We call it as a *fixed threshold* strategy. Next, we allow the defender to select a time-varying threshold while associating a cost with the threshold change. For this purpose, we present a polynomial time algorithm that computes thresholds that may depend on time. We call this approach the *adaptive threshold* strategy. We present a detailed analysis of the computational complexity and performance of both the fixed and adaptive threshold strategies. Finally, we evaluate our results using a water distribution system as a case study. Since expected damage to the system by an attack is time-dependent, the adaptive threshold strategy achieves a better overall detection delay-false positive trade-off, and consequently minimize the defender's losses. Our simulations indicate that this is indeed the case, and adaptive thresholds outperform the fixed threshold.

The remainder of this paper is organized as follows. In Sect. 2, we introduce our system model. In Sect. 3, we present our game-theoretic model and define optimal fixed and adaptive detection thresholds. In Sect. 4, we analyze both strategies and present algorithms to obtain optimal fixed and adaptive thresholds. In Sect. 5, we evaluate these algorithms using numerical example. In Sect. 6, we discuss related work on detection threshold selection in the face of strategic attacks. Finally, we offer concluding remarks in Sect. 7.

2   System Model

In this section, we present the system model. For a list of symbols used in this paper, see Table 1.

2.1   Attack Model

Let the system have a finite discrete time horizon of interest denoted by $\{1, \ldots, T\}$. Adversaries may exploit threat channels by compromising the system through a deception attack that starts at time k_a and ends at k_e, thus spanning over the interval $[k_a, k_e]$. Deception attacks are the ones that result in loss of integrity of sensor-control data, and their corresponding danger is especially profound due to the tight coupling of physical and cyber components (see [5] for details). If an attack remains undetected, it will enable the attacker to cause physical or financial damage. In order to represent the tight relation between

Table 1. List of symbols

Symbol	Description
$\mathcal{D}(k)$	Expected damage caused by an attack at timestep k
$\delta(\eta)$	Expected detection delay given detection threshold η
$FP(\eta)$	False positive rate given detection threshold is η
C	Cost of false alarms
$\mathcal{P}(\eta, k_a)$	Attacker's payoff for threshold η and attack time k_a
$\mathcal{L}(\eta, k_a)$	Defender's loss for threshold η and attack time k_a
Adaptive threshold	
$\mathcal{P}(\boldsymbol{\eta}, k_a)$	Attacker's payoff for adaptive threshold $\boldsymbol{\eta} = \{\eta_k\}$ and attack time k_a
$\mathcal{L}(\boldsymbol{\eta}, k_a)$	Defender's loss for adaptive threshold $\boldsymbol{\eta} = \{\eta_k\}$ and attack time k_a

the CPS's dynamic behavior and the expected loss incurred from undetected attacks, we model the potential damage of an attack as a function of time.

Definition 1 *(Expected Damage Function). Damage function of a CPS is a function $\mathcal{D}$: $\{1, ..., T\} \rightarrow \mathbb{R}_+$, which represents the expected damage $\mathcal{D}(k)$ incurred to a system from an undetected attack at time $k \in \{1, ..., T\}$.*

The definition above describes instant damage at a time $k \in \{1, ..., T\}$. Following this definition, expected total damage resulting from an attack that spans over some interval is defined as follows.

Definition 2 *(Expected Total Damage). Expected total damage is denoted by a function $\bar{\mathcal{D}}$: $\{1, ..., T\} \times \{1, ..., T\} \rightarrow \mathbb{R}_+$, which represents the expected total damage $\bar{\mathcal{D}}(k_a, k_e)$ incurred to a system from an undetected attack in a period $[k_a, k_e]$. Formally,*

$$\bar{\mathcal{D}}(k_a, k_e) = \sum_{k=k_a}^{k_e} \mathcal{D}(k). \tag{1}$$

2.2   Detector

We consider a defender whose objective is to protect the physical system, which is equipped with a detector. The detector's goal is to determine whether a sequence of received measurements generated through the system corresponds to the normal behavior or an attack. To implement a detection algorithm, we utilize a widely used method known as sequential change detection [11]. This method assumes a sequence of measurements that starts under the normal hypothesis, and then, at some point in time, changes to the attack hypothesis. Change detection algorithm attempts to detect this change as soon as possible.

Example (CUSUM). The Cumulative sum (CUSUM) is a statistic used for change detection. The nonparametric CUSUM statistic $S(k)$ is described by

$$S(k) = (S(k-1) + z(k))^+,$$

where $S(0) = 0$, $(a)^+ = a$ if $a \geq 0$ and zero otherwise, and $z(k)$ is generated by an observer, such that under normal behavior it has expected value of less than zero [7]. Assigning η as the detection threshold chosen based on a desired false alarm rate, the corresponding decision rule is defined as

$$d(S(k)) = \begin{cases} \text{Attack} & \text{if } S(k) > \eta \\ \text{Normal} & \text{otherwise} \end{cases}$$

Detection Delay and False Positive Rate. In detectors implementing change detection, there might be a *detection delay*, which is the time taken by the detector to raise an alarm since the occurrence of an attack.[1] Further, there might be a *false positive*, which means raising an alarm when the system exhibits normal behavior. In general, it is desirable to reduce detection delay as much as possible while maintaining an acceptable false positive rate. But, there exists a trade-off between the detection delay and the rate of false positives, which can be controlled by changing the detection threshold. In particular, by decreasing (increasing) the detection threshold, a defender can decrease (increase) detection delay and increase (decrease) false positive rate. Finding the optimal trade-off point and its corresponding *optimal threshold* is known to be an important problem [14], however, it is much more important in CPS, since expected damage incurred from undetected attack directly depends on detector's performance.

We represent detection delay by the continuous function $\delta : \mathbb{R}_+ \to \mathbb{N} \cup \{0\}$, where $\delta(\eta)$ is the detection delay (in timesteps) when threshold is η. Further, we denote the attainable false positive rate by the continuous function $FP : \mathbb{R}_+ \to [0, 1]$, where $FP(\eta)$ is the false positive rate when the detection threshold is η. We assume that δ is increasing and FP is decreasing, which is true for most typical detectors including the CUSUM detector.

3 Problem Statement

In this section, we present the optimal threshold selection problem. We consider two cases: (1) Fixed threshold, in which the defender selects an optimal threshold and then keeps it fixed; and (2) Adaptive threshold, in which the defender changes detection threshold based on time. We model this problems as conflicts between a defender and an attacker, which are formulated as two-player Stackelberg security games.

---

[1] Note that any desired definition of detection delay may be considered, for example, stationary average delay [21, 22].

3.1 Fixed Threshold

Strategic Choices. The defender's strategic choice is to select a detection threshold η. The resulting detection delay and false positive rate are $\delta(\eta)$ and $FP(\eta)$, respectively. We consider the worst-case attacker that will not stop the attack before detection in order to maximize the damage. Consequently, the attacker's strategic choice becomes to select a time k_a to start the attack. Note that we consider damage from only undetected attacks since the mitigation of non-stealthy attacks is independent of detector.

Defender's Loss and Attacker's Payoff. As an alarm is raised, the defender needs to investigate the system to determine whether an attack has actually happened, which will cost him C. When the defender selects threshold η and the attacker starts its attack at a timestep k_a, the defender's loss (i.e., inverse payoff) is

$$\mathcal{L}(\eta, k_a) = C \cdot FP(\eta) \cdot T + \sum_{k=k_a}^{k_a + \delta(\eta)} \mathcal{D}(k), \qquad (2)$$

that is, the amount of resources wasted on manually investigating false positives and the expected amount of damage caused by undetected attacks.

For the strategies (η, k_a), the attacker's payoff is

$$\mathcal{P}(\eta, k_a) = \sum_{k=k_a}^{k_a + \delta(\eta)} \mathcal{D}(k). \qquad (3)$$

that is, the total damage incurred to the system prior to the expected detection time. The idea behind this payoff function is the assumption of a worst-case attacker that has the goal of maximizing the damage.

Best-Response Attack and Optimal Fixed Threshold. We assume that the attacker knows the system model and defender's strategy, and can thus compute the detection threshold chosen by the defender. Hence, the attacker will play a *best-response* attack to the defender's strategy, which is defined below.

Definition 3 *(Best-Response Attack). Taking the defender's strategy as given, the attacker's strategy is a best-response if it maximizes the attacker's payoff. Formally, an attack starting at k_a is a best-response attack given a defense strategy η, if it maximizes $\mathcal{P}(\eta, k_a)$.*

Further, the defender must choose his strategy expecting that the attacker will play a best-response. We formulate the defender's optimal strategy as strong Stackelberg equilibrium (SSE) [12], which is commonly used in the security literature for solving Stackelberg games.

Definition 4 *(Optimal Fixed Threshold). We call a defense strategy optimal if it minimizes the defender's loss given that the attacker always plays a best-response. Formally, an optimal defense is*

$$\underset{\substack{\eta, \\ k_a \in bestResponses(\eta)}}{\arg\min} \mathcal{L}(\eta, k_a), \tag{4}$$

where bestResponses(η) is the set of best-response attacks against η.

3.2 Adaptive Threshold

Although the optimal fixed threshold minimizes the defender's loss considering attacks at critical periods (i.e., periods with high damage), it imposes a high false alarm rate at less critical periods. Adaptive threshold strategies directly address this issue. The idea of adaptive threshold is to reduce the detector's sensitivity during less critical periods (via increasing the threshold), and increase the sensitivity during more critical periods (via decreasing the threshold). As it will be shown, this significantly decreases the loss corresponding to false alarms. However, the defender may not want to continuously change the threshold, since a threshold change requires a reconfiguration of the detector that has a cost. Hence, the rational defender needs to find an *optimal adaptive threshold*, which is a balance between continuously changing the threshold and keeping it fixed.

The adaptive threshold is defined by $\eta = \{\eta_k\}_{k=1}^{T}$. The number of threshold changes is described by $N = |S|$, where $S = \{k \mid \eta_k \neq \eta_{k+1}, k \in \{1, ..., T-1\}\}$. If the system is under an undetected attack, the detection delay for each timestep k is the delay corresponding to its threshold, i.e., $\delta(\eta_k)$. We define detection time of an attack k_a as the time index at which the attack is first detected. It is given by

$$\sigma(\eta, k_a) = \{\min k \mid \delta(\eta_k) \leq k - k_a\}. \tag{5}$$

Note that the equation above represents the time index at which the attack is first detected, and not the detection delay. The detection delay for an attack k_a can be obtained by $\delta(\eta, k_a) := \sigma(\eta, k_a) - k_a$.

Strategic Choices. The defender's strategic choice is to select the threshold for each time index, given by $\eta = \{\eta_1, \eta_2, ..., \eta_T\}$. We call η to be the set of adaptive threshold. Since we consider a worst-case attacker that will not stop the attack before detection, the attacker's strategic choice is to select a time k_a to start the attack.

Defender's Loss and Attacker's Payoff. Let C_d be the cost associated with each threshold change. When the defender selects adaptive threshold η, and the attacker starts its attack at a timestep k_a, the defender's loss is

$$\mathcal{L}(\eta, k_a) = N \cdot C_d + \sum_{k=1}^{T} C \cdot FP(\eta_k) + \sum_{k=k_a}^{\sigma(\eta, k_a)} \mathcal{D}(k), \tag{6}$$

that is, the amount of resources spent on changing the threshold, operational costs of manually investigating false alarms, and the expected amount of damage caused by undetected attacks.

For the strategies $(\boldsymbol{\eta}, k_a)$, the attacker's payoff is the total damage prior to the expected detection time,

$$P(\boldsymbol{\eta}, k_a) = \sum_{k=k_a}^{\sigma(\boldsymbol{\eta}, k_a)} \mathcal{D}(k). \qquad (7)$$

Best-Response Attack and Optimal Adaptive Threshold. The definitions presented in this part are analogous to the ones discussed above for the case of optimal fixed threshold. We assume the attacker can compute the adaptive threshold, and will play a *best-response* to the defender's strategy, as defined below.

Definition 5 *(Best-Response Attack). Taking the defender's strategy as given, the attacker's strategy is a best-response if it maximizes the attacker's payoff. Formally, an attack k_a is a best-response given a defense strategy $\boldsymbol{\eta}$, if it maximizes $P(\boldsymbol{\eta}, k_a)$ as defined in (7).*

Further, the defender must choose its strategy expecting that the attacker will play a best-response.

Definition 6 *(Optimal Adaptive Threshold). We call a defense strategy optimal if it minimizes the defender's loss given that the attacker always plays a best-response with tie-breaking in favor of the defender. Formally, an optimal defense is*

$$\underset{\substack{\boldsymbol{\eta}, \\ k_a \in bestResponses(\boldsymbol{\eta})}}{\arg\min} \mathcal{L}(\boldsymbol{\eta}, k_a), \qquad (8)$$

where $bestResponses(\boldsymbol{\eta})$ is the best-response attack against $\boldsymbol{\eta}$.

4   Selection of Optimal Thresholds

In this section, we present polynomial-time algorithms to compute optimal thresholds, both for the fixed and adaptive cases.

4.1   Fixed Threshold

To compute an optimal fixed threshold, we present Algorithm 1. Here, we consider that any detection delay can be achieved by selecting a specific threshold value. Therefore, the algorithm finds an optimal detection delay, from which the optimal threshold value can be selected. To find the optimal detection delay, the algorithm iterates through all possible values of detection delay and selects the one that minimizes the defender's loss considering a best-response attack. To find a best-response attack k_a, given a delay δ, the algorithm iterates through all possible values of k_a, and selects the one that maximizes the payoff.

Algorithm 1. Algorithm for Optimal Fixed Threshold

1: **Input** $\mathcal{D}(k)$, T, C
2: **Initialize:** $\delta \leftarrow 0$, $L^* \leftarrow \infty$
3: **while** $\delta < T$ **do**
4:    $k_a \leftarrow 1$, $P' \leftarrow 0$
5:    **while** $k_a < T$ **do**
6:       $P(\delta, k_a) \leftarrow \sum_{k_a}^{k_a+\delta} D(k)$
7:       **if** $P(\delta, k_a) > P'$ **then**
8:          $P' \leftarrow P(\delta, k_a)$
9:          $L' \leftarrow P' + C \cdot FP(\delta) \cdot T$
10:       **end if**
11:       $k_a \leftarrow k_a + 1$
12:    **end while**
13:    **if** $L' < L^*$ **then**
14:       $L^* \leftarrow L'$
15:       $\delta^* \leftarrow \delta$
16:    **end if**
17:    $\delta \leftarrow \delta + 1$
18: **end while**
19: **return** δ^*

Proposition 1. *Algorithm 1 computes an optimal fixed threshold in $\mathcal{O}(T^2)$ steps.*

Proof. The obtained threshold is optimal since the algorithm evaluates all possible solutions through exhaustive search. Given a pair (δ, k_a), when computing the attacker's payoff $P(\delta, k_a)$ in Line 6, we use the payoff computed in previous iteration, and write $P(\delta, k_a) = P(\delta, k_a - 1) + \mathcal{D}(k_a - 1) + \mathcal{D}(k_a + \delta)$, which takes constant time. Therefore, the running time of the algorithm is subquadratic in the total number of timesteps T.   □

4.2 Adaptive Threshold

We present Algorithm 2 for finding optimal adaptive thresholds for any instance of the attacker-defender game, which is based on the SSE. The approach comprises (1) a dynamic-programming algorithm for finding minimum-cost thresholds subject to the constraint that the damage caused by a worst-case attack is at most a given value and (2) an exhaustive search, which finds an optimal damage value and thereby optimal thresholds. For ease of presentation, we use detection delays δ_k and the corresponding maximal thresholds η_k interchangeably (e.g., we let $FP(\delta_k)$ denote the false-positive rate of the maximal threshold that results in detection delay δ_k), and we let Δ denote the set of all attainable detection delay values.

Theorem 1. *Algorithm 2 computes an optimal adaptive threshold.*

Algorithm 2. Algorithm for Optimal Adaptive Thresholds

1: **Input** $\mathcal{D}(k)$, T, C
2: SearchSpace $\leftarrow \left\{ \sum_{k=k_a}^{k_e} D(k) \;\middle|\; k_a \in \{1, \ldots, T-1\}, \; k_e \in \{n+1, \ldots, T\} \right\}$
3: **for all** $P \in$ SearchSpace **do**
4:     $TC(P), \delta_1^*(P), \ldots, \delta_T^*(P) \leftarrow$ MINIMUMCOSTTHRESHOLDS(P)
5: **end for**
6: $P^* \leftarrow \arg\min_{P \in \text{SearchSpace}} TC(P)$
7: **return** $\delta_1^*(P^*), \ldots, \delta_T^*(P^*)$

8: **function** MINIMUMCOSTTHRESHOLDS(P)
9:     $\forall\, m \in \{0, \ldots, T-1\}, \; \delta \in \Delta : \text{COST}(T+1, m, \delta) \leftarrow 0$
10:     **for** $n = T, \ldots, 1$ **do**
11:         **for all** $m \in \{0, \ldots n-1\}$ **do**
12:             **for all** $\delta_{n-1} \in \Delta$ **do**
13:                 **for all** $\delta_n \in \Delta$ **do**
14:                     **if** $\delta_n > m$ **then**
15:                         $S(\delta_n) \leftarrow \text{COST}(n+1, m+1, \delta_n) + C \cdot FP(\delta_n)$
16:                     **else if** $\sum_{k=n-m}^{n} \mathcal{D}(k) \leq P$ **then**
17:                         $S(\delta_n) \leftarrow \text{COST}_P(n+1, \delta_n, \delta_n) + C \cdot FP(\delta_n)$
18:                     **else**
19:                         $S(\delta_n) \leftarrow \infty$
20:                     **end if**
21:                     **if** $\delta_{n-1} \neq \delta_n \wedge n > 1$ **then**
22:                         $S(\delta_n) \leftarrow S(\delta_n) + C_d$
23:                     **end if**
24:                 **end for**
25:                 $\delta^*(n, m, \delta_{n-1}) \leftarrow \arg\min_{\delta_n} S(\delta_n)$
26:                 $\text{COST}(n, m, \delta_{n-1}) \leftarrow \min_{\delta_n} S(\delta_n)$
27:             **end for**
28:         **end for**
29:     **end for**
30:     $m \leftarrow 0$, $\delta_0^* \leftarrow$ arbitrary
31:     **for all** $n = 1, \ldots T$ **do**
32:         $\delta_n^* \leftarrow \delta^*(n, m, \delta_{n-1}^*)$
33:         $m \leftarrow \min\{m+1, \delta_n^*\}$
34:     **end for**
35:     **return** $\text{COST}(1, 0, \text{arbitrary}), \delta_1^*, \ldots, \delta_T^*$
36: **end function**

Proof (Sketch). First, we prove that our dynamic-programming algorithm, called MINIMUMCOSTTHRESHOLDS in Algorithm 2, finds minimum-cost thresholds subject to any damage constraint P. Then, we show that our exhaustive search finds an optimal damage constraint P, which minimizes the defender's loss given that the attacker plays a best response.

(1) Minimum-Cost Thresholds. In the first part, we assume that we are given a damage value P, and we have to find thresholds that minimize the total cost of false positives and threshold changes, subject to the constraint that any attack against these thresholds will result in at most P damage. In order to solve this problem, we use a dynamic-programming algorithm. We will first discuss the algorithm without a cost for changing thresholds, and then show how to extend it to consider costly threshold changes.

For any two variables n and m such that $n \in \{1, \dots, T\}$ and $0 \leq m < n$, we define $\text{COST}(n, m)$ to be the minimum cost of false positives from n to T subject to the damage constraint P, given that we only have to consider attacks that start at $k_a \in \{n-m, \dots, T\}$ and that attacks are not detected prior to n. If there are no thresholds that satisfy the damage constraint P under these conditions, we let $\text{COST}(n, m)$ be ∞^2.

We can recursively compute $\text{COST}(n, m)$ as follows. For any $n < T$ and m, iterate over all possible detection delay values δ_n, and choose the one that results in the lowest cost $\text{COST}(n, m)$. If $\delta_n > m$, then no attack could be detected at timestep n, and $\text{COST}(n, m)$ would be the cost at timestep n plus the minimum cost for timesteps $\{n+1, \dots, T\}$ given that attacks may start at $\{n-m, \dots, T\} = \{(n+1) - (m+1), \dots, T\}$. On the other hand, if $\delta_n \leq m$, then some attacks could be detected at timestep n, and the worst of these attacks would start at $n - m$. Hence, if $\sum_{k=n-m}^{n} \mathcal{D}(k) \leq P$, then $\text{COST}(n, m)$ would be the cost at timestep n plus the minimum cost for timesteps $\{n+1, \dots, T\}$ given that attacks may start at $\{n+1-\delta_n, \dots, T\}$. Otherwise, there would be an attack that could cause more than P damage, so $\text{COST}(n, m)$ would be ∞ by definition since there would be no feasible thresholds for the remaining timesteps. Formally, we let

$$\text{COST}(n, m) = \min_{\delta_n} \begin{cases} \text{COST}(n+1, m+1) + FP(\delta_n), & \text{if } \delta_n > m \\ \text{COST}(n+1, \delta_n) + FP(\delta_n), & \text{else if } \sum_{k=n-m}^{n} \mathcal{D}(k) \leq P \\ \infty & \text{otherwise} \end{cases}$$

(9)

Note that in the equation above, $\text{COST}(n, m)$ does not depend on $\delta_1, \dots, \delta_{n-1}$, it depends only on the feasible thresholds for the subsequent timesteps. Therefore, starting from the last timestep T and iterating backwards, we are able to compute $\text{COST}(n, m)$ for all timesteps n and all values m. Note that for $n = T$ and any δ_T, computing $\text{COST}(T, m)$ is straightforward: if $\sum_{T-m}^{T} \mathcal{D}(k) \leq P$, then $\text{COST}(T, m) = FP(\delta_T)$; otherwise, $\text{COST}(T, m) = \infty$.

Having found $\text{COST}(n, m)$ for all n and m, $\text{COST}(1, 0)$ is by definition the minimum cost of false positives subject to the damage constraint P. The minimizing threshold values can be recovered by iterating forwards from $n = 1$ to T and again using Eq. (9). That is, for every n, we select the threshold corresponding to the delay value δ_n^* that attains the minimum cost $\text{COST}(n, m)$, where m can easily be computed from the preceding delay values $\delta_1^*, \dots, \delta_n^{*3}$.

---

[2] Note that in practice, ∞ can be represented by a sufficiently high natural number.

[3] Note that in Algorithm 2, we store the minimizing values $\delta^*(n, m)$ for every n and m when iterating backwards, thereby decreasing running time and simplifying the presentation of our algorithm.

Costly Threshold Changes. Now, we show how to extend the computation of COST to consider the cost C_d of changing the threshold. Let COST(n, m, δ_{n-1}) be the minimum cost for timesteps starting from n subject to the same constraints as before but also given that the detection delay at timestep $n-1$ is δ_{n-1}. Then, COST(n, m, δ_{n-1}) can be computed similarly to COST(n, m): for any $n < T$, iterate over all possible detection delay values δ_n, and choose the one that results in the lowest cost COST(n, m, δ_{n-1}). If $\delta_{n-1} = \delta_n$ or $n = 1$, then the cost would be computed the same way as in the previous case (i.e., similarly to Eq. (9)). Otherwise, the cost would have to also include the cost C_d of changing the threshold. Consequently, similarly to Eq. (9), we define

$$\widehat{\text{COST}}(n, m, \delta_{n-1}) = \begin{cases} \text{COST}(n + 1, m + 1, \delta_n) + FP(\delta_n) & \text{if } \delta_n > m \\ \text{COST}(n + 1, \delta_n, \delta_n) + FP(\delta_n) & \text{if } \sum_{k=n-m}^{n} \mathcal{D}(k) \leq P, \\ \infty & \text{otherwise} \end{cases}$$
(10)

and then based on the value of δ_{n-1}, we can compute COST(n, m, δ_{n-1}) as

$$\text{COST}(n, m, \delta_{n-1}) = \min_{\delta_n} \begin{cases} \widehat{\text{COST}}(n, m, \delta_{n-1}) & \text{if } \delta_n = \delta_{n-1} \vee n = 1 \\ \widehat{\text{COST}}(n, m, \delta_{n-1}) + C_d & \text{otherwise} \end{cases}.$$
(11)

Note that for $n = 1$, we do not add the cost C_d of changing the threshold. Similarly to the previous case, COST$(1, 0, \text{arbitrary})$ is the minimum cost subject to the damage constraint P, and the minimizing thresholds can be recovered by iterating forwards.

(2) Optimal Damage Constraint. For any damage value P, using the above dynamic-programming algorithm, we can find thresholds that minimize the total cost $TC(P)$ of false positives and threshold changes subject to the constraint that an attack can do at most P damage. Since the defender's loss is the sum of its total cost and the damage resulting from a best-response attack, we can find optimal adaptive thresholds by solving

$$\min_P TC(P) + P \tag{12}$$

and computing the optimal thresholds η^* for the minimizing P^* using our dynamic-programming algorithm.

To show that this formulation does indeed solve the problem of finding optimal adaptive thresholds, we use indirect proof. For the sake of contradiction, suppose that there exist thresholds η' for which the defender's loss $\mathcal{L}'$ is lower than the loss $\mathcal{L}^*$ for the solution η^* of the above formulation. Let P' be the damage resulting from the attacker's best response against η', and let TC' be the defender's total cost for η'. Since the worst-case attack against η' achieves at most P' damage, we have from the definition of $TC(P)$ that $TC' \geq TC(P')$.

It also follows from the definition of $TC(P)$ that $L^* \leq TC(P^*) + P^*$. Combining the above with our supposition $L^* > L'$, we get

$$TC(P^*) + P^* \geq L^* > L' = TC' + P' \geq TC(P') + P'.$$

However, this is a contradiction since P^* minimizes $TC(P) + P$ by definition. Therefore, η^* must be optimal.

It remains to show that Algorithm 2 finds an optimal damage value P^*. To this end, we show that P^* can be found in polynomial time using an exhaustive search. Consider the set of damage values $\bar{\mathcal{D}}(k_a, k_e)$ from all possible attacks $k_a \leq k_e$, that is, the set

$$\left\{ \sum_{k=k_a}^{k_e} \mathcal{D}(k) \; \middle| \; k_a \in \{1, \dots, T\}, k_e \in \{k_a, \dots, T\} \right\}.$$

Let the elements of this set be denoted by $P_1, P_2, \dots$ in increasing order. It is easy to see that for any i, the set of thresholds that satisfy the constraint is the same for every $P \in [P_i, P_{i+1})$. Consequently, for any i, the cost $TC(P)$ is the same for every $P \in [P_i, P_{i+1})$. Therefore, the optimal P^* must be a damage value P_i from the above set, which we can find by simply iterating over the set. □

Proposition 2. *The running time of Algorithm 2 is $\mathcal{O}(T^4 \cdot |\Delta|^2)$.*

Note that since possible detection delay values can be upper-bounded by T, the running time of Algorithm 2 is also $\mathcal{O}(T^6)$.

Proof. In the dynamic-programming algorithm, we first compute $\text{COST}(n, m, \delta_{n-1})$ for every $n \in \{1, \dots, T\}$, $m \in \{1, \dots, n-1\}$, and $\delta_{n-1} \in \Delta$, and each computation takes $\mathcal{O}(|\Delta|)$ time. Then, we recover the optimal detection delay for all timesteps $\{1, \dots, T\}$, and the computation for each timestep takes a constant amount of time. Consequently, the running time of the dynamic-programming algorithm is $\mathcal{O}(T^2 \cdot |\Delta|^2)$.

In the exhaustive search, we first enumerate all possible damage values by iterating over all possible attacks (k_a, k_e), where $k_a \in \{1, \dots, T\}$ and $k_e \in \{k_a, \dots, T\}$. Then, for each possible damage value, we execute the dynamic-programming algorithm, which takes $\mathcal{O}(T^2 \cdot |\Delta|^2)$ time. Consequently, the running time of Algorithm 2 is $\mathcal{O}(T^4 \cdot |\Delta|^2)$. □

Finally, note that the running time of the algorithm can be substantially reduced in practice by computing COST in a lazy manner: starting from $n = 1$ and $m = 0$, compute and store the value of each $\text{COST}(n, m, \delta_{n-1})$ only when it is referenced, and then reuse it when it is referenced again. Unfortunately, this does not change the worst-case running time of the algorithm.

5   Numerical Results

In this section, we evaluate our approach numerically using an example. In particular, we consider the anomaly-based detection of deception attacks in water distribution networks. In such networks, an adversary may compromise pressure sensors deployed to monitor the leakages and bursts in water pipes. By compromising sensors, adversary may alter their true observations, which can then result in physical damage and financial losses. Next, we present the system model and the simulations of our results.

System Model. Figure 1 presents hourly water demand for a water network during a day [9]. Since demand is time-dependent, the expected physical damage and financial loss caused by an attack on sensors is also time-dependent. That is, the expected disruptions at a high-demand time would be more problematic than the disruptions at a low-demand time. Therefore, for each timestep $k \in \{1, ..., 24\}$, we can define the expected damage as $\mathcal{D}(k) = \alpha \cdot d(k)$ where $d(k)$ is the demand at time k, and $\alpha \in \mathbb{R}_+$ is a fixed value for scaling (for example, water price rate). In our experiments, we let $\alpha = 2$.

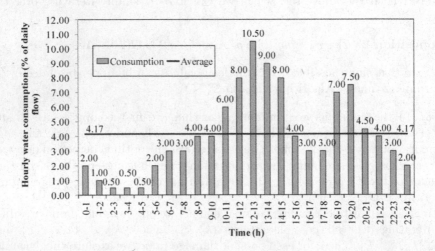

Fig. 1. Hourly water demand during a day [9].

To discover attacks, we use anomaly-based detection systems implementing sequential change detection. Based on the results presented in [7], we derive the attainable detection delays and false alarm rates for the detector as shown in Fig. 2. We observe that for the detection delay $\delta = 0$, the false positive rate is $FP(\delta) = 0.95$, and for $\delta = 23$, the false positive rate is $FP(\delta) = 0.02$. As expected, the detection delay is proportional to the threshold, and the false positive rate is inversely proportional to the threshold [6].

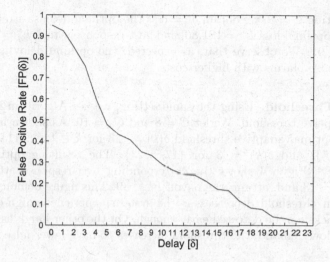

Fig. 2. Trade-off between the detection delay and the false positive rate.

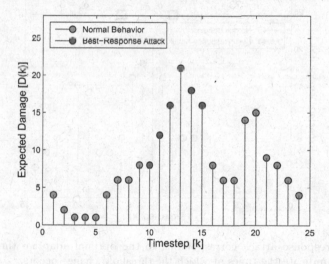

Fig. 3. Best-response attack corresponding to the optimal fixed threshold $\delta^* = 5$.

Fixed Threshold. In the case of fixed threshold, the objective is to select the strategy that minimizes the defender's loss (2) while assuming the attacker will respond using a best-response attack. Letting $C = 7$ and using Algorithm 1, we obtain $\delta^* = 5$, and the optimal loss $L^* = 171.30$. Figure 3 shows the best-response attack corresponding to this threshold value. The best-response attack starts at $k_a^* = 10$ and attains the payoff $P^* = \sum_{k=10}^{15} \mathcal{D}(k) = 91$. Note that if the attacker starts the attack at any other timestep, the damage caused before detection is less than P^*.

Next, letting $C = 8$, we obtain $\delta^* = 6$ as the optimal defense strategy, which leads to the optimal loss $L^* = 181.86$, and best-response attack $k_a^* = 9$, with the payoff $P^* = 99$. We observe that, as expected, the optimal delay is higher for the case of false alarms with higher costs.

Adaptive Threshold. Using the same setting, we use Algorithm 2 to find an optimal adaptive threshold. We let $C = 8$ and $C_d = 10$. As shown in Fig. 4, we obtain the optimal adaptive threshold $\delta(k) = 23$ for $k \in \{1, ..., 11\}$, $\delta(k) = 1$ for $\{12, ..., 15\}$, and $\delta(k) = 3$ for $\{17, ..., 23\}$. The resulting optimal loss is $L^* = 138.88$. Figure 4 shows the corresponding best-response attack, which starts at $k_a = 13$ and, attains the payoff $P^* = 39$. This figure demonstrates that the detection threshold decreases as the system experiences high-demand, so that the attacks can be detected early enough. On the other hand, as the system experiences low-demand, the threshold increases to have fewer false alarms.

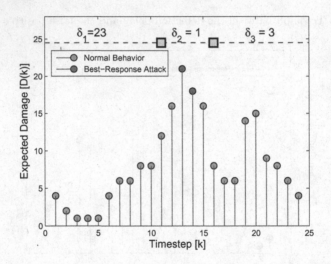

Fig. 4. Best-response attack corresponding to the optimal adaptive threshold. The yellow points indicate the times at which the threshold change occurs.

Comparison. Keeping $C = 8$ fixed, Fig. 5 shows the optimal loss as a function of cost of threshold change C_d. For small values of C_d, the optimal losses obtained by the adaptive threshold strategy are significantly lower than the loss obtained by the fixed threshold strategy. As the cost of threshold change C_d increases, the solutions of adaptive and fixed threshold problems become more similar. In the current setting, the adaptive threshold solution converges to a fixed threshold when $C_d \geq 45$.

Furthermore, letting $C_d = 8$, Fig. 6 shows optimal loss as a function of cost of false positives for fixed and adaptive threshold strategies. It can be seen that

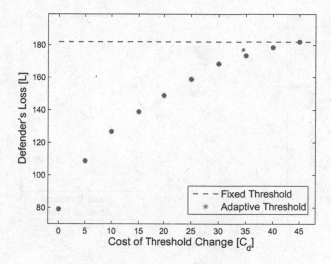

Fig. 5. The defender's loss as a function of cost of threshold change.

in both cases, the optimal loss increases as the cost of false alarms increases. However, in the case of adaptive threshold, the change in loss is relatively smaller than the fixed threshold.

6 Related Work

The problem of threshold selection for anomaly detection systems has been widely studied in the literature. Nevertheless, prior work has not particularly addressed the optimal threshold selection problem in the face of strategic attacks when the damage corresponding to an attack depends on time-varying properties of the underlying physical system.

Laszka et al. study the problem of finding detection thresholds for multiple detectors while considering time-invariant damages [14]. They show that the problem of finding optimal attacks and defenses is computationally expensive, thereby, proposing polynomial-time heuristic algorithms for computing approximately optimal strategies. Cardenas et al. study the use of physical models for anomaly detection, and describe the trade-off between false alarm rates and the delay for detecting attacks [7]. Pasqualetti et al. characterize detection limitations for CPS and prove that an attack is undetectable if the measurements due to the attack coincide with the measurements due to some nominal operating condition [18].

Signaling games are also used to model intrusion detection [8,10]. Shen et al. propose an intrusion detection game based on the signaling game in order to select the optimal detection strategy that lowers resource consumption [20]. Further, Alpcan and Basar study distributed intrusion detection as a game between an IDS and an attacker, using a model that represents the flow of information

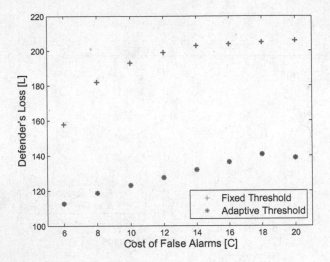

Fig. 6. The defender's loss as a function of cost of false alarms.

from the attacker to the IDS through a network [3,4]. The authors investigate the existence of a unique Nash equilibrium and best-response strategies.

This work is also related to the FlipIt literature [15,16,24]. FlipIt is an attacker-defender game that studies the problem of stealthy takeover of control over a critical resource, in which the players receive benefits proportional to the total time that they control the resource. In [19], the authors present a framework for the interaction between an attacker, defender, and a cloud-connected device. They describe the interactions using a combination of the FlipIt game and a signaling game.

In the detection theory literature, Tantawy presents a comprehensive discussion on design concerns and different optimality criteria used in model-based detection problems [23]. Alippi et al. propose a model of adaptive change detection that can be configured at run-time [2]. This is followed by [25], in which the authors present a procedure for obtaining adaptive thresholds in change detection problems.

7   Concluding Remarks

In this paper, we studied the problem of finding optimal detection thresholds for anomaly-based detectors implemented in dynamical systems in the face of strategic attacks. We formulated the problem as an attacker-defender security game that determines thresholds for the detector to achieve an optimal trade-off between the detection delay and the false positive rates. To this end, first we presented an algorithm that computes optimal fixed threshold that is independent of time. Next, we defined adaptive threshold, in which the defender is allowed to change the detector's threshold with time. We provided a polynomial time algorithm to compute optimal adaptive threshold. Finally, we evaluated our results using a case study. Our simulations indicated that the adaptive threshold strategy achieves a better overall detection delay-false positive trade-off, and consequently minimize the defender's losses, especially when the damage incurred by the successful attack varied with time.

In future work, we aim to extend this work by considering: (1) Multiple systems with different time-varying damage for each subsystem; (2) Sequential hypothesis testing detectors, in which there exits a trade-off between false alarm rate, missed detection rate, and detection delay; and (3) Moving target defense techniques based on randomized thresholds.

Acknowledgment. This work is supported in part by the the the National Science Foundation (CNS-1238959), Air Force Research Laboratory (FA 8750-14-2-0180), National Institute of Standards and Technology (70NANB15H263), Office of Naval Research (N00014-15-1-2621), and by Army Research Office (W911NF-16-1-0069).

References

1. Abrams, M., Weiss, J.: Malicious control system cyber security attack case study - Maroochy Water Services, Australia, July 2008. http://csrc.nist.gov/groups/SMA/fisma/ics/documents/Maroochy-Water-Services-Case-Study_report.pdf
2. Alippi, C., Roveri, M.: An adaptive CUSUM-based test for signal change detection. In: Proceedings of the 2006 IEEE International Symposium on Circuits and Systems (ISCAS), pp. 5752–5755. IEEE (2006)
3. Alpcan, T., Basar, T.: A game theoretic approach to decision and analysis in network intrusion detection. In: Proceedings of the 42nd IEEE Conference on Decision and Control (CDC), vol. 3, pp. 2595–2600. IEEE (2003)
4. Alpcan, T., Başar, T.: A game theoretic analysis of intrusion detection in access control systems. In: Proceedings of the 43rd IEEE Conference on Decision and Control (CDC), vol. 2, pp. 1568–1573. IEEE (2004)
5. Amin, S., Schwartz, G.A., Hussain, A.: In quest of benchmarking security risks to cyber-physical systems. IEEE Netw. **27**(1), 19–24 (2013)
6. Basseville, M., Nikiforov, I.V., et al.: Detection of Abrupt Changes: Theory and Application, vol. 104. Prentice Hall, Englewood Cliffs (1993)
7. Cárdenas, A.A., Amin, S., Lin, Z.-S., Huang, Y.-L., Huang, C.-Y., Sastry, S.: Attacks against process control systems: risk assessment, detection, and response. In: Proceedings of the 6th ACM Symposium on Information, Computer and Communications Security (ASIACCS), pp. 355–366. ACM (2011)
8. Casey, W., Morales, J.A., Nguyen, T., Spring, J., Weaver, R., Wright, E., Metcalf, L., Mishra, B.: Cyber security via signaling games: toward a science of cyber security. In: Natarajan, R. (ed.) ICDCIT 2014. LNCS, vol. 8337, pp. 34–42. Springer, Heidelberg (2014). doi:10.1007/978-3-319-04483-5_4
9. Durin, B., Margeta, J.: Analysis of the possible use of solar photovoltaic energy in urban water supply systems. Water **6**(6), 1546–1561 (2014)
10. Estiri, M., Khademzadeh, A.: A theoretical signaling game model for intrusion detection in wireless sensor networks. In: Proceedings of the 14th International Telecommunications Network Strategy and Planning Symposium (NETWORKS), pp. 1–6. IEEE (2010)
11. Kailath, T., Poor, H.V.: Detection of stochastic processes. IEEE Trans. Inf. Theor. **44**(6), 2230–2231 (1998)

12. Korzhyk, D., Yin, Z., Kiekintveld, C., Conitzer, V., Tambe, M.: Stackelberg vs. Nash in security games: an extended investigation of interchangeability, equivalence, and uniqueness. J. Artif. Intell. Res. **41**, 297–327 (2011)
13. Kushner, D.: The real story of stuxnet. IEEE Spectr. **50**(3), 48–53 (2013)
14. Laszka, A., Abbas, W., Sastry, S.S., Vorobeychik, Y., Koutsoukos, X.: Optimal thresholds for intrusion detection systems. In: Proceedings of the 3rd Annual Symposium and Bootcamp on the Science of Security (HotSoS), pp. 72–81 (2016)
15. Laszka, A., Horvath, G., Felegyhazi, M., Buttyan, L., FlipThem: modeling targeted attacks with FlipIt for multiple resources. In: Proceedings of the 5th Conference on Decision and Game Theory for Security (GameSec), pp. 175–194, November 2014
16. Laszka, A., Johnson, B., Grossklags, J.: Mitigating covert compromises. In: Chen, Y., Immorlica, N. (eds.) WINE 2013. LNCS, vol. 8289, pp. 319–332. Springer, Heidelberg (2013). doi:10.1007/978-3-642-45046-4_26
17. Lee, R.M., Assante, M.J., Conway, T.: German steel mill cyber attack. Technical report, SANS Industrial Control Systems (2014)
18. Pasqualetti, F., Dorfler, F., Bullo, F.: Attack detection and identification in cyber-physical systems. IEEE Trans. Autom. Control **58**(11), 2715–2729 (2013)
19. Pawlick, J., Farhang, S., Zhu, Q.: Flip the cloud: cyber-physical signaling games in the presence of advanced persistent threats. In: Khouzani, M.H.R., Panaousis, E., Theodorakopoulos, G. (eds.) GameSec 2015. LNCS, vol. 9406, pp. 289–308. Springer, Heidelberg (2015). doi:10.1007/978-3-319-25594-1_16
20. Shen, S., Li, Y., Xu, H., Cao, Q.: Signaling game based strategy of intrusion detection in wireless sensor networks. Comput. Math. Appl. **62**(6), 2404–2416 (2011)
21. Shiryaev, A.: The problem of the most rapid detection of a disturbance in a stationary process. Soviet Math. Dokl **2**, 795–799 (1961)
22. Srivastava, M., Wu, Y.: Comparison of EWMA, CUSUM and Shiryayev-Roberts procedures for detecting a shift in the mean. Ann. Stat. **21**, 645–670 (1993)
23. Tantawy, A.M.: Model-based detection in cyber-physical systems. Ph.D. thesis, Vanderbilt University (2011)
24. Van Dijk, M., Juels, A., Oprea, A., Rivest, R.L.: FlipIt: the game of stealthy takeover. J. Cryptol. **26**(4), 655–713 (2013)
25. Verdier, G., Hilgert, N., Vila, J.-P.: Adaptive threshold computation for cusum-type procedures in change detection and isolation problems. Comput. Stat. Data Anal. **52**(9), 4161–4174 (2008)

A Point-Based Approximate Algorithm for One-Sided Partially Observable Pursuit-Evasion Games

Karel Horák[(⊠)] and Branislav Bošanský

Faculty of Electrical Engineering, Department of Computer Science,
Czech Technical University in Prague, Prague, Czech Republic
{horak,bosansky}@agents.fel.cvut.cz

Abstract. Pursuit-evasion games model many security problems where an evader is trying to escape a group of pursuing units. We consider a variant with partial observability and simultaneous moves of all units, and assume the worst-case setup, where the evader knows the location of pursuer's units, but the pursuer does not know the location of the evader. Recent work has shown that the solution of such games is compactly representable as a collection of finite-dimensional value functions. We extend this result and propose the first practical algorithm for approximating optimal policies in pursuit-evasion games with one-sided partial observability. Our approach extends the point-based updates that exist for POMDPs to one-sided partially observable stochastic games. The experimental evaluation on multiple graphs shows significant improvements over approximate algorithms that operate on finite game trees.

1 Introduction

Pursuit-evasion games commonly arise in robotics and many security applications [1,9]. In these games, a team of centrally controlled pursuing units (a *pursuer*) aims to locate and capture an *evader*, while the evader aims for the opposite. From the theoretical perspective, a pursuit-evasion game on a finite graph corresponds to a two-player finite-state discrete-time zero-sum partially observable stochastic game with infinite horizon (POSG). We consider the concurrent setup where both players act simultaneously and we aim to find robust strategies of the pursuer against the worst-case evader. Specifically, we assume that the evader knows the positions of the pursuing units and her only uncertainty is the strategy of the pursuer and the move that will be performed in the current time step. We term these games as *one-sided partially observable pursuit-evasion games*.

Even though in POSGs with infinite horizon and discounted rewards the value of the game exists [2] and optimal strategies of the pursuer can be in general approximated, there are no known practical approximate algorithms that can be used to find these approximate strategies. Therefore, most of the existing works on pursuit-evasion games use heuristic algorithms in order to find strategies that are good in practice by relaxing some of the assumptions (e.g., [9]).

© Springer International Publishing AG 2016
Q. Zhu et al. (Eds.): GameSec 2016, LNCS 9996, pp. 435–454, 2016.
DOI: 10.1007/978-3-319-47413-7_25

Recently, a theoretic analysis has shown that a dynamic programming algorithm can be designed for such type of pursuit-evasion games [3]. However, similarly to models for sequential decision making that allow partial information, Partially Observable Markov Decision Processes (POMDPs), a direct application of this dynamic programming value-iteration algorithm is intractable in practice due to "curse of history" and "curse of dimensionality".

We extend these theoretic results to a broader class of one-sided partially observable pursuit-evasion games and present the first practical approximate algorithm based on the value iteration algorithm. We follow the approach used in POMDPs and adapt the heuristic search value iteration (HSVI) algorithm [7,8] that provably converges to approximate optimal strategies and provides an excellent scalability when solving POMDPs. The key challenge and the main difference compared to the POMDP version is a limited control of belief transitions. While beliefs over states are changed due to actions of the decision maker and known transition probabilities only in POMDPs, the opponent of the decision maker influence belief transitions in the case of a POSG.

Our paper addresses these challenges. After a brief description of POMDPs and the main idea of the HSVI algorithm we follow by the description of one-sided partially observable pursuit-evasion games. Afterwards we give all our contributions: (1) we extend the original definition of one-sided partially observable pursuit-evasion games to a more realistic setting compared to [3] by allowing the pursuer to capture the evader also on the edges of a graph, (2) we present for the first time a novel HSVI-inspired algorithm for approximating optimal strategies in one-sided partially observable pursuit-evasion games, (3) we provide an experimental evaluation of our algorithm. We demonstrate current scalability of our algorithm and compare it with a sampling algorithm for solving sequential games with a finite-horizon [4]. The results show that our novel algorithm can closely approximate games with tens of nodes and two pursuing units in couple of minutes and significantly outperforms the finite-horizon algorithm.

2 POMDPs and Heuristic Search Value Iteration (HSVI)

POMDPs represent the standard model for a planning problem with imperfect information about the current state and uncertainty about action effects. Formally, a POMDP is described by a finite set of states $\mathcal{S}$, a finite set of actions $\mathcal{A}$, a finite set of observations $\mathcal{O}$, transition probabilities $T_{a,o}(s_i, s_j)$, reward function $R(s, a)$, a discount factor $\gamma \in (0, 1)$, and an initial belief b_0. The space of beliefs corresponds to a simplex over $\mathcal{S}$ (denoted $\Delta(\mathcal{S})$).

A POMDP is typically solved by approximating optimal value function v^* that returns the optimal reward for each belief. A value-iteration algorithm starts with an initial estimate of the value function, v^0, and iteratively improves the estimate using the Bellman update in order to approximate the optimal value function. In each iteration, v^t is a piecewise linear function represented as a set of vectors $\Gamma = \{\alpha_1, \ldots, \alpha_\Gamma\}$, such that $v^t(b) = \max_{\alpha \in \Gamma}(\alpha, b)$.

Practical algorithms approximate the simplex of beliefs by considering only a finite set of belief points $\mathcal{B}$ and perform a point-based update for each such belief point. While the sampling of original point-based algorithms was close to being uniform [5], HSVI algorithm samples the belief space according to a forward simulation process that biases $\mathcal{B}$ to beliefs that are reachable from b_0 given the transition probabilities. The main idea of HSVI is to keep two estimates of the optimal value function v^* that form a lower bound (denoted $\underline{v}$ and represented using a set of α vectors Γ) and an upper bound (denoted $\overline{v}$ and represented as a convex lower envelope of a set of points Υ). In each iteration, these estimates are tightened by:

1 sampling a belief space by forward search using transition probabilities so that the belief point with the largest excess gap between bounds is targeted,
2 performing an update by (i) adding new α vectors into the lower bound set Γ, and (ii) adding new points into the upper bound set Υ.

By using the forward exploration heuristic, selected belief points are reachable and relevant to the initial belief and transition probabilities of the environment. Moreover, by preferring the largest excess gap between the bounds the algorithm is also guaranteed to converge.

Towards HSVI for One-Sided Partially Observable Pursuit-Evasion Games. Contrary to POMDPs, belief transitions in one-sided partially observable games are controlled also by the opponent – the evader. Therefore, we use the intuition behind HSVI to explore the belief space so that the convergence is guaranteed, preferring belief states with the largest gap between the two bounds, and perform a point-wise update to decrease this gap over time.

3   One-Sided Partially Observable Pursuit-Evasion Games

We now define one-sided partially observable pursuit-evasion games and summarize existing key characteristics [3]. There are two players – the *pursuer* who controls N units located in vertices of graph and aims to capture the opponent – the *evader*. The evader knows the exact position of the units of the pursuer, while the pursuer does not know the position of the evader. The pursuer aims to capture the evader as quickly as possible – the reward for the pursuer is defined as γ^t for capturing the evader in t steps, where $\gamma \in (0,1)$ is the *discount factor*.

The main result for this class of games is that to find optimal strategies it is sufficient to consider strategies for the pursuer based only on the current state of the pursuing units and the belief the pursuer has about the evader's position [3]. Specifically, it is not necessary to consider the history of the actions played by the pursuer. Next, we can formulate value functions in these games that are similar to value functions in POMDPs. These value functions are piecewise linear and convex (PWLC) in the belief about the position of the evader [3].

In the following we leverage the existing results, extend them to a more realistic setting, and design a point-based value iteration algorithm for this class of POSGs. We now formally define one-sided partially observable pursuit-evasion games followed by the formal description of the value functions and value update.

3.1 Definitions

The game is played on a finite undirected graph $\mathcal{G} = (\mathcal{V}, \mathcal{E})$, where $\mathcal{V} = \{1, \ldots, |\mathcal{V}|\}$ denotes the set of vertices and $\mathcal{E}$ stands for the set of edges. The set of vertices adjacent to v will be denoted $\operatorname{adj}(v) \subseteq \mathcal{V}$, the set of edges incident to v will be denoted $\operatorname{inc}(v) \subseteq \mathcal{E}$. We overload this notion to multisets of vertices $V = \{v_1, \ldots, v_n\}$, when $\operatorname{adj}(V) = \times_{i=1\ldots n} \operatorname{adj}(v_i)$ and $\operatorname{inc}(V) = \times_{i=1\ldots n} \operatorname{inc}(v_i)$.

Since the players move simultaneously, we talk about a *stage* in the game, where both players act. A stage corresponds to a position of the pursuing units $s_p = \{v_1, \ldots, v_N\} \in \mathcal{V}^N$ (where s_p is a N-element multiset of individual positions of pursuer's units, i.e. multiple units may be located in the same vertex) and a position of the evader $s_e \in \mathcal{V}$. The pursuer chooses his action $E = \{e_1, \ldots, e_N\} \in \operatorname{inc}(s_p)$, where $e_i = \{v_i, v_i'\}$ – i.e. makes his i-th unit traverse the edge e_i and reach vertex v_i'. Similarly the evader chooses her action $\{s_e, s_e'\} \in \operatorname{inc}(s_e)$. After the application of these actions, the evader is either caught if $s_e' \in \{v_1', \ldots, v_N'\}$, or the game continues with the next stage where the pursuer is located in a new position $\{v_1', \ldots, v_N'\}$ which we will denote as s_p^E, and the evader is in s_e'.

The imperfect information of the pursuer about the position of the evader is modeled as a *belief* – a probability distribution $\tilde{b} \in \Delta(\mathcal{V})$ over the set of vertices, where $\tilde{b}_i$ is the probability that the evader is located in vertex $i \in \mathcal{V}$. Therefore, a *game situation* $\langle s_p, \tilde{b} \rangle$ is sufficient information for the players to play optimally in the current stage [3]. In order to simplify the notation we will assume that the probability that the evader is already caught in $\langle s_p, \tilde{b} \rangle$ is zero, i.e. $\tilde{b}_i = 0$ for all $i \in s_p$. Note that the situation $\langle s_p, b \rangle$ (b denotes belief where the evader may be already caught) is strategically equivalent to the situation $\langle s_p, \tilde{b} \rangle$ since

$$\tilde{b}_i = \begin{cases} b_i/(1 - \beta) & i \notin s_p \\ 0 & \text{otherwise} \end{cases} \tag{1}$$

where $(1 - \beta)$ is a normalization term corresponding to a probability of not capturing the evader.

3.2 Value Functions and Strategies

Previous work showed that we can use a value-iteration type of algorithm to approximate optimal strategies of a game with an infinite horizon [3].

This results from an observation, that value of a variant of this game with a restricted horizon (horizon-t games, where players perform at most t moves) can be represented by a collection of *value functions*, one for each position of the pursuer. A horizon-t value function $v^t\langle s_p \rangle : \Delta(\mathcal{V}) \to \mathbb{R}$ assigns the value $v^t\langle s_p \rangle(b)$ of a horizon-t game starting in situation $\langle s_p, b \rangle$ for each belief b (where the evader might be already caught)—i.e. the value of a game where pursuer starts in s_p and the evader starts in vertex $i \in \mathcal{V}$ with probability b_i. For every $s_p \in \mathcal{V}^N$, the value function $v^t\langle s_p \rangle$ is piecewise linear and convex (PWLC) and we can represent it as a set of α-vectors. Moreover the value of every fixed strategy of the pursuer is linear in the belief.

The value functions can now be used to approximate the value of the game (for some initial situation $\langle s_p, \tilde{b} \rangle$) by a repeated application of a dynamic programming operator H such that $v' = Hv$. Formally, v' for state $\langle s_p \rangle$ is computed using the following equation:

$$v' \langle s_p \rangle (\tilde{b}) = \gamma \max_{\pi_p} \min_{\pi_e} \sum_{E \in \text{inc}(s_p)} \pi_p(E) \cdot v \langle s_p^E \rangle (b_{\pi_e}) \tag{2}$$

where, π_p and π_e are *one-step strategies* of the pursuer and the evader, respectively; s_p^E is the position reached from s_p by playing action E, and b_{π_e} is a *transformed belief* about the position of the evader after playing strategy π_e. Formally, a strategy of the pursuer is a distribution over the actions available in the current state s_p – i.e., over edges that the pursuing units can take from current vertices, $\pi_p \in \Delta(\text{inc}(s_p))$. Similarly, a strategy of the evader is a mapping $\pi_e : \mathcal{V} \to \Delta(\mathcal{V})$. We will use $\pi_p(E)$ and $\pi_e(i,j)$ for the probability of traversing edges E, or $\{i,j\}$ respectively. Strategies π_p and π_e solving the maximin in Eq. (2) constitute a Nash equilibrium of this one-step game.

Transformed belief b_{π_e} depends on the previous belief in situation $\langle s_p, \tilde{b} \rangle$ and the probability with which the evader uses edges $(i,j) \in \mathcal{E}$. Formally, probability $b_{\pi_e,j}$ of evader being in vertex j in the subsequent stage equals to

$$b_{\pi_e,j} = \sum_{i \in \mathcal{V}} \tilde{b}_i \cdot \pi_e(i,j) \tag{3}$$

3.3 Value Update

The dynamic programming operator H has a unique fixpoint v^* that satisfies $v^* = Hv^*$, and for an arbitrary set of value functions v the recursive application of H converges to v^* (i.e. $H^\infty v = v^*$) [3]. To apply the operator H, we need to solve Eq. (2) – i.e., to solve the maximin problem for the situation $\langle s_p, \tilde{b} \rangle$.

Each of the value functions $v \langle s_p \rangle$ is represented using a set of α-vectors $\Gamma \langle s_p \rangle = \{\alpha^k \mid k = 1 \ldots |\Gamma \langle s_p \rangle|\}$, $\alpha^k = (\alpha_1^k, \ldots, \alpha_N^k)$. The evader seeks a strategy π_e that minimizes the payoff of pursuer's best response to π_e. This problem can be solved by means of a linear program shown in LP 1.

Since value of every action E of the pursuer against a fixed strategy π_e of the evader equals to

$$v' \langle s_p \rangle (\tilde{b})_{\pi_e}^E = \gamma v \langle s_p^E \rangle (b_{\pi_e}) = \gamma \max_{\alpha \in \Gamma \langle s_p^E \rangle} \sum_{j \in \mathcal{V}} \alpha_j b_{\pi_e,j}, \tag{9}$$

$$\min_{V,b',\pi_e} V \tag{4}$$

$$\text{s.t.} \gamma \sum_{j \in \mathcal{V}} \alpha_j \cdot b_{\pi_e,j} \leq V \qquad \forall E \in \text{inc}(s_p), \forall \alpha \in \Gamma \langle s_p^E \rangle \tag{5}$$

$$\sum_{i \in \mathcal{V}} \tilde{b}_i \cdot \pi_e(i,j) = b_{\pi_e,j} \qquad \forall j \in \mathcal{V} \tag{6}$$

$$\sum_{j \in \text{adj}(i)} \pi_e(i,j) = 1 \qquad \forall i \in \mathcal{V} \tag{7}$$

$$\pi_e(i,j) \geq 0 \qquad \forall i,j \in \mathcal{V} \tag{8}$$

LP 1: Linear program for computing optimal evader's strategy π_e in game situation $\langle s_p, \tilde{b} \rangle$ for state s_p and belief $\tilde{b}$ (assuming the evader is currently not at the position of the pursuer).

the maximum can be rewritten using a set of inequalities – one for each $\alpha \in \Gamma \langle s_p^E \rangle$ (Constraints (5)). Constraints (6) ensure that the transformed belief b_{π_e} is computed according to Eq. (3), while constraints (7) and (8) ensure that the strategy π_e is valid (i.e., it assigns positive probabilities only to vertices adjacent to evader's current position). The strategy of the pursuer π_p is obtained from the dual linear program of LP 1 – the dual variables corresponding to the constraints (5) represent the optimal strategy of the pursuer.

4   Extended Value Update

The original approach [3] did not consider that the pursuer can capture the evader on an edge – i.e., the case when both the evader and a pursuing unit choose to use the same edge at the same time. While from the theoretic perspective the results and methodology applies, designing a practical algorithm requires us to formally extend the model since capturing on edges is a natural prerequisite.

The main difference is in the computation of the transformed belief b_{π_e} that can now also depend on the action E the pursuer played – if the game continues after playing action E, the pursuer not only knows that the evader is not in current vertices s_p^E, but also that the evader has used none of the edges in E in the previous move. We will thus denote the transformed belief as $b_{\pi_e}^E$. Formally:

$$b_{\pi_e,j}^E = \frac{1}{1-\beta} \sum_{i \mid \{j,i\} \notin E} \tilde{b}_i \cdot \pi_e(i,j) \tag{10}$$

Note that due to the exclusion of some of the edges, it is necessary to perform normalization, where β again corresponds to a probability that the evader will be captured in this step, $\beta = \sum_{\{j,i\} \in E} \tilde{b}_i \cdot \pi_e(i,j)$.

Second, we also need to update the expected value of playing action E by the pursuer, since the probability that the evader is caught has changed

$$v'\langle s_p \rangle (\tilde{b})_{\pi_e}^E = \gamma \left[\beta + (1 - \beta) \cdot v \langle s_p^E \rangle (b_{\pi_e}^E) \right]. \tag{11}$$

Now, since each of the value functions $v \langle s_p \rangle$ is represented using a finite set $\Gamma \langle s_p \rangle$ of α-vectors, its value in $b_{\pi_e}^E$ can be computed as

$$v\langle s_p \rangle (b_{\pi_e}^E) = \max_{\alpha \in \Gamma \langle s_p \rangle} \sum_{j \in \mathcal{V}} \alpha_j b_{\pi_e,j}^E = \frac{1}{1 - \beta} \max_{\alpha \in \Gamma \langle s_p \rangle} \sum_{j \in \mathcal{V}} \sum_{i | \{j,i\} \notin E} \alpha_j \cdot \tilde{b}_i \cdot \pi_e(i,j). \tag{12}$$

Value $v'\langle s_p \rangle (\tilde{b})_{\pi_e}^E$ of playing action E against evader's strategy π_e is linear in π_e since the term $(1 - \beta)$ cancels out. This value is used to replace the best-response constraints (5) in LP 1 and the resulting linear program can be used to compute evader's optimal strategy π_e in the game with capturing on edges (see LP 2). The optimal strategy of the pursuer is again obtained by solving a dual linear program to LP 2.

$$\min_{V, \pi_e} V \tag{13}$$

$$\text{s.t.} V \geq \gamma \left[\beta + \sum_{j \in \mathcal{V}} \sum_{i | \{j,i\} \notin E} \alpha_j \cdot \tilde{b}_i \cdot \pi_e(i,j) \right] \quad \forall E \in \text{inc}(s_p), \forall \alpha \in \Gamma \langle s_p^E \rangle \tag{14}$$

$$\sum_{j \in \text{adj}(i)} \pi_e(i,j) = 1 \qquad \forall i \in \mathcal{V} \tag{15}$$

$$\pi_e(i,j) \geq 0 \qquad \forall i,j \in \mathcal{V} \tag{16}$$

LP 2: Linear program for computing evader's strategy.

5   HSVI for Pursuit-Evasion Games

We are now ready to present the first approximate algorithm for solving pursuit-evasion games with one-sided partial observability and concurrent moves. We use the value update as described in Sect. 4 in a point-based manner and present a new method for selecting belief points to be updated, compared to the original version of HSVI (see Sect. 2). Specifically, we let the evader choose the new belief (induced by his possibly suboptimal one-step strategy π_e) in an optimistic way, assuming that the value of the subsequent stage might be lower than the true value. We will then either try to prove that this optimism is unjustified (i.e. the true value of the subsequent stage is higher than what the evader thinks) and hence make the evader change his strategy, or show that the strategy π_e of the evader is near-optimal. We model these optimistic values from the evader's perspective by value functions $\underline{v}$, while the optimistic values from the pursuer's perspective are modeled by $\overline{v}$. These value functions correspond to the ones used in the HSVI algorithm for POMDPs.

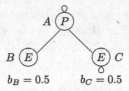

Fig. 1. Game used for demonstration of the ideas behind the algorithm. Pursuer starts in vertex A, evader starts in either B or C with uniform probability.

We describe important building blocks of the algorithm. First, we present the representation of value functions $\underline{v}$ and $\overline{v}$ used to approximate the optimal value functions v^* in Sect. 5.1. These value functions have to satisfy $\underline{v} \leq v^* \leq \overline{v}$ and hence they must be initialized properly, which is discussed in Sect. 5.2. The value functions $\overline{v}$ use the point-representation instead of the α-vector one and the linear program for strategy computation presented in Sect. 4 is not directly applicable. In Sect. 5.3 a modification of the linear program for computing strategies of the players that uses the point representation of value functions is presented. Similarly as in the case of the HSVI algorithm for POMDPs, we rely on point-based updates. We discuss these updates in Sect. 5.4. Each point corresponds to a game situation in our case. We sample these game situations using a forward exploration heuristic, discussed in Sect. 5.5. This heuristic skews the sampling process towards reachable game situations. Finally the complete algorithm is presented in Sect. 5.6.

Example. We will be illustrating the ideas behind the algorithm on a simple game with three vertices shown in Fig. 1. The pursuer starts in vertex A, while the evader starts either in vertex B or C with uniform probability. The game situation therefore corresponds to $\langle A, \{B : 0.5, C : 0.5\}\rangle$. Note that there is a loop in vertices A and C (i.e. players can wait in these vertices), while there is no loop in vertex B.

5.1   Value Functions Representation

The algorithm approximates the optimal strategies within ϵ tolerance by approximating the optimal value functions v^* corresponding to the values of infinite horizon games. Similarly as in the case of the POMDP counterpart, we will use a pair of PWLC value functions $\underline{v}$ and $\overline{v}$ to approximate v^*, however in our case we have to consider a pair of value functions $\underline{v}\langle s_p \rangle$ and $\overline{v}\langle s_p \rangle$ for each position s_p of the pursuer. The collection of *lower value functions* $\underline{v}$ bound v^* from below and the collection of *upper value functions* $\overline{v}$ bound v^* from above, i.e. $\underline{v}\langle s_p \rangle (b) \leq v^*\langle s_p \rangle (b) \leq \overline{v}\langle s_p \rangle (b)$ for every $s_p \in \mathcal{V}^N$ and every belief $b \in \Delta(\mathcal{V})$. The pair of lower and upper value functions $\underline{v}$ and $\overline{v}$ will be jointly referred to as $\hat{v}$. The goal of the algorithm is to reduce the gap between $\underline{v}$ and $\overline{v}$ at relevant belief points (i.e., those reached when optimal strategies are played). We denote

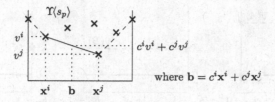

Fig. 2. A visualization of a projection to a lower envelope of a set of points in LP 3.

this gap in situation $\langle s_p, b \rangle$ by

$$\text{width}(\hat{v}\langle s_p \rangle (b)) = \overline{v}\langle s_p \rangle (b) - \underline{v}\langle s_p \rangle (b). \tag{17}$$

The lower value function $\underline{v}\langle s_p \rangle$ is represented by a set of α-vectors, the upper envelope of which forms the PWLC function $\underline{v}\langle s_p \rangle$. The set of α-vectors representing $\underline{v}\langle s_p \rangle$ is denoted $\Gamma\langle s_p \rangle$. Given the representation, equations for value updates and linear program for finding optimal strategies from Sect. 4 are directly applicable to $\underline{v}$.

The upper value function $\overline{v}\langle s_p \rangle$ is represented by a set of points $\Upsilon\langle s_p \rangle$, the lower envelope of which forms the PWLC function $\overline{v}\langle s_p \rangle$

$$\Upsilon\langle s_p \rangle = \left\{ (x_1^i, x_2^i, \ldots, x_{|\mathcal{V}|}^i) \rightarrow v^i \;\middle|\; i = 1 \ldots |\Upsilon\langle s_p \rangle| \right\}. \tag{18}$$

The value of $\overline{v}\langle s_p \rangle$ can be computed by means of the LP 3 that projects coordinates $(b_1, \ldots, b_{|\mathcal{V}|})$ on the lower envelope of points in $\Upsilon\langle s_p \rangle$. Coordinates $(b_1, \ldots, b_{|\mathcal{V}|})$ are expressed as convex combination of coordinates of points in $\Upsilon\langle s_p \rangle$ (Eqs. (20) and (21)). The minimization of the convex combination of the respective values (v^i is the value of a point with coordinates $(x_1^i, \ldots, x_{|\mathcal{V}|}^i)$) ensures that the projection lies on the lower envelope of $\Upsilon\langle s_p \rangle$ (Eq. (19)). The idea of this linear program is depicted graphically in Fig. 2.

$$\min_{\mathbf{c} \in [0,1]^k} \sum_{i=1\ldots k} c^i \cdot v^i \tag{19}$$

$$\text{s.t.} \sum_{i=1\ldots k} c^i \cdot x_j^i = b_j \qquad \forall j = 1 \ldots N \tag{20}$$

$$\sum_{i=1\ldots k} c^i = 1 \tag{21}$$

LP 3: Lower envelope projection LP.

5.2   Initialization of Value Functions

The value functions have to be initialized so that $\underline{v} \leq v^* \leq \overline{v}$ holds for every state. Each of the lower value function $\underline{v}\langle s_p \rangle$ initially corresponds to the probability

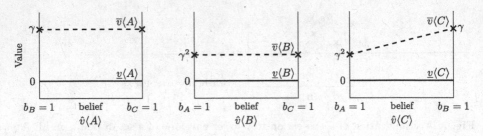

Fig. 3. Initialization of value functions

that the evader is caught in the current game situation and hence the pursuer gets the reward $\gamma^0 = 1$, i.e.

$$\underline{v}\langle s_p\rangle\,(b) = \sum_{i\in s_p} b_i. \tag{22}$$

This initialization disregards all future rewards that may be obtained by the pursuer, thus it lower bounds the optimal value $v^*\langle s_p\rangle\,(b)$.

The upper value functions $\overline{v}$ are initialized by solving a perfect-information version of the game (which is a discounted stochastic game solvable e.g. by Shapley's algorithm [6]). This is similar to the case of the HSVI for POMDPs, where the values of a perfect-information version of the POMDP, a Markov decision process, are used to initialize the value function. The values of the perfect-information game define the value of $\overline{v}$ in extreme points of belief simplices.

Example. The initialization of value functions for the example game (Fig. 1) is shown in Fig. 3. In order to provide cleaner visualization, we consider only a subset of the belief space where the evader is caught with zero probability. The lower value functions $\underline{v}$ are initialized to the probability that the evader is caught in the initial situation, which is zero in the considered part of belief space. The upper value functions are initialized by solving the perfect-information version of the game: If the pursuer starts in vertex A, he captures the evader in one move regardless of his initial position B or C (because he *knows* where she is located) and gets a reward γ. If the pursuer starts in vertex B, he will need two moves to capture the evader starting in A or C (the time needed to reach vertex C) and hence the reward of γ^2. Finally if the pursuer starts in vertex C, he will need two moves to capture the evader starting in A (with reward γ^2) and one move to capture the evader starting in B (with reward γ). These values are used to form points marked in Fig. 3.

5.3   Computing Strategies with Point-Represented Value Functions

In Sect. 4 we presented a linear program (LP 2) for computing strategies using value functions in the α-vectors representation that are applicable for updating the lower bound value functions $\underline{v}$. Here we present the modification for updating value functions $\bar{v}$ represented by sets of points $\Upsilon\langle s_p\rangle$. Specifically we modify Constraint (14) to use ideas from the projection LP 3.

Let us denote the probability that the evader was caught on edge when E was played by the pursuer as $\beta^E_{\pi_e} = \sum_{\{i,j\}\in E} \tilde{b}_i \pi_e(i,j)$. The value of playing action E in situation $\langle s_p, \tilde{b}\rangle$ was presented in Eq. (11). This leads to a set of best response constraints (corresponding to (14))

$$V \geq \gamma\left[\beta^E_{\pi_e} + \left(1 - \beta^E_{\pi_e}\right) \cdot v\langle s^E_p\rangle(b^E_{\pi_e})\right] \qquad \forall E \in \text{inc}(s_p) \tag{23}$$

The minimization over V allows us to rewrite every Constraint (23) using Constraints (19) to (21) from the LP 3:

$$V \geq \gamma\left[\beta^E_{\pi_e} + \left(1 - \beta^E_{\pi_e}\right) \sum_{(\mathbf{x},v)\in\Upsilon\langle s^E_p\rangle} c^{\mathbf{x},v}_E \cdot v\right] \tag{24}$$

$$\sum_{(\mathbf{x},v)\in\Upsilon\langle s^E_p\rangle} c^{\mathbf{x},v}_E \cdot x_j = \frac{1}{1 - \beta^E_{\pi_e}} \sum_{\{i,j\}\notin E} \tilde{b}_i \pi_e(i,j) \quad \left(= b^E_{\pi_e,j}\right) \quad \forall j \in 1\ldots N \tag{25}$$

$$\sum_{(\mathbf{x},v)\in\Upsilon\langle s^E_p\rangle} c^{\mathbf{x},v}_E = 1 \tag{26}$$

This set of constraints is however not linear. The linearity can be established by multiplying both sides of Constraints (25) by $(1 - \beta^E_{\pi_e})$ and introducing new variables $\hat{c}^{\mathbf{x},v}_E$ as a substitution for $(1 - \beta^E_{\pi_e}) \cdot c^{\mathbf{x},v}_E$. The resulting linear program is shown in LP 4. The strategy of the pursuer can be once again obtained from dual variables corresponding to Constraints (28).

$$\min_{V,\pi_e,\hat{c}} V \tag{27}$$

$$\text{s.t.}\, \gamma \sum_{(\mathbf{x},v)\in\Upsilon\langle s^E_p\rangle} \hat{c}^{\mathbf{x},v}_E \cdot v + \gamma \sum_{\{i,j\}\in E} \tilde{b}_i \pi_e(i,j) \leq V \qquad \forall E \in \text{inc}(s_p) \tag{28}$$

$$\sum_{(\mathbf{x},v)\in\Upsilon\langle s^E_p\rangle} \hat{c}^{\mathbf{x},v}_E \cdot x_j = \sum_{\{i,j\}\notin E} \tilde{b}_i \pi_e(i,j) \qquad \forall E \in \text{inc}(s_p),\, \forall j \in 1\ldots N \tag{29}$$

$$\sum_{(\mathbf{x},v)\in\Upsilon\langle s'_p\rangle} \hat{c}^{\mathbf{x},v}_E = 1 - \sum_{\{i,j\}\in E} \tilde{b}_i \pi_e(i,j) \qquad \forall E \in \text{inc}(s_p) \tag{30}$$

$$\sum_{j\in\text{adj}(i)} \pi_e(i,j) = 1 \qquad \forall i \in 1\ldots N \tag{31}$$

$$\pi_e(i,j) \geq 0 \qquad \forall i,j \in 1\ldots N \tag{32}$$

$$\hat{c}^{\mathbf{x},v}_E \geq 0 \qquad \forall E \in \text{inc}(s_p),\, \forall(\mathbf{x},v) \in \Upsilon\langle s^e_p\rangle \tag{33}$$

LP 4: LP for computing evader's strategy using point-represented $\bar{v}$.

5.4 Point-Based Updates

In the HSVI algorithm for POMDPs, the gap between value functions is reduced by performing point-based updates. The lower value function is updated by adding a new α-vector to the set Γ, obtained using the Bellman backup, and the upper value function is updated by addition of a new point into the set Υ by performing a one-step lookahead. In our algorithm, the point-based update in a game situation $\langle s_p, b \rangle$ is performed in a similar way, however unlike in the POMDP case we have to account for strategies of both players.

The newly created α-vector to add into $\Gamma \langle s_p \rangle$ corresponds to the value of optimal pursuer's strategy with respect to $\underline{v}$ in the game situation $\langle s_p, b \rangle$ where the point-based update is performed. This strategy is obtained by solving the LP 2 and its value is linear in the belief b (and thus representable as an α-vector). We will denote this α-vector $L\Gamma(s_p, b)$.

The update of the upper value function $\overline{v}\langle s_p \rangle$ requires to compute the value of the game situation $\langle s_p, b \rangle$ when the value of the subsequent stage is given by $\overline{v}$ (which is similar to the one-step lookahead). This value V is obtained by solving LP 4. A newly formed point $U\Upsilon(s_p, b) = b \rightarrow V$ is added into $\Upsilon \langle s_p \rangle$ which completes the point-based update of $\overline{v}\langle s_p \rangle$.

Example. The idea behind point-based updates will be presented on the game from Fig. 1. We will perform point based updates of value functions depicted in Fig. 3 in two game situations $\langle A, \{C : 1\}\rangle$ and $\langle A, \{B : 0.5, C : 0.5\}\rangle$ (i.e. the pursuer is located in vertex A and the evader is located either in vertex C, or she is located in either B or C with uniform probability).

The optimal strategy of the pursuer in situation $\langle A, \{C : 1\}\rangle$ with respect to $\underline{v}$ depicted in Fig. 3 is to move along the edge $\{A, C\}$. This leads to immediate capture of the evader located in vertex C and the pursuer hence gets a reward γ. If the pursuer was located in vertex B instead, the pursuer would reach a situation $\langle C, \{A : 1\}\rangle$ by playing $\{A, C\}$, where his reward according to current state of $\underline{v}$ is zero. The value of this strategy is expressed by α-vector α_1 shown in Fig. 4(Left). The value of the one-step lookahead in situation $\langle A, \{C : 1\}\rangle$ is γ which forms a new point on the upper value function $\overline{v}\langle A \rangle$.

The value of the one-step lookahead at situation $\langle A, \{B : 0.5, C : 0.5\}\rangle$ is $(\gamma + \gamma^2)/2$. A new point corresponding to this belief and to this value is added to $\Upsilon \langle A \rangle$ in Fig. 4(Middle). The optimal strategy of the pursuer to play in situation $\langle A, \{B : 0.5, C : 0.5\}\rangle$ with respect to $\underline{v}$ depicted in Fig. 4(Left) is to wait in vertex A. The strategy guarantees him reward γ when the evader starts in vertex A, and γ^2 when the evader starts in C. The value of this strategy is represented by α-vector α_2 shown in Fig. 4(Right).

Note that these two point-based updates were sufficient to reduce the gap in the initial game situation $\langle A, \{B : 0.5, C : 0.5\}\rangle$ to zero. In the next section we present a heuristic approach to identify these points.

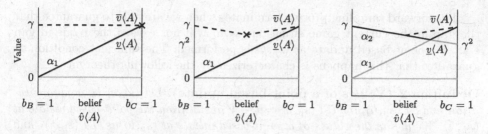

Fig. 4. Visualization of point-based updates: (Left) update at $\langle A, \{C : 1\}\rangle$ (Middle) upper value function update at $\langle A, \{B : 0.5, C : 0.5\}\rangle$ (Right) lower value function update at $\langle A, \{B : 0.5, C : 0.5\}\rangle$.

5.5   Forward Exploration

In every iteration of the HSVI algorithm, a forward sampling is used to generate a sequence of points to be updated using the point-based updates. In this section we propose a heuristic approach to direct this sampling.

The HSVI algorithm for POMDPs uses the concept of *excess* to identify observations (and consequently belief points) that contribute to the fact that the value in the initial belief point exceeds the desired precision ϵ. We define a similar concept of *excess contribution* taking pursuer's actions into account instead of observations.

Definition 1 (Excess contribution). *Let $\langle s_p, b\rangle$ be a game situation and $\underline{v}$ and $\overline{v}$ be lower and upper value functions. Let π_e be evader's strategy for the first stage of the game and E be action of the pursuer. We define excess contribution of playing π_e and E as*

$$\rho\langle s_p, b\rangle(\pi_e, E) = \gamma \left[1 - \sum_{i \in s_p} b_i\right] \cdot \left[1 - \sum_{\{i,j\} \in E} \tilde{b}_i \overline{\pi}_e(i,j)\right] \cdot \text{width}\big(\hat{v}\langle s_p^E\rangle (b_{\pi_e}^E)\big)$$

(34)

where $\tilde{b}$ is the belief updated with the information that evader is located in none of the vertices of s_p computed using Eq. (1).

The forward sampling process aims to ensure that the excess contribution in all game situations reachable from the current situation $\langle s_p, b\rangle$ gets sufficiently small in order to guarantee that the gap in the current situation width($\hat{v}\langle s_p\rangle (b)$) does not exceed $\hat{\epsilon}$. Whenever a new game situation is to be sampled in the course of forward exploration, the one with the highest *weighted excess contribution* is selected. The weighted excess contribution takes both the excess contribution reached by individual pursuer's actions E and optimal optimistic strategy π_e of the evader (solved using LP 2 with respect to $\underline{v}$) *and* the probability of playing each of the actions of the pursuer according to his optimistic strategy π_p (obtained from LP 4 using value functions $\overline{v}$) into account. First an action E with the highest weighted excess contribution $\pi_p(E) \cdot \rho\langle s_p, b\rangle (\pi_e, E)$ is selected and then the new game situation $\langle s_p^E, b_{\pi_e}^E\rangle$ is considered for further exploration.

The forward sampling process terminates when we are able to guarantee that the gap in the current game situation $\langle s_p, b \rangle$ will not exceed the required gap $\hat{\epsilon}$ after a point-based update at $\langle s_p, b \rangle$ is performed. The sufficient condition to guarantee that this happens is characterized by the following theorem.

Definition 2 (Values of a point-based update). *Let $\langle s_p, b \rangle$ be a game situation and $L\Gamma(s_p, b)$ and $U\Upsilon(s_p, b) = b \rightarrow V$ result from the point-based update at $\langle s_p, b \rangle$. We define the values of a point-based update at $\langle s_p, b \rangle$ as $(L\Gamma(s_p, b))$ and $\mathrm{val}(U\Upsilon(s_p, b))$, where $\mathrm{val}(L\Gamma(s_p, b))$ is the value of α-vector $L\Gamma(s_p, b)$ evaluated at belief point b and $\mathrm{val}(U\Upsilon(s_p, b)) = V$.*

Theorem 1. *Let $\langle s_p, b \rangle$ be a game situation and $\underline{v}$, $\overline{v}$ be a collection of lower and upper value functions corresponding to the game. Let π_e be optimal strategy of the evader with respect to $\underline{v}$. Let $\hat{\epsilon} > 0$ be chosen so that for every $E \in \mathrm{inc}(s_p)$ the following holds*

$$\rho\langle s_p, b \rangle(\pi_e, E) \leq \hat{\epsilon} \tag{35}$$

then $\mathrm{val}(U\Upsilon(s_p, b)) - \mathrm{val}(L\Gamma(s_p, b)) \leq \hat{\epsilon}$.

Proof. We will show that the values of pursuer's best responses to π_e at $\langle s_p, b \rangle$ with respect to $\underline{v}$ and $\overline{v}$ cannot differ by more than $\hat{\epsilon}$. The expected utility of playing action E against π_e with respect to $\underline{v}$ follows from Eq. (11):

$$\underline{v}_E = \sum_{i \in s_p} b_i + \gamma \left[1 - \sum_{i \in s_p} b_i \right] \cdot \left[\sum_{\{i,j\} \in E} \tilde{b}_i \pi_e(i,j) + \left(1 - \sum_{\{i,j\} \in E} \tilde{b}_i \pi_e(i,j) \right) \cdot \underline{v}\langle s_p^E \rangle(b_{\pi_e}^E) \right] \tag{36}$$

Value $\overline{v}_E$ representing the expected utility of playing E against π_e with respect to $\overline{v}$ is defined analogously. Note that

$$\max_{E \in \mathrm{inc}(s_p)} \underline{v}_E = \mathrm{val}(L\Gamma(s_p, b)) \quad \text{and} \quad \max_{E \in \mathrm{inc}(s_p)} \overline{v}_E \geq \mathrm{val}(U\Upsilon(s_p, b)). \tag{37}$$

The first equality holds because π_e is optimal with respect to $\underline{v}$, while the second inequality holds because π_e might be suboptimal with respect to $\overline{v}$. It holds for every $E \in \mathrm{inc}(s_p)$ that

$$\overline{v}_E - \underline{v}_E = \gamma \left[1 - \sum_{i \in s_p} b_i \right] \cdot \left[1 - \sum_{\{i,j\} \in E} \tilde{b}_i \pi_e(i,j) \right] \cdot \mathrm{width}(\hat{v}\langle s_p^E \rangle(b_{\pi_e}^E)) \tag{38}$$

$$= \rho\langle s_p, b \rangle(\pi_e, E) \leq \hat{\epsilon}. \tag{39}$$

Then

$$\mathrm{val}(U\Upsilon(s_p, b)) - \mathrm{val}(L\Gamma(s_p, b)) \leq \max_{E \in \mathrm{inc}(s_p)} \overline{v}_E - \max_{E \in \mathrm{inc}(s_p)} \underline{v}_E \tag{40}$$

$$\leq \max_{E \in \mathrm{inc}(s_p)} [\overline{v}_E - \underline{v}_E] \leq \hat{\epsilon}. \qquad \square$$

Example. Let us consider the game from Fig. 1 and value functions depicted in Fig. 3. The initial game situation is $\langle A, \{B : 0.5, C : 0.5\}\rangle$. The optimal optimistic strategy π_p of the pursuer in this situation (chosen according to $\overline{v}$) is to wait in vertex A, while the optimal optimistic strategy π_e of the evader (chosen according to $\underline{v}$) is to stay in the vertex C (if he started there), or to move from vertex B to A (as he has no other option). As the action $\{A, A\}$ is the only action played by the pursuer with a positive probability according to π_p, it has the highest weighted excess contribution. New game situation after action $\{A, A\}$ and strategy π_e are played is $\langle A, \{C : 1\}\rangle$. The optimal optimistic strategy of the pursuer in the newly generated situation captures the evader immediately (by moving from A to C) and hence its weighted excess contribution is zero and the forward sampling process can terminate. Note that the situations generated by the forward sampling process correspond to the situations considered in the example from Sect. 5.4.

5.6   Complete Algorithm

We can now state the complete HSVI algorithm for one-sided partially-observable pursuit-evasion games (Algorithm 1). First of all, value functions $\underline{v}$ and $\overline{v}$ are initialized on line 1. This initialization was described in Sect. 5.2.

The algorithm then iteratively performs exploration (line 3) until the required precision ϵ is reached (line 2). The goal of the exploration (line 8) is to ensure that the gap between $\underline{v}\langle s_p\rangle (b)$ and $\overline{v}\langle s_p\rangle (b)$ is bounded by $\hat{\epsilon}$.

If this goal has not been fulfilled yet, there must then be according to Theorem 1 an action E such that $\rho\langle s_p, b\rangle(\pi_e, E) > \hat{\epsilon}$. This action is selected on line 15 either using the weighted excess heuristic with probability $1 - \eta$ (line 12), or it is selected based on the excess contribution only to guarantee that all actions are eventually explored (this happens with probability η on line 14). We term the parameter η the *exploration parameter*. In order to make sure that $\rho\langle s_p, b\rangle(\pi_e, E) \leq \hat{\epsilon}$, it must hold that

$$\text{width}(\hat{v}\langle s_p^E\rangle (b_{\pi_e}^E)) \leq \hat{\epsilon}/\left(\gamma\left[1 - \sum_{i \in s_p} b_i\right] \cdot \left[1 - \sum_{\{i,j\} \in E} \tilde{b}_i \pi_e(i,j)\right]\right) \quad (41)$$

which follows from Theorem 1. This is satisfied by recursive application of the Explore procedure (line 18) on the newly generated game situation.

The optimal strategy π'_e of the evader with respect to $\overline{v}$ need not equal π_e—however the strategy π'_e is used when the point $U\Upsilon(s_p, b)$ is constructed. For this reason, point-based updates are performed at the belief point $b_{\pi_e}^{E'}$ on lines 19 to 23, where E' is chosen similarly to line 15, however a pessimistic strategy π'_e of the evader (chosen according to $\overline{v}$) is taken into account. Finally, point-based update is performed in the current game situation $\langle s_p, b\rangle$ on lines 24 and 25.

In order to guarantee the convergence of the algorithm a game situation $\langle s_p, b\rangle$ is selected randomly and a point-based update in this situation is performed on lines 4 to 6. There is a non-zero probability of sampling a point where

Data: Graph $\mathcal{G}$, initial position $\langle s_p^0, b^0 \rangle$, discount factor γ, precision $\epsilon > 0$,
exploration parameter η
Result: Set of approximate value functions $\hat{v}$

1 Initialize $\hat{v}$
2 **while** width$\left(\hat{v}\langle s_p^0 \rangle (b^0)\right) > \epsilon$ **do**
3 | Explore(s_p^0, b^0, ϵ)
4 | $\langle s_p, b \rangle \leftarrow$ random game situation
5 | $\Gamma\langle s_p \rangle \leftarrow \Gamma\langle s_p \rangle \cup \{L\Gamma(s_p, b)\}$
6 | $\Upsilon\langle s_p \rangle \leftarrow \Upsilon\langle s_p \rangle \cup \{U\Upsilon(s_p, b)\}$
7 **return** $\hat{v}$

8 **procedure** Explore$(s_p, b, \hat{\epsilon})$
9 | **if** width$(\hat{v}\langle s_p \rangle (b)) > \hat{\epsilon}$ **then**
10 | $\pi_e \leftarrow$ optimal evader's strategy at $\langle s_p, b \rangle$ with respect to $\underline{v}$
11 | **with probability** $1 - \eta$
12 | | $\pi_p \leftarrow$ optimal strategy at $\langle s_p, b \rangle$ with respect to $\overline{v}$
13 | **otherwise**
14 | | $\pi_p \leftarrow$ uniform distribution over inc(s_p)
15 | $E \leftarrow \arg\max_{E'} \pi_p(E') \cdot \rho\langle s_p, b\rangle(\pi_e, E')$
16 | $\hat{\epsilon}' \leftarrow \hat{\epsilon}/\left(\gamma\left[1 - \sum_{i \in s_p} b_i\right] \cdot \left[1 - \sum_{\{i,j\} \in E} \tilde{b}_i \pi_e(i,j)\right]\right)$ // $\hat{\epsilon}' = \infty$ in **case of division by zero**
17
18 | Explore$(s_p^E, b_{\pi_e}^E, \hat{\epsilon}')$
19 | $\pi_e' \leftarrow$ optimal evader's strategy at $\langle s_p, b \rangle$ with respect to $\overline{v}$
20 | $E' \leftarrow \arg\max_{E''} \pi_p(E'') \cdot \rho\langle s_p, b\rangle(\pi_e', E'')$
21 | **if** $\pi_p(E') \cdot \rho\langle s_p, b\rangle(\pi_e', E') > 0$ **then**
22 | | $\Gamma\left\langle s_p^{E'} \right\rangle \leftarrow \Gamma\left\langle s_p^{E'} \right\rangle \cup \{L\Gamma(s_p^{E'}, b_{\pi_e'}^{E'})\}$
23 | | $\Upsilon\left\langle s_p^{E'} \right\rangle \leftarrow \Upsilon\left\langle s_p^{E'} \right\rangle \cup \{U\Upsilon(s_p^{E'}, b_{\pi_e'}^{E'})\}$
24 | $\Gamma\langle s_p \rangle \leftarrow \Gamma\langle s_p \rangle \cup \{L\Gamma(s_p, b)\}$
25 | $\Upsilon\langle s_p \rangle \leftarrow \Upsilon\langle s_p \rangle \cup \{U\Upsilon(s_p, b)\}$

Algorithm 1. HSVI algorithm for one-sided pursuit-evasion games

the approximation can be improved, and the probability of not sampling such a point vanishes in time.

6   Experiments

We now turn to the experimental evaluation of the scalability of our algorithm. We consider games with two pursuing units played on a $m \times n$ grid and assume that the initial distance between the pursuers and the evader is maximal – i.e., the pursuer starts in the top left corner, while the evader starts in the bottom right one. The pursuer knows the initial position of the evader, however he does not observe her actions until the encounter. This scenario corresponds to a situation when a suspect is seen at some location and the goal of the pursuer is to locate and capture her. An example of a $m \times n$ grid game is shown in Fig. 5.

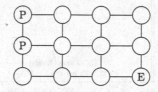

Fig. 5. 3×4 grid game. Initial positions of the pursuer are marked by P, initial position of the evader is marked by E.

6.1 Algorithm Configuration

As mentioned in Sect. 5.2, the initialization of the upper value functions can be done by the Shapley's value iteration algorithm [6] on the perfect information variant of the game. The algorithm was run for at most 20 s, unless the maximum change in the current iteration is lower than 10^{-3}. In order to guarantee that the value functions $\overline{v}$ upper bound the optimal value functions v^*, we initialized Shapley's algorithm with valuation assigning value 1 to every game state (and the values thus approach v^* from above).

For our HSVI algorithm, we set the exploration parameter to $\eta = 0.1$. The algorithm was implemented in Java and IBM ILOG CPLEX 12.6.3 was used to solve all linear programs.

6.2 Comparison with Online Outcome Sampling

Since there are, to the best of our knowledge, no other existing algorithms that approximate optimal strategies in partially observable stochastic games with infinite horizon, we compare the performance with the representative of the algorithms for approximately solving games with finite horizon (or imperfect-information extensive-form games; EFGs) – online outcome sampling algorithm (OOS) [4]. The algorithm is a modification of Monte Carlo Tree Search for imperfect-information games, so that the algorithm constructs the game tree *incrementally*. While the algorithm is designed for finite-horizon games, the combination of sampling and incremental building of the game tree allows us (in theory) to use this algorithm for infinite horizon games and extend the horizon over time. However, the size of the EFG grows exponentially in the horizon t, thus increasing the horizon to improve the approximation is typically not an option. This behavior is demonstrated on a 3×4 grid game with discount factor $\gamma = 0.95$ and horizon $t = 8$. The OOS algorithm runs out of 50GB of memory before reaching a close approximation of the value of the infinite-horizon game. Specifically, while our algorithm was able to compute 0.02-optimal strategy with value 0.769 in 25.5 s, OOS found a strategy with value of only 0.553 after 160 s. Less than 1 GB of memory was sufficient for our HSVI algorithm.

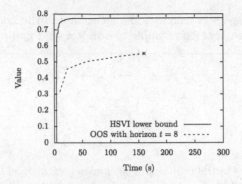

Fig. 6. Convergence of the OOS algorithm on 3×4 grid game, $\gamma = 0.95$, horizon $t = 8$

6.3 Scalability of HSVI

We now focus only on our HSVI algorithm and analyze its scalability. We show how the computation time and the quality of found strategies depend on the size of the graph, desired precision (parameter ϵ), and the discount factor γ. All of these parameters influence the length of the samples generated by the forward exploration heuristic (Sect. 5.5), the lower ϵ or the higher γ, the longer samples may have to be generated in order to satisfy property required by Theorem 1. We measure the runtime of the algorithm needed to reach 2ϵ precision on $3 \times N$ grid games with 4 independent runs for each parametrization.

Figure 7(Left) shows the computation times for increasing number of columns of the grid N and for different desired precisions ϵ with discount factor set to $\gamma = 0.95$. The runtime of the algorithm exhibits the exponential dependency on N as well as on the desired precision ϵ. While an 0.02-optimal strategy for a 3×4 game was found in 15.1 s on average, sufficiently accurate strategy was not found after 3 h in the case of the 3×8 game (the average gap in the initial situation reached 0.0275 after 3 h). Figure 7(Right) displays the runtime dependency on the discount factor γ. The higher the discount factor γ, the more important future rewards are and hence longer plays have to be considered.

While on smaller graphs our algorithm can compute strategies that closely approximate optimal strategies, the current approach has its limitations and the convergence has a long tail on larger graphs. We illustrate such an example on a game obtained from a 4×8 grid where six randomly selected edges were contracted to a final graph with 26 vertices. The convergence of the lower and upper bounds in the initial game situation is depicted in Fig. 8(Left). The long-tail convergence is caused by two factors: (1) the increasing time of each iteration as the number of α-vectors and points in the representation of the value functions grows (shown in Fig. 8; Right) and (2) the necessity to play near-optimal strategies in distant stages that are less likely to be reached using the forward exploration heuristic.

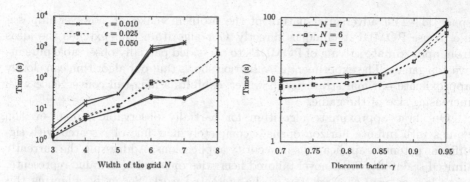

Fig. 7. Time needed to reach 2ϵ precision on a $3 \times N$ grid game. (Left) scalability in the width of the grid N ($\gamma = 0.95$) (Right) scalability in the discount factor γ ($\epsilon = 0.025$)

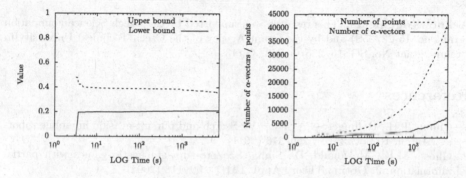

Fig. 8. Game with 26 vertices. (Left) convergence plot (Right) size of value functions

The algorithm was able to approximate the optimal value of the infinite game within 0.147 tolerance in 2 h. The larger the representation of value functions is (in terms of α-vectors and points), the larger the linear programs are. While iterations took 1–2 s when value functions were represented using roughly 1000 vectors and 4000 points, the iteration time increases to 20–25 s when the number of vectors reaches 7000 and the number of points reaches 40000.

7   Conclusions

The class of stochastic games with one-sided partial observability is a well-suited model for decision making in security domains and designing robust control strategies. We focus on pursuit-evasion games where the evader knows the position of the pursuer, but the pursuer does not know evader's position.

We present the first approximate algorithm for solving games from this class. The algorithm builds upon recent theoretic results showing that the value functions exhibit similar properties to their counterparts in Partially Observable Markov Decision Processes (POMDPs); namely the convexity in the belief space [3]. The adversarial nature of the problem makes it however impossible to use POMDP techniques directly. Our algorithm thus extends the ideas from approximate solving of POMDPs to one-sided partially observable pursuit-evasion games. The experimental evaluation shows that our algorithm can closely approximate optimal strategies, however, exhibits a long-tail convergence with increasing size of the game.

Our first approximate algorithm for partially observable pursuit-evasion games with infinite horizon opens a completely new line of research with significant potential applications in security applications. Addressing the currently limited scalability with novel tailored heuristics or pruning in value representations is a natural continuation of the presented work. Secondly, adapting this approach for other game models relevant in security domain (e.g., sequential security games) is another promising direction.

Acknowledgements. This research was supported by the Czech Science Foundation (grant no. 15-23235S) and by the Grant Agency of the Czech Technical University in Prague, grant No. SGS16/235/OHK3/3T/13.

References

1. Chung, T.H., Hollinger, G.A., Isler, V.: Search and pursuit-evasion in mobile robotics. Auton. Robot. **31**(4), 299–316 (2011)
2. Ghosh, M.K., McDonald, D., Sinha, S.: Zero-sum stochastic games with partial information. J. Optim. Theory Appl. **121**(1), 99–118 (2004)
3. Horak, K., Bosansky, B.: Dynamic programming for one-sided partially observable Pursuit-Evasion games. arXiv preprint arXiv:1606.06271 (2016)
4. Lisý, V., Lanctot, M., Bowling, M.: Online Monte Carlo counterfactual regret minimization for search in imperfect information games. In: Proceedings of the 14th International Conference on Autonomous Agents and Multiagent Systems (AAMAS), pp. 27–36 (2015)
5. Pineau, J., Gordon, G., Thrun, S., et al.: Point-based value iteration: an anytime algorithm for POMDPs. IJCAI **3**, 1025–1032 (2003)
6. Shapley, L.S.: Stochastic games. Proc. Natl. Acad. Sci. **39**(10), 1095–1100 (1953)
7. Smith, T., Simmons, R.: Heuristic search value iteration for POMDPs. In: Proceedings of the 20th Conference on Uncertainty in Artificial Intelligence, pp. 520–527. AUAI Press (2004)
8. Smith, T., Simmons, R.: Point-based POMDP algorithms: improved analysis and implementation. arXiv preprint arXiv:1207.1412 (2012)
9. Vidal, R., Shakernia, O., Kim, H.J., Shim, D.H., Sastry, S.: Probabilistic pursuit-evasion games: theory, implementation, and experimental evaluation. IEEE Trans. Robot. Autom. **18**(5), 662–669 (2002)

Consensus Algorithm with Censored Data for Distributed Detection with Corrupted Measurements: A Game-Theoretic Approach

Kassem Kallas[✉], Benedetta Tondi, Riccardo Lazzeretti, and Mauro Barni

Department of Information Engineering and Mathematics,
University of Siena, Via Roma 56, 53100 Siena, Italy
kassem.kallas@unisi.it, benedettatondi@gmail.com,
riccardo.lazzeretti@gmail.com, barni@dii.unisi.it

Abstract. In distributed detection based on consensus algorithm, all nodes reach the same decision by locally exchanging information with their neighbors. Due to the distributed nature of the consensus algorithm, an attacker can induce a wrong decision by corrupting just a few measurements. As a countermeasure, we propose a modified algorithm wherein the nodes discard the corrupted measurements by comparing them to the expected statistics under the two hypothesis. Although the nodes with corrupted measurements are not considered in the protocol, under proper assumptions on network topology, the convergence of the distributed algorithm can be preserved. On his hand, the attacker may try to corrupt the measurements up to a level which is not detectable to avoid that the corrupted measurements are discarded. We describe the interplay between the nodes and the attacker in a game-theoretic setting and use simulations to derive the equilibrium point of the game and evaluate the performance of the proposed scheme.

Keywords: Adversarial signal processing · Consensus algorithm · Distributed detection with corrupted measurements · Data fusion in malicious settings · Game theory

1 Introduction

In distributed detection applications a group of nodes in a network collect measurements about a certain phenomenon [1]. In centralized architectures, the measurements are sent to a central processor, called fusion center (FC), which is responsible of making a global decision. If needed, the result of the decision is then transmitted to all the nodes. Though attractive for the possibility of adopting an optimum decision strategy based on the entire set of measurements collected by the network, centralized solutions present a number of drawbacks, most of which related to the security of the network. For instance, the FC represents a single point of failure or a bottleneck for the network, and its failure may compromise the correct behavior of the whole network. In addition, due to

© Springer International Publishing AG 2016
Q. Zhu et al. (Eds.): GameSec 2016, LNCS 9996, pp. 455–466, 2016.
DOI: 10.1007/978-3-319-47413-7_26

privacy considerations or power constraints, the nodes may prefer not to share the gathered information with a remote device. For the above reasons, decentralized solutions have attracted an increasing interest. Consensus Algorithm is a fusion decentralized algorithm in which the nodes locally exchange information to reach a final agreement about the phenomenon of interest [2,3]. Consensus algorithm have been proven to provide good performance in many applications like cognitive radio [4], social networks or experimental sociology [5], and many others like flocking, formation control, load-balancing network, wireless sensor networks, etc. [2].

Despite the benefits of decentralized solutions using consensus algorithm, their nature makes them vulnerable to many security threats: for instance, attacks that emulate the phenomenon of interest to have an exclusive benefit from the resource, i.e., the Primary User Emulation Attack (PUEA) in cognitive radio applications [6], or data (measurements) falsification attacks [7], in which the attacker tries to induce a wrong decision by injecting forged measurements. This kind of attack can be launched in one of two ways: either the attacker can directly access the programmable device or, more simply, attack the physical link between the phenomenon and the nodes. In the first case, the attacker has full control over the nodes, and many effective solutions are proposed [8–11] whereas in the second case, the attacker cannot control the node and then he is not part of the network. In this paper, we focus on this second case.

In this attack scenario, when centralized systems are considered, by relying on the observation of the entire set of measurements, the fusion center can easily 'detect' the corrupted values and discard them, as long as their number remains limited. In this way, a reliable decision can still be done, see [12–15]. Attempts have been made to defend against those attacks in decentralized networks that employ a consensus algorithm to make a decision [16–18], when the attacker is assumed to control the nodes. Other solutions based on network control theory are proposed in [8–11]. However, all these methods do not consider the possibility that the attackers are aware of the defense mechanism adopted by the network and hence have their own countermeasures.

In this paper, by focusing on the measurement falsification attack with corruption of the physical link, we propose a game theoretical framework to distributed detection based on consensus algorithm. Specifically, we propose to include a preliminary isolation step in which each node may discard its own measurement based on the available a priori knowledge of the measurements statistics under the tho hypotheses. Then, the algorithm proceeds as usual, with the nodes that continue to receive and dispatch messages from their neighbors. Under some assumptions on network topology, that prevents that isolation step causes the network to disconnect, the convergence of the consensus algorithm is preserved. By following the principles of adversarial signal processing [19], we assume that in turn the attacker may adjust the strength of the falsification attack to avoid that the fake measurements are discarded. We then formalise the interplay between the network designed and the attacker as a zero-sum competitive game and use simulations to derive equilibrium point of the game.

The rest of this paper is organized as follows. In Sect. 2, we introduce the network model and describe the consensus algorithm. In Sect. 3, we introduce the measurement falsification attack against the detection based on consensus showing its powerfulness. In Sect. 4, we propose a refinement of the consensus algorithm to make it robust to the measurement falsification attack. Then, the interplay between the attacker and the network designer is casted into a game-theoretic framework in Sect. 5. The equilibrium point is found numerically in Sect. 6. Then, we conclude the paper in Sect. 7 with some final remarks.

2   Distributed Detection Based on Consensus Algorithm

In this section, we describe the distributed detection system considered in this paper, when no adversary is present and introduce the consensus algorithm the detection system relies on.

2.1   The Network Model

The network is modeled as an undirected graph $\mathcal{G}$ where the information can be exchanged in both directions between the nodes. A graph $\mathcal{G} = (\mathcal{N}, \mathcal{E})$ consists of the set of nodes $\mathcal{N} = \{n_1, ..., n_N\}$ and the set of edges $\mathcal{E}$ where $(n_i, n_j) \in \mathcal{E}$ if and only if there is a common communication link between n_i and n_j, i.e., they are neighbors. The neighborhood of a node n_i is indicated as $\mathcal{N}_i = \{n_j \in \mathcal{N} : (n_i, n_j) \in \mathcal{E} \}$. For task of simplicity, we sometimes refer to $\mathcal{N}_i$ as the set of indexes j instead than directly of nodes. The graph $\mathcal{G}$ can be represented by its adjacency matrix $A = \{a_{ij}\}$ where $a_{ij} = 1$, if $(n_i, n_j) \in \mathcal{E}$, 0 otherwise.

The degree matrix D of $\mathcal{G}$ is a diagonal matrix with $d_{ii} = a_{i1} + a_{i2} + ... + a_{in}$, $d_{ij} = 0, \forall i, j \neq i$ [20].

2.2   The Measurement Model

Let S be the status of the system under observation: we have $S = 0$, under hypothesis H_0 and $S = 1$ under hypothesis H_1. We use the capital letter X_i to denote the random variable describing the measurement at node n_i, and the lower-case letter x_i for a specific instantiation. By adopting a Gaussian model, the probability distribution of each measurement x_i under the two hypothesis is given by:[1]

$$P_X(x) = \begin{cases} \mathcal{N}(-\mu, \sigma), & \text{under } H_0, \\ \mathcal{N}(\mu, \sigma), & \text{under } H_1, \end{cases} \tag{1}$$

where, $\mathcal{N}(\mu, \sigma)$ is the Normal Distribution with mean μ and variance σ^2.

Let us denote with U the result of the final (binary) decision. An error occurs if $u \neq s$. By assuming that the measurements are conditionally independent, that

---

[1] We are assuming that the statistical characterization of the measurement at all the nodes is the same.

is that are independent conditioned to the status of the system, the optimum decision strategy consists in computing the mean of the measurements, $\bar{x} = \sum_i x_i/N$ and comparing it with a threshold λ which is set based on the a-priori probability ($\lambda = 0$ in the case of equiprobable system states). In a distributed architecture based on consensus, the value of $\bar{x}$ is computed iteratively by means of a proper message exchanging procedure between neighboring nodes, the final decision is made at each single node by comparing $\bar{x}$ with λ.

In this paper we consider the case of equiprobable system states. It is worth observing that our analysis, included the game formulation in Sect. 5, can be extended to the general case in which this assumption does not hold.

2.3   The Consensus Algorithm

Consensus algorithm for distributed detection is a protocol where the nodes locally exchange information with their neighbors in order to converge to an agreement about an event or a physical phenomenon [2], e.g. the existence of a transmission signal in cognitive radio applications [4]. It consists of three phases: the initial phase, the state update phase and the decision phase.

1. Initial phase: the nodes collect their initial measurement $x_i(0)$ about the phenomenon they are monitoring, and exchange the measurement with their neighbors.
2. State update phase: at each time step k, each node updates its state based on the information received from its neighbors. Then, at step $k + 1$ we have:

$$x_i(k + 1) = x_i(k) + \epsilon \sum_{j \in \mathcal{N}_i} (x_j(k) - x_i(k)) \tag{2}$$

where, $0 < \epsilon < (\max_i \mathcal{N}_i)^{-1}$ is the update step parameter. This phase is iterated until they reach the consensus value $\bar{x}(\mathcal{N}) = \frac{1}{N} \sum_{i \in \mathcal{N}} x_i(0)$, which corresponds to the mean of the initial measurements. It is proven that, with the above choice for ϵ, the consensus algorithm converges to $\bar{x}$ regardless of the network topology [3].
3. The final decision phase: this is the last phase in which all nodes compare the consensus value $\bar{x}$ to a threshold λ to make the final decision u:

$$u = \begin{cases} 1, & \text{if } \bar{x} > \lambda. \\ 0, & \text{otherwise,} \end{cases} \tag{3}$$

In the symmetric setup considered in this paper $\lambda = 0$.

3   Measurement Falsification Attack Against Consensus-Based Detection

In this section, we consider an adversarial setup of the setting described in the previous section and show that even a single false measurement can result in a wrong decision.

3.1   Consensus Algorithm with Corrupted Measurements

In the binary decision setup we are considering, the objective of the attacker is inducing one of the two decision errors (or both of them): decide that $S = 0$ when H_1 holds (False Alarm), decide that $S = 1$ when H_0 holds (Missed Detection). For simplicity, we optimistically make the worst case assumption that the attacker knows the true system state. In this case, he can try to push the network toward a wrong decision by replacing one or more measurements so to bias the average computed by using the consensus algorithm. Specifically, for any corrupted node, the attacker forces the measurement to a positive value Δ_0 under H_0 and to a negative value Δ_1 under H_1. For the symmetric setup, reasonably, $\Delta_1 = -\Delta_0 = \Delta > 0$. In the following we assume that the attacker corrupts a fraction α of the nodes, that is the number of attacked nodes is $N_A = \alpha N$.

Given the initial vector of measurements, the consensus value the network converge to because of the attack is:

$$\tilde{x} = \frac{1}{N} \sum_{i \in \mathcal{N}_H} x_i(0) + \frac{N_A \Delta}{N}, \tag{4}$$

where $\mathcal{N}_H$ is the set of the uncorrupted nodes ($|\mathcal{N}_H| = N - N_A$).

By referring to the model described in Sect. 2.2, it is easy to draw a relation between Δ, α and the probability p that the attacker induces a decision error. By exploiting the symmetry of the considered setup we can compute p by considering the behavior under one hypothesis only, that is we have $p = P(U = 1|H_0) = P(\tilde{X} > 0|H_0)$.

In the following we indicate with $\bar{X}(\mathcal{N})$ the average of the measurements made by the nodes in a set $\mathcal{N}$.

The error probability p for a given N_A can be written as:

$$p = P(\tilde{X} > 0|H_0) = P\left(\frac{N - N_A}{N}\bar{X}(\mathcal{N}_H) > -\frac{N_A \Delta}{N}\Big|H_0\right) \tag{5}$$

$$= P\left(\bar{X}(\mathcal{N}_H) > \frac{N}{N - N_A}\left(-\frac{N_A \Delta}{N}\right)\Big|H_0\right) = \int_{-\frac{N_A \Delta}{N - N_A}}^{\infty} \mathcal{N}(-\mu, \sigma/\sqrt{N - N_A}).$$

Clearly, if there is no limit to the value of Δ, the attacker will always succeed in inducing a wrong decision (see for example Fig. 1).

This shows how harmful the attack can be against distributed detection based on consensus algorithm. Therefore, the issue of securing distributed detection with consensus algorithm must be studied, at the purpose to increase the robustness of the consensus algorithm to intentional attacks.

4   Consensus Algorithm with Censored Data

With centralized fusion is quite easy to detect false measurements, since they assume outlier values with respect to the majority of the measurements. In a

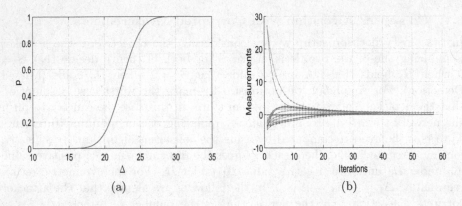

Fig. 1. (a): Success probability of the attack versus Δ in the adversarial setup $N = 20$, $\mu = 2.5$, $\sigma = 1$, $N_A = 2(\alpha = 0.1)$. (b): Effect of the attack on the convergence of the consensus algorithm for $\Delta = 27$, the network decides for H_1 even if H_0 is true.

distributed setting, however, this is not easy since, at least in the initial phase (see Sect. 2.3), each node sees only its measurement and has no other clue about the system status.

In contrary with most of proposed solutions in the literature [8,18], in this paper we propose to tackle with the problem of the measurement falsifications at the initial phase of the consensus algorithm (see for instance [2]), by letting each node discard its measurement if it does not fall within a predefined interval containing most of the probability mass associated to both H_0 and H_1, being then a suspect measurement. In the subsequent phase the remaining nodes continue exchanging messages as usual according to the algorithm, whereas the nodes which discarded their measurements only act as receivers and do not take part in the protocol. Due to the removal, the measurements exchanged by the nodes follows a censored gaussian distribution, i.e. the distribution which results by constraining the (initial) gaussian variable to stay within an interval [21]. Specifically, the nodes discards all the measurements whose absolute values are large than a removal threshold η. By considering the results shown in Fig. 1a, we see that, in the setup considered, if we let by letting $\eta = 17.5$ the error probability drops to nearly zero since the attacker must confine the choice of Δ to values lower than 17.5. The proposed strategy is simple, yet effective, and allow us to use a game theoretical approach to set the parameters (see Sect. 5).

For our analysis, we consider conditions on the network topology, such that the connectivity of the network is preserved and then the algorithm converges to the average of measurements which have not been discarded. For a given graph, this fact is characterized by the node connectivity, namely, the maximum number of nodes whose removal does not cause a disconnection [22]. Convergence is guaranteed for instance in the following cases (see [23] for an extensive analysis of the connectivity properties for the various topologies): Fully-connected graph; Random Graph [24], when the probability of having a connection between two nodes is large enough; Small-World Graph [25] when the neighbour list in ring formation is large and the rewiring probability is large as well; Scale-Free Graph [26], for sufficiently large degree of the non-fully meshed nodes.

We now give a more precise formulation of the consensus algorithm based on censored data. Let us denote with $\mathcal{R}$ the set of all the remaining nodes after the removal, that is

$$\mathcal{R} = \{n_j \in \mathcal{N} : -\eta < x_j < \eta\}, \tag{6}$$

and let $\mathcal{R}_i$ be the 'active' neighborhood of node i after the isolation, $i \in \mathcal{R}$ (i.e. the set of the nodes in the neighborhood of i which take part in the protocol). The update rule for node $i \subset \mathcal{R}$ can be written as:

$$x_i(k+1) = x_i(k) + \epsilon \sum_{j \in \mathcal{R}_i} (x_j(k) - x_i(k)), \tag{7}$$

where $0 < \epsilon < (\max_i \mathcal{N}_i)^{-1}$, and the degree refers to the network after the removal of the suspect nodes, that is to the graph $(\mathcal{R}, \mathcal{E})$ (instead of $(\mathcal{N}, \mathcal{E})$).

Under the above conditions on the network topologies, the consensus algorithm converges to the average value computed over the measurements made by the nodes in $\mathcal{R}$, namely $\bar{x}(\mathcal{R})$. Otherwise, disconnection may occur and then it is possible that different parts of the network (connected components) converge to possibly different values.

5   Game-Theoretic Formulation

The consensus based on censored data is expected to be robust in the presence of corrupted measurements. On the other hand, we should assume that the attacker is aware that the network takes countermeasures and removes suspect measurements in the initial phase, hence he will adjust the attack strength Δ to avoid that the false measurement is removed. We model the interplay between the attacker and the network as a two-player zero sum game where each player will try to maximize its own payoff. Specifically, we assume that the network designer, hereafter referred as the defender (D), does not know the attack strength Δ, while the attacker (A) does not know the value of the removal threshold η adopted by the defender.

With these ideas in mind, the Consensus-based Distributed Detection game $\mathcal{CDD}(\mathcal{S}_\mathcal{A}, \mathcal{S}_\mathcal{D}, v)$ is a two-player, strategic game played by the attacker and the defender, defined by the following strategies and payoff.

• The space strategies of the defender and the attacker are respectively

$$\mathcal{S}_\mathcal{D} = \{\eta \in [0, \infty)\}$$
$$\mathcal{S}_\mathcal{A} = \{\Delta \in [0, \infty)\}; \tag{8}$$

The reason to limit the strategies of D to values larger than by λ is to avoid removing correct measurements at the defender side and to prevent to vote for the correct hypothesis at the attacker side.

- The payoff function is defined as the final error probability,

$$v = P_e = P(U \neq S) = P(\bar{X} > 0/H_0), \tag{9}$$

where $\bar{X} = \bar{X}(\mathcal{R})$, that is the mean computed over the nodes that remain after the removal. The attacker wishes to maximize v, whereas the defender wants to minimize it.

Note that according to the definition of the $\mathcal{CDD}$ game, the sets of strategies of the attacker and the defender are continuous sets. We remind that, in this paper, we consider situations in which the network remains connected after the isolation and then convergence of the algorithm is preserved. Notice that, with general topologies, when disconnection may occur, the payoff function should be redefined in terms of error probability at the node level.

In the next section, we use numerical simulations to derive the equilibrium point of the game under different settings and to evaluate the payoff at the equilibrium.

6   Simulation Results

We run numerical simulations in order to investigate the behavior of the $\mathcal{CDD}$ game for different setups and analyze the achievable performance when the attacker and the defender adopt their best strategies with parameters tuned following a game-theoretic formalization. Specifically, the first goal of the simulations is to study the existence of an equilibrium point for the $\mathcal{CDD}$ game and analyze the expected behavior of the attacker as well as the defender at the equilibrium. The second goal is to evaluate the payoff at the equilibrium as a measure of the best achievable performance of distributed detection with the consensus algorithm based on censored data.

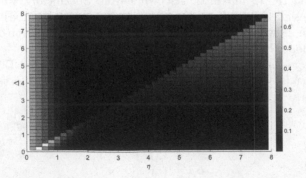

Fig. 2. Payoff matrix of the game with $N = 20$, $\alpha = 0.1$ and $\mu = 1$ (SNR = 4).

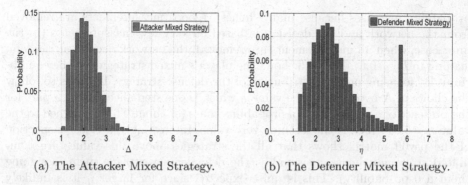

(a) The Attacker Mixed Strategy.   (b) The Defender Mixed Strategy.

Fig. 3. Equilibrium strategies in the following setup: $N = 20$, $\alpha = 0.1$, $\mu = 1$, (SNR = 4). Payoff at the equilibrium: $v = 0.0176$.

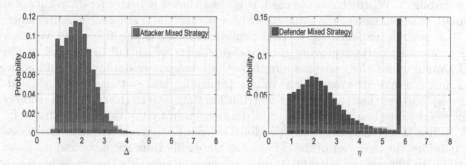

Fig. 4. Equilibrium strategies in the following setup: $N = 20$, $\alpha = 0.2$, $\mu = 1$ (SNR = 4). Payoff at the equilibrium: $v = 0.1097$.

To perform our experiments, we quantize the values of η and Δ with step 0.2 and then we consider the following sets: $\mathcal{S}_D = \{\eta \in \{0, 0.2, ...\}\}$ and $\mathcal{S}_A = \{\Delta \in \{0, 0.2, ...\}\}$. Simulations were carried out according to the following setup. We considered a network with $N = \{20, 50\}$ nodes where the measurement of each node is corrupted with probability $\alpha \in \{0.1, 0.2\}$. We assume that the probability that the measurement of a node is corrupted does not depend on the other nodes (independent node corruption). According to the model introduced in Sect. 2.2, the measurements are drawn according to Gaussian distribution with variance $\sigma^2 = 1$ and mean $-\mu$ and μ under H_0 and H_1 respectively. In our tests, we take $\mu = \{1, 2\}$. For each setting, we estimated the error probability of the decision based on the censored data over 10^5 trials. Then, we find the mixed strategies Nash equilibrium by relying on the minimax theorem [27] and then solving two separate linear programming problems.

Figure 2 shows the payoff matrix in gray levels for the game with $\alpha = 0.1$ and $\mu = 1$ (i.e., $SNR = 4$). Notice that the stepwise behavior of the values of the payoff in correspondence of the diagonal, which is due to the hard isolation (for each Δ, when $\eta < \Delta$ all the corrupted measurements are kept, while they are removed for $\eta \geq \Delta$). When very low values of η are considered, the error probability increases because many 'honest' (good) measurements are removed from the network and the decision is based on very few measurements (in the limit case, when all measurements are removed, the network decides at random, leading to $P_e = 0.5$). Figure 3 shows the player's mixed strategies at the equilibrium. By focusing on the distribution of the defense strategy, D seems to follow the choice of A by choosing the value η which is one step ahead of Δ, a part for the presence of a peak, that is a probability mass (of about 0.075) assigned to the value $\eta = 5.6$, which is the last non-zero value. Interestingly, a closer inspection of the payoff matrix shows that all the strategies above this values are dominated strategies; hence, reasonably, the defender never plays them (assigning them a 0 probability). This is quite expected since for larger η it is unlikely that an observation falls outside the range $[-\eta, \eta]$ and then the 'censoring' does not significantly affect the 'honest' measurements (i.e. $\mathcal{R} = \mathcal{N}$ with very high probability). When this is the case, it is clear that it is better for D to choose η small, thus increasing the probability of removing the corrupted measurements.

A possible explanation for the peaked behavior is the following. When η decreases, D starts removing good measurements which fall in the tail of the Gaussian under the corresponding hypothesis, whose values are not limited to Δ, but can take arbitrarily large values. Depending also on the setup considered, it may happen that the positive contribution they give to the correct decision is more relevant than the negative contribution given by the values introduced by A. When this is the case, it is better for the defender to use all the measurements. Therefore, the behavior of the defender at the equilibrium has a twofold purpose: trying to remove the corrupted measurements on one side (by choosing η one step ahead of Δ) and avoiding to rule out the large good measurements on the other (by selecting the critical η). The error probability at the equilibrium is 0.0176 thus showing that the proposed scheme allows to get correct detection with high probability despite the data corruption performed by A.

Figure 4 shows the equilibrium strategies for $\alpha = 0.2$. Since the removal of the large good measurements has more impact when α is large, a bit higher weight is associated in this case to the peak. The error probability at the equilibrium is

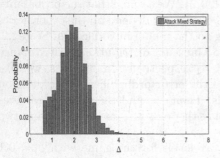 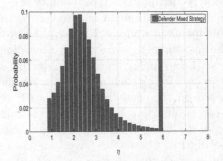

Fig. 5. Equilibrium strategies in the following setup: $N = 50$, $\alpha = 0.2$, $\mu = 2$ (SNR = 4). Payoff at the equilibrium $v = 0.0556$.

$v = 0.1097$. Finally, Fig. 5 shows the equilibrium mixed strategies for D and A when $N = 50$, $\alpha = 0.2$ and $\mu = 2$.

7 Conclusion

We proposed a consensus algorithm based on censored data which is robust to measurement falsification attacks. Besides, we formalized the interplay between the attacker and the network in a game-theoretic sense, and we numerically derive the optimal strategies for both players and the achievable performance in terms of error probability in different setups. Simulation results show that, by adopting the proposed scheme, the network can still achieve correct detection through consensus, despite the presence of corrupted measurements.

As a future work, we would like to extend the game-theoretic approach to include the graph disconnection as a part of the defender payoff and then apply our analysis to general topologies. In addition, we would like to extend the analysis to more complicated statistical models for the measurements, e.g. the case of the chi-square distribution, and to consider more complicated versions of the game, e.g. by allowing the players to adopt randomized strategies.

References

1. Varshney, P.K., Burrus, C.S.: Distributed Detection and Data Fusion. Springer, New York (1997)
2. Olfati-Saber, R., Fax, J.A., Murray, R.M.: Consensus and cooperation in networked multi-agent systems. Proc. IEEE 95, 215–233 (2007)
3. Sardellitti, S., Barbarossa, S., Swami, A.: Optimal topology control and power allocation for minimum energy consumption in consensus networks. IEEE Trans. Signal Process. 60, 383–399 (2012)
4. Li, Z., Yu, F., Huang, M.: A distributed consensus-based cooperative spectrum-sensing scheme in cognitive radios. IEEE Trans. Veh. Technol. 59, 383–393 (2010)
5. Mossel, E., Schoenebeck, G.: Reaching consensus on social networks. In: Innovations in Computer Science, pp. 214–229 (2010)
6. Chen, R., Park, J.-M., Reed, J.: Defense against primary user emulation attacks in cognitive radio networks. IEEE J. Sel. Areas Commun. 26, 25–37 (2008)
7. Chen, R., Park, J.-M., Bian, K.: Robust distributed spectrum sensing in cognitive radio networks. In: 2008 IEEE Proceedings INFOCOM - The 27th Conference on Computer Communications (2008). doi:10.1109/infocom.2007.251
8. Sundaram, S., Hadjicostis, C.N.: Distributed function calculation via linear iterative strategies in the presence of malicious agents. IEEE Trans. Autom. control 56, 1495–1508 (2011)
9. Pasqualetti, F., Dorfler, F., Bullo, F.: Attack detection and identification in cyber-physical systems. IEEE Trans. Autom. Control 58, 2715–2729 (2013)
10. Leblanc, H.J., Zhang, H., Koutsoukos, X., Sundaram, S.: Resilient asymptotic consensus in robust networks. IEEE J. Sel. Areas Commun. 31, 766–781 (2013)
11. Vaidya, N.H., Tseng, L., Liang, G.: Iterative approximate Byzantine consensus in arbitrary directed graphs. In: Proceedings of the 2012 ACM Symposium on Principles of Distributed Computing - PODC 2012 (2012). doi:10.1145/2332432.2332505

12. Barni, M., Tondi, B.: Multiple-observation hypothesis testing under adversarial conditions (2013). doi:10.1109/wifs.2013.6707800

13. Abrardo, A., Barni, M., Kallas, K., Tondi, B.: A game-theoretic framework for optimum decision fusion in the presence of Byzantines. IEEE Trans. Inf. Forensics Secur. **11**, 1333–1345 (2016)

14. Abrardo, A., Barni, M., Kallas, K., Tondi, B.: A game-theoretic approach. In: 53rd IEEE Conference on Decision and Control. doi:10.1109/cdc.2014.7039431

15. Rawat, A.S., Anand, P., Chen, H., Varshney, P.K.: Collaborative Spectrum sensing in the presence of Byzantine attacks in cognitive radio networks. IEEE Trans. Signal Process. **59**, 774–786 (2011)

16. Yu, F.R., Tang, H., Huang, M., Li, Z., Mason, P.C.: Defense against spectrum sensing data falsification attacks in mobile ad hoc networks with cognitive radios. In: 2009 IEEE Military Communications Conference (MILCOM) (2009). doi:10.1109/milcom.2009.5379832

17. Liu, S., Zhu, H., Li, S., Li, X., Chen, C., Guan, X.: An adaptive deviation-tolerant secure scheme for distributed cooperative spectrum sensing. In: 2012 IEEE Global Communications Conference (GLOBECOM) (2012). doi:10.1109/glocom.2012.6503179

18. Yan, Q., Li, M., Jiang, T., Lou, W., Hou, Y.T.: Vulnerability, protection for distributed consensus-based spectrum sensing in cognitive radio networks. In: 2012 Proceedings of IEEE International Conference on Computer Communications (INFOCOM) (2012). doi:10.1109/infcom.2012.6195839

19. Barni, M., Pérez-González, F.: Advances in adversary-aware signal processing. In: 2013 IEEE International Conference on Acoustics, Speech and Signal Processing (2013). doi:10.1109/icassp.6639361

20. Godsil, C.D., Royle, G.: Algebraic Graph Theory. Springer, New York (2001)

21. Helsel, D.R., et al.: Nondetects and Data Analysis. Statistics for Censored Environmental Data. Wiley-Interscience, Hoboken (2005)

22. Bollobás, B.: Modern Graph Theory. Springer, New York (2013)

23. Jamakovic, A., Uhlig, S.: On the relationship between the algebraic connectivity and graph's robustness to node and link failures. In: 3rd EuroNGI Conference on Next Generation Internet Networks, Trondheim, pp. 96–102 (2007). doi:10.1109/NGI.2007.371203

24. Bollobás, B.: Random Graphs. Cambridge University Press, Cambridge (2001)

25. Watts, D.J., Strogatz, S.H.: Collective dynamics of 'small-world' networks. Nature **393**(1998), 440–442 (1998)

26. Barabási, A.: Emergence of scaling in random networks. Science **286**, 509512 (1999)

27. Osborne, M.J., Rubinstein, A.: A Course in Game Theory. MIT Press, Cambridge (1994)

Posters Abstracts

A Game-Theoretical Framework for Industrial Control System Security

Edward J.M. Colbert[1]([✉]), Quanyan Zhu[2], and Craig G. Rieger[3]

[1] US Army Research Laboratory, Adelphi, MD 20783, USA
edward.j.colbert2.civ@mail.mil
[2] New York University, New York, NY 11201, USA
quanyan.zhu@nyu.edu
[3] Idaho National Laboratory, Idaho Falls, ID 83415, USA
craig.riger@inl.gov

Abstract. A simplistic cyber encounter between an attacker and a defender (security engineer) can be described by a zero-sum game between two players who both have complete information about the cyber system and their opponent. The rational moves of the two players are well-defined by saddle-points (Nash equilibria points) once the costs and awards are defined over all game strategies. A reasonable question to ask is whether this simplistic game model can be reasonable adapted to Industrial Control System (ICS) security. In our presentation, we describe some important differences between ICS attack scenarios and this simplistic game, and then elaborate on sample modifications to the security game model.

First, ICSs are not merely cyber networks. They are connected to physical systems and are affected by the physical systems. Attacks focused on the physical system can penetrate into the cyber network. In addition, the operator of the control process is an important player to consider. He or she dutifully monitors critical elements of the process and makes optimal choices to maintain system operability given policy constraints dictated by the system owner. There are clearly more players and more systems to consider than the attacker and defender in the simplistic game. We propose a three-game model in which defender and attacker play in the cyber regime, physical control devices and perturbations (intentional or accidental) play in the physical regime, and operator and system owner play in an abstracted process regime. All three regimes and all players can affect each other in this complex game.

Next, one nominally assumes that if all of the information in the game is readily available to the players, the players will choose the optimal path so that they suffer the least cost. If cost is a monetary measure, this may not be true, especially for state-sponsored attackers. Their defensive opponent however may indeed act to minimize monetary costs. Even if cost measures were completely known of all players, players are inefficient and often not rational. They can be coerced by psychological affects or swayed by political demands of their peers and supervisors. For extended attacks, multiple humans may play the part of a single actor. Human behavior can be modelled in some circumstances so that these uncertainties can be taken into account.

© Springer International Publishing AG 2016
Q. Zhu et al. (Eds.): GameSec 2016, LNCS 9996, pp. 469–470, 2016.
DOI: 10.1007/978-3-319-47413-7

Lastly, assuming costs and behavior can be modelled well, the attacker will often not have complete knowledge of the three regimes when they begin their attack. They may have done some reconnaissance work, but will be missing important pieces of information. This lack of information will affect their instantaneous strategy, and their path taken through attack space may be highly dependent on the amount of information available. Incomplete-information game models using Bayesian methods can be used to accommodate this effect.

We describe a game-theoretical framework that incorporates these effects, and some preliminary results using that framework.

Keywords: Industrial control system · ICS · SCADA · Game theory · Cyber security

Risk Minimization in Physical Surveillance: Playing an Uncertain Cops-and-Robbers Game

Stefan Schauer[1]([⊠]), Sandra König[1], Stefan Rass[2],
Antonios Gouglidis[3], Ali Alshawish[4], and Hermann de Meer[4]

[1] Digital Safety and Security Department,
AIT Austrian Institute of Technology GmbH, Klagenfurt, Austria
{stefan.schauer, sandra.koenig}@ait.ac.at
[2] Institute of Applied Informatics, System Security Group,
Universität Klagenfurt, Klagenfurt, Austria
stefan.rass@aau.at
[3] School of Computing and Communications,
Lancaster University, Lancaster, UK
a.gouglidis@lancaster.ac.uk
[4] Faculty of Computer Science and Mathematics,
Universität Passau, Passau, Germany
{ali.alshawish, hermann.demeer}@uni-passau.de

Keywords: Risk minimization · Physical surveillance · Distribution-valued payoff

Many security infrastructures incorporate some sort of surveillance technologies to operate as an early incident warning or even prevention system. The special case of surveillance by cameras and human security staff has a natural reflection in game-theory as the well-known "Cops-and-Robbers" game (a.k.a. graph searching). Traditionally, such models assume a deterministic outcome of the gameplay, e.g., the robber is caught when it shares its location with a cop. In real life, however, the detection rate is far from perfect (as models assume), and thus required to play the game with uncertain outcomes. This work applies a simple game-theoretic model for the optimization of physical surveillance systems in light of imperfect detection rates of incidents, minimizing the potential damage an intruder can cause. We explicitly address the *uncertainty* in assessing the potential damage caused by the intruder by making use of *empirical data* (i.e., diverging expert opinions, inaccuracies of detection mechanisms, etc.). This particularly aids standardized risk management processes, where decision-making is based on qualitative assessments (e.g., from "low damage" to "critical danger") and nominally quantified likelihoods (e.g., "low", "medium" and "high"). The unique feature of our approach is threefold: 1) it models the *practical imperfections* of surveillance systems accounting for the *subjectivity of expert opinions,* 2) it treats the uncertainty in the outcome as a *full-fledged categorical distribution* (rather than requiring numerical data to optimize characteristic measures), and 3) it optimizes the whole distribution of randomly suffered damages, thus *avoiding information loss* due to data aggregation (required in many standard game-theoretic models using numbers for their specification). The resulting optimal security strategies provide risk managers with the information they need to make better decisions.

© Springer International Publishing AG 2016
Q. Zhu et al. (Eds.): GameSec 2016, LNCS 9996, p. 471, 2016.
DOI: 10.1007/978-3-319-47413-7

Game-Theoretic Framework for Integrity Verification in Computation Outsourcing

Qiang Tang and Balázs Pejó[✉]

University of Luxembourg,
6, Rue Richard Coudenhove-Kalergi, L-1359 Esch-sur-alzette, Luxembourg
{qiang.tang,balazs.pejo}@uni.lu

Abstract. Nowadays, in order to avoid computational burdens, many organizations tend to outsource their computations to third-party cloud servers [2]. In order to protect service quality, the integrity of computation results need to be guaranteed. We define a game, where the client wants to outsource some computation to the server and verify the results. We provide a strategy for the client to minimize its own cost and force a rational server to execute the computation task honestly, e.g. *not-cheat* is a dominant strategy for the server. The details of our work appear in the full paper [1]. We give a sketch below.

The Settings. In our model we have two entity: the client, who wants to outsource some computation and the server, who will actually perform the computation. Both of them have they own strategies: the server sets ρ percent of the results to be random numbers while the client chooses σ percent of the outputs to verify by recomputing them. Based on these values, a detection rate P_d can be defined. However, ρ is unknown to the client, it is infeasible to calculate $P_d(\sigma, \rho)$. To tackle this, we define a threshold cheating toleration rate θ. Now, if the server sets ρ above this threshold then it will be caught at least with probability $P_d(\sigma, \theta)$.

The Game. We define a two-player Stackelberg game, where the client makes an offer W (e.g. how much it willing to pay for the computation) to the server. If the server rejects this, the game terminates. If the offer is accepted, then the server carries out the computation with some level of cheating (ρ). Then the client verifies the results and in case of detected cheating it refuses to pay.

Results. Our analysis showed, that the only condition that must be satisfied to make the *not-cheat* a dominant strategy for the server is the following: $W^{-1} < P_d(\sigma, \theta)$. In other words, the inverse of the payment is the lower limit for the detection rate. Furthermore, this rate corresponds to a verification cost V_d which is the other part of the client's cost (besides the payment W). So for each possible payment there is a corresponding verification cost. By searching exhaustively amongst these pairs $(W + V_d)$, the client is able to determine the one with the lowest sum, which is the game's Nash Equilibrium.

Use Cases. We apply our model to two recommender algorithm (Weighted Slope One and Stochastic Gradient Descent Matrix Factorization) with two real world dataset (Movilens 1M and Netflix). We show that the payment in the equilibrium is only slightly bigger than the cost of the calculation.

© Springer International Publishing AG 2016
Q. Zhu et al. (Eds.): GameSec 2016, LNCS 9996, pp. 472–473, 2016.
DOI: 10.1007/978-3-319-47413-7

References

1. Tang, Q., Pejó, B.: Game-Theoretic Framework for Integrity Verification in Computation Outsourcing. http://eprint.iacr.org/2016/639
2. Vaidya, J., Yakut, I., Basu, A.: Efficient integrity verification for outsourced collaborative filtering. In: IEEE International Conference on Data Mining, pp. 560–569. IEEE (2014)

On the Vulnerability of Outlier Detection Algorithms in Smart Traffic Control Systems

Rowan Powell[1], Bo An[2], Nicholas R. Jennings[3], and Long Tran-Thanh[1(✉)]

[1] University of Southampton, Southampton, UK
l.tran-thanh@soton.ac.uk
[2] Nanyang Technological University, Singapore, Singapore
[3] Imperial College London, London, UK

With the rise of interest in the development of smart traffic control systems in the recent years [1], comes the increasing need to focus on their cyber security challenges. In particular, a typical smart traffic control system would collect data from the participants in the traffic, based on which it would make intelligent traffic light control, or provide efficient routing plans to the participants. To this end, these systems typically consist of many devices with limited computational and power resources, as well as easy physical accessibility, thus, making them easy targets against cyber attack [2]. In this work, we focus on a data manipulation based attack scenario, motivated by the following example. Consider an attacker who is interested in directing the traffic from route A towards route B (a possible reason is that the attacker wants to use route A as an escaping route, and thus, the emptier it is, the faster he can escape). To do so, he wants to manipulate the traffic data of so that for the control system, which provides route plans to the vehicles within the traffic, route B seems to be a faster route than A (this attack is feasible as it can be done by, e.g., using an infected vehicle cloud [3]). However, this data manipulation has to be carefully done, as smart traffic control systems typically use some outlier detection algorithms to identify suspicious data, which then could easily pick up the malicious behaviour (i.e., data is being manipulated) if the attack is not well designed. Therefore, an interesting research question is that whether it is possible to manipulate the data such that the manipulation will not be identified by outlier detectors.

In this work-in-progress, we first show that standard outlier detectors are indeed vulnerable against data manipulation. In particular, we propose an efficient data manipulation technique that will not be detected by standard outlier detection algorithms, and is provably optimal in minimising the number of manipulated data. We then show how this data attack can be detected by using a combination of outlier and change point detection. In the second part, we further develop the previous attack algorithm by adding a randomisation module to the method, making it resistant against the abovementioned change point detection based defense. As future work, we aim to propose a security game based defence mechanism that combines the previous defence methods with some (costly) data verification approach to identify whether a data stream is under attack. The goal of this approach is to calculate the (minimax) solution of the game, and thus, to identify a robust defence strategy for the defender.

© Springer International Publishing AG 2016
Q. Zhu et al. (Eds.): GameSec 2016, LNCS 9996, pp. 474–475, 2016.
DOI: 10.1007/978-3-319-47413-7

References

1. Aslam et al.: Congestion-aware traffic routing system using sensor data. IEEE ITSC (2012)
2. Castelluccia et al.: On the difficulty of software-based attestation of embedded devices. In: ACM CCS 2009 (2009)
3. Yan et al.: Security challenges in vehicular cloud computing. IEEE TITS **14**(1), 284–294 (2013)

Author Index

Printed in the United States
By Bookmasters